MW00754439

Construction Management

Fourth Edition

Daniel W. Halpin
Purdue University

Bolivar A. Senior
Colorado State University

WILEY

John Wiley & Sons, Inc.

VP & Executive Publisher:	Don Fowley
Acquisitions Editor:	Jennifer Welter
Editorial Assistant:	Alexandra Spicehandler
Marketing Manager:	Christopher Ruel
Designer:	Seng Ping Ngien
Production Manager:	Janis Soo
Senior Production Editor:	Joyce Poh

Cover image: Digital Vision/Getty Images, Inc.

This book was set in 10/12 Times Roman by Thomson Digital and printed and bound by Courier Digital Solutions. The cover was printed by Courier Digital Solutions.

This book is printed on acid free paper.

Founded in 1807, John Wiley & Sons, Inc. has been a valued source of knowledge and understanding for more than 200 years, helping people around the world meet their needs and fulfill their aspirations. Our company is built on a foundation of principles that include responsibility to the communities we serve and where we live and work. In 2008, we launched a Corporate Citizenship Initiative, a global effort to address the environmental, social, economic, and ethical challenges we face in our business. Among the issues we are addressing are carbon impact, paper specifications and procurement, ethical conduct within our business and among our vendors, and community and charitable support. For more information, please visit our website: www.wiley.com/go/citizenship.

Copyright © 2011, 2006, 1998, 1981 John Wiley & Sons, Inc. All rights reserved. No part of this publication may be reproduced, stored in a retrieval system or transmitted in any form or by any means, electronic, mechanical, photocopying, recording, scanning or otherwise, except as permitted under Sections 107 or 108 of the 1976 United States Copyright Act, without either the prior written permission of the Publisher, or authorization through payment of the appropriate per-copy fee to the Copyright Clearance Center, Inc. 222 Rosewood Drive, Danvers, MA 01923, website www.copyright.com. Requests to the Publisher for permission should be addressed to the Permissions Department, John Wiley & Sons, Inc., 111 River Street, Hoboken, NJ 07030-5774, (201)748-6011, fax (201)748-6008, website http://www.wiley.com/go/permissions.

Evaluation copies are provided to qualified academics and professionals for review purposes only, for use in their courses during the next academic year. These copies are licensed and may not be sold or transferred to a third party. Upon completion of the review period, please return the evaluation copy to Wiley. Return instructions and a free of charge return shipping label are available at **www.wiley.com/go/returnlabel**. Outside of the United States, please contact your local representative.

Library of Congress Cataloging-in-Publication Data

Halpin, Daniel W.
  Construction management / Daniel W. Halpin, Bolivar Senior. – 4th ed.
    p. cm.
  Includes index.
  ISBN 978-0-470-44723-9 (hardback)
  1. Construction industry–United States–Management.   2. Construction industry–Law and legislation–United States.
I. Senior, Bolivar A. II. Title.
  HD9715.U52H324 2011
  624.068–dc22   2010021340

Printed in the United States of America
10 9 8 7 6 5 4

Dedicated to the loving memory
of Maria Kirchner Halpin
1937–2009

Preface

INTRODUCTION

This book has been used widely and has become a standard course text at many universities around the U.S. and abroad. The first three editions have enjoyed wide success as an introductory treatment of the subjects which are critical to success in the construction industry. This fourth edition preserves the features that have been well received by users throughout the years and adds new material suggested by instructors and students through formal surveys and informal feedback to the authors.

Construction Management requires high levels of expertise in a wide range of specialties. The analogy which has been cited in previous editions is that

> " . . . a construction manager is like an Olympic decathlon athlete who must show great competence in a multitude of areas ranging from design of construction operations to labor relations."

Experience teaches that it is very important to understand how these various construction specialties relate to one another and which activities have priority in a given situation. On some days estimating and bidding are the controlling factors for job success. On other days, labor relations are the key to success. A thorough understanding of these topics is critical to being a successful construction manager.

The construction professional has to be a "jack of all trades, and master of all." This text covers a wide range of subjects, reflecting the breadth of knowledge needed to understand the dynamics of this large and complex industry.

ORGANIZATION

This text is arranged in relatively short chapters with well defined, self-contained scope. This strategy results in a compact and direct delivery of the intended material, easy to teach and read. Chapters are sequenced in the logical flow described below.

- **Overall context.** Chapter 1 (*History and Basic Concepts*) presents an overview of the construction industry, including a historical framework.
- **Procurement and execution.** Chapters 2 (*Preparing the Bid Package*) and 3 (*Issues During Construction Phase*) discuss the procurement and administration of construction projects.
- **Company organization and management.** Chapters 4 (*Construction Contracts*), 5 (*Legal Structure*) and 6 (*Impact of Taxes*) look at issues affecting the company as a whole.
- **Planning and scheduling.** Chapters 7 (*Project Planning*), 8 (*Project Scheduling*) and 9 (*PERT Networks and Linear Operations*) address these central aspects of the construction industry.
- **Managing money.** Chapters 10 (*The Mathematics of Money*), 11 (*Project Cash Flow*) and 12 (*Project Funding*) deal with money as a company-wide resource to be planned and managed.
- **Construction equipment.** Chapters 13 (*Equipment Ownership*) and 14 (*Equipment Productivity*) address the cost and productivity of the equipment used in the construction industry.
- **Labor.** Chapter 15 (*Construction Labor*) discusses another basic construction resource. It discusses the changes in labor organization and legislation, alongside an explanation of the factors included in the computation of labor costs.
- **Cost estimating and control.** Chapter 16 (*Estimating Process*) provides an overview of the process used in construction to estimate the cost of a project. Chapter 17 (*Cost Control*) addresses the development of the systems required to keep track of a project's cost and to compare it against its progress.
- **Materials management.** Chapter 18 (*Materials Management*) offers a close-up of the procurement and on-site management of the materials used in a construction project.
- **Safety.** Chapter 19 (*Safety*) offers a concise and practical guide for safety considerations.
- **Construction operations.** Chapter 20 is on the book website at www.wiley.com/college/halpin.

NEW MATERIAL IN THIS EDITION

This edition introduces extended coverage in the scheduling area to address more advanced and practice oriented procedures such as Start to Start, Finish to Finish, and similar relationships between activities in a network schedule.

To provide increased coverage of topics related to the financial side of the house, two new chapters have been added addressing the impact of taxes on the operation of a company and the evaluation of borrowing, lending, and other financial opportunities using the mathematical concepts of engineering economy. Chapters from the previous edition have been rearranged to improve the flow of topics covered.

In addition, new problems and exercises have been added to every chapter. Figures, tables and references have been updated. New chapter openers provide summaries of emerging techniques, including Building Information Modeling, Lean Construction, Life Cycle Costs, and LEED certification.

As in previous editions, this edition provides a historical perspective of construction and related topics, an unique feature that instructors have valued over the years.

STUDENT RESOURCES

The following resources are available from the book website at www.wiley.com/college/halpin. Visit the Student section of the website.

- Information about and support materials for two leading scheduling software programs: Primavera, and Microsoft Project.
- Chapter 20, covering Construction Operations. This chapter involves the actual placement of construction in the field, using resources such as equipment and labor. This material covers the business of placing the construction physically in the field, what is done to actually build the project (e.g. putting concrete in place with a crane and bucket and pump) and the sequence of when tasks need to be done. The chapter discusses the concept of sequence and technical logic. It addresses the questions "How and in what sequence am I going to build?" and "What resources will I use?"
- The book website includes a comprehensive Simulation Homepage. This home-page includes material regarding construction process simulation and describing glossary/definition terms specific to the CYCLONE modeling format.
- A web-based program, Web CYCLONE permits simulation of construction operations. The website also includes extensive material regarding the CYCLONE construction modeling system.
- Examples showing how to build models (e.g. a masonry model, an asphalt paving model and a concrete supply model.)

INSTRUCTOR RESOURCES

All instructor resources are available from the book website at www.wiley.com/college/halpin, available only to instructors who adopt the text:

- Solutions Manual.
- Text figures in PowerPoint format.
- Image Gallery of Text Figures.
- All resources from Student section of the website.

These resources are password-protected. Visit the Instructor section of the book website to register for a password to access these materials.

ACKNOWLEDGMENTS

The authors would like to recognize Professor Emeritus Ronald Woodhead of the University of New South Wales in Sydney, Australia, for the major role he played in the realization of the first two Editions of this book. His experience and insights were critical to the success of these earlier editions.

We would also like to thank the many colleagues and numerous students who have provided very useful feedback regarding various aspects of this book over the past 30 years.

In particular, the following academic colleagues and industry practitioners provided important insights and relevant material for this book:

Dulcy Abraham and Joe Sinfield, Purdue University
Bob Bowen, Bowen Engineering, Indianapolis, IN
Peter Dozzi, U. of Alberta, Edmonton, Canada
Jimmie Hinze, U. of Florida, Gainesville, FL
E. Paul Hitter Jr., Messer Construction, Cincinnati, OH
Mike Kenig, Holder Construction, Atlanta, GA
Jerry Kerr, Construction Consultant, Indianapolis, IN

Finally, we would like to recognize the continuing support and understanding of our families throughout the process of writing, updating and preparing this text for publication.

Contents

Chapter 1

History and Basic Concepts

CONTENTS

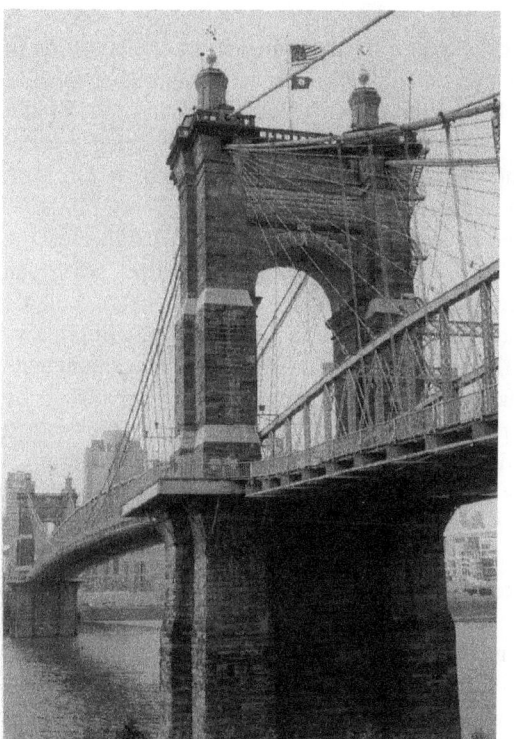

Bridges and History

Water crossings have always been seen as great engineering challenges. Since Roman times, bridges and various river crossings have been linked with great engineering achievements. Apollodorus was chief engineer for the Emperor Trajan and built a bridge across the Danube River in the second century A.D. This bridge allowed Trajan to invade Dacia and annex the territory of modern-day Romania.

The length of clear span bridging was greatly increased by the development of the cable supported suspension bridge. The oldest vehicular steel cable suspension bridge in the world in continuous use was built by John A. Roebling in Cincinnati, Ohio, during the Civil War. It is still one of the major arteries connecting Cincinnati with Covington, Kentucky.

When construction started in 1856, the charter authorizing the construction required a clear span of 1,000 feet between two towers, with the deck located a minimum of 100 feet above the water's surface. The bridge was completed in December 1866. The 1,057-foot main span was, at the time, the longest in the world. It was one of the first suspension bridges to use both vertical suspenders and diagonal cable stays that radiated from

John A. Roebling Bridge Covington, Kentucky Side

the top of each tower. This innovative use of cable stays gave the bridge great rigidity and resistance to movement during high winds. Roebling used this same concept later when building the Brooklyn Bridge.

The bridge was upgraded to its present configuration in 1894. A second set of 10.5-inch cables was added to carry heavier decks. This reconstruction increased the carrying capacity of the bridge to a 30-ton limit. As a native of Covington, one of the authors rode both trolley (street) cars and electrically powered buses hundreds of times to the transit terminal in Cincinnati located at the north end of the bridge. In 1984, the bridge was renamed the John A. Roebling Bridge.

Bridges Today

World famous bridges have become a symbol of civil engineering. The Golden Gate Bridge in San Francisco has not only been hailed a tremendous engineering achievement but also a beautifully balanced aesthetic achievement. Plans are now underway to bridge the famous Straits of Messina between the toe of Italy and the island of Sicily. This bridge will have a clear span of almost 2 miles, approximately 10 times the span of the Roebling Bridge in Cincinnati. It will also be designed to resist hurricane-force winds. Construction of this bridge will rival the construction of the Channel Tunnel connecting England and France.

1.1 HISTORICAL PERSPECTIVE

Construction and the ability to build things is one of the most ancient of human skills. In prehistoric times, it was one of the talents that set *Homo sapiens* apart from other species. Humans struggled to survive and sought shelter from the elements and the hostile environment that surrounded them by building protective structures. Using natural materials such as earth, stone, wood, and animal skins, humans were able to fabricate housing that provided both shelter and a degree of protection.

As society became more organized, the ability to build things became a hallmark of the sophistication of ancient civilizations. The wonders of the ancient world reflect an astounding ability to build not only structures for shelter but also monuments of gigantic scale. The pyramids and Greek temples such as the Parthenon (Figure 1.1) are impressive testimony to the building skills of the civilizations of antiquity. Great structures punctuate the march of time, and many of the structures of ancient times are impressive even by modern standards. The great Church of Hagia Sophia in Constantinople, constructed during the sixth century, was the greatest domed structure in the world for nine centuries. It is an impressive example of the ingenuity of the builders of that time and their mastery of how forces can be carried to the ground using arches in one dimension and in three dimensions as domes.

In modern times, the Brooklyn Bridge and the Panama Canal stand as legendary feats of engineering achievement. They are also testimonies to the fact that realizing a construction project involves solving a multitude of problems, many of which are not technical. In both the Brooklyn Bridge and Panama Canal projects, people problems requiring great innovation and leadership were just as formidable as the technical problems encountered. To solve them, the engineers involved accomplished "heroic" feats.

Figure 1.1 The Parthenon in Athens.

1.2 GREAT CAPTAINS OF CONSTRUCTION

The Roebling family as a group can be credited with building the Brooklyn Bridge during the period from 1869 to 1883. It was the greatest project of its time and required the use of technology at a scale never before tried. The concept of a cable-supported suspension bridge was literally invented by John A. Roebling (Figure 1.2). Roebling was born in Germany and was the favorite student of the famous philosopher Georg Wilhelm Friedrich Hegel. Roebling was a man of tremendous energy and powerful intellect. He built a number of suspension bridges, notably the John A. Roebling Bridge in Cincinnati (which is still in daily use), that demonstrated the cable-supported concept prior to designing the Brooklyn Bridge. Upon his death (precipitated by an accident that occurred during the initial survey of the centerline of the bridge), his son, Washington, took charge.

Washington Roebling (Figure 1.3) was a decorated hero of the Civil War who had received his training in civil engineering at Rensselaer Polytechnic Institute. Like his father he was a man of great vision and courage. He refined the concepts of caisson construction and solved numerous problems as the great towers of the bridge rose above New York City (Figure 1.4). Since he would not require anyone to work under unsafe conditions, he entered the caissons and supervised the work personally. He ultimately suffered from a mysterious illness related to the fact that the work was carried out under elevated air pressure in the caissons. We now know that this illness, called "the bends," was caused by the absorption and rapid exit of nitrogen from the bloodstream when workers entered and exited the pressurized caissons.

Although incapacitated, Washington continued to supervise the work from an apartment that overlooked the site. At this point, Emily, Washington's wife and the sister of an army general, entered the picture (Figure 1.5). Emily carried information to Roebling's supervising engineers on the site. She became the surrogate chief engineer and gave directives in the name of her husband. She was able to gain the confidence and

Figure 1.2 John A. Roebling, designer of the Brooklyn Bridge. (American Society of Civil Engineers)

respect of the site engineers and was instrumental in carrying the project through to successful accomplishment. The tale of the building of the great bridge—see *The Great Bridge* by David McCullough, 1972—is one of the most extraordinary stories of technical innovation and personal achievement in the annals of U.S. history.

Figure 1.3 Washington A. Roebling, chief engineer of the Brooklyn Bridge. (Special Collections and University Archives, Rutgers University Libraries)

Figure 1.4 Brooklyn Bridge under construction 1881. (Museum of the City of New York, Print Archives, Gift of the Essex Institute)

1.3   PANAMA CANAL

The end of the 19th century was a time of visionaries who conceived of projects that would change the history of humankind. Since the time Vasco Núñez de Balboa crossed Panama and discovered a great ocean, planners had conceived of the idea of a water link between the Atlantic and the Pacific Oceans. Having successfully connected the Mediterranean with

Figure 1.5 Emily Warren Roebling, wife of Washington Roebling. (Special Collections and University Archives, Rutgers University Libraries)

the Red Sea at Suez, in 1882 the French began work on a canal across the narrow Isthmus of Panama, which at that time was part of Colombia. After struggling for 9 years, the French were ultimately defeated by the formidable technical difficulties as well as the hostile climate and the scourge of yellow fever.

Theodore Roosevelt became president during this period and his administration decided to take up the canal project and carry it to completion. Using what would he referred to as "gun-boat" diplomacy, Roosevelt precipitated a revolution that led to the formation of the Republic of Panama. Having clarified the political situation with this stratagem, the famous "Teddy" then looked for the right man to actually construct the canal. That right man turned out to be John F. Stevens, a railroad engineer who had made his reputation building the Great Northern Railroad (Figure 1.6). Stevens proved to be the right man at the right time.

Stevens understood the organizational aspects of large projects. He immediately realized that the working conditions of the laborers had to be improved. He also understood that measures had to be taken to eradicate the fear of yellow fever. To address the first problem, he constructed large and functional camps for the workers in which good food was available. To deal with the problem of yellow fever he enlisted the help of an army doctor named William C. Gorgas. Prior to being assigned to Panama, Gorgas had worked with Dr. Walter Reed in wiping out yellow fever in Havana, Cuba. He had come to understand that the key to controlling and eliminating this disease was, as Reed had shown, the control of the mosquitoes that carried the dreaded infection and the elimination of their breeding places (see *The Microbe Hunters* by Paul DeKruif). Gorgas was successful in effectively controlling the threat of yellow fever, but his success would not have been possible without the total commitment and support of Stevens.

Having established an organizational framework for the project and provided a safe and reasonably comfortable environment for the workers, Stevens addressed the technical problems presented by the project. The French had initially conceived of a canal built at sea level and similar to the Suez Canal. That is, the initial technical concept was to build a canal

Figure 1.6 John F. Stevens, chief engineer of the Panama Canal. (National Archives, Washington, D.C.)

at one elevation. Because of the high ground and low mountains of the interior portion of the isthmus, it became apparent that this approach would not work. To solve the problem of moving ships over the "hump" of the interior, it was decided that a set of water steps, or locks, would be needed to lift the ships transiting the canal up and over the high ground of Central Panama and down to the elevation of the opposite side. The immense size of a single lock gate is shown in Figure 1.7.

The construction of this system of locks presented a formidable challenge. Particularly on the Atlantic side of the canal, the situation was complicated by the presence of the wild Chagres River, which flowed in torrents during the rainy season and dropped to a much lower elevation during the dry season. The decision was made to control the Chagres by constructing a great dam that would impound its water and allow for control of its flow. The dam would create a large lake that would become one of the levels in the set of steps used to move ships through the canal. The damming of the Chagres and the creation of Lake Gatun itself was a project of immense proportions, which required concrete and earthwork structures of unprecedented size.

The other major problem had to do with the excavation of a great cut through the highest area of the canal. The Culebra Cut, as this part of the canal was called, required the excavation of earthwork quantities that even by today's standards stretch the imagination. Stevens viewed this part of the project as the construction of a gigantic railroad system that would operate continuously (24 hours a day) moving earth from the area of the cut to the Chagres dam construction site. The material removed from the cut would provide the fill for the dam. It was an ingenious idea.

To realize this system, Stevens built one of the greatest rail systems of the world at that time. Steam-driven excavators (shovel fronts) worked continuously loading railcars. The excavators worked on flexible rail spurs that could be repositioned by labor crews to maintain contact with the work face. In effect, the shovels worked on sidings that could be moved many times each day to facilitate access to the work face. The railcars passed continuously under these shovels on parallel rail lines.

Stevens' qualities as a great engineer and leader were on a level with those of the Roebling's. As an engineer, he understood that planning must be done to provide a climate and environment for success. Based on his railroading experience, he knew that a project of this magnitude could not be accomplished by committing resources in a piecemeal fashion.

Figure 1.7 Work in progress on the Great Gatun lock gates. (© Hulton Archive/Getty Images, Inc.)

He took the required time to organize and mass his forces. He also intuitively understood that the problem of disease had to be confronted and conquered. Some credit for Stevens' success must go to President Roosevelt and his secretary of war, William Howard Taft. Taft gave Stevens a free hand to make decisions on the spot and, in effect, gave him total control of the project. Stevens was able to be decisive and was not held in check by a committee of bureaucrats located in Washington (i.e., the situation present before he took charge of the job).

Having set the course that would ultimately lead to successful completion of the canal. Stevens abruptly resigned. It is not clear why he decided not to carry the project through to completion. President Roosevelt reacted to his resignation by appointing a man who, as

Roosevelt would say, "could not resign." Roosevelt selected an army colonel and West Point graduate named George Washington Goethals to succeed Stevens. Goethals had the managerial and organizational skills needed to push the job to successful completion. Rightfully so, General Goethals received a great deal of credit for the construction of the Panama Canal. However, primary credit for pulling the job "out of the mud," getting it on track, and developing the technical concept of the canal that ultimately led to success must be given to Stevens—a great engineer and a great construction manager.

1.4   OTHER HISTORIC PROJECTS

Much can he learned from reading about and understanding projects like the Brooklyn Bridge and the Panama Canal. David McCullough's books *The Great Bridge* and *The Path between the Seas* are as exciting and gripping as any spy novel. They also reflect the many dimensions of great and small construction projects. Other projects, such as the building of the Hoover Dam on the Colorado River, have the same sweep of adventure and challenge as the construction of the Panama Canal. The construction of the Golden Gate Bridge in San Francisco was just as challenging a project as the construction of the Brooklyn Bridge in its time.

The construction of the Empire State Building in only 18 months is another example of a heroic engineering accomplishment. Realization of great skyscrapers, such as the Empire State Building and the Chrysler Building in New York, was made possible by the development of technologies and techniques in the construction of previous projects. The construction of the Eiffel Tower in Paris and the towers of the "Magnificent Mile" in Chicago in the early 1900s demonstrated the feasibility of building tall structures supported by steel frames. Until the advent of the steel frame with its enclosing curtain walls, the height of buildings had been limited based on the strength of materials used in the bearing walls, which carried loads to the ground.

The perfection of the concept of steel-frame-supported structures and the development of the elevator as a means of moving people vertically in tall buildings provided the necessary technologies for the construction of the tall buildings that we take for granted today. Modern-day city skylines would not have been possible without these engineering innovations.

More recently, a project of historical proportions was realized with the completion of the Eurotunnel connecting the British Isles and France. This project was dreamed of for many centuries. Through the skill and leadership of a large team of engineers and managers, it has now become a reality. Great projects are still being proposed and constructed. For the interested reader, brief coverage of many historical projects is given in *The Builders—Marvels of Engineering* published by the National Geographic Society (edited by Elizabeth L. Newhouse, 1998).

1.5   CONSTRUCTION VERSUS MANUFACTURING PROCESSES

Construction is the largest product-based (as opposed to service-oriented) industry in the United States. The dollar volume of the industry is on the order of $1 trillion ($1,000 billion) annually. The process of realizing a constructed facility such as a road, bridge, or building, however, is quite different from that involved in manufacturing an automobile or a television set.

Manufactured products are typically designed and produced without a designated purchaser. In other words, products (e.g., automobiles or TV sets) are produced and then presented for sale to any potential purchaser. The product is produced on the speculation that a purchaser will be found for the item produced. A manufacturer of bicycles, for instance, must determine the size of the market, design a bicycle that appeals to the potential purchaser, and then manufacture the number of units that market studies indicate can be sold. Design and production are done prior to sale. To attract possible buyers, advertising is required and is an important cost center.

Many variables exist in this undertaking, and the manufacturer is at risk of failing to recover the money invested once a decision is made to proceed with design and production of the end item. The market may not respond to the product at the price offered. Units may remain unsold or sell at or below the cost of production (i.e., yielding no profit). If the product cannot be sold so as to recover the cost of manufacture, a loss is incurred and the enterprise is unprofitable. When pricing a given product, the manufacturer must not only recover the direct (i.e., labor, materials, etc.) cost of manufacturing but also the so-called indirect and general and administrative (G&A) costs such as the cost of management and implementation of the production process (e.g., legal costs, marketing costs, supervisory costs, etc.). Finally, unless the enterprise is a non-profit, the desire of the manufacturer is to increase the value of the firm. Therefore, profit must be added to the direct, indirect, and G&A costs of manufacturing.

Manufacturers offer their products for sale either directly to individuals (e.g., by mail order or directly over the Web), to wholesalers who purchase in quantity and provide units to specific sales outlets, or to retailers who sell directly to the public. This sales network approach has developed as the framework for moving products to the eventual purchaser. (See if you can think of some manufacturers who sell products directly to the end user, sell to wholesalers, or sell to retail stores.)

In construction, projects are sold to the client in a different way. The process of purchase begins with a client who has need for a facility. The purchaser typically approaches a design professional to more specifically define the nature of the project. This leads to a conceptual definition of the scope of work required to build the desired facility. Prior to the age of mass production, purchasers presented plans of the end object (e.g., a piece of furniture) to a craftsman for manufacture. The craftsman then proceeded to produce the desired object. If King Louis XIV desired a desk at which he could work, an artisan would design the object, and a craftsman would be selected to complete the construction of the desk. In this situation, the purchaser (King Louis) contracts with a specialist to construct a unique object. The end item is not available for inspection until it is fabricated. That is, because the object is unique, it is not sitting on the show room floor and must be specially fabricated.

Because of the one-of-a-kind unique nature of constructed facilities, this is still the method used for building construction projects. The purchaser approaches a set of potential contractors. Once agreement is reached among the parties (i.e., client, designer, etc.) as to the scope of work to be performed, the details of the project or end item are designed and constructed. Purchase is made based on a graphical and verbal description of the end item, rather than the completed item itself. This is the opposite of the speculative process where design and manufacture of the product are done prior to identifying specific purchasers. A constructed facility is not commenced until the purchaser has been identified. It would be hard to imagine, for instance, building a bridge without having identified the potential buyer. (Can you think of a construction situation in which the construction is completed prior to identifying the buyer?)

The nature of risk is influenced by this process of purchasing construction. For the manufacturer of a refrigerator, risk relates primarily to being able to produce units at a competitive price. For the purchaser of the refrigerator, the risk involves mainly whether the appliance operates as advertised.

In construction, because the item purchased is to be produced (rather than being in a finished state), there are many complex issues that can lead to failure to complete the project in a functional or timely manner. The number of stakeholders and issues that must be dealt with prior to project completion leads to a complex level of risk for all parties

Figure 1.8 Manufacturing versus Construction Process.

involved (e.g., designer, constructor, government authorities, real estate brokers, etc.). A manufactured product is, so to say, "a bird in the hand." A construction project is "a bird in the bush."

The risks of the manufacturing process to the consumer are somewhat like those incurred when a person goes to the store and buys a music CD. If the recording is good and the disk is serviceable, the risk is reduced to whether the customer is satisfied with the musical group's performance. The client in a construction project is more like a musical director who must assemble an orchestra and do a live performance hoping that the recording will be acceptable. The risks of a failure in this case are infinitely greater. A comparative chronological diagram of the events involved in the manufacturing process versus those in the project construction project process is shown schematically in Figure 1.8.

1.6   PROJECT FORMAT

In contrast to other manufacturing industries that fabricate large numbers of units, such as automobiles or television sets, the construction industry is generally focused on the production of a single and unique end product. That is, the product of the construction industry is a facility that is usually unique in design and method of fabrication. It is a single "one-off" item that is stylized in terms of its function, appearance, and location. In certain cases, basically similar units are constructed—as in the case of town houses or fast-food restaurants. But even in this case, the units must be site adapted and stylized to some degree.

Mass production is typical of most manufacturing activities. Some manufacturing sectors make large numbers of similar units or batches of units that are exactly the same. A single item is designed to be fabricated many times. Firms manufacture many repetitions of the same item (e.g., cell phones, thermos bottles, etc.) and sell large numbers to achieve a profit. In certain cases, a limited number or batch of units of a product is required. For instance, a specially designed transformer or hydropower turbine may be fabricated in limited numbers (e.g., 2, 3, or 10) to meet the special requirements of a specific client. This production of a limited number of similar units is referred to as batch production.

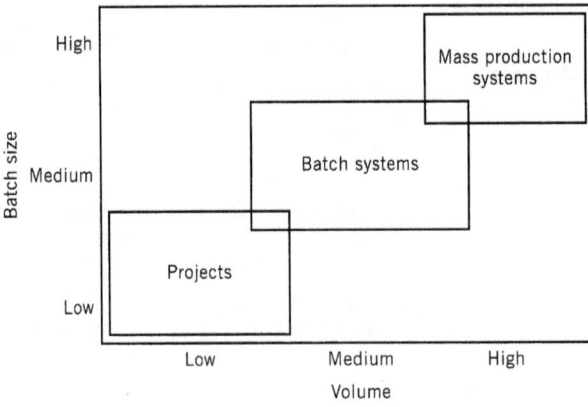

Figure 1.9 Comparison of production systems.

Mass production and batch production are not typical of the construction industry. Since the industry is oriented to the production of single unique units, the format in which these one-off units are achieved is called the *project format*. Both the design and production of constructed facilities are realized in the framework of a project. Figure 1.9 compares the batch size against the number of units produced (i.e., *volume*) for project, batch, and mass production systems.

The focus of construction management is the planning and control of resources within the framework of a project. This is in contrast to other manufacturing sectors that are interested in the application of resources over the life of an extended production run of many units.

1.7   PROJECT DEVELOPMENT

Construction projects develop in a clearly sequential or linear fashion. The general steps involved are as follows:

1. A need for a facility is identified by the client.
2. Initial feasibility and cost projections are developed.
3. The decision to proceed with conceptual design is made and a design professional is retained.
4. The conceptual design and scope of work is developed to include an approximate estimate of cost.
5. The decision is made to proceed with the development of final design documents, which fully define the project for purposes of construction.
6. Based on the final design documents, the project is advertised and proposals to include quotations for construction of the work are solicited.
7. Based on proposals received, a constructor is selected and a notice to the constructor to proceed with the work is given. The proposal and the acceptance of the proposal on the part of the owner constitute the formation of a contract for the work.
8. The process of constructing the facility is initiated. Work is completed and the facility is available for acceptance, occupancy and utilization.
9. In complex projects, a period of testing decides if the facility operates as designed and planned. This period is typical of industrial projects and is referred to as project start-up.

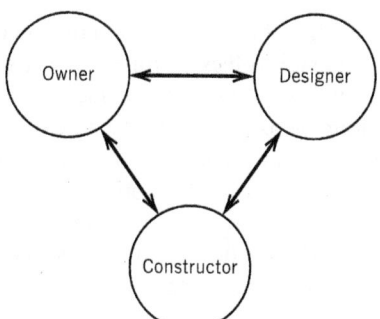

Figure 1.10 Relationship between owner, designer, and constructor.

10. The facility operates and is maintained during a specified service life.

11. The facility is disposed of if appropriate or maintained in perpetuity.

These steps must be modified on a case-by-case basis to address the special aspects of a given project. Topics relating to items 1 through 8 will be discussed in detail in Chapters 2 and 3.

The key players in this developmental sequence are:

- The owner.
- The designer or design professional.
- The constructor.

The interaction of these three major entities is shown in Figure 1.10.

Although other entities such as regulators, subcontractors, materials vendors, and so forth are important supporting players in this sequence, the major development of the project revolves about these three major entities. The legal definition of this interaction is established in the general conditions of the contract. This interaction will be described in detail in the following chapters.

1.8   CONSTRUCTION TECHNOLOGY AND CONSTRUCTION MANAGEMENT

The study of construction as a discipline can be broadly structured into two general themes:

1. Construction technology

2. Construction management

As the name implies, *construction technology* relates to the methods or techniques used to place the physical materials and elements of construction at the job site. The word *technology* can be broken into two subwords—technical from *techno* and *logic*. *Logic* addresses the concept of sequence or procedure. That is, logic addresses the order of things; something is done first, another thing second, and so on until a result is achieved. Adding *technical* to this leads to the idea that technology has to do with the technical sequence in which something is done to produce an end result. It is possible to talk about a technology that applies to placing concrete, cladding a building, excavating a tunnel, and so on.

Once a project has been defined, one of the most critical questions facing the construction manager is "What construction technique or method should be selected?" The types of methods for placing construction are diverse. New methods are continuously being perfected and a construction manager must weigh the advantages and disadvantages of a given method or technique.

In contrast to construction technology, construction management addresses how the resources available to the manager can be best applied. Typically, when speaking of resources for construction, we think of the four M's of construction: **m**anpower, **m**achines, **m**aterials, and **m**oney. Management involves the timely and efficient application of the four M's to construct a project. Many issues must be considered when managing a project and successfully applying the four M's. Some are technical (e.g., design of formwork, capacities of excavators, weather tightness of exterior finishes, etc.). Many issues, however, are more qualitative in nature and deal with the motivation of workers, labor relations, the form of contracts, legal liability, and safety on the job site. As noted in discussing the Panama Canal, organizational issues can be critical to the success of any project. This book will focus mainly on the topic of construction management. Therefore, we will be talking about the four M's and subjects that relate to management and the timely and cost-effective realization of a project.

1.9 CONSTRUCTION MANAGEMENT IS RESOURCE DRIVEN

The job of a construction manager is to efficiently and economically apply the required resources to realize a constructed facility of acceptable quality within the time frame and budgeted cost specified. Among the many watch words within the construction industry is the expression "on time and within budget." More recently, the concept of quality as a requirement has become an increasingly important aspect of the construction process. So this old adage can he expanded to say "a quality facility on time and within budget."

The construction manager is provided with resources such as labor, equipment, and materials and is expected to build a facility that meets the specifications and is consistent with the drawings provided for the project. The mission of construction is constrained in terms of the available time and amount of available money. The challenge faced by the construction manager is to apply the resources of workers, machines, and materials within the limited funding (money) and time available. This is the essence of construction.

The manager must be clever and innovative in the utilization of resources available. Somewhat like a general in battle, the manager must develop a plan of action and then direct and control forces (resources) in a coordinated and timely fashion so that the objective is achieved.

This requires a variety of skills. A high level of competency is needed in a broad range of qualitative and quantitative subjects. A manager must be like a decathlon athlete. A strong ability in many areas is a necessity. Being outstanding in one area (e.g., engineering) but weak in a number of others (e.g., interpersonal relationships, contract law, labor relations, etc.) is not enough to be a successful construction manager. A strong performance across the board is required.

1.10 CONSTRUCTION INDUSTRY

The construction industry has been referred to as the engine that drives the overall economy. It represents one of the largest economic sectors in the United States. Until the early 1980s the construction industry accounted for the largest percentage of the gross domestic product (GDP) and had the highest dollar turnover of any U.S. industry. Presently, construction is still the largest manufacturing (i.e., focus on an end item of production) industry in the United States. New construction accounts for approximately 8% of the GDP and retrofit projects contribute an additional 5%. As noted previously, the total annual volume of activity in the construction sector is estimated to be on the order of $1 trillion. More than a million firms operate in the construction sector, and the number of people employed in construction is estimated to be 10 million.

The industry consists of large and small firms. The largest firms may have an annual work volume in excess of $20 billion, and consist of thousands of employees. Many of the largest firms work both domestically and in the international market. In contrast to the large companies, statistics indicate that over two-thirds of construction-related firms have less than five employees. The spectrum of work ranges from the construction of large power plants and interstate highways costing billions of dollars to the construction of single-family houses and the paving of driveways and sidewalks. The high quality of life available in the United States is possible in large part because of the highly developed infrastructure. The U.S. infrastructure, which consists of the roads, tunnels, bridges, communications systems, power plants and distribution networks, water treatment systems, and all of the structures and facilities that support daily life, is without peer. The infrastructure is constructed and maintained by the construction industry. Without it, the country would not be able to function.

1.11   STRUCTURE OF THE CONSTRUCTION INDUSTRY

Since the construction sector is so diverse, it is helpful to look at the major types of projects typical of construction to understand the structure of the industry. Construction projects can be broadly classified as (a) building construction, (b) engineered construction, and (c) industrial construction, depending on whether they are associated with housing, public works, or manufacturing processes.

The building construction category includes facilities commonly built for habitational, institutional, educational, light industrial (e.g., warehousing, etc.), commercial, social, and recreational purposes. Typical building construction projects include office buildings, shopping centers, sports complexes, banks, and automobile dealerships. Building construction projects are usually designed by architects or architect/engineers (A/Es). The materials required for the construction emphasize the architectural aspects of the construction (e.g., interior and exterior finishes).

Engineered construction usually involves structures that are planned and designed primarily by trained professional engineers (in contrast to architects). Normally, engineered construction projects provide facilities that have a public function relating to the infrastructure and, therefore, public or semipublic (e.g., utilities) owners generate the requirements for such projects. This category of construction is commonly subdivided into two major subcategories; thus, engineered construction is also referred to as (a) highway construction and (b) heavy construction.

Highway projects are generally designed by state or local highway departments. These projects commonly require excavation, fill, paving, and the construction of bridges and drainage structures. Consequently, highway construction differs from building construction in terms of the division of activity between owner, designer, and constructor. In highway construction, owners may use in-house designers and design teams to perform the design so that both owner and designer are public entities.

Heavy construction projects are also typically funded by public or quasi-public agencies and include sewage plants, flood protection projects, dams, transportation projects (other than highways), pipelines, and waterways. The owner and design firm can be either public or private depending on the situation. In the United States, for instance, the U.S. Army Corps of Engineers (public agency) has in the past used its in-house design force to engineer public flood protection structures (i.e., dams, dikes) and waterway navigational structures (i.e., river dams, locks, etc.). Due to the trend toward downsizing government agencies, more design work is now being subcontracted to private design engineering firms. Public electrical power companies use private engineering firms to design their power plants. Public mass-transit authorities also call on private design firms for assistance in the engineering of rapid-transit projects.

Industrial construction usually involves highly technical projects in manufacturing and processing of products. Private clients retain engineering firms to design such facilities. In some cases, specialty firms perform both design and construction under a single contract for the owner/client.

Figure 1.11 Breakdown of construction industry segments.

1.12   DIFFERING APPROACHES TO INDUSTRY BREAKDOWN

Figure 1.11 represents one of many ways in which the industry can be divided into a number of sectors. This breakdown includes single-family houses within the residential construction sector. In some breakdowns, one- and two-family houses are considered to be a separate industry, and this single family residential activity is not reported as part of the construction industry. As can be seen from the pie chart, residential and building construction accounts for 70 to 75% of the industry. Industrial construction and heavy engineering construction (which are more closely related to the infrastructure) account for 25 to 30% of industry activity.

A slightly different approach to project classification is used by the *Engineering News Record (ENR)* magazine, which reflects the weekly dynamics of the construction industry in the United States. This breakdown of construction identifies three major construction categories:

1. Heavy and highway

2. Nonresidential building

3. Multiunit housing

The nonresidential building category includes building and industrial construction as defined previously. These overall categories are further dissected to reflect the major areas of specialization within the construction industry. The *ENR* publishes a Web-based update of information based on this set of construction categories each week.

1.13   MANAGEMENT LEVELS OF CONSTRUCTION

Organizational considerations lead to a number of hierarchical levels that can be identified in construction. This derives from the project format. Decision making at levels above the project relate to company management considerations. Decisions within the project relate to operational considerations (e.g., selection of production methods) as well as the application of resources to the various construction production processes and work tasks selected to realize the constructed facility. Specifically, four levels of hierarchy can be identified as follows:

1. **Organizational.** The organizational level is concerned with the legal and business structure of a firm, the various functional areas of management, and the interaction between head office and field managers performing these management functions.

2. **Project.** *Project-level* vocabulary is dominated by terms relating to the breakdown of the project for the purpose of time and cost control (e.g., the project activity and the project cost account). Also, the concept of resources is defined and related to the activity as either an added descriptive attribute of the activity or for resource scheduling purposes.

3. **Operation** (and **Process**). The construction operation and process level is concerned with the technology and details of how construction is performed. It focuses on work at the field level. Usually a construction operation is so complex that it encompasses several distinct processes, each having its own technology and work task sequences. However, for simple situations involving a single process, the terms are synonymous.

4. **Task.** The task level is concerned with the identification and assignment of elemental portions of work to field units and work crews.

The relative hierarchical breakout and description of these levels in construction management are shown in Figure 1.12. It is clear that the organizational, project, and

Figure 1.12 Management levels in construction.

Figure 1.13 Schematic of concourse building.

activity levels have a basic project and top management focus, whereas the operation, process, and work task levels have a basic work focus.

To illustrate the definitions given in this section, consider a glazing subcontract for the installation of glass and exterior opaque panels on the concourses of Hartsfield-Jackson International Airport in Atlanta, Georgia. This was a project requiring the installation of five panels per bay on 72 bays of each of the concourses. Figure 1.13 shows a schematic diagram of the project. A breakout of typical items of activity at each level of hierarchy is given in Table 1.1. At the project level, activities within the schedule relate to the glass and panel installation in certain areas of the concourses. At the work task level, unloading, stripping, and other crew-related activity is required.

Table 1.1 Example of Hierarchical Terms

Project	Installation of all exterior glass and panel wall construction on the Concourses of the Hartsfield International Airport, Atlanta, GA
Activity	Glass and panel installation on Concourse A, Bays 65–72
Operation	Frame installation to include preparation and installation of five panel frames in each concourse bay; column cover plate installation
Process	Sill clip placement; mullion strips installation
	Glass placement in frame; move and adjust hanging scaffold
Work task	Locate and drill clip fastener; unload and position mullion strips; strip protective cover from glass panel; secure scaffold in travel position

REVIEW QUESTIONS AND EXERCISES

1.1 Look up the names of the largest contractors reported by *ENR* in its three main categories: Heavy and highway, non-residential building, and multiunit housing. Notice that the list also includes their ranking for the previous years. Were the contractors the largest the previous year?

1.2 There have been many construction marvels in human history, of which this chapter only mentioned a few. Comment on three historical projects of your choosing not included here. Examples include the Great Wall of China, the pyramids, the Suez Canal, the Eiffel Tower, and the Golden Gate Bridge, among many others. Why were they built? What makes them unique? How were they built? Who paid for them?

1.3 What advantages do you see in consolidating the roles of owner, designer, and constructor shown in Figure 1.10? What disadvantages could it have?

1.4 Give examples of the management levels in construction shown in Figure 1.12.

Chapter 2

Preparing the Bid Package

CONTENTS

Online Plan Rooms

Plan rooms are typically conveniently located and can be visited by contractors in large cities. These facilities provide access to bidding documents in order to facilitate the bidding process. Sets of project plans, specifications, bidding information, and general contractor lists for jobs being let for bid are available at plan rooms. However, utilizing these plan room services can be time consuming and costly because of the need to travel to the physical location of the plan room. Even though many plan rooms mail their

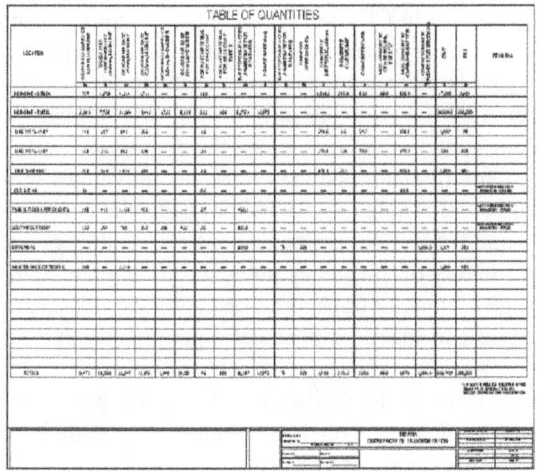

members a weekly newsletter to update projects, a visit to the facility is required to check out the project information in detail.

Online plan rooms enable their members to conveniently view plan room projects 24 hours a day, 7 days a week from the member's home or office location via the Internet. Members can access, study, and download useful information for various construction projects, and even print out actual plans using plotters or printers. Contractors and estimators can get information and prepare for a bid in a timely and efficient manner. In addition, establishing an online plan room enables owners, architects, and contractors to control and manage their bid process more effectively.

Additional features augmenting the online plan room services include Internet access to the plan room via secure login, computerized estimating tools, worksheet build-up function, real-time bidder notification system (e-mail/fax), project search with specific criteria, and blueprint measurement on screen, etc.

Utilization of these online plan rooms provides prospective bidders with 24 hour accessibility. They also make the bidding process efficient and well organized, which results in reduced costs.

2.1   PROJECT CONCEPT AND NEED

Since the constructed environment in which we live is realized in a project format, the construction process can be best understood by examining the steps required to realize a complete project. In this chapter and the next, we will examine the step-by-step development of a project. A schematic flow diagram of the sequential actions required to realize a project is shown in Figure 2.1. The framework for this discussion will be the development

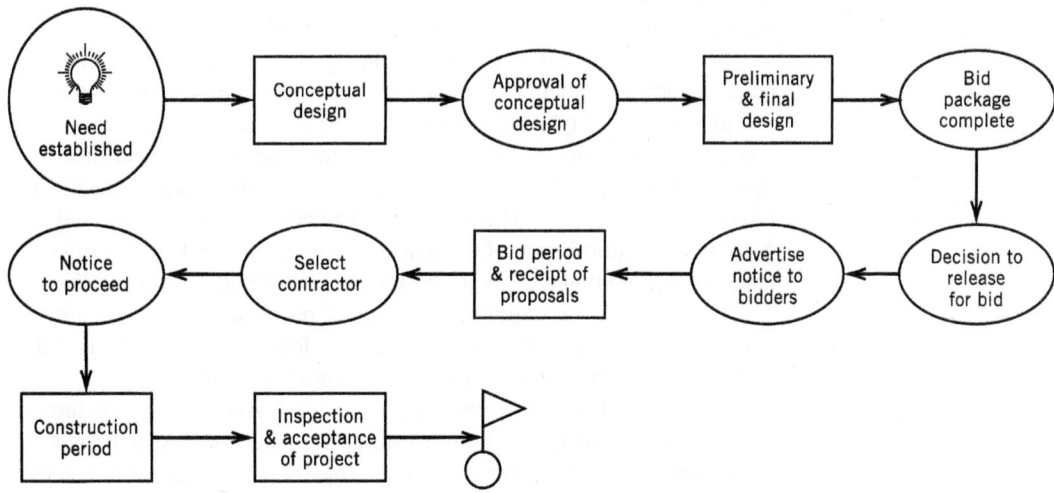

Figure 2.1 Project development cycle (new).

of a project for competitive bid. As we will see in Chapter 4, this is the delivery system characteristic of publicly contracted work. This approach requires that a full set of project documents be developed before the project is offered for bid and construction.

Each project has a life cycle triggered by the recognition of a need that can best be addressed with the construction of a facility. In a complex society, the number of entities generating needs that will shape the built environment is diverse. Private individuals seek to construct housing that is functional and comfortable (e.g., home or residential construction). Public entities such as city, state, and federal governments construct buildings and required public structures to enhance the quality of life. Many public projects relate to the development of the infrastructure. Bridges, tunnels, transportation facilities, dikes, and dams are typical of public projects designed to meet the needs of a community and the society in general.

Private entities such as commercial firms build facilities that provide goods and services to the economy. These entities are typically driven by the objective of realizing a profit. Facilities constructed by private owners include manufacturing plants, hospitals, research laboratories, hotels and commercial buildings, communications networks, and a host of other project types.

2.2   ESTABLISHING NEED

The first step in any project is the establishment of a need and a conceptual definition and refinement of the facility that will meet that need. If the need has a commercial basis, it is normally defined in terms of a market analysis that establishes the profitability of the proposed project. For instance, if the need relates to the construction of a chemical plant in Spain, the firm constructing the plant will want to establish that a market exists that can be profitably accessed once the plant is in operation.

The economic basis for the plant must be established based on market studies projecting the demand for the plant's product mix across the planning horizon under consideration. In many cases, these studies recommend optimal time frames for the plant construction to meet the market in advance of competition. Plant size, site location, availability of labor and supporting resources, such as energy, water, and shipping connections, are considered. This study is sometimes referred to as a *feasibility study*.

This type of information must be developed so that planning decisions by senior management within the company can be made. Typically, feasibility information and supporting cost analyses are submitted to the board of directors. The board then must decide whether the investment required to build a plant is justified.

Similar analysis is necessary for any project. If a group of entrepreneurs decides to build a hotel in Phoenix, Arizona, the basic economic considerations to determine the potential profitability of this venture must be examined. If the economic study supports the idea of a hotel, a need is established. In this case, the financial institutions that lend the money for the development of the hotel typically require certain justification before providing the financing. Therefore, the structure of the feasibility study is dictated, in large part, by the requirements of the lending institution. The types of information required for developing a commercial building project will be addressed in Chapter 12.

Public and community-service-related projects do not typically involve profit and, therefore, are triggered by other considerations. If the church board of the Smallville Methodist Church decides to add a wing to provide a larger area for the Sunday school, this decision is based on improving the quality of services provided by the church. Since funds must be developed for such an addition, the church board will seek the assistance of a consultant (e.g., architect or architect/engineer [A/E]) to better define the scope of the new addition. Design and cost information are required to approach a bank or lending agency regarding financing. Based on information from design and cost consultants, the church board must decide whether to proceed to development of final design documents or place the project on hold.

Public entities, such as city, state, and federal governments, are continuously reviewing societal needs. The annual cycle of activity for public agencies looks at the changing demands of constituents with the objective of developing a plan (e.g., a set of projects) that will improve services. State highway departments, for instance, have annual budgets based on existing strategic plans. These master plans envision the construction of new roads and bridges and the maintenance of existing infrastructure. Such plans are reviewed annually, and projects to repair and enhance the transportation network of each state are budgeted. In this situation, the needs of the state are under continuous review. A balance between funds available and transportation needs must be maintained.

2.3   FORMAL NEED EVALUATION

In deciding whether or not to proceed with preliminary and final design of a given project, three items should be developed during the conceptual portion of the project cycle. The following elements provide input to the decision process:

1. Cost/benefit analysis.
2. Graphical representation of the project (e.g., sketch or artist's rendering), a layout diagram of the facility and/or a three-dimensional (3-D) depiction that may be part of a building information model (BIM).
3. Cost estimate based on the conceptual-level information available.

These documents assist the key decision maker(s) in deciding whether to proceed with a proposed project.

The cost/benefit analysis in the case of commercial or profit-based projects is simply a comparison of the estimated cost of the project against the revenues that can be reasonably expected to be generated. In public and other non-profit-based projects (e.g., monuments, churches, and museums), development of the benefit to be achieved is more difficult to pin down.

For instance, if a dam is to be constructed on the Colorado River, part of the benefit will be tangible (i.e., developed in dollars) and part will be intangible (i.e., related to the quality of life). If power is to be generated by the dam, the sale of the electricity and the revenues generated are tangible and definable in dollars amounts. Much of the benefit may, however, derive from control of the river and the changing of the environment. This dam will prevent flooding of downstream communities and form a lake that can be used as a recreational resource.

The recreational aspects of the project and the protection of communities from flooding are difficult to characterize in dollars and cents. They can be viewed as intangible benefits related to improvement of the quality of life. Protocols for converting intangible aspects of a dam project into benefits have been developed by the Bureau of Reclamation and the Army Corps of Engineers, both which are government agencies involved in water resource development. At best, however, evaluating intangibles is a judgment call and subject to review and criticism.

2.4   CONCEPTUAL DRAWINGS AND ESTIMATES

In seeking funding for entrepreneurial projects such as hotels, apartment buildings and complexes, shopping malls, and office structures, it is common practice to present conceptual documentation to potential funding sources (e.g., banks and investors). In addition to a cost/benefit analysis, graphical information to include architect's renderings or sketches as well as layout drawings, 3-D computer models and BIM assist the potential investor in better understanding the project. For this reason, such concept drawings and models are typically part of the conceptual design package. A cost estimate based on the conceptual drawings and other design information (e.g., square footage of roof area, floor space, and size of heating and air conditioning units) is prepared.

Government projects at the federal level require similar supporting analysis and are submitted with budget requests each year for congressional action. Supporting documentation includes layout sketches and outline specifications such as those shown in Figure 2.2. The supporting budget for this project is shown in Figure 2.3. These projects are included as line items in the budget of the government agency requesting funding. In this case, the requestor would be the post engineer, Fort Campbell, Kentucky. This request would be consolidated with requests at the Army and Department of Defense level and forwarded to the Bureau of the Budget to be included in the budget submitted to Congress.

It is of interest to note that because the post office project will not be built for at least a year (assuming it is approved), a projection of cost to the future date on which construction will begin is required. The projection is made using the *Engineering News Record (ENR)* indexes of basic construction cost. Construction cost indexes such as the *ENR* index allow estimators to project costs into the future. The building and construction cost indexes through March 2009 are shown in Figure 2.4.

The summary on the last page of the estimate indicates that the baseline cost of the project will be $909,050. The reserve for contingency is 10% of the base cost, or $90,900. The amount budgeted for supervision of the work by the Corps of Engineers is $54,800. The design cost is projected to be $70,000.

The amount of conceptual design documentation varies based on the complexity of the project. Fairly simple building projects, such as the Sunday school addition or the military post office, can be conceptually defined in terms of drawings, guide specifications, and a cost estimate such as those shown in Figures 2.2 and 2.3. Large and complex projects, such as petrochemical plants and power generation facilities, require expanded documentation (e.g., hundreds of pages) to define the scope of work. For this reason, the number of engineering man-hours required to develop conceptual design documentation for the chemical plant in Spain would be significantly greater than that required for a small commercial building.

2.5   PRELIMINARY AND DETAIL DESIGN

Once the concept of the project has been approved, the owner desiring the construction retains an engineer or an architect. These functions may be combined into a single stakeholder, called an architect/engineer (A/E). The end product of the design phase of project development is a set of plans and specifications that define the project to be constructed. The drawings are a graphical or schematic indication of the work to be accomplished. The specifications are a verbal or word description of what is to be constructed and to what levels of quality. When completed, they are included as legally binding elements of the contract. The production of the plans and specifications usually proceeds in two steps. The first step is called *preliminary design* and offers the owner a pause in which to review construction before detail design commences.

A common time for this review to take place is at 40% completion of the total design. The preliminary design extends the concept documentation. In most projects, a design team leader concept is utilized. The design team leader coordinates the efforts of architects and engineers from differing disciplines. The disciplines normally identified are architectural, civil and structural, mechanical, and electrical. The architect or A/E, for instance, is responsible for the development of floor plans and general layout drawings as well as

Figure 2.2 Project proposal: layout sketch and outline specifications.

25

TO: Chief of Engineers Department of the Army Washington, D.C.	FROM: Louisville District Corps of Engineers Louisville, KY
Fiscal Year 2XXX	Date Prepared: 14 Oct 2XXX
Name and Address of A.E. N.A.	
Basis of Estimate Budget Sketch & 1391	A.E. Fee N.A.
Name and Location of Installation Ft. Campbell, Kentucky	Type of Construction Permanent
Status of Design Preliminary 0% complete	Final 0% complete
Line Item Number 224	Description of Facility Post Office
Final Design Completion Date Not Authorized	

Description	Quantity	Unit	Unit Price	Totals ($000)
1. *Building*				
General construction	13,725	Sq ft	$42.24	$579.7
Plumbing	13,725	Sq ft	2.42	33.2
Heating and ventilating	13,725	Sq ft	2.68	36.8
Air conditioning (50-ton)	13,725	Sq ft	7.62	104.6
Electrical	13,725	Sq ft	5.76	79.0
Subtotal	13,725	Sq ft	60.72	833.3
2. *Utilities*				
a. *Electrical*				
Transformers	112.5	kVA	50	5.6
Poles with X-arms, pins, insulation, etc.	4	Each	720	2.9
Dead ends	6	Each	80	.5
Down guys and anchors	4	Each	180	.7
Fused cutouts and L.A.	6	Each	60	.4
#6 Bar Cu. conductor	2,400	lin ft	.30	0.7
#3/0 Neoprene covered service	160	lin ft	1.48	0.2
Parking area lights on aluminum pole 3C #8 DB 600-V	7	Each	933.41	6.5

TO: Chief of Engineers
Department of the Army
Washington, D.C.

FROM: Louisville District
Corps of Engineers
Louisville, KY

Figure 2.3 Current working estimate for budget purposes.

Description	Quantity	Unit	Unit Price	Totals ($000)
3-in. duct conc.				
encased U.G.	100	lin ft	4.75	0.5
Subtotal				19.3
b. *Water*				
3-in. Water line	365	lin ft	8.60	3.1
3-in. Gate valve and				
box	1	Each	200.00	.2
Fire hydrants	2	Each	1000.00	2.0
Connections to				
existing lines	3	Each	500.00	1.5
Subtotal				6.8
c. *Sewer*				
6-in. sanitary sewer	215	lin ft	12.00	2.6
8-in. sanitary sewer	375	lin ft	14.00	5.3
Manhole	2	Each	1000.00	2.0
Connection to exist.				
manhole	1	Each	250.00	.25
Subtotal				10.15
d. *Gas*				
$1\frac{1}{4}$ in. gas line	1,000	lin ft	6.00	6.0
$1\frac{1}{4}$ in. plug valve				
and box	1	Each	200.00	0.2
Connect to existing	1	Each	237.60	0.2
street and parking				
area crossing	280	lin ft	2.0	0.6
Subtotal				7.0
3. *Site Work*				
Clearing and grubbing	2.4	Acre	500	1.2
Borrow excavation	10,000	cu yd	8.00	80.0
Remove B. T. paving	1,070	sq yd	4.00	4.3
Subtotal				85.3
4. *Paving*				
Paving-$1\frac{1}{2}$ A.C. and				
8-in. stab.	3,950	sq yd	10.00	39.5
aggr. base	2,250	lin ft	8.50	19.2
6-in. P.C. concrete				
paving	380	sq yd	20.00	7.6
3-in. painted parking				
lines	1,680	lin ft	8.50	0.9
Concrete sidewalk	440	sq yd	18.00	7.9
Subtotal				75.1
5. *Storm Damage*				
15-in. concrete Cl. II pipe	40	lin ft	18.00	0.7
15-in. concrete Cl. III pipe	20	lin ft	20.00	0.4
Reinf. drainage structure				
concrete	8	cu yd	300.00	2.4
C.I. grates and frames	1,900	lb	1.00	1.9
Subtotal				5.4
6. *Landscaping*				

Figure 2.3 (*Continued*)

Description	Quantity	Unit	Unit Price	Totals ($000)
Sprigging and seeding	1.6	Acre	$1500.00	2.4
Landscaping		Job		3.8
Subtotal				6.2
7. *Communications*				
a. Telephone		LS	$1400.00	1.4
b. Support (within building)				
100 Pr. DB Pic Cable	600	LF	1.26	0.8
51 Pr. DB Pic Cable	550	LF	0.72	0.4
Splicing sleeves and material		LS	$900.00	0.9
Labor		LS		2.5
Subtotal				6.0
Total estimated cost (excluding design, but including reserve for contingencies and super-vision and administration (S&A)				1054.75
1. Estimated contract cost				909.05
2. Reserve for contingencies	10	percent		90.90
3. Supervision and administration (S&A); total estimated cost (excluding design, but including reserve for contingencies and supervision and administration)				54.8
				1054.75
4. *Design*				
District expenses (preliminary and final)				70.0
Subtotal				70.0

Figure 2.3 (*Continued*)

considerations such as building cladding, exterior effects, and interior finish. The mechanical engineer is concerned with the heating, ventilating, and air conditioning (HVAC), as well as service water systems. At preliminary design, decisions regarding size and location of air conditioning and heating units as well as primary water distribution components (e.g., pumps) are made. Similar decisions regarding the electrical system are made at this point by the electrical engineers. The structural and civil engineers develop the preliminary design of the structural frame and the subsurface foundation support. All of these designs are interlinked. The architectural layout impacts the weight support characteristics of the floor structure and, hence, the selection of structural system. The structural superstructure influences the way in which the foundation of the structure can be handled. The floor plan also determines the positioning of pipes and ducts and the space available for service mains.

(a) Construction Cost Index History (1919-2009)

How ENR Builds the index: 200 hours of common labor at the 20-city average of common labor rates, plus 25 cwt of standard structural steel shapes at the mill price prior to 1996 and the fabricated 20-city price from 1996, plus 1.128 tons of portland cement at the 20-city price, plus 1,088 board-ft of 2x4 lumber at the 20-city price.

ANNUAL AVERAGE

Year	Val	Year	Val	Year	Val	Year	Val
1919	198	1938	236	1957	724	1976	2401
1920	251	1939	236	1958	759	1977	2576
1921	202	1940	242	1959	797	1978	2776
1922	174	1941	258	1960	824	1979	3003
1923	214	1942	276	1961	847	1980	3237
1924	215	1943	290	1962	872	1981	3535
1925	207	1944	299	1963	901	1982	3825
1926	208	1945	308	1964	936	1983	4066
1927	206	1946	346	1965	971	1984	4148
1928	207	1947	413	1966	1019	1985	4182
1929	207	1948	461	1967	1074	1986	4295
1930	203	1949	477	1968	1155	1987	4406
1931	181	1950	510	1969	1269	1988	4519
1932	157	1951	543	1970	1381	1989	4615
1933	170	1952	569	1971	1581	1990	4732
1934	198	1953	600	1972	1753		
1935	196	1954	628	1973	1895		
1936	206	1955	660	1974	2020		
1937	235	1956	692	1975	2212		

	JAN.	FEB.	MAR.	APRIL	MAY	JUNE	JULY	AUG.	SEPT.	OCT.	NOV.	DEC.	ANNUAL AVG.
1991	4777	4773	4772	4766	4801	4818	4854	4892	4891	4892	4896	4889	4835
1992	4888	4884	4927	4946	4965	4973	4992	5032	5042	5052	5058	5059	4985
1993	5071	5070	5106	5167	5262	5260	5252	5230	5255	5264	5278	5310	5210
1994	5336	5371	5381	5405	5405	5408	5409	5424	5437	5437	5439	5439	5408
1995	5443	5444	5435	5432	5433	5432	5484	5506	5491	5511	5519	5524	5471
1996	5523	5532	5537	5550	5572	5597	5617	5652	5683	5719	5740	5744	5620
1997	5765	5769	5759	5799	5837	5860	5863	5854	5851	5848	5838	5858	5826
1998	5852	5874	5875	5883	5881	5895	5921	5929	5963	5986	5995	5991	5920
1999	6000	5992	5986	6008	6006	6039	6076	6091	6128	6134	6127	6127	6059
2000	6130	6160	6202	6201	6233	6238	6225	6233	6224	6259	6266	6283	6221
2001	6281	6272	6279	6286	6288	6318	6404	6389	6391	6397	6410	6390	6334
2002	6462	6462	6502	6480	6512	6532	6605	6592	6589	6579	6578	6563	6538
2003	6581	6640	6627	6635	6642	6694	6696	6733	6741	6771	6794	6782	6695
2004	6825	6861	6957	7017	7064	7109	7126	7188	7298	7314	7312	7308	7115
2005	7297	7298	7309	7355	7398	7415	7422	7479	7540	7563	7630	7647	7446
2006	7660	7689	7692	7695	7691	7700	7721	7723	7763	7883	7911	7888	7751
2007	7880	7880	7856	7865	7942	7939	7959	8007	8050	8045	8092	8089	7967
2008	8090	8094	8109	8112	8141	8185	8293	8362	8557	8623	8602	8551	8310
2009	8549	8533	8534										

Base: 1913=100

(b) Building Cost Index History (1923-2009)

How ENR Builds the index: 68.38 hours of skilled labor at the 20-city average of bricklayers, carpenters and structural ironworkers rates, plus 25 cwt of standard structural steel shapes at the mill price prior to 1996 and the fabricated 20-city price from 1996, plus 1.128 tons of portland cement at the 20-city price, plus 1,088 board-ft of 2×4 lumber at the 20-city price.

ANNUAL AVERAGE

Year	Val	Year	Val	Year	Val	Year	Val
1923	186	1941	211	1959	548	1977	1545
1924	186	1942	222	1960	559	1978	1674
1925	183	1943	229	1961	568	1979	1819
1926	185	1944	235	1962	580	1980	1941
1927	186	1945	239	1963	594	1981	2097
1928	188	1946	262	1964	612	1982	2234
1929	191	1947	313	1965	627	1983	2384
1930	185	1948	341	1966	650	1984	2417
1931	168	1949	352	1967	676	1985	2425
1932	131	1950	375	1968	721	1986	2483
1933	148	1951	401	1969	790	1987	2541
1934	167	1952	416	1970	836	1988	2598
1935	166	1953	431	1971	948	1989	2634
1936	172	1954	446	1972	1048	1990	2702
1937	196	1955	469	1973	1138	1991	2751
1938	197	1956	491	1974	1205		
1939	197	1957	509	1975	1306		
1940	203	1958	525	1976	1425		

	JAN.	FEB.	MAR.	APRIL	MAY	JUNE	JULY	AUG.	SEPT.	OCT.	NOV.	DEC.	ANNUAL AVG.
1992	2784	2775	2799	2809	2828	2838	2845	2854	2857	2867	2873	2875	2834
1993	2886	2886	2915	2976	3071	3066	3038	3014	3009	3016	3029	3046	2996
1994	3071	3106	3116	3127	3125	3115	3107	3109	3116	3116	3109	3110	3111
1995	3112	3111	3103	3100	3096	3095	3114	3121	3109	3117	3131	3128	3112
1996	3127	3131	3135	3148	3161	3178	3190	3223	3246	3284	3304	3311	3203
1997	3332	3333	3323	3364	3377	3396	3392	3385	3378	3372	3350	3370	3364
1998	3363	3372	3368	3375	3374	3379	3382	3391	3414	3423	3424	3419	3391
1999	3425	3417	3411	3421	3422	3433	3460	3474	3504	3505	3498	3497	3456
2000	3503	3523	3536	3534	3558	3553	3545	3546	3539	3547	3541	3548	3539
2001	3545	3536	3541	3541	3547	3572	3625	3605	3597	3602	3596	3577	3574
2002	3581	3581	3597	3583	3612	3624	3652	3648	3655	3651	3654	3640	3623
2003	3648	3655	3649	3652	3660	3677	3684	3712	3717	3745	3766	3758	3694
2004	3767	3802	3859	3908	3955	3996	4013	4027	4103	4129	4128	4123	3984
2005	4112	4116	4127	4168	4189	4195	4197	4210	4242	4265	4312	4329	4205
2006	4335	4337	4330	4335	4331	4340	4356	4360	4375	4431	4462	4441	4369
2007	4432	4432	4411	4416	4475	4471	4493	4515	4533	4535	4558	4556	4486
2008	4557	4556	4571	4574	4599	4640	4723	4733	4827	4867	4847	4797	4691
2009	4782	4765	4767										

Base: 1913=100

Figure 2.4 *Engineering News Record* Construction Cost Indexes (a) Building Cost Index (b) Construction Cost Index (Reprinted from Engineering News-Record, Copyright McGraw-Hill Companies Inc., 23 March 2009. All rights reserved.)

Once the preliminary design has been approved by the owner, final or detail design is accomplished. This is the second step in the production of the plans and specifications. For the A/E this focuses on the interior finishes, which include walls, floors, ceilings, and glazing. Details required to install special finish items are designed. Precise locations and layout of electrical and mechanical systems as well as the detail design of structural members and connections are accomplished by the appropriate engineers. As noted, the detail design phase culminates in the plans and specifications that are given to the constructor for bidding purposes. In addition to these detailed design documents, the A/E produces a final *owner's estimate* indicating the total job cost minus markup. This estimate should achieve approximately ±3% accuracy, since the total design is now available. The owner's estimate is used (a) to ensure the design produced is within the owner's financial resources to construct (i.e., the A/E has not designed a gold-plated project) and (b) to establish a reference point in evaluating the bids submitted by the competing contractors. In some cases, when all contractor bids greatly exceed the owner's estimate, all bids are rejected and the project is withdrawn for redesign or reconsideration. Once detailed design is completed, the owner again approves the design prior to advertising the project to prospective bidders.

2.6 NOTICE TO BIDDERS

The document announcing to prospective bidders that design documents are available for consideration and that the owner is ready to receive bids is called the *notice to bidders*. Because of his commitment to the owner to design a facility that can be constructed within a given budget and at an acceptable level of quality, the A/E wants to be sure that the lowest bid price is achieved. To ensure this, the job is advertised to those contractors who are capable of completing the work at a reasonable price. All A/E firms maintain computer-based lists or other methods of contact with qualified bidders. When design is complete, a notice to bidders, such as the one shown in Figure 2.5, is sent either electronically or by regular mail to all prospective bidders. The notice to bidders contains information

---

NOTICE TO BIDDERS
FOR
CONSTRUCTING SEWERAGE SYSTEM IMPROVEMENTS
CONTRACT "B"
CENTRAL STATE HOSPITAL
FOR THE
GEORGIA BUILDING AUTHORITY (HOSPITAL)
STATE CAPITOL—ATLANTA, GEORGIA

Sealed proposals will be received for Constructing Sewerage System Improvements, Contract "B," for the Georgia Building Authority (Hospital), State Capitol, Atlanta, Georgia, at Room 315, State Health Building, 47 Trinity Avenue, S.W., Atlanta, Georgia, until 2:00 P.M., E.S.T., February, 18 __, at which time and place they will be publicly opened and read. Bidding information on equipment in Section No. 10 shall be submitted on or before February 4 __.

Work to Be Done: The work to be done consists of furnishing all materials, equipment, and labor and constructing:

Division One. Approximately 12,400 L.F. 36″ Sewer Pipe, 5,650 L.F. 30″ Sewer Pipe, 7,300 L.F. 24″ Sewer Pipe, 1,160 L.F. 15″ Sewer Pipe, 3,170 L.F. 12″ Sewer Pipe, 300 L.F. 8″ Sewer Pipe, 418 L.F. 36″ C.I. Pipe Sewer, 324 L.F. 30″ C.I. Pipe Sewer, 1,150 L.F. 30″ C.I. Force Main, 333 L.F. 24″ C.I. Force Main, 686 L.F. 24″ C.I. Pipe Sewer, and all other appurtenances for sewers.
Division Two. One Sewage Pumping Station—"Main Pump Station."
Division Three. One Sewage Pumping Station—"Fishing Creek Pump Station."
Division Four. One Sewage Pumping Station—"Camp Creek Pump Station."

Bids may be made on any or all Divisions, any of which may be awarded individually or in any combination.

Figure 2.5 Notice to bidders (Courtesy of Georgia Building Authority).

Proposals. Proposals shall contain prices, in words and figures, for the work bid on. All Proposals must be accompanied by a certified check, or a bid bond of a reputable bonding company authorized to do business in the State of Georgia, in an amount equal to at least five (5%) percent of the total amount of the bid.

Upon the proper execution of the contract and required bonds, the checks or bid bonds of all bidders will be returned to them.

If Proposals are submitted via mail rather than delivery they should be addressed to Mr. Smith, Director, Department of Administration and Finance, Georgia Department of Public Health, Room 519, State Health Building, 47 Trinity Avenue, S.W., Atlanta, Georgia 30334.

Performance and Payment Bonds: A contract performance bond and payment bond, each in an amount equal to one hundred (100%) percent of the contract amount, will be required of the successful bidder.

Withdrawal of Bids: No submitted bid may be withdrawn for a period of sixty (60) days after the scheduled closing time for the receipt of bids.

Plans, Specifications, and Contract Documents: Plans, Specifications and Contract Documents are open to inspection at the Office of the Georgia Building Authority (Hospital), State Capitol, Atlanta, Georgia, or may be obtained from Wiedeman and Singleton, Engineers, P.O. Box 1878, Atlanta, Georgia 30301, upon deposit of the following amounts.

Division One: $45.00 for Plans and Specifications.
Divisions Two, Three and Four (Combined). $50.00 for Plans and Specifications.
Divisions One to Four, Inclusive. $75.00 for Plans and Specifications.
All Divisions. $20.00 for Specifications only.

Upon the return of all documents in undamaged condition within thirty (30) days after the date of opening of bids, one-half of the deposit will be refunded. No refunds will be made for plans and documents after thirty (30) days.

Wage Schedule: The schedule of minimum hourly rates of wages required to be paid to the various laborers and mechanics employed directly upon the site of the work embraced by the Plans and Specifications as determined by the Secretary of the U.S. Department of Labor, Decision No. AI-971, is included in the General Conditions of the Specifications. This decision, expiring prior to the receipt of bids, will be superceded by a new decision to be incorporated in the Contract before the award is made to the successful bidder.

Acceptance or Rejection of Bids: The right is reserved to accept or reject any or all bids and to waive informalities.

THIS PROJECT WILL BE FINANCED IN PART BY A GRANT FROM THE FEDERAL WATER POLLUTION CONTROL ADMINISTRATION AND WILL BE REFERRED TO AS PROJECT WPC-GA-157.

BIDDERS ON THIS WORK WILL BE REQUIRED TO COMPLY WITH THE PRESIDENT'S EXECUTIVE ORDERS NO. 11246 and NO. 11375. THE REQUIREMENTS FOR BIDDERS AND CONTRACTORS UNDER THESE ORDERS ARE EXPLAINED IN THE SPECIFICATIONS.

GEORGIA BUILDING AUTHORITY (HOSPITAL)

By: _____
Secretary–Treasurer

Figure 2.5 (*Continued*)

regarding the general type and size of the project, the availability of plans and specifications for review, and the time, place, and date of the bid opening. Normally, sets of plans and specifications are available electronically for perusal at online plan rooms. Such facilities provide plans and specifications for a large number of jobs being let for bid. They afford contractors the opportunity to electronically review a large number of jobs available for bid within a given time frame and geographical area. The expenditure of effort on the part of the contractor to review available jobs is modest. Should a contractor decide to bid on a particular job, his or her commitment increases sharply in terms of money and time invested.

Contractors have other methods of learning about jobs that are available for bid. In some large cities, a builder's exchange may operate to serve the contracting community and keep it appraised of the status of design and bid activity within a given area. These exchanges have Web sites with information such as that shown in Figure 2.6. These reports indicate what jobs are available for bidding, and A/Es make use of such facilities to gain maximum coverage in advertising their jobs.

Nationwide services such as the Dodge Reporting System also provide information on projects being let for bid. For a subscription fee, such services provide information regarding jobs categorized by type of construction, geographical location, job size, and other parameters directly to the contractor. The information announcements indicate whether the job is under design, ready for bid, or awarded. In the cases of jobs that have been awarded, the low bid and other bid prices submitted are furnished so that the contractor can detect bidding trends in the market. A typical Web-Based Invitation to Bid is shown in Figure 2.7.

2.7   BID PACKAGE

The documents that are available to the contractor and on which he must make a decision to bid or not to bid are those in the *bid package*. In addition to the *plans* and *technical specifications*, the bid package prepared by the design professional consists of a *proposal* form, *general conditions* that cover procedures common to all construction contracts, and *special conditions*, which pertain to procedures to be used that are unique to a given project. All supporting documents are included by reference in the proposal form. A schematic representation of the bid package layout is shown in Figure 2.8.

The proposal form as designed and laid out by the A/E is the document that, when completed and submitted by the contractor, indicates the contractor's desire to perform the work and the price at which he or she will construct the project. A typical example of a proposal is shown in Figure 2.9.

The proposal form when completed and submitted establishes intent on the part of the contractor to enter into a contract to complete the work specified at the price indicated in the proposal. It is an offer and by itself is not a formal contract. If, however, the owner responds by awarding the contract based on the proposal, an acceptance of the offer results and a contractual relationship is established. The prices at which the work will be constructed can be stated either as lump-sum or as unit-price figures. A portion of the price schedule for the example project described in Figure 2.5 is shown in Figure 2.9. As shown in the figure, two methods (lump sum and unit price) of quoting price are illustrated. Items 1 to 4 require the bidder to specify unit price (i.e., dollar per unit) for the guide quantities specified. Therefore, if the contractor will do the rock excavation for $80.00 per cubic yard, this price is entered along with the total price (550 × $80.00 = $44,000). Items 5, 6, and 7 require lump- or stipulated-sum quotations. Therefore, the contractor states a single price for the access road, finish grading, and so on.

bids

Replace Roof, Building 3245 Ft Benning, GA

---

2 PM August 20

---

bids

Family Life Center for First Assembly of God Griffin, GA

---

2 PM August 22

---

bids

Misc Elect, Bldg 2752 Ft Benning, GA

---

2 PM August 22

REPLACE ROOF, BLDG 3245 (B-0037), FT BENNING, GA (CHATTAHOOCHEE CO)
Submit Bids To .... Directorate of Contracting, Bldg 6 (Meloy Hall), Mailroom 207, ATZB-KTD, Ft Benning GA 31905-5000 (Isaac D Larry, contact) 706/545-2193 2 PM August 20
Plans Available From .... Owner
Bond Info .... 20% bid bond. 50% pymt bond. 100% perf bond.
Div 2 .... demol
Div 5 .... misc mtl
Div 6 .... rough carp
Div 7 .... roof insul, b/u rfg, sht mtl wk, caulking & sealants
Div 9 .... ptg
PLANS ON FILE .... ATL Bin 56 - #107293

FAMILY LIFE CENTER FOR FIRST ASSEMBLY OF GOD (2000 W McIntosh Rd) GRIFFIN, GA (SPALDING CO)
Submit Bids To .... First Assembly of God, Griffin, GA 2 PM August 22
Archt .... Rardin & Carroll Architects, 6105 Preservation Drive, Ste A, Chattanooga, TN 37416 423/(894-2839) 894-3242
Civil Engr .... Breelove Land Planners, Atlanta, GA
Struct Engr .... Robinson & Associates, Atlanta, GA
Mech/Elect/Plbg Engr ...... Brewer & Skala Engineers, Atlanta, GA
Plans Available From .... Archt $100 plan dep to GC's, one set ref to bona fide GC's within 10 days, all others non-ref cost of reprod.
Bond Info .... 5% bid bond. 100% pymt/perf bond.
Est Cost $1,400,000
Scope .... Approx 26,000 s.f. 1 story classrooms, offices & gym
Div 2 .... sel demol, site clrg & prep, earthwk, termite control
Div 3 .... cast-in-place conc, cementitious wood fiber decking
Div 4 .... unit masonry, precast conc window stool
Div 5 .... struct stl, open web joists, composite roof deck assemb, light gauge stl frmg, mtl handrails
Div 6 .... rough carp, struct glued laminated timber, finish carp & millwk
Div 7 .... waterprfg, dampprfg, bldg insul, masonry insul, EIFS, roof insul over wood deck, firestpg, fiberglass shingles, vinyl siding, rubber membrane rfg, flash & s/m, int caulking, ext sealant
Div 8 .... hol mtl wk, alum doors & frames, wood doors, counter doors, alum windows, glass & glzg
Div 9 .... gyp wallbd, acoust ceilings, poured in place synth sports flrg, resil flrg, cpt (owner provided), ptg, vinyl-coated fabric
Div 10 .... laminated plastic toilet partns, solid phenolic core partns, toilet & bath acces, fire extin & cabinets
Div 11 .... athletic equip
Div 15 .... hvac, ductwk & acces, computerized damper (VVT) sys, louvers-grilles-registers & diffusers, unitary exhaust & supply fans & vents, split sys a/c, refrigerant piping, pkg rooftop htg & vent units, auto controls, high efficiecy gas-fired duct heater, kitchen vent equip, test & balance, plbg, natl gas piping sys, fire prot, roof curbs
Div 16 .... elect, lighting, fire alarm sys
PLANS ON FILE .... ATL Bin 38 - #107182

MISCELLANEOUS ELECTRICAL REPAIRS, BLDG 2752 (B-0051), FT BENNING, GA (CHATTAHOOCHEE CO)
Submit Bids To .... Directorate of Contracting, Bldg 6 (Meloy Hall), Mailroom 207, ATZB-KTD, Ft Benning GA 31905-5000 (Sabra A Boynton, contact) 706/545-2221 2 PM August 22
Plans Available From .... Owner
Bond Info .... 20% bid bond. 50% pymt bond. 100% perf bond.
Div 2 .... demol
Div 16 .... int elect wk
PLANS ON FILE .... ATL Bin 55 - #107292

Figure 2.6 Typical Daily Building Report.

In the proposal form shown in Figure 2.9, the contract duration is also specified, although this is not always the case. In many instances, the project duration in working or calendar days is specified in the *special conditions* portion of the bid package. This proposal form indicates also that the contractor is to begin work within 10 calendar days after receipt of written *notice of award* of contract. Award of contract is usually communicated to the contractor in the form

> **THE FORT COLLINS LAFAYETTE SCHOOL DISTRICT**, 400 E La Porte Avenue Fort Collins CO 80522, (hereinafter referred to as "The School") will receive sealed bids prior to the date and time stated for the Bid Opening for contracts for the construction of the **Lafayette High School Additions {Contract LHS-118)**. The contract will involve the addition of a Gym and Arts wing to existing facilities of the **Lafayette High School, 406 E Jupiter Drive, Ft Collins CO 80522.DIEK AND BACO, LLC,** a Colorado Limited Liability Company, is the project's Engineer, with business offices at: 122 W Straw Circle, Fort Collins, CO 80522, (970)-555-4021.
>
> A **bid package, plan holders list, and other contract documents relating to the project** may be obtained beginning February 9, 20XX through: Blue Helios, Inc., 620 Laurel Avenue, Fort Collins, CO 80523 (970)-555-1010. A deposit of $200.00 check or money order only is required for the first set of plans and any additional plans may be purchased through Blue Helios, Inc. A **pre-bid conference** will be held at 8:30 a.m. {MST) on Tuesday, February 16, 20XX at Room 214 of the project's Engineer business office. Response to any questions submitted will be provided publicly at the pre-bid conference or through addenda. Correspondence must be addressed to the attention of: Jorge L. Lopez, jorge.lopez@dbllc.com. **Sealed bids must be received no later than 2:00 p.m. on March 9, 20XX**, with a public opening at such time on such date, at the project's Engineer business office. **Bid security** in the amount of five percent (5%) of bid amount must accompany each bid in accordance with the instructions to Bidders.

Figure 2.7 Typical Web-Based Invitation to Bid.

Figure 2.8 Bid package documents.

of a *notice to proceed*. Response by the owner to the contractor's proposal with a notice to proceed establishes a legally binding contractual relationship. This relationship must be supported by the legal signatures of individuals empowered to represent (i.e., commit contractually) the firm making the proposal.[1]

2.8   GENERAL CONDITIONS

Certain stipulations regarding how a contract is to be administered and the relationships between the parties involved are often the same for all contracts. An organization that enters into a large number of contracts each year normally evolves a standard set of stipulations that establishes these procedures and applies them to all construction contracts. This set of provisions is normally referred to as the *general conditions*. Large government contracting organizations such as the U.S. Army Corps of Engineers, the Bureau of

---

[1] The Electronic Signatures in Global and National Commerce Act of 2000 ("E-Sign Act") made legally binding the use of any "electronic sound, symbol, or process attached to or logically associated with a contract or other record and executed or adopted by a person with the intent to sign the record."

PROPOSAL
TO THE GEORGIA BUILDING AUTHORITY (HOSPITAL)
STATE CAPITOL
ATLANTA, GEORGIA

Submitted: _____(date)_____, 2XXX

The undersigned, as Bidder, hereby declares that the only person or persons interested in the Proposal as principal or principals is or are named herein and that no other person than herein mentioned has any interest in this Proposal or in the Contract to be entered into; that this Proposal is made without connection with any other person, company, or parties making a bid or Proposal; and that it is in all respects fair and in good faith without collusion or fraud.

The Bidder further declares that he has examined the site of the work and informed himself fully in regard to all conditions pertaining to the place where the work is to be done; that he has examined the plans and specifications for the work and contractual documents relative thereto, and has read all Special Provisions and General Conditions furnished prior to the opening of bids; and that he has satisfied himself relative to the work to be performed.

The Bidder proposes and agrees, if this Proposal is accepted, to contract with the Georgia Building Authority (Hospital), Atlanta, Georgia, in the form of contract specified, to furnish all necessary material, equipment, machinery, tools, apparatus, means of transportation, and labor, and to finish the construction of the work in complete accordance with the shown, noted, described, and reasonable intended requirements of the plans and specifications and contract documents to the full and entire satisfaction of the Authority with a definite understanding that no money will be allowed for extra work except as set forth in the attached General Conditions and Contract Documents, for the following prices:

CAMP CREEK PUMP STATION

Section 1: Unit Price Work

(For part payment—except rock excavation—by unit prices, to establish price for variation in quantities. Include balance of quantities for these items—except rock excavation—in lump sum bid for Section 2.)

Item Number	Quantity	Unit	Description	Unit Price	Total Amount
1.	550	cubic yard (cu yd)	Rock excavation (for structures and pipes only)	$_____	$_____
2.	50	linear foot (lin ft)	8″ C.I. force main	$_____	$_____
3.	20	cubic yard (cu yd)	Trench excavation for pipes	$_____	$_____
4.	200	square yard (sq yd)	Paving	$_____	$_____

Subtotal, Section 1, Item Nos. 1 to 4, Inclusive _____

_____ Dollars ($_____)

Figure 2.9 A typical proposal (Courtesy of Georgia Building Authority).

Section 2: Lump Sum Work

Item No.	Description		Total Amount
5.	Excavation and Fill		
	(a) Access Roadway	$_____	
	(b) Structure Excavation and Backfill	$_____	
	(c) Finish Grading	$_____	
	Total for Item No. 5		$_____
6.	Paving		
	(a) Access Roadway	$_____	
	(b) Station Area	$_____	
	Total for Item No. 6		$_____
7.	Concrete Work		$_____

The Bidder further proposes and agrees hereby to commence work under his contract, with adequate force and equipment, on a date to be specified in a written order of the Engineer, and shall fully complete all work thereunder within the time stipulated, from and including said date, in 300 consecutive calendar days.

The Bidder further declares that he understands that the quantities shown in the Proposal are subject to adjustment by either increase or decrease, and that should the quantities of any of the items of work be increased, the Bidder proposes to do the additional work at the unit prices stated herein; and should the quantities be decreased, the Bidder also understands that payment will be made on the basis of actual quantities at the unit price bid and will make no claim for anticipated profits for any decrease in quantities, and that actual quantities will be determined upon completion of the work, at which time adjustment will be made to the Contract amount by direct increase or decrease.

The Bidder further agrees that, in case of failure on his part to execute the Construction Agreement and the Bonds within ten (10) consecutive calendar days after written notice being given of the award of the Contract, the check or bid bond accompanying this bid, and the monies payable thereon, shall be paid into the funds of the Georgia Building Authority (Hospital), Atlanta, Georgia, as liquidated damages for such failure, otherwise the check or bid bond accompanying his Proposal shall be returned to the undersigned.

Attached hereto is a bid bond by the _____
_____ in the amount of
_____
_____ Dollars ($_____)
made payable to the Georgia Building Authority (Hospital), Atlanta, Georgia, in accordance with the conditions of the Advertisement for Bids and the provisions herein.

Submitted: _____ L.S.

By: _____ L.S.

Title: _____

Note: If the Bidder is a corporation, the Proposal shall be signed by an officer of the corporation; if a partnership, it shall be signed by a partner. If signed by others, authority for signature shall be attached.

ADDRESS: _____
_____

Figure 2.9 (*Continued*)

Table 2.1 Topics Typically Addressed in General Conditions

1. Definitions
2. Preliminary matters
3. Contract documents
4. Bonds and insurance
5. Contractor's responsibilities
6. Owner's responsibilities
7. Engineer's responsibilities
8. Changes in the work
9. Change of contract price
10. Change of contract times
11. Tests and inspections
12. Payments to contractor and completion
13. Suspension of work and termination
14. Dispute resolution

Reclamation, and the General Services Administration (Public Building Service) have a standard set of general provisions. For those organizations that enter into construction contracts on a less frequent basis, professional and trade organizations publish standards that are commonly used in the industry. A committee for engineer documents has been formed jointly by the American Consulting Engineers Council, the National Society of Professional Engineers, and the American Society of Civil Engineers to prepare standard contract documents. The committee is called the Engineers Joint Contract Documents Committee (EJCDC). These documents have been endorsed by the Associated General Contractors (AGC) of America and the Construction Specifications Institute (CSI). Some topics that are typically considered in the general conditions are shown in Table 2.1. The rights, privileges, and responsibilities that accrue to the primary contractual parties in any construction contract are also defined in the general conditions. Therefore, sections pertaining to the (a) owner, (b) architect (or A/E), (c) contractor, and (d) subcontractors are typically found in the general conditions. Most contractors become thoroughly familiar with the standard forms of general conditions (i.e., Corps of Engineers) and can immediately pick up any additions or changes. Each of the provisions of a standard set of general conditions has legal implications, and the wording cannot be changed without careful consideration. The contract language embodied in the general conditions has been hammered out over the years from countless test cases and precedents in both claims and civil courts. The wording has evolved to establish a fair and equitable balance of protection for all parties concerned. In cases where a contractor finds considerable deviation from the standard language, he or she may decline the opportunity to bid fearing costs of litigation in clarifying contractual problems. In areas in which small deviation is possible, the language of a given standard form may tend to be protective of one party (e.g., the architect) and to hold others responsible when gray areas arise. Predictably, the AGC standard sub-contract protects the contractor in areas in which responsibility is unclear or subject to interpretation.

2.9 SUPPLEMENTARY CONDITIONS

Those aspects of the contractual relationship that are peculiar or unique to a given project are given in the *supplementary conditions*. Items, such as the duration of the project, additional instructions regarding commencement of work, owner-procured materials, mandatory wage rates characteristic of the local area, format required for project progress reporting (e.g., a network schedule), and amount of liquidated damages, are typical of the provisions included in the supplementary conditions. Items contained in supplementary conditions are of two types:

1. Modifications to the basic articles of the general condition in the form of additions, deletions, or substitutions.
2. Additional articles of a contractual-legal nature that may be desirable or necessary for a particular project.

Because some of the provisions are extensions or interpretations of the general conditions, some of the major paragraph titles are similar to those used in the general conditions. The contents of a typical set of supplementary or special conditions for a Corps of Engineers channel improvement project are shown in Figure 2.10.

2.10   TECHNICAL SPECIFICATIONS

The contract documents must convey the requirements of the project to potential bidders and establish a legally precise picture of the technical aspects of the work to be performed. This is accomplished visually through the use of drawings. A verbal description of the technical requirements is established in the *technical specifications*. These provisions pertain in large part to the establishment of quality levels. Standards of workmanship and material standards are defined in the specifications. For materials and equipment, this is often done by citing a specific brand name and model number as the desired item for installation. In government procurement, in which competitive procurement must take place, a similar approach is used. Government specifications usually cite a specific brand or model and then establish the requirement that this or an equal item be used. The fact that equality exists must be established by the bidder.

Often the quality required will be established by reference to an accepted practice or quality specification. The American Concrete Institute (ACI), the American Welding Society (AWS), the American Association of State Highway and Transportation Officials (AASHTO), the American Society for Testing and Materials (ASTM), as well as federal procurement agencies publish recognized specifications and guides. A list of some typical references is given in Figure 2.11. The organization of the technical specifications section usually follows the sequence of construction. Therefore, specifications regarding concrete placement precede those pertaining to mechanical installation. A typical index of specifications for a heavy construction project might appear as follows:

Section

1. Clearing and grubbing
2. Removal of existing structures
3. Excavation and fill
4. Sheet steel piling
5. Stone protection
6. Concrete
7. Miscellaneous items of work
8. Metal work fabrication
9. Water supply facilities
10. Painting
11. Seeding

Figure 2.10 Special conditions: typical index of special conditions (Courtesy of the Army Corps of Engineers).

American Concrete Institute

211.1-91 Standard Practice for Selecting Proportions for Normal, Heavyweight, and Mass Concrete (Reapproved 2009)

211.2-98 Standard Practice for Selecting Proportions for Structural Lightweight Concrete (Reapproved 2004)

214R-02 Evaluation of Strength Test Results of Concrete

301-05 Specifications for Structural Concrete

318-08 Building Code Requirements for Structural Concrete and Commentary

421.2R-07 Seismic Design of Punching Shear Reinforcement in Flat Plates

ITG-7-09 Specification for Tolerances for Precast Concrete

SP-211 Large-Scale Structural Testing

ACI 305.1-06—Specification for Hot Weather Concreting

ACI 306.1-90 Standard Specification for Cold Weather Concreting (Reapproved 2002)

ACI 530/530.1—Building Code Requirements & Specifications for Masonry Structures, 2005

American Society for Testing and Materials

A185/A185M-07 Standard Specification for Steel Welded Wire Reinforcement, Plain, for Concrete

A934/A934M-07 Standard Specification for Epoxy-Coated Prefabricated Steel Reinforcing Bars

C270-08a Standard Specification for Mortar for Unit Masonry

C55-09 Standard Specification for Concrete Building Brick

C109/C109M-08 Standard Test Method for Compressive Strength of Hydraulic Cement Mortars

C1116/C1116M-09 Standard Specification for Fiber-Reinforced Concrete

C1329-05 Standard Specification for Mortar Cement

C140-09 Standard Test Methods for Sampling and Testing Concrete Masonry Units and Related Units

C33/C33M-08 Standard Specification for Concrete Aggregates

C39/C39M-09 Standard Test Method for Compressive Strength of Cylindrical Concrete Specimens

C42/C42M-04 Standard Test Method for Obtaining and Testing Drilled Cores and Sawed Beams of Concrete

C91-05 Standard Specification for Masonry Cement

C94/C94M-09a Standard Specification for Ready-Mixed Concrete

C977-03 Standard Specification for Quicklime and Hydrated Lime for Soil Stabilization

C1328-05 Standard Specification for Plastic (Stucco) Cement

C141/C141M-09 Standard Specification for Hydrated Hydraulic Lime for Structural Purposes

American Welding Society

AWS D1.4/D1.4M: 2005 Structural Welding Code—Reinforcing Steel

American Society of Civil Engineers

ASCE/SEI 7-05 Minimum Design Loads for Buildings and Other Structures

Building Seismic Safety Council

NEHRP Recommended Provisions and Commentary for Seismic Regulations for New Buildings and Other Structures. 2003 Edition

National Ready Mix Concrete Association

Concrete Plant Standards of Concrete Plant Manufacturers Bureau (15th Revision, 2007)

Truck Mixer, Agitator and Front Discharge Concrete Carrier (15th Revision, 2001)

Concrete Reinforcing Steel Institute Reinforcement

Reinforcement, Anchorages and Splices, 5th Edition 2008

Figure 2.11 Typical references to structural inspection and testing standards.

As with the general conditions, most contractors are familiar with the appearance and provisions of typical technical specifications. A contractor can quickly review the specifications to determine whether there appears to be any extraordinary or nonstandard aspects that will have an impact on cost. These clauses or nonstandard provisions are underlined or highlighted so that they can be studied carefully.

2.11   ADDENDA

The bid package documents represent a description of the project to be constructed. They also spell out the responsibilities of the various parties to the contract and the manner in which the contract will be administered. These documents establish the basis for determining the bid price and influence the willingness of the prospective bidder to bid or enter into a contract. It is important, therefore, that the bid package documents accurately reflect the project to be constructed and the contract administration intentions of this owner or of the owner's representative. Any changes in detail, additions, corrections, and contract conditions that arise before bids are opened that are intended to become part of the bid package and the basis for bidding are incorporated into the bid package through *addenda* (whose singular is *addendum*).

An addendum thus becomes part of the contract documents and provides the vehicle for the owner (or the owner's representative) to modify the scope and detail of a contract before it is finalized. It is important therefore that addenda details be rapidly communicated to all potential bidders prior to bid submission. Since addenda serve notice on the prospective bidder of changes in the scope or interpretation of the proposed contract, steps must be taken to ensure that all bidders have received all issued addenda. Consequently, addenda delivery must be documented and confirmed at the time of bid submission. This is done by the submission of a signed document listing receipt of each duly identified addendum.

Once a contract has been signed, future changes in the scope or details of a contract may form the basis for a new financial relationship between contracting parties. The original contract, in such cases, can no longer be accepted as forming the basis for a full description of the project. Such changes are referred to as *change orders* (see Section 3.5).

2.12   THE ESTIMATE AND THE DECISION TO BID

After investigating the plans and specifications, the contractor must make a major decision as to whether or not to bid the job. This is an important financial decision because it implies incurring substantial cost that may not be recovered. Bidding the job requires a commitment of man-hours by the contractor for the development of the estimate.

Estimating is the process of looking into the future and trying to predict project costs and various resource requirements. The key to this entire process is the fact that these predictions are made based on past experiences and the ability of the estimator to sense potential trouble spots that will affect field costs. The accuracy of the result is a direct function of the skill of the estimator and the accuracy and suitability of the method by which these past experiences were recorded. Since the estimate is the basis for determining the bid price of a project, it is important that the estimate be carefully prepared. Studies reveal the fact that the most frequent causes of contractor failure are incorrect and unrealistic estimating and bidding practices.

The quantities of materials must be developed from the drawings by an expert. The process of determining the required material quantities on a job is referred to as *quantity takeoff* or *quantity surveying*. Once quantities are established, estimators who have access to pricing information use these quantities and their knowledge of construction methods and productivities to establish estimates of the *direct costs* of performing each construction task.

They then add to the totaled project direct costs those *indirect costs* that cannot be assigned directly to a particular physical piece of the project but which must be distributed across the entire project (e.g., project supervising costs). Finally, the bid price is established by adding the *management and overhead* costs, allowances for contingencies, and a suitable profit margin. Appendix A gives typical considerations affecting the decision to bid.

The cost of the time and effort expended to develop a total bid price and submit a proposal is only recovered in the event the contractor receives the contract. A common rule of thumb states that the contractor's estimating cost will be approximately 0.25% of the total bid price. This varies, of course, based on the complexity of the job. Based on this rule, a cost of $25,000 can be anticipated for estimating a job with a total bid cost in the vicinity of $10 million. Expending this amount of money to prepare and submit a bid with only a probability of being awarded the work is a major monetary decision. Therefore, most contractors consider it carefully. To recover bidding costs for jobs not awarded, contractors place a charge in all bids to cover bid preparation costs. This charge is based on their frequency of contract award. That is, if a contractor, on the average, is the selected bidder on one in four of the contracts bid, he or she will adjust the bid cost included in each proposal to recover costs for the three in four jobs not awarded. In addition to the direct costs of bid preparation, the contractor may be required to pay certain fees for accessing a full set of plans for the project. Other costs include telephone and computer charges related to obtaining quotations from subcontractors and vendors and the administrative costs of getting these quotations in writing. A small fee must be paid for a bid bond (see Section 2.15) and the administrative aspects of submitting the proposal in conformance with the *instructions to bidders*.

2.13 PREQUALIFICATION

In some cases the complexity of the work dictates that the owner must be certain that the selected contractor is capable of performing the work described. Therefore, before considering a bid, the owner may decide to prequalify all bidders. This is announced in the instructions to bidders. Each contractor interested in preparing and submitting a bid is asked to submit documents that establish his firm's expertise and capability in accomplishing similar types of construction. In effect, the owner asks the firm to submit its résumé for consideration. If the owner has doubts regarding the contractor's ability to successfully complete the work, the owner can simply withhold qualification.

This is helpful to both parties. The contractor does not prepare a bid and incur the inherent cost unless qualified. On the other hand, the owner does not find himself or herself in the position of being under pressure to accept a low bid from a firm he or she feels cannot perform the work. In the extreme case, a small firm with experience only on single-family residential housing may bid low on a complex radar-tracking station. If the owner feels the contractor will not be able to successfully pursue the work, prequalification can be withheld.

2.14 SUBCONTRACTOR AND VENDOR QUOTATIONS/CONTRACTS

As already noted, estimating section personnel will establish costs directly for those items to be constructed by the prime contractor with in-house forces. For specialty areas such as electrical work, interior finish, and roofing, the prime contractor solicits quotations from subcontractors with whom he or she has successfully worked in the past. Material price quotations are also developed from vendors. These quotations are normally taken electronically and included in the bid. It is good business practice to use a sub-contractor/vendor bid quotation form that includes a legal signature as well as the bid quotation. The contractor soliciting the quotation should request that the bid be faxed with

2.15 Bid Bond **43**

the quotation and a signature binding the bidder to the quoted price. Alternatively, signed documentation of the quotation can be submitted as an e-mail attachment.

The contractor integrates these quotations into the total bid price. Until the contractor has a firm subcontract or purchase order, the only protection against the vendor or subcontractor reneging on or changing the quoted price is submitted by fax or signed e-mail attachment quotation form. Prior to the availability of electronically transmitted methods of submission, the contractor had only a telephone quote and the subcontractor's word that he or she would sign a formal agreement at the stated price. The use of electronically transmitted documents and signed quotations has greatly improved this situation and eliminated potential misunderstandings.

Following award of the contract, the prime contractor has his purchasing or procurement group move immediately to establish subcontracts with the appropriate specialty firms. Both the American Institute of Architects (AIA) and the Engineers Joint Contract Documents Committee (EJCDC) have standard forms for this purpose.

2.15 BID BOND

Various defaults are possible in the relationship between owner and contractor. The concept of a bond allows one party to protect itself against default in a relationship with a second party. A third party, referred to as the *surety*, provides protection such that, if a default between two parties occurs that results in damage (e.g., loss of money or other value), the surety protects the damaged party. This protection is typically in the form of offsetting or covering the damage involved. A construction bond, therefore, involves a relationship between three parties—the principal or party who might default, the obligee or party who may be damaged or lose some advantage, and the surety who will offset the damage or loss of advantage. For this reason, a bond is referred to as a three-party instrument establishing a three-party relationship (see Figure 2.12).

A similar relationship exists if a person of limited means attempts to borrow money at the bank. If the bank is concerned about the ability of the borrower to pay back the borrowed money, it may require that a separate individual cosign the note or instrument of the loan. The cosigner is required to pay back the borrowed money in the event that the primary borrower defaults in repayment. In this situation, the bank is in the position of the obligee, the borrower is the principal, and the cosigner acts as the surety.

During the bidding process, the owner usually requires a *bid security*. The security is required to offset the harm or *damage* occurring in the event that the firm selected fails to begin the project as directed. This may occur in the event that the selected bidder realizes that he has underbid the project and that pursuing the work will result in a financial loss.

In such cases, the owner would incur a damage, since the next lowest bidder would have to be selected. If, for instance, a contractor bidding $3 million refuses to enter into contract, and the next low bid is $3,080,000, the owner is damaged in the amount of

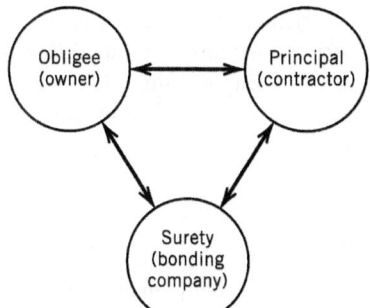

Figure 2.12 Bonding relationship (three-party).

$80,000. The responsibility of the surety is indemnified (covered) by the principal. If the principal fails to enter into contract,

the Principal shall pay to the Obligee the difference not to exceed the penalty hereof between the amount specified in said bid and such larger amount for which the Obligee may in good faith contract with another party to perform the work.

If the principal is unable to pay this amount, the surety must step in and cover the damage.

In most cases, the surety firm will not issue a bid bond unless it is sure the assets of the principal will offset any default occurring due to failure to enter into a contract. Therefore, if issued at all, the bid bond is issued for a small administrative fee. From the bonding company's point of view, the importance of the bid bond is not the fee paid by the contractor for its issuance, but instead its implication that if the contract is awarded, the surety will issue performance and payment bonds. The bid bond is a "lead parachute," which pulls these two bonds out of the main pack. Typical performance and payment bonds are shown in Appendix B. The performance bond protects the owner against default on the contractor's part in completing the project in accordance with the contract documents. The payment bond protects the owner against failure on the part of the prime contractor to pay all subcontractors or vendors having outstanding charges against the project. If the surety fails to issue these bonds (required in the contract documents), the contractor is prevented from entering into contract and the surety could be forced to cover damages resulting from this default.

As an alternative to a bid bond, the owner will sometimes specify (in the notice to bidders or the proposal) acceptance of a cashier's check in a specified amount made out to the owner to secure the bid. If the contractor fails to enter into the contract, he or she forfeits this check and the owner can use it to defray the cost of entering into contract with the second lowest bidder at a higher bid price. This method of bid security is indicated in the notice to bidders in Figure 2.5. The notice to bidders states:

All Proposals must be accompanied by a certified check, or a bid bond of a reputable bonding company authorized to do business in the State of Georgia, in an amount equal to at least five (5%) percent of the total amount of the bid.

This procedure is further explained in the proposal form (Figure 2.9):

The bidder further agrees that, in case of failure on his part to execute the Construction Agreement and the Bonds within ten (10) consecutive days after written notice being given of the award of the Contract, the check or bid bond accompanying this bid, and monies payable thereon, shall be paid into the funds of the {owner}.

Competitively bid government construction contracts require a bid bond that usually must cover a higher percentage of the bid price than the percentage stipulated in private construction bids (in fact, some private bids do not require a bid bond). For this reason, residential and commercial construction contractors are different from public construction contractors to a surety. A contractor failing to enter into contract after acceptance of his or her low bid in public construction places a surety at greater risk because of the larger bid bond for government contracts.

2.16   PERFORMANCE AND PAYMENT BONDS

If the contractor is awarded the contract, performance and payment bonds are issued. A performance bond is issued to a contractor to guarantee the owner that the contract work will be completed and that it will comply with project specifications. In other words, a performance bond protects the owner against default on the part of the contractor in performing the project as required. If the contractor fails to perform the work as required, the surety must provide for completion of the project in compliance with the plans and specifications at the price originally quoted by the defaulting contractor.

A payment bond is issued to guarantee the owner protection against any liens or charges against the project that are unpaid as a result of the contractor's default. That is, if the contractor fails to pay outstanding liens and charges against the project occurring as a result of the construction work, the surety will pay these debts. If the contractor does not pay subcontracts or suppliers, the surety must protect the owner from their claims. Standard bonding forms for performance and payment bonds are shown in Appendix B.

Because of the potential cost and trouble involved in taking over the work of a contractor about to default, the surety may elect to negotiate short-term financing for a contractor who has current liquidity problems. The surety may grant loans directly or assist the contractor in getting additional loans for the construction. In the event of default, there is no surety payment until the contractor's funds are completely exhausted. Then the surety will normally rebid the job, at a cost to itself and delay to the owner. For these reasons, a surety will often seek to assist a contractor overcome temporary cash shortages.

Surety companies typically have a list of troubleshooters who have a proven record of quickly taking over projects that are in trouble and bringing them to successful completion. In certain cases, the surety will replace the defaulting contractor's management team with a team of troubleshooters and attempt to complete the job with the existing work force. In other cases, the surety may negotiate with a second contractor to complete the work at a fixed price that is acceptable under the circumstances. The owner's interest is that the loss of time and disruption occurring because of default is minimized.

2.17   COST AND REQUIREMENTS FOR BONDS

Performance and payment bonds are issued for a service charge. A common rate is 1% or $10 per $1,000 on the first $200,000 of contract cost. At higher contract costs, the rate is reduced incrementally. Based on the size of the project and past performance of the contractor, rates fluctuate between 0.5% and 3%. Normally, the surety is not at any great risk because the bond includes an indemnity agreement on the part of the contractor. In other words, the contracting corporation, partnership, or proprietorship must pledge to pay back any monies expended by the surety on its behalf. Since the contractor or principal in a construction bond is required to indemnify the surety against any loss as a result of the bond, the protection of personal wealth and assets of corporate stockholders typical of closely held corporations is not in force. Key personnel may be required to sign that they will back the bond with their personal wealth in the case of closely held corporations or limited partnerships.

The Miller Act (enacted in 1935) establishes the level of bonding required for federally funded projects. Performance bonds must cover 100% of the contractor amount, whereas payment bonds are required based on the following sliding scale:

- 50% if the contract is $1 million or less
- 40% if the contract is between $1 and $5 million
- Fixed amount of $2.5 million if the contract is greater than $5 million

A surety seeks to keep itself well informed of a contractor's progress on bonded projects and with the contractor's changing business and financial status. To help with this, the contractor makes periodic reports on the work in progress with particular emphasis on costs, payments, and disputes associated with uncompleted work. Based on these reports, the contractor's bonding capacity can be determined. This is calculated as a multiple of the net quick assets of the contractor as reflected in the company balance sheet. The *net quick assets* are the contractor's assets that can quickly be converted to cash or negotiable instruments to cover the cost of default. Such items as cash on hand, demand deposits, accounts receivable, and similar highly *liquid* assets are available to the surety in case of a default.

The multiple to determine bonding capacity is based on the contractor's performance over the years. New contractors with no track record will have a low multiple such as 5 or 6. Old and reliable contractors may have a multiple of 40 or greater. Based on a multiple of 40, a reliable firm with net quick assets of $500,000 would have a bonding capacity of $20 million. In this case, the surety would provide bonding on work in progress (new jobs plus remaining value in jobs under way) up to the amount of $20 million. In other words, the firm will be able to pursue work for which bonds are required up to a total amount of $20 million.

Bonding companies have experienced larger than normal losses since the year 2000, and have raised rates and reduced the dollar amount available for writing new bonds. Many surety companies have raised premium rates by 10 to 30% since 2002 on performance and payment bonds. This has made it much more difficult for construction firms to find bonding and has impacted bid costs for firms working in markets which require bonds.

REVIEW QUESTIONS AND EXERCISES

2.1   What are the three major types of construction bonds? Why are they required? Name three items that affect bonding capacity.

2.2   In what major section of the contract is the time duration of the project normally specified?

2.3   Who are the three basic parties involved in any construction bonding arrangement?

2.4   What type of bond guarantees that if a contractor goes broke on a project the surety will pay the necessary amount to complete the job?

2.5   What is the purpose of the following documents in a construction contract?

    a.   General conditions

    b.   Special conditions

    c.   Addenda

    d.   Technical specifications

2.6   Why is the contractor normally required to submit a bid bond when making a proposal to an owner on a competitively bid contract?

2.7   What is the Miller Act and what does it specify regarding government contracts?

2.8   What is the purpose of the notice to bidders?

2.9   List the various specialty groups that are normally involved in the design of a high-rise building project.

2.10   How much money is the contractor investing in an advertised project available for bid at the time of:

    a.   Going to the A/E's office or checking the web to look at the plans and specifications?

    b.   Deciding to acquire a set of the drawings for further consideration?

    c.   Deciding to make initial quantity take-off?

    d.   Full preparation of bid for submittal?

2.11   What are the major parameters to be considered in the prequalification assessment of a contractor? Investigate the local criteria used in the prequalification of both small housing and general contractors.

2.12   Obtain sample specification clauses relating to the quality of finish of an item such as face brick, exterior concrete, or paint surfaces. Who has the major responsibility for the definition, achievement in the field, and paid acceptance?

2.13   Read those clauses of the general conditions of the contract for construction that refer to the owner, architect, contractor, and subcontractor. Then list the major responsibilities of these agents with respect to the following:

    a.   The definition, or attention to, the scope of the project

    b.   The financial transactions on the project

    c.   The finished quality of the work

2.14   Look up the percentage of the total cost of a building that *RS Means Building Construction Costs*[TM] (a compendium of cost data, discussed later in this book) estimates for its architectural design.

2.15   An architect asks $100,000 for the architectural design of a new building. However, after the preliminary design is completed, the owner and the architect decide to stop the design process. How much of the $100,000 would it be fair to pay to the architect? Why? Ask an experienced contractor to comment on your answer. Do not consider any legal aspects that would complicate your answer.

2.16   Visit the McGraw-Hill Dodge Construction Reports[TM] Website (http://dodge.construction.com) and find the three largest projects currently being bid in your state in the categories of engineering, commercial, houses, and utilities. Report the provided general description ("type"), estimated cost ("valuation"), and their owner.

2.17   The AIA has standard contract documents similar to the ones provided in the Appendixes of this book, developed by the EJCDC. What differences in the content of their respective General Conditions would you expect to find? Discuss at least five of the topics shown in Table 2.1.

2.18   The blueprints for a one-story condominium project indicate that it will use high-performance engineered wood joists. The specifications, however, indicate that the joists will consist of cheaper, traditional TJI® joists. Which of the two items would have priority, if the mistake is not detected prior to awarding the construction contract?

Chapter 3

Issues During Construction Phase

CONTENTS

Project Rework Reduction

Rework, and particularly field rework, continues to be one of the major sources of unplanned cost growth on industrial construction projects. Rework occurs when the installed work does not comply with or meet required specifications. According to Construction Industry Institute research, if field rework alone can be significantly reduced, or even eliminated, as much as 10% of overall project costs can be saved. The savings are expected to be substantially greater across an entire project cycle, which includes engineering and procurement.

Field rework is not caused solely by construction site activities in isolation. For the field rework to be reduced, a substantial effort must be made to improve the effectiveness of the prior project phases with a view to preventing all too frequent "catch-up" work during the site construction and commissioning phases. A number of computer-based tools have been developed to combat project-wide reworks.

3.1 ACCEPTANCE PERIOD/WITHDRAWAL

In formal competitive bid situations, the timing of various activities has legal implications. The issuance of the notice to bidders opens the bidding period. The date and time at which the bid opening is to take place marks the formal end of the bidding period. Usually either a physical or virtual bid box is established at some central location. Bids that have not been received at the bid box by the appointed date and hour are late and are normally disqualified. Prior to the close of the bidding (i.e., bid opening), contractors are free to withdraw their bids without penalty. If they have noted a mistake, they can also submit a correction to their original bid. Once bid opening has commenced, these prerogatives are no longer available. If bids have been opened and the low bidder declares a mistake in bid, procedures are available to reconcile this problem. If it can be clearly established that a mathematical error has occurred, the owner usually will reject the bid. However, if the mistake appears contrived to establish a basis for withdrawal of the bid, the owner will not reject. Then the contractor must enter into contract or forfeit bid security. The chronology of the bid procedure is shown in Figure 3.1.

The bid security protects the owner from failure by the contractor to enter into a formal construction agreement. The contractor is protected by the acceptance period. The notice to bidders specifies a period following bid opening during which the proposed bids are to remain in force. The indication is that if the owner does not act in this period to accept one of the bids, then the contractors can withdraw or adjust their bids. This is indicated in the notice to bidders (see Figure 2.5) as follows:

> **Withdrawal of Bids:** *No submitted bid may be withdrawn for a period of sixty (60) days after the scheduled closing time for the receipt of bid.*

This is designed to protect the bidder because otherwise the owner could hold the contractors to their bids for an unspecified amount of time. If the expected financing or appropriation for the project does not materialize, the owner could, in theory, say "Wait until next year, and I will enter into contract with you at this price." This, of course, would be potentially disadvantageous to the bidder. Therefore, the owner must send written notice of award (e.g., notice to proceed) to the selected contractor during the acceptance period or the bidders are released from their original proposals.

3.2 AWARD OF CONTRACT/NOTICE TO PROCEED

Notification of award of contract is normally accomplished by a letter indicating selection and directing the contractor to proceed with the work. This notice to proceed consummates the contractual relationship from a legal viewpoint despite the fact that a formal agreement has not been signed. The contractual legal protocol for accepting proposals (offers) is satisfied by the issuance of this letter. The letter also implies that the site is free of encumbrances and that the contractor can occupy the site for work purposes. Provisions of the contract usually direct that selected bidders commence work on the site within a specified period of time, such as 10 days.

Figure 3.1 Chronology of bid procedure.

The notice to proceed has an additional significance. The date of the notice to proceed establishes the reference date from which the beginning of the project is calculated. Therefore, based on the stipulated duration of the project as specified in the supplementary conditions, the projected end of the project can be established. As will be discussed, time extensions may increase the duration of the project, but the end of project beyond which damages will be assessed for failure to complete the project on time is referenced to the date of the notice to proceed. This might be specified as follows:

Work shall be completed no later than one thousand fifty (1,050) calendar days after the date of receipt by the Contractor of Notice to Proceed.

Calendar days are used because they simplify the calculation of the end-of-project date. In certain cases, the duration of the project is specified in working days. The general conditions normally specify working days as Monday through Friday. Therefore, each week contains 5 working days.

In some projects, all encumbrances to entry of the construction site have not been reconciled. Therefore, the "notice to proceed" cannot be issued since the contractor is not authorized to enter the site. In such cases, to indicate selection and acceptance of a proposal, the owner may send the selected bidder a *letter of intent*. This letter will indicate the nature of encumbrance and establish the owner's intent to enter into contracts as soon as barriers to the site availability have been removed.

3.3   CONTRACT AGREEMENT

Although the issuance of the notice to proceed establishes the elements of a contract, this is formalized by the signing of a *contract agreement*. In a legal sense, the formal contract agreement is the single document that binds the parties and by reference describes the work to be performed for a consideration. It pulls together under one cover all documents to include (a) the drawings, (b) the general conditions, (c) the supplementary conditions, (d) the technical specifications, and (e) any addenda describing changes published to these original contract documents. As with other bid package components, standard forms for the contract agreement for a variety of contractual formats are available from various professional organizations. Forms for the *stipulated* (lump) *sum* and *negotiated* (cost of work plus a fee) type contract are given in Appendixes C and D.

3.4   TIME EXTENSIONS

Once the formal contract has been signed, certain aspects of the contractor's activity during construction must be considered. Often circumstances beyond the contractor's control, which could not have been reasonably anticipated at the time of bidding, lead to delays. These delays make it difficult or impossible to meet the project completion date. In such cases the contractor will request an extension of time to offset the delay. These time extensions, if granted, act to increase the duration of the project. Procedures for dealing with time extensions are established in the general conditions of the contract. Claims for extension of time must be based on delays that are caused by the owner or the owner's agents or on delays due to Acts of God. Delays that result from design errors or changes are typical of owner-assignable delays and are not uncommon. A study of delay sources on government contracts indicated that a large percentage of all delays can be traced to the reconciliation of design-related problems (see Table 3.1). Weather delays are typical of the so-called act of God type delay. Normal weather, however, is not justification for the granting of a time extension. Most general conditions state

Table 3.1 Average Percent Extension by Extension Type

Facility	Design Problem	Owner Modification	Weather	Strike	Late Delivery	Other
Airfield paving/lighting	7.2	1.3	2.3	0.0	10.5	4.9
Airfield buildings	12.1	2.3	3.7	3.2	0.8	29.9
Training facilities	6.2	20.8	2.9	0.0	0.6	4.6
Aircraft maintenance facilities	12.0	2.0	8.4	1.0	2.2	0.2
Automotive maintenance facilities	12.9	2.3	3.4	1.4	0.7	0.4
Hospital buildings	16.0	3.4	2.6	0.6	0.6	0.9
Community facilities	6.7	5.4	2.3	1.7	1.5	0.3

Source: From D. W. Halpin and R. D. Neathammer, "Construction Time Overruns," Technical Report P-16, Construction Engineering Research Laboratory, Champaign, IL, August 1973.

specifically that only "adverse weather conditions not reasonably anticipatable" qualify as a basis for time extensions. This means that a contractor working in Minnesota in January who requests a 15-day time extension due to frozen ground that could not be excavated will probably not be granted a time extension. Since frozen ground is typical of Minnesota in January, the contractor should have "reasonably anticipated" this condition and scheduled around it. Weather is a continuing question of debate, and many contractors will submit a request for time extension automatically each month with their progress pay request, if the weather is the least bit out of the ordinary.

Time extensions are added to the original duration so that if 62 days of time extension are granted to an original duration of 1,050 days, the project must be completed by 1,112 calendar days after notice to proceed. If the contractor exceeds this duration, liquidated damages (see Section 3.9) are assessed on a daily basis for each day of overrun. The question of what constitutes completion can be answered as follows:

> The Date of Substantial Completion of the Work or designated portion thereof is the Date certified by the owner's representative when construction is sufficiently complete, in accordance with the Contract Documents, so the Owner can occupy, or utilize the Work or designated portion thereof for the use for which it is intended.

This date is often referred to as the beneficial occupancy date, or BOD. Once the owner occupies the facility, he or she relinquishes a large portion of the legal leverage available to make the contractor complete outstanding deficiencies. Usually, a mutually acceptable date is established when substantial completion appears to have been reached. On this date an inspection of the facility is conducted. The owner's representative (normally the design professional) and the contractor conduct this inspection recording deficiencies that exist and representing items for correction. Correction of these items will satisfy the owner's requirement for substantial completion. This deficiency list is referred to in the industry as the punch list. Theoretically, once the contractor satisfactorily corrects the deficiencies noted on the punch list, the owner will accept the facility as complete. If the rapport between owner and contractor is good, this phase of the work is accomplished smoothly. If not, this turnover phase can lead to claims for damages on both sides.

An indication of the amounts of time extension granted for various reasons on some typical government projects is given in Table 3.1. The types of delay sources categorized were due to (a) design problems, (b) owner modification, (c) weather, (d) strike, (e) late delivery, and (f) others. The percentages presented were calculated as % extension = (no. of days of time extension granted ÷ originally specified project duration) × 100.

3.5   CHANGE ORDERS

Since the contract documents are included by reference in the formal agreement, the lines on the drawings, the words in the technical specifications, and all other aspects of the contract documentation are legally binding. Any alteration of these documents constitutes an alteration of the contract. As will be discussed in Chapter 4, certain contractual formats such as the unit-price contract have a degree of flexibility. However, the stipulated or lump-sum contract has virtually no leeway for change or interpretation. At the time it is presented to the bidders for consideration (i.e., is advertised), it represents a statement of the project scope and design as precise as the final drawings for an airplane or a violin. Changes that are dictated, for any reason, during construction represent an alteration of a legal arrangement and, therefore, must be formally handled as a modification to the contract. These modifications to the original contract, which themselves are small augmenting contracts, are called *change orders*.

Procedures for implementing change orders are specified in the general conditions of the contract. Since change orders are minicontracts, their implementation has many of the elements of the original contract bid cycle. The major difference is that there is no competition, since the contractor has already been selected. Normally, a formal communication of the change to include scope and supporting technical documents is sent to the contractor. The contractor responds with a price quotation for performing the work, which constitutes his or her offer. The owner can accept the offer or attempt to negotiate (i.e., make a counteroffer). This is, of course, the classical contractual cycle. Usually, the contractor is justified in increasing the price to recover costs due to disruption of the work and possible loss of job rhythm. If the original contract documents were poorly scoped and prepared, the project can turn into a patchwork of change orders. This can lead to a sharpening of the adversary roles of the contractor and the owner and can substantially disrupt job activities.

3.6   CHANGED CONDITIONS

Engineering designs are based on the project site conditions as they are perceived by the architect/engineer (A/E) or design professional. For structural and finish items as well as mechanical and electrical systems above ground, the conditions are constant and easily determined. Variation in wind patterns leading to deviation from original design criteria may pose a problem. But normally, elements of the superstructure of a facility are constructed in a highly predictable environment.

This is not the case when designing the subsurface and site topographical portions of the project. Since the ability to look below the surface of the site is limited, the designer relies on approximations that indicate the general nature of the soil and rock conditions below grade. The designer's "eyes" in establishing the design environment are the reports from subsurface investigations. These reports indicate the strata of soil and rock below the site based on a series of bore holes. These holes are generally located on a grid and attempt to establish the profile of soil and rock. The ability of the below-grade area to support weight may be established by a grid of test piles. The money available for this design activity (i.e., subsurface investigation) varies, and an inadequate set of bore logs or test piles may lead to an erroneous picture of subsurface characteristics. The engineer uses the information provided by the subsurface investigation to design the foundation of the facility. If the investigation is not extensive enough, the design can be inadequate.

The information provided from the subsurface investigation is also the contractor's basis for making the estimate of the excavation and foundation work to be accomplished. Again, if the investigation does not adequately represent the site conditions, the contractor's estimate will be affected. The topographic survey of the site is also a basis for estimate and, if in error, will impact the estimate and price quoted by the contractor. If the contractor feels the work conditions as reflected in the original investigation made available for bidding purposes are not representative of the conditions "as found," he or she can claim a changed condition. For instance, based on the boring logs, a reasonable estimate may indicate 2,000 cubic yards of soil excavation and 500 cubic yards of rock. After work commences, the site may be found to contain 1,500 cubic yards of rock and only 1,000 cubic yards of soil. This, obviously, substantially affects the price of excavation and would be the basis for claiming a changed condition.

In some cases, a condition may not be detected during design, and the assumption is that it does not exist. For instance, an underground river or flow of water may go undetected. This condition requires dewatering and a major temporary-construction structure to coffer the site and to construct the foundation. If this condition could not reasonably have been foreseen by the contractor, there would be no allowance for it in his or her bid. The failure of the bid documents to reflect this situation would cause the contractor to claim a changed condition.

If the owner accepts the changed condition, the extended scope of work represented will be included in the contract as a change order. If the owner does not accept the changed condition claim, the validity of the claim must be established by litigation or arbitration.

3.7   VALUE ENGINEERING

Value engineering (VE) was developed during World War II in the United States. It began as a search for alternative product components due to a shortage of critical items during the war. Innovation was required. It was discovered that a process of "function analysis" produced low-cost products without impacting functional characteristics or reducing quality. This initiative showed that innovation can yield products can be developed that cost less but maintain the expected levels of performance. In this case, "necessity was the mother of invention."

In the early 1960s, this concept of value was introduced in the construction industry through directives from the Navy and Army Corps of Engineers relating to facility procurement. Other major government agencies (e.g., Public Building Service) joined this movement by introducing incentive clauses in facility procurement (construction) contracts, which provided rewards to contractors for value proposals that led to reduced construction costs while maintaining the functionality of the completed facility. These clauses are structured generally as follows:

> *The Contractor is encouraged to develop, prepare, and submit value engineering change proposals (VECPs) voluntarily. The Contractor shall share in any net acquisition savings realized from accepted VECPs, in accordance with the incentive sharing rates specified in the contract.*

In Army Corps of Engineers contracts, the VE incentive clause allows the construction contractor to share 50% or more of the net savings in firm fixed-price contracts. For example, if a contractor is constructing bridge towers supported (in the original design) by drilled pile foundations, and the contractor can redesign the foundations as spread mat footers with a savings of $400,000, a portion of the savings (usually 50%) is distributed to the contractor. The construction contractor must prepare a value engineering change proposal (VECP) which will be reviewed and then accepted or rejected by the owner. A potential reward to the contractor (in this case) of $200,000 is available if the proposal is accepted.

The VECP procedure allows the owner to harvest new and innovative ideas from the construction contractor. This overcomes, to some degree, the factors which obstruct the transfer of information from the contractor to the designer in classical design-bid-build (DBB) contracts. In such contracts, information flow is impeded by the "friendly enemy" attitude, which is often characteristic of the relationship between the design professional and the contractor in competitively bid contracts. Also, due to the sequence of design and construction in the DBB format, contractors seldom have input to the design process.

Construction contractors are typically more knowledgeable about field conditions and construction methods than design engineers. The construction methods used to realize a given design in the field have a great impact on cost. Contractors are in a better position to know what materials are easiest to install and which designs are most constructable. This knowledge can greatly influence cost. The VECP process allows this expertise to be transferred to the owner yielding a cost saving.

The idea behind VE is the improvement of design by encouraging the contractor to make suggestions during construction. This is in contrast to the implementation of VE during the design phase, which involves the designer or design professional in a systematic program of "value analysis." If at any point following selection, the contractor feels that a proposal to improve cost effectiveness of the design as transmitted at the time of bidding is appropriate, there is a monetary incentive to submit such a proposal. Again, if the contactor makes a suggestion that cuts the cost of the air-conditioning system by $60,000 and an equal sharing VE clause is in the contract, $30,000 is received for this VECP if accepted by the owner. The guiding principle in making the suggestion is that the cost is reduced while the functionality is maintained or improved.

VE can also be implemented during the design phase of project development. This aspect of VE uses various procedures such as brainstorming, prioritization, research, matrix analysis, and scoring systems to evaluate design alternatives. Criteria evaluation can be used to assess multiple factors such as aesthetics, performance, safety, and so on. A weighted analysis is used to do the final analysis. In the weighting process, a criteria matrix such as that shown in Figure 3.2 is used. All criteria to be considered are listed and compared. In comparing two criteria, preference for one over the other is scored as follows:

1 = slight preference

2 = average preference

3 = above average preference

4 = major preference

In the figure, five criteria are listed as A through E. In the triangular portion of the matrix, *cost* (A) is compared with *aesthetic* (B). Cost has an "average preference" so A-2 is placed in the diamond linking these two criteria. B and C are compared because they are equally preferred (no preference); B/C is entered in the diamond linking these two criteria. Now the A-2 diamond is compared to the B/C diamond and the preference of A-2 prevails and is noted in the diamond linking these two composite preferences (A to B and B to C). This process is continued until all criteria have been compared. The right most diamond shows that the most preferred criteria are A and E. They are effectively equally preferred.

In the lower part of the matrix however, E is given a raw score of 5 whereas A receives a slightly higher score of 6. These raw scores are converted to weights as shown. Two alternatives are compared with the original design solution. Each of the factors (A to E) are given a score between 1 (poor) and 5 (excellent). These scores are multiplied by the weight of importance. For the original solution, factor E receives a score of 3 times 9 or 27. The original solution score is 27 + 12 + 4 + 4 + 40 or 87. This compares with scores for alternative 1 and 2 of 99 and 86 respectively. Based on this weighted analysis, alternative 1 would be selected.

Figure 3.2 Criteria Evaluation Matrix.

This structured approach to evaluating alternatives during design helps to formalize consideration of various alternatives. A more comprehensive discussion of this topic is beyond the scope of this section. Government agencies have led the way in the in the use of VE incentive clauses in construction contracts. This allows the owner (i.e., the government) to enhance the flow of cost saving information between the designer and the contractor during the construction phase. Success in the use of VECP clauses by public agencies has led to widespread implementation of such incentive clauses in construction contracts used by private owners.

3.8 SUSPENSION, DELAY, OR INTERRUPTION

The standard general conditions utilized for many government contracts provide that:

The Contracting officer may order the Contractor in writing to suspend, delay, or interrupt all or any part of the work for such period of time as he or she may determine to be appropriate for the convenience of the government.

Interrupting or suspending work for an extended period of time may be costly to the contractor, since he or she must go through a demobilization-remobilization cycle and may confront inflated labor and materials at the time of restarting. In such cases, within the provisions of the contract, the owner (i.e., the government) is required to pay an adjustment for "unreasonable" suspensions as follows:

An adjustment shall be made for any increase in cost of performance of this contract (excluding profit) necessarily caused by such unreasonable suspension, delay, or interruption and the contract modified in writing accordingly.

The amount of this adjustment is often contested by the contractor and can lead to lengthy litigation. Normally, the owner will attempt to avoid interruptions. Difficulties in obtaining continuing funding, however, are a common cause for these suspensions.

3.9 LIQUIDATED DAMAGES

Projects vary in their purpose and function. Some projects are built to exploit a developing commercial opportunity (e.g., a fertilizer plant in Singapore) while others are government funded for the good and safety of the public (e.g., roads or bridges). In any case, the purpose and function of a project are often based on the completion of the project by a certain point in time. To this end, a project duration is specified in the contract document. This duration is tied to the date the project is needed for occupancy and use. If the project is not completed on this date, the owner may incur certain damages due to lacking the availability of the facility.

For instance, assume an entrepreneur is building a shopping center. The project is to be complete for occupancy by 1 October. The projected monthly rental value of the project is $30,000. If all space is rented for occupancy on 1 October and the contractor fails to complete the project until 15 October, the space cannot be occupied and half a month's rental has been lost. The entrepreneur has been "damaged" in the amount of $15,000 and could sue the contractor for the amount of the damage. Contracts provide a more immediate means of recourse in liquidating or recovering the damage. The special conditions allow the owner under the contractual relationship to charge the contractor for damages for each day the contractor overruns the date of completion. The amount of the liquidated damage to be paid per day is given in the special or supplementary conditions of the contract. The clause SC-2 of the Special Conditions in Figure 2.10 is typical of the language used, and reads:

Liquidated Damages In case of failure on the part of the Contractor to complete the work within the time fixed in the contract or any extensions thereof, the Contractor shall pay the owner as liquidated damages the sum of $3000 for each calendar day of delay until the work is completed or accepted.

The amount of the liquidated damage to be recovered per day is not arbitrary and must be a just reflection of the actual damage incurred. The owner who is damaged must be able, if challenged, to establish the basis of the figure used. In the rental example given, the basis of the liquidated damage might be as follows:

Rental loss: $30,000 rent/month/30 days = $1,000/day

Cost of administration and supervision of contract = $200/day

Total amount claimed as liquidated damage = $1,200/day

If a project overruns, the owner not only incurs costs due to lost revenues but also must maintain a staff to control and supervise the contract. This is the $200 cost for supervision. The point is that an owner cannot specify an arbitrarily high figure such as $20,000 per day to scare the contractor into completion without a justification. The courts have ruled that such unsupported high charges are in fact not the liquidation of a damage but instead a penalty charge. The legal precedent established is that if the owner desires to specify a penalty for overrun (rather than liquidated damages), he must offer a bonus in the same amount for every day the contractor brings the project in early. That is, if the contractor completes the project 3 days late he must pay a penalty of $60,000 (based on the preceding figure). On the other hand, if the contractor completes the project 3 days early, he would be entitled to a bonus of $60,000. This has discouraged the use of such penalty-bonus clauses except in unusual situations.

Establishing the level of liquidated damages for government projects is difficult, and in most cases the amount of damage is limited to the cost of maintaining a resident staff and main office liaison personnel on the project beyond the original date of completion. In a claims court, it is difficult to establish the social loss in dollars of, for instance, the failure to complete a bridge or large dam by the specified completion date.

3.10   PROGRESS PAYMENTS AND RETAINAGE

During the construction period, the contractor is reimbursed on a periodic basis. Normally, at the end of each month, the owner's representative (e.g., project or resident engineer) and the contractor jointly make an estimate of the work performed during the month and the owner agrees to pay a progress payment to cover the contractor's expenditures and fee or markup for the portion of the work performed. The method of making progress payments is implemented in language as follows:

> *At least ten days before the date for each progress payment established in the Owner Contractor Agreement, the Contractor shall submit to the owner's representative an itemized Application for Payment, notarized if required, supported by such data substantiating the Contractor's right to payment as the Owner may require, and reflecting retainage, if any, as provided elsewhere in the Contract Documents.*

Retainage is considered in greater detail in Chapter 11 in discussing cash flow. The owner typically retains or holds back a portion of the monies due the contractor as an incentive for the contractor to properly complete the project. The philosophy of retainage is that if the project is nearing completion virtually the entire bid price has been paid out, the contractor will not be motivated to do the small closing-out tasks that inevitably are required to complete the project. By withholding or escrowing a certain portion of the monies due the contractor as retainage, the owner has a "carrot," which can be used at the end of a project. The owner can say essentially, "Until you have completed the project to my satisfaction, I will not release the retainage." Retainage amounts are fairly substantial, and therefore, the contractor has a strong incentive to complete small finish items at the end of the project.

The amount of retainage is stated in the contract documents using wording similar to the following:

> *In making progress payments, there shall be retained 10 percent of the estimated amount until final completion and acceptance of the work.*

Various retainage formulas can be used based on the owner's experience and policy. If work is progressing satisfactorily at the 50% completion point, the owner may drop the retainage requirement. This may be specified in terms similar to the following:

If the owner's representative (architect/engineer) at any time after 50 percent of the work has been completed finds that satisfactory progress is being made, he may authorize any of the remaining progress payments to be made in full.

If a project has been awarded at a price of $1,500,000 and 10% retainage is withheld throughout the first half of the job, the retained amount is $75,000. This is a formidable incentive and motivates the contractor to complete the details of the job in a timely fashion.

3.11   PROGRESS REPORTING

Contracts require the prime contractor to submit a schedule of activity and periodically update the schedule reflecting actual progress. This requirement is normally stated in the general conditions as follows:

Progress Charts The contractor shall within 5 days or within such time as determined by the owner's representative, after the date of commencement of work, prepare and submit to the owner's representative for approval a practicable schedule, showing the order in which the contractor proposes to carry on the work, the date on which he will start the several salient features (including procurement of materials, plant, and equipment) and the contemplated dates for completing the same. The schedule shall be in the form of a progress chart of suitable scale to indicate appropriately the percentage of work scheduled for completion at any time. The contractor shall revise the schedule as necessary to keep it current, shall enter on the chart the actual progress at the end of each week or at such intervals as directed by the owner's representative, and shall immediately deliver to the owner's representative three copies thereof. If the contractor fails to submit a progress schedule within the time herein prescribed, the owner's representative may withhold approval of progress payment estimates until such time as the contractor submits the required progress schedule.

This provision is fairly broad and could well be interpreted to require only grossly defined S-curves or bar charts. These bar charts may be based either on activities or percentage completion of the various work categories such as concrete, structural, electrical, and mechanical work. These reports are used at the time of developing the monthly progress payments and to ensure the contractor is making satisfactory progress. Figures 3.3 and 3.4 indicate sample reporting methods involving bar charts for work activities and S-curves of overall percentages complete.

Network methods provide greater detail and have the advantage during planning and scheduling of being oriented to individual activities and their logical sequence. From the owner's viewpoint they allow a more precise review of logic and progress during construction and acceptance periods. If the contractor is behind schedule on critical activities, a simple bar chart or S-curve will not highlight this. The Critical Path Method (see Chapter 8) provides greater early warning of the impact of delays on total project completion.

3.12   ACCEPTANCE AND FINAL PAYMENT

Final acceptance of the project is important to all parties concerned. As noted previously, it is particularly important to the contractor, since final acceptance means the release of retainage. Final acceptance of the project is implemented by a joint inspection on the part of the owner's representative and the contractor. The owner's representative notes deficiencies that should be corrected, and the contractor makes note of the deficiencies. These are generally detail items, and the list generated by the joint inspection is called the deficiency, or punch, list. It becomes the basis for accepting the work as final and releasing final payment (to include retainage) to the contractor. A similar procedure is used between the prime contractor and the subcontractors. When the subcontractor's work is complete,

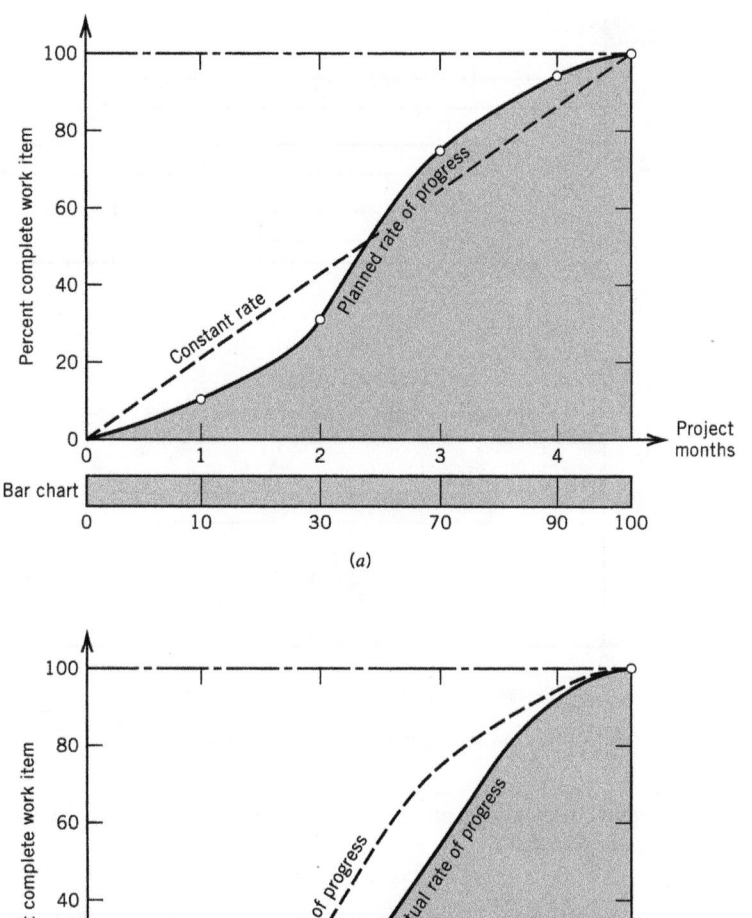

Figure 3.3 Bar chart planning and control models: (*a*) planned rate of progress and (*b*) actual rate of progress.

representatives of the prime and subcontractor "walk the job" and compile the deficiencies list for final acceptance of subcontract work. An example of a punch list between prime contractor and subcontractor is shown in Figure 3.5.

3.13   SUMMARY

This chapter has presented an overview of the cycle of activity that moves a project from the bid award stage through construction to acceptance by the owner or client. It is necessarily brief but provides a general frame of reference indicating how the contractor receives the project and some of the contractual considerations that must be addressed during construction. The competitively bid type of contract and the bid sequence peculiar to this contractual format have been used as the basis for this presentation. Other forms of

Figure 3.4 Bar chart project models: (*a*) bar chart schedule (plan focus) and (*b*) bar chart updating (control focus).

May 20, 2XXX
Punch List Items Acme Plastering Co.
Barfield-400 Project

Larger Building
　1. Caulking required between stucco and brick on the lower level.
　2. Streaks and cracking on stucco must be remedied.

Smaller Building
　1. Very noticeable line of stucco in rear of building.
　2. Streaks and cracking on stucco must be repaired.
　3. Exterior bridge entrances: patch stucco must be made uniform.
　4. Areas of excess spalling must be corrected.

Figure 3.5 Typical punch list.

contract will be discussed in Chapter 4. However, the basic chronology of events is the same. Having established this general mapping of the construction process, the following chapters develop the details of the contractor's role in the construction team.

REVIEW QUESTIONS AND EXERCISES

3.1 What is the difference between liquidated damages and a penalty for late completion of the contract?

3.2 What is the purpose of retainage?

3.3 During what period can a contractor withdraw the bid without penalty?

3.4   As a contractor you have built a 100-unit apartment complex that rents for $450 per unit a month. For late completion, you were assessed $2,000 per day. Would you qualify the assessment as liquidated damages or as a penalty? If the contract had included a bonus of $500 per day for early completion, would you expect to gain any assessment from court action? Why?

3.5   Describe the procedures to be followed for the receiving and opening of bids. If possible attend a bid opening and determine the number of bids that were submitted. For several unsuccessful bids determine the dollar amounts by which they exceeded the winning bid. Then calculate (relative to the winning bid) the percentages by which they exceed the winning bid. What do these figures tell about the strength of the current estimating and market environments? How much did the winning bidder "leave on the table"?

3.6   Scan a typical stipulated sum contract and identify those clauses that prescribe, modify, or are related to time considerations. Then develop a time strip map (similar to Figure 3.1) for the contract that locates the times (or time zones) for which each of the clauses are relevant. Which clauses rigorously fix time constraints for the contract and which are dependent on Acts of God or the owner for relevance?

3.7   Describe the procedure to be followed by the contractor who wishes to claim a time extension. What sort of documentation do you think is necessary to either refute or defend a time extension claim due to unusual weather? What sort of records do local contractors keep of weather conditions?

3.8   Must a contractor accept and perform all the work involved in each contract change order? Is there a limit to the number or magnitude of change orders that can be applied to a contract? When can a contractor refuse to accept a change order?

3.9   List the common causes of changed conditions in a building contract. What typical contract clauses bear on the problems caused by changed conditions? Suppose separate contracts are let for the building foundations and all remaining work. If you are the second contractor and you find that the foundations are incorrectly located, either in plan or elevation, would you be able to claim a changed condition?

3.10   Prepare a punch list of deficiencies or repairs that you consider necessary for your room, garage, or classroom. Can any of these items be related back to the original acceptance of the facility?

3.11   How would you go about either documenting a claim for a contractor's progress payment or its verification by the contract administrator for a typical building project in your locality?

3.12   Fudd Associates, Inc. submits a bid for $20 million. Mr. Fudd discovers at bid opening that all other bids are over $25 million. It is obvious that there is a mistake in his cost estimate. Could he withdraw his bid after all bids are opened, but no winner has been declared? Why? What circumstances would make you change your answer?

3.13   A contractor signs a contract for the construction of a $150 million ship harbor expansion. Before the Notice to Proceed is issued, a hurricane passes near the area and substantially changes the shape of the sea floor. The contractor claims that different drilling and dragging equipment will be required for the changed shape and requests a bid price adjustment. The owner responds that the passing of a hurricane is common in the area, and therefore, the contractor should have included the event as a contingency cost in his bid. Who is right? How do you think that this real case was settled?

3.14   Alright Construction, Inc., signs a contract with Casa Del Monte, Inc., to build five new 20-room modules for a large tourist resort in a Caribbean country. Casa Del Monte establishes liquidated damages of $30,000 for each day over the contractual deadline of December 1. The damages are considered reasonable, based on a net income of $300 for each of the 100 new rooms. However, political instability in the country keeps tourists off the resort. It has an occupancy rate of just 20% during December. Alright Construction does have a delay of 15 days. Can this contractor claim that the liquidated damages should be limited to 20% × $30,000 = $6,000 per day?

Chapter 4

Construction Contracts

CONTENTS

Web-Based Contracts

Web-based contracting provides an environment supporting scheduling, controlling, regulating, analyzing, and auditing the procurement and delivery of materials and services for construction in an electronic format. By enabling online competitive bidding and improving the record-keeping associated with the purchasing process, buyers will be able to quickly and easily compare product offerings from different manufacturers, as well as solicit pricing and availability. This puts buyers in a position to make optimal price- and time-of-delivery decisions.

A number of companies are developing network-based contracting services. These networks are adopting such solutions to improve their subcontract bidding process. With this system, construction companies can easily submit bid documents and specifications to solicit competitive bids for subcontracted work. They are able to route request for quotations and request for proposals (RFQs/RFPs) to approved contractors or search for matching contractors according to attributes, such as Construction Specifications Institute (CSI) classification, geographic location, specialty, minority status, licensing, and bonding. Eligible contractors and suppliers are notified of pending project bids and may then respond electronically. Their responses are automatically organized into bid summary spreadsheets for review and award. Buyers can anticipate an immediate return on investment from reduced costs associated with the processing of purchase orders, more competitive prices, and overall supply-chain efficiencies. Sellers can find new opportunities to increase sales by expanding their customer base and effectively communicating their product line and pricing. These online services help construction companies manage their complex supply chains, while enabling the project owner to make trade-off decisions about construction costs versus lifetime operational costs.

4.1 CONTRACT ENVIRONMENT

Construction is a product-oriented activity that has many dimensions. One of these dimensions is the business side of construction. The business world is structured by contractual relationships, and the business aspects of construction require the establishment of contractual relationships with a wide range of parties. The central role played by contracts is reflected by the fact that construction firms are referred to as "contractors." In addition to the contractual relationship with the owner/client, construction managers supervise contracts with subcontractors, specialty firms such as scheduling services, labor unions, as well as equipment and materials vendors. Insurance and bonds as well as the documents establishing the legal structure of a company have the elements of contractual requirements. This chapter investigates the major contractual forms used to establish contracts for the construction of projects.

An agreement between two or more competent parties to do something for a consideration establishes the basis for a contract. What is agreed to be performed (or refrained to be performed) cannot be impossible or enjoined by law. For example, a contract cannot bind a party to perform an illegal act or one that would be virtually impossible such as completing a 20-story building in a week.[1]

The courts are often called upon to determine:

1. Who are the parties to a contract?
2. What are their promises?
3. Other aspects of the contractual agreement.

A whole body of law has grown up around the many facets of contractual relationships. Because these issues remain constant for most construction situations, contract language in the construction industry has been normalized over many years and a variety of standard contract forms have developed.

4.2 PROCESS OF PURCHASING CONSTRUCTION

As noted in Chapter 1, construction contracts structure the way in which construction is "purchased." It is interesting to compare the construction purchasing system with the way in which we would buy a new lawn mower or a set of living room furniture. Consumers who need to purchase something go to a store, look at the range of product choices, and then pay a single supplier (the store owner) for the item of interest. If, for instance, we need a refrigerator, we go to an appliance store, inspect the various models, check prices, select one for purchase, pay for it, and the store owner sends it to our house or apartment within the next few days.

---

[1] What is *virtually impossible*? The concept is relative and frequently at the center of construction litigation.

Two major aspects of this process contrast with the way in which we "buy" construction.

1. We have the finished product available for our inspection, and we can decide whether it meets our requirements. That is, the manufactured product is available for our inspection prior to purchase.

2. Because the final product is available, we purchase it from a single individual or source.

In construction the facility is purchased before it is "manufactured" based on a set of drawings and work descriptors. Also the end item requires the purchaser to coordinate many entities to include designer(s), contractor(s), specialty subcontractors, and vendors. It is as if to buy a refrigerator we must develop a drawing of the refrigerator, purchase the materials required, and then coordinate 10 different entities who build it for us. Typically, none of the entities building the end item will warrant the proper operation of the refrigerator. They will only warrant the work that they provide.

In the building of the Brooklyn Bridge referred to in Chapter 1, Washington Roebling ordered a gigantic wooden box (e.g., a caisson) that was built by a local shipyard. The shipyard required payment in advance and would only warrant that the box was built according to plans that Roebling provided. They would not guarantee that it would perform adequately as a caisson because they did not know what a caisson was or how it was to be used.

Ideally, we would like to go to a single source and purchase the construction project as a finished unit. Obviously, in construction this is seldom possible. Traditional contract formats address this problem by focusing on the purchase of the design from a single entity (e.g., the design professional) and the construction of the facility by a general contractor who purchases the needed materials and services and coordinates the work of all the entities building the facility. As has been noted, even this three-party purchasing relationship (e.g., owner, designer, and constructor) can lead to an adversarial relationship between the parties.

Owners would, in general, like to work with a single source and be able to purchase the facility as-built (e.g., fully constructed and ready for occupancy). That single source would warrant the operation of the facility and act as a single point of contact to reconcile all problems with the product. This approach is used in single-family housing to a large extent. The home builder builds a house on a speculative basis without a buyer on hand. The house is presented to the public as a finished product and the contractor is a single point of contact for purchase and warranty.

Tunnels, bridges, and most large construction projects are not first built and then presented to the public for purchase. These are not built "on spec" (i.e., on a speculative basis) simply because the risk of building a "white elephant" that cannot be sold is too great. For this reason, the ability to present a finished product to a prospective buyer of construction is not feasible for most construction projects (one exception being single-family housing).

Project delivery systems have been developed to provide the construction buyer (i.e., the client) with a single point of contact or source of purchase. These contract formats have gained popularity over the past 25 years and are still evolving. The two major varieties of contract formats designed to provide the client with the construction equivalent of "one-stop shopping" are (a) design-build contracts and (b) construction management contracts. Before discussing these more recent developments in construction contracting, it is important to understand the contract types that have been used most widely over the past 60 years.

4.3 MAJOR CONSTRUCTION CONTRACT TYPES

The most widely used format of contract is the competitively bid contract. For a number of reasons, almost all contracts that involve public funds are awarded using competitively bid contracts. A competitively bid contract is used because it yields a low and competitive price that ensures taxpayers that their monies are being equitably and cost-effectively disbursed. The basic sequence of events associated with this type of contract has been described in Chapters 2 and 3. The two main categories of competitively bid contracts are (a) the lump-, or stipulated, sum contract and (b) the unit-price contract. The names of both of these contract formats refer to the method in which the price for the work is quoted.

The second most widely used contract format is the negotiated contract. This form of contract is also referred to as a cost-plus contract, although this refers to the method of payment rather than the nature of the selection process. The contractor is reimbursed for the cost of doing the work plus a fee. In this type of contract, the contractor risk is greatly reduced because the requirement of completing the work at a fixed price is not present. The owner has the flexibility to select the contractor based on considerations other than lowest price quotation. The method of selection involves the identification of a group of qualified contractors who are invited to prepare proposals based on the project documentation available. The proposals present the credentials of the firm and an approximate estimate of cost based on the project data available. The estimate includes not only the "bricks-and-mortar" direct costs but also estimates of the cost of supervision by the contractor's personnel and the level of fee requested. The proposal is often presented in a semiformal interview framework in which the contractor meets with the client and his or her representatives. Selection of the contractor is based on the preferences of the owner and the strengths and weaknesses of the contractor's proposal. This format is not well suited to public projects since favoritism can play a major part in determining which contractor is selected. In recent years, these issues have been addressed with innovative stipulations that reduce these risks.

4.4 COMPETITIVELY BID CONTRACTS

The mechanism by which competitively bid contracts are advertised and awarded has been described in Chapters 2 and 3. Essentially, the owner invites a quote for the work to be performed based on complete plans and specifications. The award of contract is generally made to the lowest responsible bidder. The word *responsible* is important since the contractor submitting the lowest bid may not, in fact, be competent to carry out the work. Once bids have been opened and read publicly (at the time and place announced in the notice to bidders), an "apparent" low bidder is announced. The owner then immediately reviews the qualifications of the bidders in ascending order from lowest to highest. If the lowest bidder can be considered responsible based on his or her capability for carrying out the work, then further review is unnecessary.

The factors that affect whether a contractor can be considered responsible are the same as those used in considering a contractor for prequalification:

1. Technical competence and experience.
2. Current financial position based on the firm's balance sheet and income statement.
3. Bonding capacity.
4. Current amount of work under way.
5. Past history of claims litigation.
6. Defaults on previous contracts.

Because of shortcomings in any of these areas, a contractor can be considered a risk and, therefore, not responsible. Owners normally verify the bidder's financial status by consulting the *Dun and Bradstreet Credit Reports* (Building Construction Division) or similar credit reporting system to verify the financial picture presented in the bid documents.

Generally, the advantages that derive from the use of competitively bid contracts are twofold. First, because of the competitive nature of the award, selection of the low bidder ensures that the lowest responsible price is obtained. This is only theoretically true, however, since change orders and modifications to the contract tend to offset or negate this advantage and increase the contract price. Some contractors, upon finding a set of poorly defined plans and specifications will purposely bid low (i.e., zero or negative profit) knowing that many change orders will be necessary and will yield a handsome profit. That is, they will bid low to get the award and then negotiate high prices on the many change orders that are issued.

The major advantage, which is essential for public work, is that all bidders are treated equally and there are no favorites. This is important because in the public sector political influence and other pressures could bias the selection of the contractor. Presently, public design contracts are not awarded by competitive bidding. The practice of negotiating design contracts is traditional and supported by engineering professional societies (e.g., the American Society of Civil Engineers and the National Society of Professional Engineers). Nevertheless it has been challenged by the U.S. Department of Justice.

Recent rulings indicate that competitive procedures for award of design contracts are gaining support, and this may become common in the design field as it is now in the construction field.

The competitive method of awarding construction contracts has several inherent disadvantages. First, the plans and specifications must be totally complete prior to bid advertisement. This leads to a sequentiality of design followed by construction and breaks down feedback from the field regarding the appropriateness of the design. Also, it tends to extend the total design-build time frame since the shortening of time available by designing and constructing in parallel is not possible. In many cases, the owner wants to commence construction as quickly as possible to achieve an early completion and avoid the escalating prices of labor and materials. The requirement that all design must be complete before construction commences preempts any opportunity for commencing construction while design is still under way.

4.5 STIPULATED-SUM CONTRACTS

A lump-sum, or stipulated-sum, contract is one in which the contractor quotes one price, which covers all work and services required by the contract plans and specifications. In this format, the owner goes to a set of firms with a complete set of plans and specifications and asks for a single quoted price for the entire job. This is like a client going to a marine company with the plans for a sailboat or catamaran and requesting a price. The price quoted by the boat builder is the total cost of building the vessel and is a lump-sum price. Thus, the lump sum must include not only the contractor's direct costs for labor, machines, and so forth but also all indirect costs such as field and front office supervision, secretarial support, and equipment maintenance and support costs. It must also include profit.

In stipulated-sum contracts the price quoted is a guaranteed price for the work specified in the plans and supporting documents. This is helpful for the owner since he knows the exact amount of money that must be budgeted for the project, barring any contingencies or change of contractual documents (i.e., change orders).

In addition, the contractor receives monthly progress payments based on the estimated percentage of the total job that has been completed. In other contract forms, precise field measurement of the quantity of work placed (e.g., cubic yards of concrete) must be made continuously because the contractor is paid based on the units placed rather than on the percentage of job completed. Since the percentage of the total contract completed is an estimate, the accuracy of the field measurements of quantities placed need only be accurate enough to establish the estimated percentage of the project completed. This means that the number and quality of field teams performing field quantity measurements for the owner can be reduced. The total payout by the owner cannot exceed the fixed or stipulated price for the total job. Therefore, rough field measurements and observations, together with some "Kentucky windage," are sufficient support for establishing the amount of progress payment to be awarded.

In addition to the disadvantage already noted (i.e., the requirement to have detailed plans and specifications complete before bidding and construction can begin), the difficulties involved in changing design or modifying the contract based on changed conditions are an important disadvantage. The flexibility of this contract form is limited. Any deviation from the original plans and specifications to accommodate a change must be handled as a change order (see Section 3.5). This leads to the potential for litigation and considerable wrangling over the cost of contract changes and heightens the adversary relationship between owner and contractor.

The stipulated-sum form of contract is used primarily in building construction in which detailed plans and specifications requiring little or no modification can be developed. Contracts with large quantities of earthwork or subsurface work are not normally handled on a lump-sum basis since such contracts must be flexible enough to handle the imponderables of working below grade. Public contracts for buildings and housing are typical candidates for lump-sum competitively bid contracts.

4.6   UNIT-PRICE CONTRACTS

In contrast to the lump-sum, or fixed-price, type of contract, the unit-price contract allows some flexibility in meeting variations in the amount and quantity of work encountered during construction. In this type of contract, the project is broken down into work items that can be characterized by units such as cubic yards, linear and square feet, and piece numbers (e.g., 16 window frames). The contractor quotes the price by units rather than as a single total contract price. For instance, he or she quotes a price per cubic yard for concrete, machine excavation, square foot of masonry wall, etc. The contract proposal contains a list of all work items to be defined for payment. Items 1 to 4 in Section 1 of Figure 2.9 provide a typical listing for unit-price quotation. This section is reprinted for reference.

Item Number	Quantity	Unit	Description	Unit Price	Total Amount
1	550	cubic yard (cu yd)	Rock excavation (for structures and pipes only)	$_____	$_____
2	50	linear foot (lin ft)	8-in. C.I. force main	$_____	$_____
3	20	cubic yard (cu yd)	Trench excavation for pipes	$_____	$_____
4	200	square yard (sq yd)	Paving	$_____	$_____

Four items of unit-price work are listed. A guide quantity is given for each work item. The estimated amount of rock excavation, for example, is 550 cu yd. Based on this quantity of work, the contractor quotes a unit price. The total price is computed by multiplying the unit price by the guide quantity. The low bidder is determined by summing the total amount for each of the work items to obtain a grand total. The bidder with the lowest grand total is considered the low bidder. In true unit-price contracts, the entire contract is divided into unit-price work items. Those items that are not easily expressed in units such as cubic yards are expressed in the unit column as "one job."

Unit-price quotations are based on the guide quantity specified. If a small quantity is specified, the price will normally be higher to offset mobilization and demobilization costs. Larger quantities allow economies of scale, which reduce the price per unit. That is, if 100 square feet of masonry brick wall is to be installed, the cost per square foot would normally be higher than the cost for 5,000 square feet. Mobilization and demobilization costs are spread over only 100 units in the first case, whereas in the second case these costs are distributed over 5,000 units, reducing the individual unit cost.

Most unit-price contracts provide for a price renegotiation in the event that the actual field quantity placed deviates significantly from the guide quantity specified. If the deviation exceeds 10%, the unit price is normally renegotiated. If the field quantity is over 10% greater than the specified guide quantity, the owner or the owner's representative will request a price reduction based on economies possible due to the larger placement quantity. If the field quantity underruns the guide quantity by more than 10%, the contractor will usually ask to increase the unit price. He or she will argue that to recover mobilization, demobilization, and overhead, prices must be increased since they were based on the larger guide quantities provided in the bid schedule. That is, there are fewer units across which to recover these costs and, therefore, the unit price must be adjusted upward. In developing the unit-price quotation, the contractor must include not only direct costs for the unit but also indirect costs such as field and office overheads as well as a provision for profit.

In unit-price contracts, the progress payments for the contractor are based on precise measurement of the field quantities placed. Therefore, the owner should have a good indication of the total cost of the project based on the grand total price submitted. However, deviations between field-measured quantities and the guide quantities will lead to deviations in overall job price. Therefore, one disadvantage of the unit price contract form is that the owner does not have a precise final price for the work until the project is complete. In other words, allowances in the budget for deviations must be made. In addition, the precision of field measurement of quantities is much more critical than with the lump-sum contract. The measured field quantities must be exact because they are, in fact, the payment quantities. Therefore, the owner's quantity measurement teams must be more careful and precise in their assessments since their quantity determinations establish the actual cost of the project.

Unit-price contracts can also be manipulated using the technique called *unbalancing the bid*. The relationship between the contractor's expenditures and income across the life of a typical project is shown schematically in Figure 4.1. Because of delays in payment and retainage as described in Section 3.10, the revenue curve lags behind the expenditure curve and leads the contractor to borrow money to finance the difference. The nature and amount of this financing is discussed in detail in Chapter 11.

The shaded area in Figure 4.1 gives an approximate indication of the amount of overdraft the contractor must support at the bank pending reimbursement from the client. To reduce this financing as much as possible, the contractor would like to move the income curve as far to the left as possible.

One way to achieve this is to unbalance the bid. Essentially, for those items that occur early in the construction, inflated unit prices are quoted. For example, hand excavation that in fact costs $50 per cubic yard will be quoted at $75 per cubic yard. Foundation piles that cost $40 per linear foot will be quoted at $60 per linear foot. Since these items are overpriced, to remain competitive, the contractor must reduce the quoted prices for latter

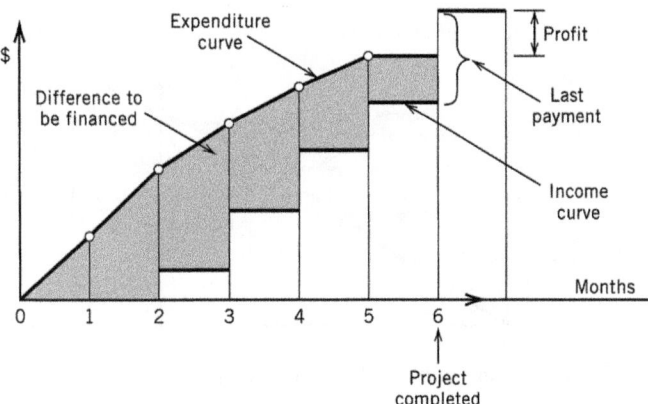

Figure 4.1 Project expense/income curves.

bid items. "Close-out" items such as landscaping and paving will be quoted at lower-than-cost prices. This has the effect of moving reimbursement for the work forward in the project construction period. It unbalances the cost of the bid items leading to front-end loading.

The amount of overdraft financing is reduced, as shown by the revenue and expense profiles in Figure 4.2. Owners using the unit-price contract format are usually sensitive to this practice by bidders. If the level of unbalancing the quotations for early project bid items versus later ones is too blatant, the owner may ask the contractor to justify his or her price or even reject the bid.

Some contracts obviate the need to unbalance the bid by allowing the contractor to quote a "mobilization" bid item. This essentially allows the bidder to request front money from the owner. The mobilization item moves the income curve to the left of the expense curve (see Figure 4.3). The contractor in the normal situation (Figure 4.1) will bid the cost (e.g., interest paid) of financing the income/expense difference into his prices. Therefore, the owner ultimately pays the cost of financing the delay in payments of income. If the owner's borrowing (i.e., interest) rate at the bank is better than that of the contractor, money can be saved by providing a mobilization item, thereby offsetting the contractor's charge for interim financing. Large owners, for instance, are often able to borrow at the prime rate (e.g., 3 or 4%), whereas contractors must pay several percent above the prime rate (e.g., 6 or 9%). By providing a mobilization item, the owner essentially assumes the overdraft financing at his rate, rather than having the contractor charge for financing at the higher rate.

In addition to the flexibility in accommodating the variation in field quantities, the unit-price contract has the added advantage to the contractor that quantity estimates developed as part of the bidding process need only verify the guide quantities given in the

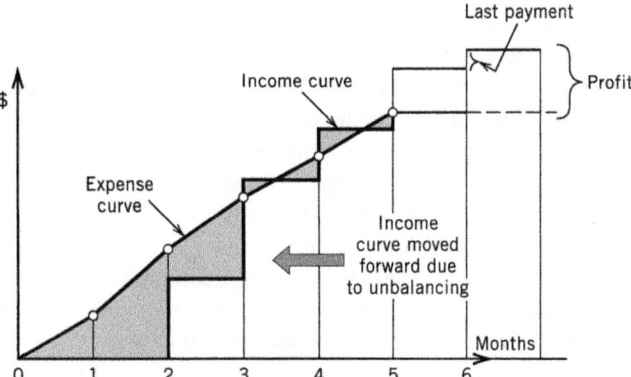

Figure 4.2 Unbalanced bid income profile.

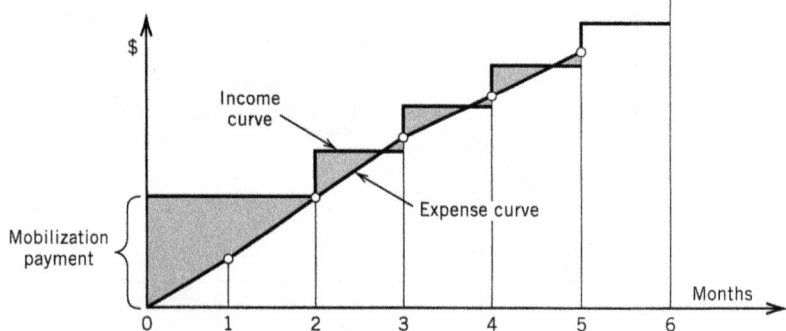

Figure 4.3 Income profile with mobilization payment.

bid item list (i.e., bid schedule). Therefore, the precision of the quantity takeoff developed for the contractor's estimate prior to construction need not be as exact as that developed for a fixed-price (lump-sum) contract. The leeway for quantity deviation about the specified guide quantities also normally reduces the number of change orders due to the automatic allowance for deviation.

Because of its flexibility, the unit-price contract is almost always used on heavy and highway construction contracts where earthwork and foundation work predominate. Industrial rehab work can also be contracted using the unit-price contract form with a bid item list for price quotation. Major industrial facilities are typically bid using the negotiated contract format.

4.7   NEGOTIATED CONTRACTS

An owner can enter into contract with a constructor by negotiating the price and method of reimbursement. A number of formats of contract can be concluded based on negotiation between owner and contractors. It is possible, for example, to enter into a fixed-price or unit-price contract after a period of negotiation. In some cases, public owners will negotiate with the three low bidders on prices, materials, and schedule.

The concept of negotiation pertains primarily to the method by which the contractor is selected. It implies flexibility on the part of the owner to select the contractor on a basis other than low bid. Therefore, a contractor competing for award of contract in the negotiated format cannot expect to be selected solely on the basis of low bid. This affects the bid cycle and the completeness of plans and specifications that must be available at the time of contractor selection. The owner invites selected contractors to review the project documentation available at the time of negotiation. This documentation may be total and complete design documentation as in the case of competitively bid contracts or only concept-level documentation. Based on the documentation provided, the contractor is invited to present his qualifications to perform the work and to indicate his projected costs and fee for completing the work. Because the level of the design documentation can vary from total detail to preliminary concept drawings, the accuracy of the cost projections will also vary. Within this presentation format, the owner evaluates the experience, reputation, facilities, staff available, charge rates, and fee structures of the various bidders participating. Based on this evaluation, the field is reduced to two or three contractors, and negotiations are opened regarding actual contract form and methods of reimbursement.

Since in most cases, the design documentation is not complete at the time of negotiation, the most common form of contract concluded is the *cost + fee*. In this type of contract, the contractor is reimbursed for expenses incurred in the construction of the contracted facility. The contract describes in detail the nature of the expenses that are reimbursable. Normally, all direct expenses for labor, equipment, and materials as well as overhead charges required to properly manage the job are reimbursable. In addition, the contractor receives a fee for his expertise and the use of his plant in support of the job. The fee is essentially a profit or markup in addition to the cost reimbursement. The level and amount of fee in addition to the charge schedule to be used in reimbursement of the direct costs are major items of discussion during negotiation. Various formulas are used for calculating the fee and strongly influence the profitability of the job from the contractor's standpoint.

As in the case of competitively bid contracts, the contractor does the financing of the project and is reimbursed by periodic (e.g., monthly) progress payments. Both parties to the contract must agree to and clearly define the items that are reimbursable. Agreement regarding the accounting procedures to be used is essential. Areas of cost that are particularly sensitive and must be clearly established are those relating to home office overhead charges. If the owner is not careful, he may be surprised to find out he has agreed to pay for the contractor's new computer network. Other activities that must be clearly defined for purposes of reimbursement are those pertaining to award and control of subcontracts as well as the charges for equipment used on the project.

Four types of fee structure are common. They lead to the following cost-plus types of reimbursement schemes:

1. Cost + percent of cost
2. Cost + fixed fee
3. Cost + fixed fee + profit-sharing clause
4. Cost + sliding fee

The oldest form of fee structure is the percentage of cost form. This form is lucrative for the contractor but is subject to abuse. There is little incentive to be efficient and economical in the construction of the project. Just to the contrary, the larger the cost of the job, the higher the amount of fee that is paid by the owner. If the cost of the job is $40 million and the fee is 2%, then the contractor's fee is $800,000. If the costs increase to $42 million the contractor's fee increases by $40,000. Abuse of this form of contract has been referred to as the "killing of the goose that laid the golden egg." Therefore, this form of reimbursement is seldom used.

To offset this flaw in the percent-cost approach, the fixed-fee formula was developed. In this case, a fixed amount of fee is paid regardless of the fluctuation of the reimbursable cost component. This is usually established as a percentage of an originally estimated total cost figure. This form is commonly used on large multiyear industrial plant projects. If the projected cost of the plant is $500 million, a fixed fee of 1% of that figure is specified and does not change due to variation from the original estimated cost. Therefore, the contractor's fee is fixed at $5 million. This form gives the contractor an incentive to get the job done as quickly as possible to recover his or her fee over the shortest time frame. Because of the desire to move the job as quickly as possible, however, the contractor may tend to use expensive reimbursable materials and methods to expedite completion of the project.

The fixed-fee plus profit-sharing formula provides a reward to the contractor who controls costs, keeping them at a minimum. In this formula it is common to specify a target price for the total contract. If the contractor brings the job in under the target, the savings are divided or shared between owner and contractor. A common sharing formula provides that the contractor shares by getting 25% of this underrun of the target. If, for instance, the target is $15 million and the contractor completes the job for $14.5 million, a bonus of $125,000 is received. The projection of this underrun of the target and the percent bonus to be awarded the contractor are used by some construction firms as a measure of the job's profitability. If the contractor exceeds the target, there is no profit to be shared.

In some cases, the target value is used to define a guaranteed maximum price (GMP). This is a price that the contractor guarantees will not be exceeded. In this situation, any overrun of the GMP must be absorbed by the contractor. The GMP may be defined as the target plus some fraction of the target value. In the previous example, if the target is $15 million, a GMP of $16 million might be specified.

In this form, a good estimate of the target is necessary. Therefore, the plans and concept drawings and specifications must be sufficiently detailed to allow determination of a reasonable target. The incentive to save money below the target provides an additional positive factor to the contractor. The owner tends to be more ready to compromise regarding acceptance of the project as complete if the job is under target. Additional work on punch list costs the contractor 25 cents, but it costs the owner 75 cents. The quibbling that is often present at the time the punch list is developed is greatly reduced to the contractor's advantage.

A variation of the profit-sharing approach is the sliding fee, which not only provides a bonus for underrun but also penalizes the contractor for overrunning the target value. The amount of the fee increases as the contractor falls below the target and decreases as he or she overruns the target value. One formula for calculating the contractor's fee based on a sliding scale is

$$\text{Fee} = R(2T - A) \text{ where } T = \text{target price}$$
$$R = \text{base percent value}$$
$$A = \text{actual cost of the construction}$$

Negotiated contracts are most commonly used in the private sector, where the owner wants to exercise a selection criterion other than low price alone. The negotiated contract is used only in special situations in the public sector since it is open to abuse in cases in which favoritism is a factor. Private owners are also partial to the negotiated format of contracting because it allows the use of phased construction in which design and construction proceed simultaneously. This allows compression of the classical "design first-then construct" or design-bid-build sequence. Since time is literally money, every day saved in occupying the facility or putting it into operation represents a potentially large dollar saving. The cost of interest alone on the construction financing of a large hotel complex can run as high as $50,000 a day. Financial costs generated by delays on large power facilities are estimated at between $250,000 and $500,000 a day. Quite obviously, any compression of the design-build sequence is extremely important.

Large and complex projects have durations of anywhere from to 2 to 3 up to 10 years. For such cases, cost-plus contracts are the only feasible way to proceed. Contractors will not bid fixed prices for projects that continue over many years. It is impossible to forecast the price fluctuations in labor, material, equipment, and fuel costs. Therefore, negotiated cost-plus-fee contracts are used almost exclusively for such complex long-duration projects.

4.8   PROJECT DELIVERY METHODS

In ancient times, great structures were constructed by "master builders" who developed the project concept, designed the appearance and technical details of the finished building or monument, and mobilized the resources needed to realize the final structure. This classical approach was used to build the pyramids, the great castles and churches of the middle Ages, and the civil engineering infrastructure of the industrial revolution. Master builders designed and constructed facilities acting as a single point of contact for the client.

Over the past 100 years, the processes of designing and building were gradually separated. Design and construction were viewed as separate endeavors. A design professional prepared the project plans, and a separate firm was contracted to perform the actual construction of a facility. This separation of activities also led to a sequencing of activities in which design was completed before construction commenced. This became the "traditional" sequence and is now referred to as Design-Bid-Build (DBB). This contracting procedure has been the basis for our discussions in Chapters 2 and 3. In contrast, master builders conducted design and construction simultaneously.

Over the past 30 years, a number of new concepts for project delivery have been developed to compress the time required to realize a constructed facility. It has been recognized that the DBB method of project delivery, with its sequential emphasis, leads to longer-than-necessary project time frames. It is advantageous from a time perspective to have design and construction proceed simultaneously. This has led to a reconsideration of the master builder concept and a discussion of what is meant by "Project Delivery Systems or Methods."

The topic of project delivery methods addresses "the organization or the development of the framework relating the organizations required to complete or deliver a project and the establishment of the formal (i.e., contractual) and the informal relationships between these organizations." In the DBB approach, for instance, the owner holds a contract with the designer or architect/engineer (A/E) for the development of the plans and specifications and a separate contract with the construction contractor for the building of the facility. In other delivery systems, the owner contracts with a single group or entity for both the design and construction of a facility. Another accepted definition is as follows:

> *A project delivery method is the comprehensive process of assigning the contractual responsibilities for designing and constructing a project (Associated General Contractors [AGC], 2004).*

Based on guidance given by the AGC, the concept of project delivery addresses two critical issues:

1. Is the responsibility to the owner/client for project design and construction tied to a single entity (e.g., a performing group) or multiple entities? In other words, does the owner deal with a single entity or multiple entities when pursuing design and construction of a project?

2. Is the award based on lowest cost or on other criteria?

In the framework of this discussion, competitively bid contracts require multiple entities and the contract award is based on lowest quotation or cost to the owner. Negotiated contracts also involve multiple entities (e.g., architects, design professionals, and construction contractors) and may be based on lowest cost, although, in certain cases other considerations take precedence over cost (e.g., expertise, previous performance, and ability to react quickly). Two relatively new contract formats have been introduced that emphasize simplifying the project delivery process. Low cost is not as central to these formats as in the competitive or negotiated types of contracts. The emphasis is on optimizing parameters other than cost (e.g., quality, time of completion, meeting market needs, and safety). Design Build (DB) and Construction Management (CM) contracts differ from the traditional DBB format in terms of how they address the two critical issues of project delivery methods previously stated. They also facilitate the use of "phased construction" or "fast-tracking" based on design and construction occurring in parallel (i.e., at the same time) in contrast to the sequential nature of the DBB approach.

4.9   DESIGN-BUILD CONTRACTS

As noted in Section 4.2, it is advantageous from the client's point of view to have a single contractor provide the entire project as a single contract package. In the 1970s, large firms began to offer both design and construction services to provide the client with a single source for project delivery. This approach of providing both design and construction services can be viewed as a natural evolutionary step beyond the negotiated contract. It has been common practice in industrial construction to use the design-build (DB) approach for complex projects that have tight time requirements. In such cases, it is advantageous for the client to have a single firm providing both design and construction services.

This system has the advantage that differences or disputes between the design team or group and the construction force are matters internal to a single company. This eliminates the development of an adversary relationship between two or more firms involved in realizing the project (i.e., it eliminates disputes between designer and constructor). Normally, the management of the DB contractor is motivated to reconcile disputes or differences between design and construction in an expeditious and efficient manner. If such problems are not addressed, they can lead to loss of profit and potential dismissal of the contractor for poor performance.

Coordination between design and construction is also enhanced by having both functions within the same firm. This system improves the communication between designers and the field construction force and assists in designing a facility that is not only functional but is also efficient to construct. The firm is driven by the profit motive to optimize the design both for the functional life-cycle use of the building as well as to design for construction, thus enhancing the efficiency of the construction process. This can be compared to the manufacturing design of the refrigerator referred to in Section 4.2. If a firm manufactures a refrigerator, it designs the appliance to be efficient for use in the home. It also designs the item so that it can be manufactured in the most cost-efficient and timely manner so as to reduce production costs.

DB contracts also have the advantage that design and construction can be done concurrently. That means that work can be started in the field before a complete design is available. This allows for "phased construction," or a "fast track," approach as described previously and a compression of the schedule because design must not be totally complete prior to commencement of construction. During the 1970s this type of contract was used mainly on large and complex projects (e.g., petrochemical plants, industrial complexes, and power plants) to improve the flow of information between the design team and the construction people in the field. Usually, only firms with large design and construction capabilities were able to provide DB services. Projects built with a single DB contractor were often referred to as "turnkey" projects because the owner dealt with only one contractor and that contractor was charged with the completion of the facility so that the project was ready to be placed in operation with the "turn of a key." That is, this owner signed a single contract and said "Call me when you have the project complete and you want me to turn the key to start it up."

4.10   DESIGN-BUILD IN A CONSORTIUM FORMAT

In the past 20 years, the use of DB contracts has become more common in the building construction sector. A number of firms have marketed this project delivery approach to private entrepreneurs (e.g., owners building hotels, and apartment and office structures) in the building sector as way to receive the best product in the most timely way at the best price. Since most building contractors do not have an in-house design capability, lead contractors typically form a team or consortium of designers and specialty contractors who work together to meet the needs of the client. The owner/client contracts with the consortium as a single group providing the total project package (e.g., design, construction, procurement, etc.).

Each member of the consortium is at risk and is motivated to work with other members to minimize delays and disputes. In effect, a group of designers and constructors form a consortium to build a project based on conceptual documentation provided by the owner. They agree to work together to achieve the project and, therefore, implicitly agree to avoid developing an adversary relationship between one another.

The attraction of this consortium-based approach is the fact that the owner/client is given a stipulated-sum price for the project after 30 to 40% of the design of the project is completed. Barring major changes to the project, the consortium locks in the final price at the end of the preliminary design phase. This is attractive to the owner because financing for the overall project can be lined up based on a definitive cost figure developed early in the design development process. This reduces the need for contingency funds and is attractive to the lender since the cost and, therefore, the amount of the borrowing is locked in.

The members of the consortium are motivated to be innovative and avoid disputes since failure to achieve the stipulated price quoted at the end of preliminary design will result in a loss to all of the members of the consortium. Again, the adversary relationship typical of designer and constructors is largely eliminated since bickering and lack of cooperation among members of the team can lead to significant losses. Incentive to avoid disputes and to develop innovative solutions to field problems is inherent in this type of contract.

This consortium-based DB contract has gained such wide acceptance in the private building construction community that it is now being used by the federal government on a number of building projects. A number of large Internal Revenue Service (IRS) facilities have been built using DB contracts. In its application in the private sector, stiff competition among a number of consortia for the same project has not been the normal case. In most cases, the owner/entrepreneur has worked with one or two lead contractors who form a DB team to meet the customized requirements of the client. Competition among competing consortia in the private building construction sector has not been a major issue.

With the advent of the use of this method in the public sector, competition has been a major factor. Selection of the winning consortium is based on competitive review of proposals from each consortium. In this format, the consortia not selected incur substantial losses based on the cost of organizing and developing a competitive proposal.

4.11 CONSTRUCTION MANAGEMENT CONTRACTS

In construction-management (CM) type contracts, one firm is retained to coordinate all activities from concept design through acceptance of the facility. The firm represents the owner in all construction management activities. In this type of contract, construction management is defined as that group of management activities related to a construction program, carried out during the predesign, design, and construction phases, which contributes to the control of time and cost in the construction of a new facility. The CM firm's position in the classical relationship linking owner, contractor, and A/E is as shown in Figure 4.4.

This firm has the function of a traffic cop or enforcer, controlling the flows of information among all parties active on the project. The CM firm establishes the procedures for award of all contracts to architect/engineers, principal vendors, and the so-called trade or specialty contractors. Once contractual relationships are established, the CM firm controls not only the prime or major contractor but all subcontractors as well as major vendors and off-site fabricators. Major and minor contractors on the site are referred to as trade contractors. In this control or management function, the CM firm uses the project schedule as a road map or flight plan to keep things moving forward in a timely and cost-effective manner. The major functions carried out by the CM firm vary depending on whether the project is in the (a) predesign, (b) design, or (c) construction phase.

Figure 4.4 Traditional versus phased construction.

CM contracts are particularly attractive for organizations that periodically build complex structures (e.g., hospital authorities, municipalities, and transit authorities) but do not desire to maintain a full-time construction staff to supervise projects on a recurring basis. In such cases an owner can retain a firm as construction manager to plan, develop, and coordinate the activities of one or more design professionals, trade contractors, vendors, and other interested parties such as licensing and control bodies.

4.12 CONSTRUCTION MANAGEMENT AT-RISK

Agency construction managers are coordinators working on behalf of the client and are not contractually liable for the successful completion of the work. In addition, problems arising out of decisions made as a part of their oversight of the project often lead to legal issues that are difficult to resolve. For instance, if the agency construction manager implements safety procedures on behalf of the owner, which are found to be faulty and lead to unsafe construction operations, the owner, not the agency construction manager, is normally viewed as liable for legal suites or claims arising out of these procedures. Therefore, the client, through lack of experience in dealing with construction problems and protocols, may be put at risk by poor decisions made by the agency construction manager while supervising and controlling the project.

To close this loophole, a version of the CM contract format has become popular in which the construction manager not only coordinates the project but also assumes responsibility for the construction phase of the work. In the CM at-risk contract, the CM firm assumes the same risk that a construction contractor in the DBB format would assume for the successful completion of the project. In this situation, the CM-at risk signs all contracts related to the construction phase of the work. The design and other preconstruction contracts and responsibilities are signed by or remain with the owner. Prior to the commencement of construction, the CM at-risk provides services similar to those provided by an agency construction manager. The CM at-risk contract is similar to the DBB format in that design and construction are separate contracts. It differs from the DBB format, however, in that the selection criteria are based on issues other than lowest total construction cost.

4.13 COMPARING PROJECT DELIVERY METHODS

From the project delivery perspective, competitively bid contracts are required to be design-bid-build (DBB) contracts. Negotiated contracts can be viewed as DBB contracts although it

Table 4.1 Project Delivery Method (PDM)

	Contract Type	Single or Multiple Contracts to Owners	Selection Criteria	Phased Construction
DBB	Competitively Bid	Multiple-Design Contract & Construction Contract	Low Construction Cost	No
	Negotiated	Multiple Design & Construction	Low Cost or Other	Possible
DB	DB	Single Contract with DB Firm	Usually not Low Cost—Based on Performance	Yes
CM	CMa	Contracts held by Owner—CM, Design, Construction, and Vendors	Based on Performance Expectations	Yes
	CM at Risk	Same as CMa above except CM and Construction Contracts are Combined	Based on Performance Expectations	Yes

is not unusual for design and construction to proceed in parallel (i.e., simultaneously) in a given situation. In other words, it is possible to use "phased construction" when working with negotiated construction contracts. A constructor involved in construction of a hotel building on a cost reimbursable basis can begin construction of the site excavation and subbasements while the roof-top restaurant is still being designed. For both competitively bid and negotiated contracts, the owner holds separate contracts with the designer or design group and with the construction contractor.

In the Design Build (DB) format, the owner enters into contract with a single entity—the design builder. The basis of selection of the DB firm or consortium is normally on the basis of considerations other than least cost. Fast-tracking or phased construction is typical of DB contracts.

In both forms of the CM format, the owner holds multiple contracts. In the case of the Agency Construction Management format, the owner signs a management contract with the CMa, but holds contracts directly with the design and construction firms involved. Selection of the Agency CM firm is based on issues other than total construction cost (e.g., quality, schedule performance) Fast-track construction is usual when using this format.

The CM at-risk format requires separate contracts for the design team and the CM at-risk firm (similar to the DBB format). Low total construction cost is not the basis for selection of the construction manager at-risk. Fast-track construction is possible when using this format. The major types of project delivery methods are summarized in Table 4.1.

REVIEW QUESTIONS AND EXERCISES

4.1 Name and briefly describe each of the two basic types of competitively bid construction contracts. Which type would be most likely used for building the piers to support a large suspension bridge? Why?

4.2 If you were asked to perform an excavation contract competitively with limited boring data, what type of contract would you want and why?

4.3 Name three ways the construction contract can be terminated.

4.4 Name two types of negotiated contracts and describe the method of payment and incentive concept.

4.5 What is meant by unbalancing g a bid? What type of contract is implied? Give an example of how a bid is unbalanced.

4.6 Why is cost plus a percentage of cost type of contract not used to a great extent?

4.7 Under what circumstances is a cost-plus contract favorable to both owner and contractor?

4.8 Valid contracts require an offer, an acceptance, and a consideration. Identify these elements in the following cases:

  (a) The purchase of an item at the store

  (b) The hiring of labor

  (c) A paid bus ride

  (d) A construction contract

  (e) The position of staff member in a firm

4.9 Suppose you are a small local building contractor responsible for the construction of the small gas station in Appendix H. List the specialty items that you would subcontract.

4.10 Visit a local building site and ascertain the number and type of subcontracts that are involved. How many subcontracts do you think may be needed for a downtown high-rise building? Why would there be more subcontractors in a building job as opposed to a heavy construction job?

4.11 From the point of view of the owner's contract administrator, each different type of contract places different demands on supervision. List the significant differences that would impact the complement (number) of field personnel required to monitor the contract.

4.12 Visit a local contractor and determine the proportion of contracts that are negotiated against those that are competitively bid awards. Is this percentage likely to change significantly with small building contractors? Is there a difference between building contractors and heavy construction contractors?

4.13 The first progress payment received by the contractor is signified by the first vertical segment of the income curve in Figures 4.1, 4.2, and 4.3. Why is this first payment shown at the end of the second month instead of at the end of the first month?

4.14 Some people feel that lump-sum contracts can foster a more contentious work environment than any of the other types of contracts discussed in this chapter. What is the basis for this perception? Do you think that this perception is true?

4.15 An architect and a general contractor join forces and win a DB project contract. What risks could this arrangement have, from the owner's perspective? What advantages could this ad-hoc enterprise have, again from the owner's viewpoint?

4.16 Comment on the advantages of the DB approach to project delivery that the Design-Build Institute of America (DBIA) provides on its Website (www.dbia.org).

Chapter 5

Legal Structure

CONTENTS

Hoover Dam (http://www.usbr.gov)

Channel Tunnel

Joint Venturing

Larger and complex projects often exceed the capability of a single firm or contractor to do the work. In such cases, a team or group of contractors will combine their resources to bid and execute the work. A famous example of this situation is the construction of the Hoover Dam during the 1930s, in which a group of six of the largest contractors in the United States banded together to build this project. The approach of a group of firms or professionals establishing a team to complete a project is referred to as a *joint venture*. Joint ventures are also referred to as *consortia* (a single team or group is a *consortium*).

There are many reasons why firms will decide to legally combine for a specific period of time to pursue a given project. A given project may be so large that the financial resources of a number of companies are required to bond the project. For instance, the construction of an addition to the McCormick Exhibition Center in Chicago was so large that bonding companies were not prepared to provide performance and payment bonds to a single company. Being prudent, the bonding companies wanted to spread their risk.

This was also a consideration in the financial structure for the construction of the Channel Tunnel, which led to the establishment of a large consortium of companies (both French and English) that allied for the specific purpose of constructing the "Chunnel." From a financial, political and technical perspective, it would not have been possible for one company to build this epic project.

A joint venture is a business relationship undertaken by two or more companies to form a legal entity for the purpose of performing a specific work item, or in the case of construction, a given project. A team of firms may be involved in both the design and construction of a project or only (as in the Hoover Dam) the construction phase of a project. In any case, the owner contracts with the joint venture.

Therefore, the joint venture must be legally established in a rigorous fashion so that the contract required is binding. A major benefit to the owner/client when working with a joint venture is that the owner deals with a single entity as opposed to a number of different companies. From a management point of view, the owner has a single point of contact or a single contractor consisting of many subentities (e.g., companies).

Joint ventures differ from proprietorships, partnerships, and corporations in that they exist for a fixed period of time defined by the duration of the project being undertaken. Therefore, as a legal entity they exist to perform a given objective and are then dissolved.

The partners in a joint venture must each bring important contributions or capabilities to the undertaking. Each firm brings special abilities, which may include technical expertise, financial resources, or special knowledge—all of which are key to successful completion of the project being pursued. In the Design Build (DB) contract format discussed in Chapter 4, some of the partners are involved with the design aspects of the project while others are focused purely on the construction phase of the work. The legal aspects of joint venture formation and operation are typically unique to each project and will vary based on special aspects of the team partners (e.g., is the consortium multinational or not, etc.) and the location and nature of the project being constructed.

5.1   TYPES OF ORGANIZATION

One of the first problems confronting an entrepreneur who has decided to become a construction contractor is that of deciding how best to organize the firm to achieve the goals of profitability and control of business as well as technical functions. When organizing a company, two organizational questions are of interest. One relates to the *legal* organization of the company and the second focuses on the *management* organization. The legal structure of a firm in any commercial undertaking, be it construction or dairy farming, is extremely important because it influences or even dictates how the firm will be taxed, the distribution of liability in the event the firm fails, the state, city, and federal laws that govern the firm's operation, and the firm's ability to raise capital. Management structure establishes areas and levels of responsibility in accomplishing the goals of the company and is the road map that determines how members of the firm communicate with one another on questions of common interest. The types of company legal organization will be considered in this chapter.

5.2   LEGAL STRUCTURE

At the time an entrepreneur decides to establish a company, one of the first questions to be resolved is which type of legal structure will be used. The nature of the business activity may point to a logical or obvious legal structure. For instance, if the entrepreneur owns a truck and decides to act as a free agent in hauling materials by contracting with various customers, the entrepreneur is acting alone and is the proprietor of his or her own business. In situations in which a single person owns and operates a business activity and makes all of the major decisions regarding the company's activity, the company is referred to as a *proprietorship*. If the business prospers, the entrepreneur may buy additional trucks and hire drivers to expand the fleet, thereby increasing business. The firm, however, remains a proprietorship even if the entrepreneur has 1,000 employees, so long as the individual retains ownership and sole control of the firm.

If a young engineer with management experience and a job superintendent with field experience decide to start a company together, this firm is referred to as a *partnership*. The size of a partnership is not limited to two persons and may consist of any number of partners. Law firms as well as other professional companies (e.g., accounting firms) are often organized as partnerships consisting of as many as 10, 12, or more partners. If two or three individuals decide to form a partnership, the division of ownership is decided by the initial contribution to the formation of the company on the part of each partner. The division of ownership may be based solely on the monetary or capital assets contributed by each partner. Therefore, if three individuals form a partnership with two contributing $20,000 and a third contributing $10,000, the division of ownership among the partners is 40, 40, and 20%. In other cases, one of the partners may bring a level of expertise that is recognized in the division of ownership. For instance, in the example just cited, if the partner contributing the $10,000 was the expert in the area of business activity to be pursued, his expertise could be valued at the nominal level of $10,000, making his overall contribution to the firm $20,000. Therefore, ownership would be equally divided among the three partners. The actual division of ownership is usually specified in the charter of the partnership. If no written charter exists, and the partnership was concluded by verbal mutual agreement only, the assumption is that the division of ownership among the partners is equal.

In some business activity the risk of failure or exposure to damage claims may be such that a corporate structure is deemed appropriate. This form of ownership recognizes the company itself as a legal entity and makes only those assets that belong to the firm attachable for settlement of claims in the event of bankruptcy or damage claims. This allows principals or stockholders in a corporation to protect their personal and private assets from being called in to settle debts or claims arising out of the firm's operation or insolvency. Therefore, if a stockholder in a corporation has private assets of $1 million and the corporation declares bankruptcy, the $1 million cannot be attached to settle debts of the corporation.[1] Other desirable features of corporate structure that cause firms to select this legal structure will be discussed later in this chapter.

Two types of corporations are commonly encountered. Corporations in which a small number of persons hold all of the stock in the firm are referred to as close or closely held corporations. This form of ownership is common in the construction industry since it offers risk protection and also allows a small group of principals to control company policies and functions. A public corporation, in contrast to a closely held corporation, allows its stock to be bought and sold freely. The actual ownership of the stock varies daily as the stock is traded by brokers, in the case of large corporations, on the stock market.

Another form of organization that has legal implications is the joint venture. This is not a form of ownership but a temporary grouping of existing firms defined for a given period to accomplish a given task or project. A joint-venture organizational structure is used when a large project is to be constructed and requires the pooling of resources or expertise from several companies. Typically the companies establish a basis for division of responsibility on the job and cooperate toward the end of successfully completing the project. They are bound together for a period of cooperation by a legal agreement that defines the nature of the relationship. Joint venturing first became popular during the construction of large dams such as the Grand Coulee and Hoover Dam in the western United States and has since been used for a wide variety of large construction tasks.

---

[1] In certain situations, stockholders may by ancillary agreement, such as a bond, waive some of the protection offered by the corporate structure and find that their personal assets are subject to attachment.

5.3 PROPRIETORSHIP

The simplest form of legal structure is the proprietorship. In this form of business ownership, an individual owns and operates the firm, retaining personal control. The proprietor makes all decisions regarding the affairs of the firm. The assets of the firm are held totally by one individual and augment the individual's personal worth. All revenue to the firm is personal cash revenue to the proprietor, and all losses or expenses incurred by the firm are personal expenses to the proprietor. The proprietor is, therefore, taxed as an individual and there is not separate taxation of the firm. Consider Uncle Fudd, who has a small contracting business. The firm generated $187,000 in total volume during the calendar year. The firm has $100,000 in expenses, so that the before-tax income of the firm is $87,000. Uncle Fudd declares this income on his personal income tax return. Assuming this is his total income (i.e., he received no further income from other sources) and that he has $17,000 in deductions and exemptions, his taxable income is $70,000.

Since the owner's capital and that of the firm are one and the same, the credit that the firm can obtain and its ability to generate new capital are limited by the personal assets of the proprietor. Furthermore, any losses incurred by the firm must be covered from the personal assets of the proprietor. Any liabilities incurred by the firm are the owner's liability, and he must cover them from his personal fortune. Therefore, bankruptcy of the firm is personal bankruptcy. Since there is no limitation of liability, high-risk businesses do not normally use the proprietorship form of structure.

The life of the proprietorship corresponds to that of the owner. Upon the death of the owner, the proprietorship ceases to exist. Assets of the proprietorship are normally divided among the heirs to the proprietor's estate.

5.4 PARTNERSHIP

The partnership is similar to the proprietorship in the sense that liabilities of the firm are directly transmitted to the partners. That is, there is no limitation of liability. However, in this case, since there are two or more partners, the liability is spread among several principals. The reason for forming a partnership is based on the principle of division of risk and pooling of management and financial resources. The ownership of the firm is shared among the partners to a degree defined in the initial charter of the partnership. Since several persons come together to form a partnership, the capital base of the firm is broadened to include the personal assets of the partners involved. This increase in assets increases the line of credit available to a partnership as opposed to a proprietorship. Control of the firm, however, is divided among the principals, who are called *general partners*. Partners share the profits and losses of the firm according to their degree of ownership as defined in the partnership agreement, but because the liability of each of the partners is not limited, one partner may carry more liability in the case of a major loss. Assume that Carol, Joan, and Bob are partners in a small contracting business. The personal fortunes and percentage of ownership of the three principals are as follows:

Carol	$1,400,000	40% ownership
Joan	$800,000	30% ownership
Bob	$100,000	30% ownership

The firm loses $1,000,000 and must pay this amount to creditors. The proportionate shares of this loss are:

Carol	$400,000
Joan	$300,000
Bob	$300,000

However, since Bob can only cover $100,000, the remaining $900,000 must be carried by Carol and Joan in proportion to their ownership share.

A *limited partnership*, as the term implies, provides a limit to the liability that is carried by some partners. This concept allows the general partners to attract capital resources to the firm. The *limited partner* is liable only to the extent of his or her investment. Assume that Tom comes into the partnership described previously as a limited partner. He makes $200,000 available for the capitalization of the firm. The percentages of ownership are redefined to provide Tom with 15% ownership. He, therefore, shares in the profit and loss of the firm in this proportion. Nevertheless, his level of loss is limited to the $200,000 he has invested. No amount beyond this investment can be attached from his personal fortune to defray claims against the firm. This provides the general partners with a mechanism to attract wealthy investors who desire liability limitation but profit participation. Limited partners have the position of a stockholder in a corporation in that loss is limited to the amount of their investment.

Limited partners have no voice in the management of the firm. Therefore, the general partners retain the same level of control but increase the capital and credit bases of the firm by bringing in limited partners. There must be at least one general partner in any partnership. The limited form of partnership (i.e., a partnership that includes limited partners) is more difficult to establish and subject to more regulation by state chartering bodies (usually the office of the Secretary of State of the state in which the partnership is chartered). This is because limited partnerships realize some of the advantages available in the corporate legal structure. Corporations are subject to close control by state chartering bodies.

The contribution made by the limited partner must be tangible. That is, the limited partner cannot contribute a patent, copyright, or similar instrument. The contribution must have a tangible asset value (i.e., equipment, cash, notes, and shares of stock in a corporation).

Any partnership is terminated in the event of the death of one of the partners. However, arrangements can be made to provide for the continuity of the partnership should one of the partners die. An agreement can be made among the partners that in the event of the death of a partner the remaining principals will purchase the ownership share of the deceased partner. Usually a formula that recognizes the fluctuating worth of the partnership is adopted in this agreement. The remaining partners pay this amount to the estate of the deceased partner.

General partners who are actively involved in the day-to-day management of the firm may decide to pay themselves a salary. In this way, the time and level of expertise contributed to the operation of the partnership are recognized. This level of day-to-day participation may be different from the level of initial contribution made in capitalizing the firm. In the case of Carol, Joan, and Bob, the levels of ownership were 40, 30, and 30%, respectively. If Bob is most active in the management of the partnership, he may be paid a full-time salary to recognize his commitment. Carol and Joan being active only on a part-time basis will be paid proportionately smaller or part-time salaries. Taxation, in any case, will be on both salary and earnings deriving from the operation of the partnership.

The action of one partner is binding on all partners. For instance, in the partnership described, if Joan enters into a contract to construct a building for the client, this agreement binds Bob and Carol as well. In this sense, a partnership is a "marriage," and any partner must be able to live with any commitment made on behalf of the partnership by another partner. On the other hand, it is not proper for a partner to sell or mortgage an asset of the partnership without the consent of the other partners. If the partner sells the asset, the income accrues to the partnership. If the partner uses a partnership asset to secure a personal note or loan, the other partners could advise the noteholder that they contest the use of this asset as security.

5.5   CORPORATION

A corporation is a separate legal entity and is created as such under the law of a state in which it chartered. In most states corporations are established by applying to the office of the secretary of state or similar official. This office issues a chartering document and approves the initial issuance of shares of stock in the corporation to establish the level of ownership of initial stockholders. As in the case of a partnership, the initial stockholders contribute financial capital and expertise as well as other intangible assets such as patents and royalty rights. The level of contribution is recognized by the number of shares of stock issued to each of the founding stockholders. If, in the partnership just described, Carol, Joan, and Bob decided to incorporate and the level of ownership were to remain the same, shares in the proper proportion would be issued to each principal. The number of shares and the share value defined at the initialization of a corporation are arbitrary and are selected to facilitate the recognition of ownership rather than actual value of the corporate assets. If the Carol-Joan-Bob (CJB) Corporation is established by the issuance of 1,000 share of stock, Carol would receive 400 shares (40%), and Joan and Bob would receive 300 shares each (30%). For simplicity, each share could have a par value of one dollar. This assignment of $1 per share simplifies the unit (i.e., share value) used to recognize ownership. On the other hand, the initial capital contributed to the formation of the corporation might have been $100,000. Therefore, the book value of each share of stock would be $100 per share. The book value of each share of a corporation is the net worth of the corporation divided by the number of shares issued. In this case, 1,000 shares are issued and the total asset value is $100,000. Therefore, each share has a book value of $100.

In addition to the par and book values associated with a share of stock in a corporation, each share has a traded or market value. This is the value that is listed on stock exchanges for those publicly traded corporation shares and that is printed in the newspaper or on the Web. It indicates what the general public or stock traders are willing to pay for a share of ownership in the corporation. If the future looks good, traders will anticipate an increase in the value of the corporation's stock and will pay to own a stock that is increasing in value. If the corporation is about to experience a loss, the market price of the stock may indicate this by declining in value. To illustrate, if CJB, Inc., wins a contract that promises to net the corporation an after-tax profit of $100,000, the market price of the stock will tend to move up. In fact, as already noted, most construction firms hold their stock closely and do not trade it publicly. Therefore, the market value of the stock is of interest primarily to the giant construction firms that are publicly traded.

Because of the legal procedures required, the corporation is the most complicated form of ownership to establish. A lawyer is normally retained to prepare the proper documents, fees must be paid to cover actions by the chartering body (e.g., office of the secretary of state), printed stock is prepared, and formal meetings by the principals are required. Since the corporation can sell further stock to raise capital, it has an advantage in this respect over the proprietorship and the partnership. This power to sell stock can be and has been abused. Once a corporation is established, it may sell stock to unsuspecting buyers based on an idea or concept that is not properly presented or explained. For this reason and others, the corporation is closely controlled by the chartering agency in regard to its issuance and sale of additional stock. Federal law also dictates certain aspects of the presentation of corporate stock for sale.

The most desirable aspect of the corporate structure to businesses that are exposed to high risk such as the construction industry is its limitation of liability. Since the corporation is a legal entity of itself, only the assets of the corporation are subject to attachment in the settling of claims against and losses incurred by the corporation. This means that stockholders in a corporation can lose the value of their investment in stock, but that is the limit of their potential loss. Other assets that they own outside of the corporation cannot be impounded to offset debts against the corporation.

One disadvantage associated with the corporation is the double-taxation feature. Since the corporation is a legal entity, it is subject to taxation. The same profit that is taxed within the corporation is taxed again when it is distributed to stockholders as a dividend. This distributed profit becomes taxable as personal income to the individual stockholders. Assume that CJB Corporation has a before-tax profit (e.g., revenue − expenses) of $100,000 during the corporation's first year of operation. Let us assume the CJB Corporation is taxed by the IRS at the rate of 34% of profit provided income is in excess of $75,000. The corporation would be taxed $34,000 for $100,000 of before-tax profit.[2] The after-tax income would be $66,000. Assume the CJB decides to distribute $30,000 to the three stockholders. That is, Carol, Joan, and Bob as directors of their closely held corporation distribute $30,000 to themselves and retain $36,000 of these earnings within the corporation as working capital. In this case Carol, the major stockholder, receives a dividend of $12,000. Joan and Bob would receive $9,000 each. If we assume that each stockholder pays approximately 25% on personal taxable income, Carol will pay $3,000 in tax on this dividend, and Joan and Bob will pay $2,250. In other words, the federal tax at the corporation and stockholder levels combined will be $34,000 plus $7,500, or $41,500.

The double-taxation feature does not always prove to be a disadvantage. Returning to the situation of Uncle Fudd who is organized as a proprietorship, assume his before-tax income with the proprietorship is $147,000.[3] Assume that Uncle Fudd decides to incorporate his proprietorship and become Fudd Associates, Inc. As president of this corporation, Uncle Fudd pays himself a salary of $85,000. At this salary level, Uncle Fudd is taxed at 21% of his taxable income (i.e., his gross income minus deductions and exemptions). In the proprietorship format, his tax would be 25% of $147,000 minus $12,000 in deductions and exemptions.[4] He will pay 25% of $135,000, or $33,750 in tax. In the corporate format, Uncle Fudd's tax will be:

Uncle Fudd's tax will be:

$147,000
−$85,000   (Fudd's Salary, which is an expense)
= $62,000   Gross income of corporation
Corporate tax = 0.25(62,000) = $15,500 (See footnote 4)
Personal tax = (0.21){$85,000 − 12,000 (deductions and exemptions)} = $15,330

Therefore, Uncle Fudd's tax in the corporate format will be $15,500 + $15,330 = $30,830. In this case, the corporate form of ownership yields a lower tax payment despite the double taxation. For this reason, a good tax consultant is a valuable advisor when deciding which form of ownership is most appropriate.

Certain states provide for a special corporate structure that avoids the double-taxation feature of a normal corporation but retains the protection of limited liability. This is referred to as a subchapter "S" corporation. In a subchapter S corporation, the principals are taxed as if they were members of a partnership. That is, corporate income is taxed only

---

[2] Corporate taxation levels vary over time due to changes in federal and state legislation.
[3] This example is different from the previous situation in which the taxable income was $87,000.
[4] The corporate rate for less than $75,000 taxable income is assumed to be 25%.

once as personal income. The corporate shareholders are, however, still protected and their loss is limited to the value of the stock they possess.

As noted previously, the corporation is advantageous when attraction of additional capital is of interest. Figure 5.1 shows a typical stock certificate as issued at the time of incorporation. The certificate indicates that 250 shares of stock are represented. In addition, the corporation has authority to issue a total of 50,000 shares. Therefore, the directors of a corporation can decide to raise money for capital expansion by selling stock rather than borrowing money. This provides for the generation of additional capital by distributing ownership. It has the advantage that the money generated is not subject to repayment and therefore is not a liability on the company balance sheet.

The corporation also has a continuity that is independent of the stockholders. Unlike the proprietorship or partnership in which the firm is terminated on the death of one of the principals, the corporation is perpetual. Unless the corporation is bankrupt or the corporate charter lapses, the corporation continues in existence until all stockholders agree to dissolve it. In most states, clauses can be included in the corporate charter, which in effect, allow the control of sale of stock outside of the circle of present stockholders. That is, any stockholder who wishes to sell a block of stock must first offer the stock for sale to the other stockholders. They have an option to purchase it before it is sold to others. This allows the closed nature of a closely held corporation to be maintained. If a stockholder should die, the stockholder's heirs are committed to offer it to the present stockholders before selling it to others. The heirs can, of course, decide simply to retain the stock.

Two disadvantages that are inherent in the corporate form of ownership are the reduced level of control exercised in management decision making and certain restrictions that can be placed on the corporation when operating outside of its state of incorporation. The larger a corporation becomes the more decentralized the ownership becomes. On questions of dividend levels, the issuance of stock to generate capital, and other critical operational decisions, agreement of all stockholders must be obtained. In large corporations, this leads to involved balloting to establish the consensus of the ownership. This process is cumbersome and greatly reduces the speed with which corporations can respond to developing situations. In small closely held corporations, however, this presents no more of a problem than it does in a partnership.

When a corporation operates in a state other than the one in which it is incorporated, it is referred to as a *foreign* corporation. For instance, a corporation incorporated in Delaware is considered a foreign corporation in Indiana. Corporations in certain industries may encounter restrictive laws when operating as a foreign corporation. They must establish legal representation in states in which they operate as foreign corporations. Restrictive legislation of this type cannot be applied to proprietorships and partnerships, since these entities consist of individuals who are legally recognized. The individual is protected by equal treatment under the Constitution, and what is a legal restriction when placed on a corporation is illegal when applied to a proprietorship or a partnership.

5.6   COMPARISON OF LEGAL STRUCTURES

The decision to choose a particular legal structure for the firm hinges on seven major considerations. The pluses and minuses of each type of structure are summarized in Table 5.1. These considerations have already been introduced in general form. Specifically, an owner contemplating a legal structure for the firm must consider:

1. Taxation.
2. Costs associated with establishing the firm.
3. Risk and liability.

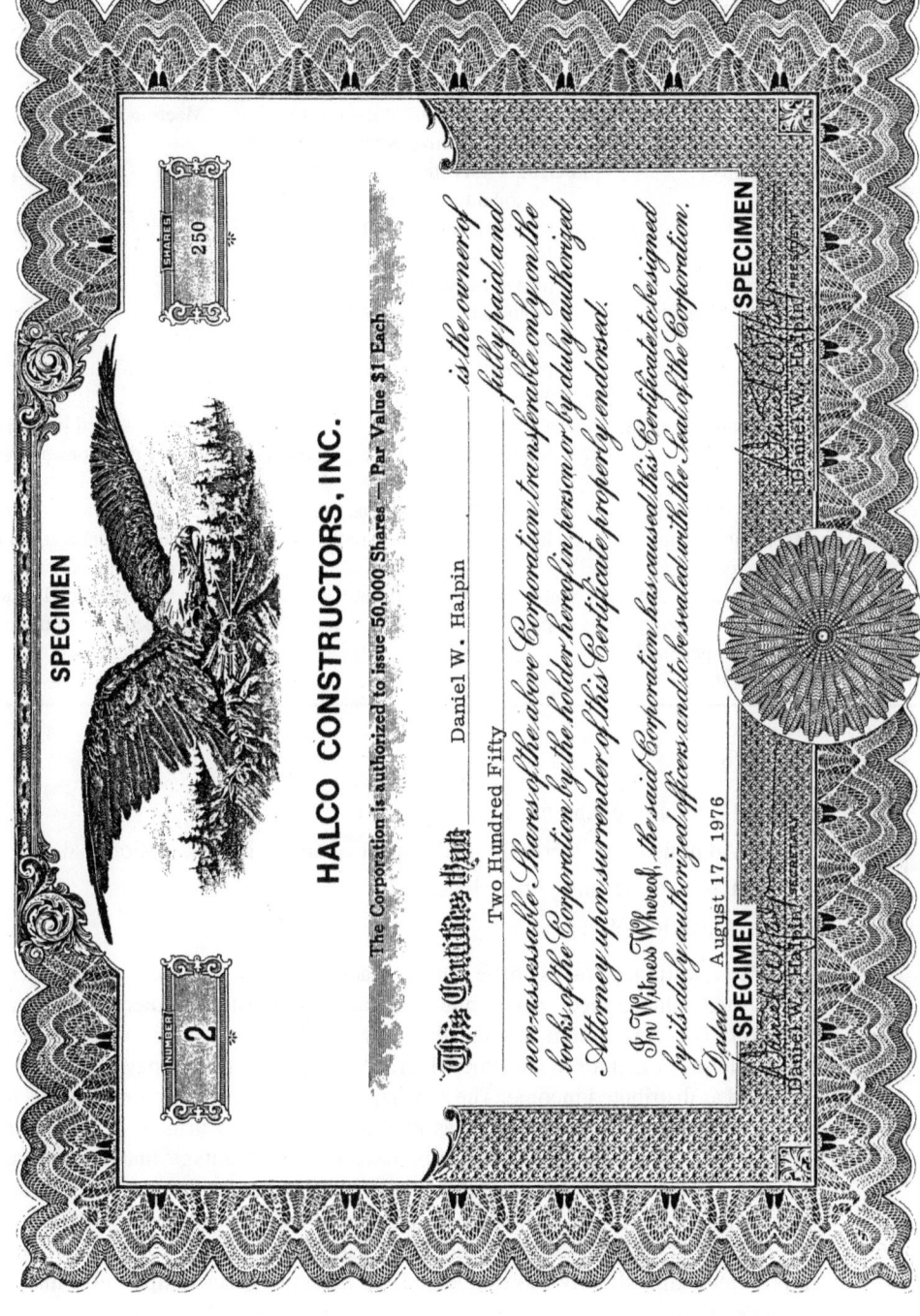

Figure 5.1 Forms of legal ownership in the construction industry. (Study by T. Gibb, Georgia Institute of Technology, 1975).

Table 5.1 Considerations in Choosing Legal Structure

	Proprietorship	**Partnership**	**Corporation**
Tax	Tax on personal income; tax on earnings whether or not they are withdrawn	Tax on personal salary and earnings	Lower taxes in some cases.[a] Dividends are not deductible; double taxing. Taxes on dividends, that is, money actually received
Costs and procedures in starting	No special legal procedure; apply for licenses; register with IRS	General: Easy—oral agreement Limited: More difficult—must closely adhere to state law	More complex and expensive. Meeting must be held
Size of risk	Personal liability	Personal liabilities: Extent of personal fortune. Limited: each partner is protected; loss of limited partner cannot exceed initial investment	Limited to assets of corporation
Continuity of the concern	No continuity on death of proprietor	Dissolution: No continuity on death of partner. Surviving partners can buy share if in agreement.	Perpetual (charter can expire)
Adaptability of administration	Simplicity of organization; direct control	Decisions and policies implemented by oral agreement	Directors—good if involved. Policy decisions predefined by by-laws
Influences of applicable laws	Laws are well defined; no limit on doing business in various states	Laws are also well defined; a license may be required	Foreign corporation status; requires legal counsel on permanent basis
Attraction of additional capital	Limited potential for capital expansion. Borrowing; line of credit; personal fortune investment	Better; more capital; limited partner concept	Issue securities; collateral provided by corporate assets; issue stock

[a]See Fudd Associates, Inc., example in text.

4. Continuity of the firm.

5. Administrative flexibility and impact of structure on decision making.

6. Laws constraining operations.

7. Attraction of capital.

The question of taxes to be paid in each organizational format is mixed, and the anticipated balance sheet and cash flow of each firm must be studied to arrive at a "best" solution. The corporation has the disadvantage that the firm is taxed twice, once on corporate profit and a second time when the stockholders must pay tax on the dividends received as distributed income. The subchapter S type of firm circumvents this to a degree in that the stockholders are taxed as individuals as if the firm were a partnership. The normal proprietorship and partnerships have the disadvantage that all income is taxed whether or not it is withdrawn from the firm. Thus, as in the example of Uncle Fudd, incorporating yields a benefit despite the double-taxation feature.

Costs and procedures associated with establishing the firms are generally minimal for a proprietorship, slightly more involved for a simple partnership, and a major financial consideration for limited partnerships and corporations. Normally whatever costs and procedures are associated with local, state, and federal tax registration and the purchasing of a license are all that must be considered in establishing a proprietorship or simple partnership. These as well as significant legal costs ($2,000 to $5,000) must be considered in establishing limited partnerships and corporations. These costs may be justified, however, based on the limitation of liability achieved and the benefits of medical, health, and insurance plans that can be implemented in a corporate format.

Corporations and limited partnerships limit the level of loss in the event of a default or bankruptcy to the level of investment. That is, stockholders cannot lose more than the value of their shares. The loss of a limited partner cannot exceed the amount of his investment. If he or she initially invested $20,000, no more than this amount can be lost, and other assets cannot be attached in the event of bankruptcy or default. The assets of stockholders in a subchapter S corporation are similarly protected. Personal assets are used to pay creditors in the proprietorship or simple partnership form of ownership. This can lead to personal bankruptcy.

Proprietorships also have the disadvantage that they terminate when the proprietor dies. This may present a problem, particularly if the firm as an asset must be divided among several heirs. It can be circumvented in part by willing the firm and its market and "goodwill" to one heir (who will carry on the business) and providing that the heir will compensate the other heirs for their share. If a partner dies, the partnership is dissolved. Again, however, provisions in the partnership agreement can provide the means for surviving partners to purchase the deceased partner's share from his estate. Corporations are perpetual and the stock certificates are transferred directly as assets to heirs of the estate.

Policy decisions are relatively simple in the proprietorship and partnership formats. Principals make all decisions. In the corporate format, certain decisions must be approved by the stockholders, which may impact the corporation's ability to react to a developing need or situation. In closely held corporations, however, this is no problem since the partners are able to call ad hoc board meetings to react quickly. Corporations with large numbers of shareholders are not as flexible in this regard. The chief operating office or president handles the day-to-day decision making. A board of directors is charged with intermediate-range and strategic planning and decision making. Major decisions, however, such as stock expansion and acquisition of other firms or major assets, must often be approved by all stockholders in a formal vote.

Local laws may encourage the formation of small and local businesses by placing restrictive constraints and burdensome additional cost on out-of-state, or foreign, corporations. These discriminating practices must be investigated when bidding construction work in a state other than the one in which the construction corporation is chartered. Special licenses and fees are sometimes required of foreign corporations. Proprietorships and partnerships that consist of individuals are protected against these discriminatory practices by the Constitution and enabling "equal rights" legislation.

In raising capital, proprietorships and simple partnerships must rely on the personal borrowing of the principals to generate capital for expansion. The unique feature of the corporation that permits it to sell stock allows corporate entities to attract new capital by further distributing ownership. The corporate assets as well as future projections of business provide a collateral basis to attract new stockholders. This mechanism is not always viable, however. From time to time, corporations may be unable to sell large issues of stock for capital expansion and will be forced to go to the commercial banks to borrow. In periods of economic uncertainty, sale of stock as a method of attracting capital may be limited.

Good information regarding the advantages and disadvantages of various legal forms of organization are contained in the Small Business Administration management guides available from the Government Printing Office and other government sources.

REVIEW QUESTIONS AND EXERCISES

5.1 Name the three principle forms of business ownership in construction and state the liability limits of the owners in each case.

5.2 Which legal structure is most difficult to establish and why?

5.3 Name three types of partnerships.

5.4   Describe briefly two advantages and two disadvantages of a corporate form of business organization as compared to a partnership.

5.5   Jack Flubber, who owns Sons of Flubber Construction Co., and runs it as a proprietorship, had gross profits last year of $80,000. His personal and family expenses are $52,000 and he has $7,000 in exemptions and deductions. He paid $17,000 in taxes. If he paid himself a salary of $55,000 taxed at 20%, would it be advantageous for him to incorporate as a closely held corporation? Explain.

5.6   What is meant by the term *foreign corporation*?

5.7   What would be the advantages of organizing as a subchapter S corporation?

5.8   Is it possible to characterize the legal structures of local contractors using the Yellow Pages as a guide?

5.9   What steps must be taken to set up a partnership? How can a partnership be dissolved?

5.10   In problem 5.5, what taxes would Jack pay if he organized as a closely held corporation (as described) and, after paying his salary, also issued himself a dividend of $10,000?

5.11   What is the difference between par and book values of corporate stock? If an incorporated construction company wins a large cost plus fixed-fee contract, what impact might this have on the market value of the company's stock?

5.12   Uncle Fudd has decided to sell his ownership in the Cougar Construction, Inc., to Cousin Elmer. How would the legal firm handling this transaction determine a fair price per share?

5.13   What is a limited liability corporation? Discuss its advantages and disadvantages. You can search the Internet for the brief IRS Publication 3402, *Tax Issues for Limited Liability Companies*, which includes a section describing this newer form of business organization.

5.14   Many banks and similar lending institutions require that the chief executive officer (CEO) of small corporations (which usually are owned by the CEO and a few relatives) cosign any loan made to the corporation. If the corporation goes into bankruptcy, to what extent is the CEO responsible for repaying the loan? What assets can the CEO lose? Assume that the loan is used exclusively by the corporation.

5.15   A stock market, such as the New York Stock Exchange (NYSE) or Nasdaq, trades shares (stock) of public corporations. What are some advantages of becoming public, from a corporation's perspective? What are some of the disadvantages?

Chapter 6

Impact of Taxes

CONTENTS

The Sphinx and Pyramid II. (© Karen Moller/iStockphoto).

Taxation in Ancient Egypt, Greece, and Rome

It has been said that nothing in life is sure except death and taxes. Taxes are not normally viewed in a positive light, but large political entities have historically relied on taxation to support and finance governmental operations and infrastructure.

The first documented tax system was in ancient Egypt. Farmers paid their taxes in kind to scribes appointed as tax collectors. The tax owed was determined in proportion to the area and location of the farm. Manufactured goods such as oil, beer and tanned skins were also subject to taxation, and individuals holding office paid their taxes in gold or silver. In the latter periods of the empire, individuals were allowed to declare their possessions for tax purposes. Scribes could spot-check these declarations, and failing the tax audit was punished with death. There were also safeguards for taxpayers. Receipts for paid taxes were normally issued, and convicted corrupt tax collectors were punished. Citizens were able to pay their taxes by performing corveé duty, which consisted in working in public works (such as the Pharaoh's burial monument) for a period of time. Interestingly, slaves were not accepted as payment for taxes.

The city-states of ancient Greece were more minimalist in their approach to taxation. Greeks were exceptional in the ancient world by their clear preference for free enterprise and limited government intervention. Citizens had to pay indirect taxes, the equivalent of today's sales taxes, on everyday goods such as houses, herds, wine, and even slaves. Higher taxes were imposed on foreigners. Wealthy citizens were levied direct taxes (i.e., according to their wealth), but only in times of national emergencies such as war. Moreover, and unique to the Greek culture, very wealthy citizens were required to contribute to public works through special-purpose taxes called *liturgias*. Some liturgias were prestigious, such as contributions to build a gymnasium or to fund a chorus, and frequently attracted volunteers

In their early days, the Romans followed the example of the Greeks in limiting the taxes levied on their citizens. The total combined taxes paid by Roman citizens has been estimated at a modest 1% of the total value of their wealth, including real estate, animals, slaves, and other possessions. This rate could climb to as much as 3% in war time, but if the collected tax turned out to be more than the actual extraordinary expenses, the excess was returned to the taxpayers. Eventually, the Romans conquered much of the known world of their time. Taxes were then virtually abolished for Roman citizens, and the funds required to govern the empire were almost entirely collected from the provinces.

The state began providing free grain, and in time, bread, to all citizens; the emperor paid for public entertainment which led to increasingly heavier taxes. Rome needed to collect greater taxes from the provinces. Collection of these taxes was performed by wealthy individuals who bid for the position and frequently remitted in advance the tax proceedings to Rome. These advances were welcomed by the cash-strapped government, but this system also guaranteed that these private collectors would go to any length to recover money from farmers and other taxpayers in their territory. This tax system proved to be unsustainable and was a significant factor in the eventual downfall of the empire as provinces grew restless with the harsh treatment.

6.1   INTRODUCTION

The profitability of any construction company is significantly affected by the amount of taxes it is required to pay. As noted in Chapter 5, the legal structure of a company greatly influences the taxation incurred by a given business or company. Understanding the rationale and conceptual framework of the current tax system is, therefore, of paramount importance to any contractor. Financial choices affect the amount of taxes paid and taxes affect the financial decisions that a construction company makes. This chapter provides a brief introduction to this topic. It must be kept in mind, however, that tax law changes *constantly* and is continuously in a state of flux. Therefore, this cursory discussion is designed to present only the conceptual outline of the U.S. tax system as it pertains to business organization and individuals.

6.2   WHY TAXES?

Taxes are primarily designed to finance government operations and to meet societal needs. In modern societies, taxes have also been implemented to influence taxpayer behavior. For example, high taxes operate to increase the price of tobacco, which presumably causes fewer individuals to use tobacco products. Tax deductions often are granted to reward individuals and companies for undertaking actions that are viewed as being positive for the economy or society in general. Charitable gifts are usually either partially or fully deductible from taxable income.

Some deductions may be given to individuals for the cost of insulating a residence with the intent of motivating individuals to make their homes more energy efficient.

6.3   TYPES OF TAXES

A tax levied on the net income realized by a company or individual is called an *income tax*. Income taxes are referred to as direct taxes. Indirect taxes, in contrast, are levied on the cost, price, or value of products or services (e.g., sales tax assessed at the time of purchase). The U.S. government relied almost exclusively on taxing resources other than income until the 16th Amendment to the Constitution was passed in 1913 (previous attempts to impose income taxes were short-lived). Although the focus of this chapter is on income taxes, everyone also pays indirect taxes. Some forms of indirect taxes will be discussed.

6.4   INCOME TAX SYSTEMS

Business entities pay tax on taxable income, which is the revenue earned by the company minus the expense of doing business and any allowable deductions. In the case of individuals, taxable income is defined as gross income minus deductions. As was noted in Chapter 5, the corporate form of legal organization leads to double taxation because the corporation is taxed as a business entity and the stockholders are taxed separately as individuals for profits distributed to them as dividends. Stockholders are also subject to further taxation at the time of sale of stock if the stock has gained in value during the time between purchase and sale.

$$Income = Revenue - Expenses$$

The concept of deductions, however, allows for this amount to be further reduced for tax purposes through the subtraction of deductions authorized by the Internal Revenue Service (IRS) and the appropriate tax law. Therefore, the taxable amount of income becomes:

$$Taxable\ Income = Revenue - Expenses - Deductions$$

For individuals required to pay federal income tax, income is what a person receives from wages, gifts, inheritance, sale of property, and similar sources. Expenses are not a major issue for most individuals, since outside of proprietorships, individuals normally don't expend funds to generate revenues. Therefore, what is reported is purely income (i.e. wages, interest at the bank, etc.) and not revenue minus expenses. As a practical matter, unless an individual operates a business, what is being reported on the standard U.S. Individual Income Tax Return (Form 1040 or 1040A) is income minus deductions and other allowances discussed in this chapter.

If, in fact, an individual does expend funds to generate a revenue stream (as would be the case in a proprietorship), a special schedule of revenues versus expenditures (*Schedule C*) must be submitted to the IRS along with the standard tax return. Business activities by an individual require the submittal of a *Profit and Loss from Business*, form, which provides a section for reporting expense.

Individuals are authorized to use a *standard deduction* (the amount of which is adjusted each year by tax law) to reduce their taxable income. The other option available to individual tax payers is to itemize deductions using a separate form called *Schedule A— Itemized Deductions*. Itemizing deductions normally is justified if an individual has major expenditures such as (a) medical and dental expenses, (b) taxes paid to entities other than the federal government, (c) interest paid on home mortgages, or (d) gifts to charity. Certain major losses due to theft or other causes specified by the tax law as well as job expenses (e.g., non-reimbursed travel, job education, and so on) can also be deducted when itemizing deductions.

6.5   TAXATION OF BUSINESSES

In business organizations, the definition of revenues and expenses becomes more important and must conform to guidelines established by tax law and IRS publications. The concept of revenue is fairly straight forward. Revenue is, in effect, the summation of all amounts due and payable to a company from clients or other financial sources for work performed, sales and services supplied, and other financial transactions.[1]

The definition of expenses is more complex. Construction companies have many expenses that are directly incurred in the physical construction of projects. Other expenses are required to support construction and provide for the management of site activities and for company administration. These are categorized as direct and indirect expenses.

1. Direct expenses: These expenses are incurred in operations that lead to the realization of the physical facility being constructed. Typical direct expenses are job payroll, project-related equipment costs, materials used for construction of the project, and subcontractor payments.

2. Indirect expenses: This term relates to costs incurred for the management support of the job site and for home office general and administrative costs (e.g., senior management and staff salaries, office equipment costs such as computer and copying charges, and costs related to the ownership or rental of office space). These types of indirect expenses are often referred to as *overhead* or *overhead expenses.*

To illustrate this, it is helpful to examine one of the key documents required for fiscal control of a company, namely the Income Statement (also called the *Profit and Loss Statement*). The Income Statement shows the company's actual net profit or loss for a given period of time (e.g., a month, quarter, or year) Revenues measure the inflow of value to the firm achieved by selling goods or providing services. In the case of a construction company, revenues reflect cash inflow generated by progress payment requests on various jobs and cash receipts received through other operations (e.g., equipment rental, consulting fees, etc.)

Expenses measure the expenditure of funds used to generate revenues (e.g., payroll, subcontractor payments, etc.) As such, they can be deducted from revenues when determining taxable income. As noted above, expenses may be direct or indirect. The entire amount of direct expenses is subtracted from the revenue to determine gross profit. In general, most indirect expenses (e.g., field and home office overheads) are also subtracted from revenues to establish the taxable income of a company. However, certain overhead expenses may not be fully deductible. Based on existing tax law, travel, meals, and entertainment cannot be fully deducted. The existing tax law limits these expenses as follows "Generally, you can deduct only 50% of your business meal and entertainment expenses, including meals incurred while away from home or business." This, of course, means that these indirect expenses are *not* fully deductible.

To better understand how a firm addresses revenues and expenses to determine its taxable income, consider the Income Statement shown in Figure 6.1.

The Income Statement in Figure 6.1 lists the revenues and expenses in an organized fashion. In this case, revenues are referred to as *Net Sales.* The gross profit is simply calculated as sales minus cost of sales (i.e., revenues minus expenses). Gross profit is reduced in the amount of *Job Related Indirect Charges* ($562,660) and *Total Fixed Overheads*" ($1,265,820). After subtracting these two overhead costs ($1,828,480), net profits before taxes are shown as $1,454,810. At this point, taxes are subtracted yielding a net profit of $945,627. This amount will either be retained (i.e., held in the company as *retained earnings*) or distributed as dividends to the owners or stock holders of the company.

---

[1] When using accrual accounting systems, taxes must be paid on revenues that are payable but possibly not yet received.

Income Statement, Fudd Construction Co. 3ʳᵈ Qtr, 2XXX

Net Sale (Revenues)		$21,427,610
Cost of Sales (Direct Expenses)		
1. Materials	10,139,130	
2. Labor (Includes all payroll taxes & fringe)	5,482,710	
3. Subcontracts	514,500	
Total Cost of Sales		$18,159,320
Gross Profit		$ 3,283,290
Job Related Indirect Charges		
1. Salaries	276,640	
2. Job Office Costs	190,430	
3. Transportation (Travel, Meals, etc.)	80,710	
4. Interest Charges (Job Carrying Costs)	12,750	
5 Miscellaneous	2,130	
Total Job Indirect Charges		$ 562,660
Fixed (Home Office) Overheads		
1. Salaries – Home Office Staff	234,180	
2. Salaries – Officers	384,100	
3. Home Office Admin Cost (Equipment, etc.)	516,950	
4. Legal and Audit	23,150	
5. Rent	25,060	
6. Depreciation	82,380	
Total Fixed Overheads		$ 1,265,820
Total Overheads		$ 1,828,480
Net Profit before Taxes		$ 1,454,810
State and Federal Taxes		$ 509,183
Net Profit		$ 945,627

Figure 6.1 Income Statement for Fudd Associates, Inc.

Two items in this report are of interest from a tax standpoint. Under job-related indirect charges, item 3 reports Transportation (Travel, Meals, etc.) costs. As noted previously, this amount cannot be fully deducted from revenues. Only half of these costs are available as a deduction. Therefore, before calculating federal tax, half of this amount (i.e., $30,000) must be added to net profit before taxes. That is, because the total amount of Item 3 is not deductible, the taxable amount of profit becomes:

$$\text{Taxable Income} = \$1,454,810 + \$30,000 = \$1,484,810.$$

On the other hand, the amount shown as Item 6 under Fixed Overhead is a major deduction allowed by tax law as a way for each firm to recover the loss of value of equipment because of obsolescence over time. This will be discussed in the next section and later in discussing the ownership cost of equipment in Chapter 13. The subtraction, then, of this amount when calculating net profit provides a major way for a company to reduce its amount of taxable income.

6.6 BUSINESS DEDUCTIONS IN GENERAL

Most firms try to reduce their amount of taxable income as much as legally possible. Deductions allow companies to reduce the amount of income subject to taxation. They can be viewed as a gift from the government recognizing both direct and indirect, as well as, latent costs of doing business. Some deductions, as noted previously, are designed to provide incentives to trigger certain actions on the part of business entities.

Probably the most significant deduction available to construction contractors is that given for the loss in value of company equipment over time. For instance, an off-highway truck or hauler will decline in value over time. After 4 years of use, a $100,000 hauler may have a market value of only $40,000. This decrease or decline in value is viewed as a cost of doing business (i.e., a cost of generating revenues). Therefore, the IRS allows this loss of value to be recovered through a deduction called *depreciation*.

Construction costs (direct costs plus field overhead) and general (home office) overhead costs associated to the operation of a construction company are deductible independent of the company's form of business organization. However, some expenses are difficult to classify and others, as noted, are not fully deductible (e.g., travel, meal and entertainment costs).

Therefore, the calculation of taxable income and the determination of what items can be deducted and to what degree must be carefully determined. Deductions, such as depreciation, typically lead to a company paying taxes on a lower amount than the net profit before income taxes appearing in a company's income statement. The net amount used to compute a company's tax liability is called its tax base or taxable income.

6.7   TAXABLE INCOME: INDIVIDUALS

Taxes levied on proprietorships are paid directly by the owner, and their computation is linked to individual tax considerations. The taxable income of individuals depends not only on the amount of money earned, but also on their civil status (e.g., married, head of household, etc.) and how they choose to report their deductions.

For tax purposes, individuals are classified into the following filing statuses:

- Single: The unmarried individual (not falling into the head of household category described below) generally pay the highest amounts of tax compared to the other categories.

- Married filing separately: A married couple may decide to pay taxes separately, filing their tax returns as essentially two single individuals. The current tax code results in a so-called marriage penalty for married couples in which both spouses work and earn higher than average, comparable incomes. In these cases, the couple pays more taxes by submitting a single joint return than by filing separately.

- Married filing jointly: This filing status comprises the traditional married couple and children, choosing to use a single tax return for the couple and dependents. Historically, tax codes protected this category, although such advantages have all but disappeared in the current code.

- Head of household: This is the one category that is given some tax benefits. It includes persons contributing more than half of the support to a household. These persons are considered unmarried and have at least one dependent person. These requirements encompass single parents, single children supporting their parents, and similar individuals frequently under financial duress.

6.8   ITEMIZED DEDUCTIONS, STANDARD DEDUCTIONS, AND PERSONAL EXEMPTIONS

Every individual (or family, if filing jointly) can deduct certain items from their income, thus reducing the tax basis on which their taxes will be computed. Contributions to charity, medical expenses over a certain threshold, the interest paid on the mortgage loan for the individual's residence, and other deductions can add up to a substantial amount of money. One of the most important deductions available to construction contractors is the depreciation of property. This item-by-item reporting is called the *itemized deduction approach*. Alternatively, individuals can choose to subtract a standard deduction from their income instead of performing an itemized deduction calculation.

Table 6.1 Standard Deductions and Personal Exemptions, 2008

Filing Status	Standard Deduction
Single	$5,450
Married filing jointly	$10,900
Married filing separately	$5,450
Heads of household	$8,000
Personal Exemption	$3,500

The amount of the standard deduction is specified by the IRS and varies depending on the filing status. Furthermore, it is adjusted every year. An advantage of using the standard deduction is that no confirming documentation is needed. If the itemized deduction approach is used, each deduction must be documented with some form of proof (typically called a voucher). Both approaches cannot be used simultaneously. If a person takes advantage of the standard deduction, no other deduction can be claimed except for a few exceptions, such as contributions to individual retirement accounts (IRAs). The standard deductions for 2008 are given in Table 6.1.[2]

The IRS also allows a personal exemption amount to be deducted for the taxpayer and each person reported as a dependent (e.g., spouse and children) on the tax return. For example, if a married couple with two children files a joint return, they can claim four personal exemptions. The amount specified for each personal exemption does not vary with the filing status, but it is adjusted every year. As can be seen in Table 6.1, the amount for each exemption for 2008 is $3,500. The personal exemptions for the family of four previously mentioned would total $4 \times \$3,500 = \$14,000$. This amounts to an actual reduction in tax for a taxpayer in the 25% tax range ($0.25 \times \$14,000 = \$3,500$).

There are several minor deductions applicable to businesses or individuals not discussed here (e.g., college tuition and fees and the interest paid on student loans). The taxable income of proprietorship owners and individuals results from applying all deductions and personal exemptions to reported income.

6.9   THE TAX SIGNIFICANCE OF DEPRECIATION

Deductions benefit the tax payer in that each dollar that can be deducted results in a reduction in the amount of tax paid. This reduction in tax amounts to a per dollar savings of $\$1.00 \times$ (tax rate). If a tax payer in the 28% marginal tax bracket and can deduct $1,000 for whatever reason, this reduces the amount of tax by $280.

Most tax agencies accept the concept that property declines in value over time and is a legitimate cost of doing business, which can be claimed as depreciation. Depreciation reflects the loss of value and obsolescence of property (e.g., equipment) involved in the operation of a business and can amount to a significant reduction in tax paid. It results in a lowering of taxes (e.g., $0.34 on the dollar for businesses based on a tax rate of 34%).

Depreciation applies to all property required for the operation of a business. Computers, office furniture, heavy equipment, and real estate are all depreciable assets. Most construction companies, particularly heavy and highway constructors, own a substantial amount of equipment that can be depreciated over time. Contractors look at equipment as small "profit centers" and will apply any depreciation associated with the piece of equipment to offset the profit generated by that machine or equipment item.

It is important to align the declaration of depreciation so that it can be used to offset taxable income. If a company has $100,000 of depreciation that can be taken in a given

---

[2] As previously noted, proprietorships can choose to use their standard deduction for personal income, independently of the costs deducted from their revenue to calculate business income (or loss) reported in their Schedule C. A proprietor can pay himself or herself a salary, which is an expense to the proprietorship. The standard deduction can be applied to this salary. Why would it be advantageous for a proprietor to accept a salary applying it as an expense to the proprietorship?

year, but only $80,000 in taxable income in that year, $20,000 in tax savings are lost. That is, depreciation cannot be carried over for use in a future tax year. In general, depreciation must be applied in the tax year in which it occurs.

Calculating Depreciation

The method by which depreciation is calculated for tax purposes must conform to standards established by the IRS. The bases available for calculation of depreciation are (a) the straight line or linear method, (b) the accelerated method and (c) a variant of the straight line method called the *production method*. The straight line method provides for the prorating of the depreciation amount for a given asset in equal amounts across the asset's service life. Federal law has implemented the use of fixed percentages for calculation of year by year depreciation when using the accelerated method. These percentages are given in Table 6.2 for various classes of property. The production method allows the depreciation associated with an asset to be recovered on a per-unit-of-output basis. These methods are discussed in detail in Chapter 13.

The accelerated method based on tabular percentages is called the *modified accelerated cost recovery system* or MACRS. This system defines various classes of property ranging from small items (e.g., computers, file cabinets, and so on) with a 3-year service life up to a 20-year class for real property. Most construction equipment items fall into the 3-, 5-, or 7-year service life categories.

For example, light trucks (less than 6.5 tons) are considered to be 3-year property. Most average-weight construction equipment is classified as 5-year equipment. Some heavy construction equipment, such as dredging barges, are depreciated over a 10-year life.

Assume that you wish to calculate the depreciation for an off-highway hauler with a net first cost of $100,000 and a service life of 5 years. Although this is a 5-year property class, it is depreciated out across a 6-year period. The depreciation percents and amounts are shown in the following table. The sixth year is assumed to be a half year.

Table 6.2 MACRS Table for Accelerated Depreciation

Recovery Year	Annual Recovery (Percent of Original Depreciable Basis)					
	3-Year Class (200% d.b.)	5-Year Class (200% d.b.)	7-Year Class (200% d.b.)	10-Year Class (200% d.b.)	15-Year Class (150% d.b.)	20-Year Class (150% d.b.)
1	33.00	20.00	14.28	10.00	5.00	3.75
2	45.00	32.00	24.49	18.00	9.50	7.22
3	15.00	19.20	17.49	14.40	8.55	6.68
4	7.00	11.52	12.49	11.52	7.69	6.18
5		11.52	8.93	9.22	6.93	5.71
6		5.76	8.93	7.37	6.23	5.28
7			8.93	6.55	5.90	4.89
8			4.46	6.55	5.90	4.52
9				6.55	5.90	4.46
10				6.55	5.90	4.46
11				3.29	5.90	4.46
12					5.90	4.46
13					5.90	4.46
14					5.90	4.46
15					5.90	4.46
16					3.00	4.46
17						4.46
18						4.46
19						4.46
20						4,46
21						2.25

Year	Accelerated Percentage	Depreciation	Straight Line Amount
1	20.0	$20,000	$20,000
2	32.0	$32,000	$20,000
3	19.2	$19,200	$20,000
4	11.52	$11,520	$20,000
5	11.52	$11,520	$20,000
6	5.76	$5,760	

Note that all categories are depreciated over $n + 1$ years, n being the stipulated property's service life. This depreciation occurs at an accelerated rate. That is, the depreciation is not evenly distributed across the 6-year period. More depreciation can be taken in the early years.

In the straight line or linear method, equal amounts of depreciation are taken in each year. Using the straight-line method the service life is 5 years, and the annual depreciation amount available is $100,000/5 or $20,000 per year for 5 years.

6.10   MARGINAL TAX RATES

The U.S. income tax system is progressive. This term is used to indicate that tax rates do not stay flat or constant across all income levels. In a flat system, all taxpayers are subject to the same tax as a percentage of their income, regardless of the amount earned. In a progressive system, the percentage of income tax paid by taxpayers with high income is higher than the percentage paid by individuals with lower income. There are a few instances of regressive taxes, which result in lower income individuals paying more, as a percentage of their income, than higher income taxpayers. The excise tax on gas is an example: The tax levied to each gallon is more significant as a percentage of earnings to lower income than to upper income individuals.

The income tax depends on two main factors: the filing status of the taxpayer (i.e., corporation, single individual, married filing jointly, etc.), and the taxable income on which the tax is computed. Each filing status has a corresponding tax table, similar to the one shown in Table 6.3. The table is divided into income brackets, and each income bracket has a different tax percentage called the marginal tax rate for the bracket. The column titled Lump Sum is derived from other data, which will be discussed. The brackets are adjusted by the IRS every year. Changing tax percentages, however, is much more difficult because percentage changes require an act of Congress.

Taxes are computed incrementally. Referring to Table 6.3, a single person with taxable income of $10,000 will pay a 10% tax over the first $8,025 and 15% on the amount above $8,025.

$$\text{Tax for } \$10,000 \text{ taxable income} = 10\% \times 8,025 + 15\% \times (10,000 - 8,025)$$
$$= \$1,098.75$$

Table 6.3 Tax Table for Single Taxpayers, 2008

Bracket Floor	Lump Sum	Marginal Tax Rate
8,025	0	10%
32,550	802.50	15%
78,850	4,481.25	25%
164,550	16,056.20	28%
357,700	40,052.20	33%
No limit	103,791.75	35%

Table 6.4 Individual Taxpayer Income Brackets and Tax Tates, 2008

Single	Married Filing Jointly or Qualified Widow(er)	Married Filing Separately	Head of Household	Marginal Tax Rate
$0	$0	$0	$0	10%
$8,025	$16,050	$8,025	$11,450	15%
$32,550	$65,100	$32,550	$43,650	25%
$78,850	$131,450	$65,725	$112,650	28%
$164,550	$200,300	$100,150	$182,400	33%
$357,700	$357,700	$178,850	$357,700	35%

Table 6.5 Corporate Income Brackets and Tax Rates, 2008

Bracket Floor	Lump Sum	Marginal Tax Rate
$0	$0	15%
$50,000	$7,500	25%
$75,000	$13,750	34%
$100,000	$22,250	39%
$335,000	$113,900	34%
$10,000,000	$3,400,000	35%
$15,000,000	$5,150,000	38%
$18,333,333	$6,416,667	35%

Similarly,

$$\text{Tax for } \$50,000 \text{ taxable income} = 10\% \times 8{,}025 + 15\% \times (32{,}550 - 8{,}025)$$
$$+ 25\%^*(50{,}000 - 32{,}551) = \$8{,}843.50$$

Notice that for a taxable income of $60,000, only the last term would be different from the previous computations, that is, $25\% \times (60{,}000 - 32{,}550)$. In fact, only the last term of the tax computations for a taxable income differs from the computations for any other taxable income within the same bracket. To simplify the computations, most tax tables include a lump sum equal to the amount of tax paid up to the bracket base.

The lump sum for the first brackets of Table 6.3 can be found as follows:

$0 to $8,025:   Lump sum = $0, since there are no preceding brackets
> $8,025 to $32,550:   Lump sum $= 10\% \times 8{,}025 = \$802.50$
> $32,550 to $78,850:   Lump sum $= 10\% \times 8{,}025 + 15\% \times (32{,}550 - 8{,}025)$
                                $= \$4{,}481.25$
> $78,850 to $164,550: Lump sum $= 10\% \times 8{,}025 + 15\% \times (32{,}550 - 8{,}025)$
                            $+ 25\% \times (78{,}850 - 32{,}550) = \$16{,}056.20$

Using the lump-sum column, tax computations are straightforward:

$$\text{Tax for } \$10{,}000 \text{ taxable income} = 802.50 + 15\% \times (10{,}000 - 8{,}025) = \$1{,}098.75$$
$$\text{Tax for } \$50{,}000 \text{ taxable income} = 4{,}481.25 + 25\% \times (50{,}000 - 32{,}550) = \$8{,}843.50$$
$$\text{Tax for } \$60{,}000 \text{ taxable income} = 4{,}481.25 + 25\% \times (60{,}000 - 32{,}550) = \$11{,}343.75$$

A summary table for other individual taxpayer filing statuses is shown in Table 6.4. Notice that although lump sums are not included, the table is perfectly usable. Table 6.5 includes the brackets and rates for corporations.

6.11 TAX CREDITS

A tax credit is an allowance that can be deducted from the tax as computed in the previous section, instead of from the individual's or corporation's taxable income. This is the case, for example, of the Hope credit and the Lifetime Learning credits, which can be claimed by college students complying with given requisites. Other examples are the credit for child and dependent care expenses and the residential energy credit. As the examples show, these allowances usually subsidize a reputedly desirable behavior (i.e., college education and energy saving investments) or to help individuals under trying financial circumstances (i.e., child and dependent care). Similarly, corporations can claim credits. They tend to encourage business activities in a region (e.g., the American Samoa Economic Development Credit), reimburse taxes paid overseas, or foster a desirable behavior (e.g., the Rehabilitation Tax Credit).

It must be repeated that credits are subtracted from the computed tax and not from the income. To emphasize the effect of a tax credit, consider the case of a person owing $10,000 in federal income tax who decided to purchase a hybrid car in 2008. At the time, the IRS allowed a credit of up to $3,000 to foster the use of this technology. By purchasing a model qualified for the full $3,000 credit, the person would have lowered the tax to $7,000 instead of the original $10,000. In other words, a tax credit results in a reduction in the tax paid whereas a deduction results in a reduction of the amount that is taxable. Therefore a deduction of $3,000 for a person in the 28% tax bracket reduces the tax actually paid by only $3,000 × 0.28 = $840, not $3,000. Clearly, a tax credit is more desirable.

6.12   TAX PAYROLL WITHHOLDING

Employers are required to retain an amount from each paycheck that is roughly enough to cover the employee's tax liability at the end of the year. This withholding ensures that amounts are set aside on a paycheck-by-paycheck basis. In effect, the employee accumulates the amount required for end of year tax payment on a pay-as-you-go basis. The amount held back is essentially an escrow account held for the employee by the employer. At the time of payment, this escrow or withheld amount provides the employee with a basis for paying tax. As many taxpayers know, it often occurs that the total retention exceeds the total tax owed, in which case the IRS refunds the amount of overpayment to the individual taxpayer.

6.13   TAX PAYMENT SCHEDULES

Corporations and some individuals must pay their taxes in quarterly installments, although their deadlines are slightly different. The requirement to make quarterly payments is similar, in concept, to withholding taxes for the individual employee discussed in the previous section. Installments are intended to avoid the situation of a company being confronted with a large payment at the end of the tax period. It ensures that companies and individuals pay as they go and spread tax payments over the four quarters of the tax year. This avoids the shock of paying a single large balloon payment at the end of the tax period.

Corporations can select a fiscal year, which does not need to coincide with the calendar year. Whereas a calendar year ends on December 31st, a corporate fiscal year can end on any arbitrary date, usually at the end or the middle of a month. Many retailers, for example, do not like a December 31st fiscal year end because they are still busy with the Christmas season. Nevertheless, about two-thirds of U.S. corporations choose to have their fiscal year ending with the calendar year. A corporation with this calendar must estimate their profit and submit installments on the 15th of April, June, September, and December. Proprietorships and other organizational forms without an independent tax standing, as well as individuals with substantial income that is not subject to withholding, use the same calendar for the first three installments (15th of April, June, and September) but pay the last installment on January 15th of the following year. Oddly enough, this means that there are 2 months between the 1st and 2nd payments, 3 months between the 2nd and 3rd payments, and 4 months between the 3rd and last payments.

6.14 MARGINAL, AVERAGE, AND EFFECTIVE TAX RATES

Three different tax rates are of interest. As already discussed, the marginal tax rate is related to the tax bracket into which the taxable income falls. A progressive system imposes a higher tax percentage to higher levels of income. However, it is also important to view taxes as a percentage of the money earned by the taxpayer. Therefore, two other rates need to be considered. The average tax rate compares the tax paid to the taxable income.

$$\text{Average Tax Rate} = \text{Tax Paid}/\text{Taxable Income}$$

The effective tax rate is the ratio between tax paid and income prior to subtraction of deductions and exemptions.

$$\text{Effective Tax Rate} = \text{Tax Paid}/\text{Gross Income}[3]$$

Assume that Elmer Fudd of Fudd Construction has a total income (before taking exemptions and deductions) of \$200,000. His taxable income is \$160,000. He runs his company as a proprietorship and is married filing a joint return. Mr. Fudd falls into the 28% marginal tax bracket (i.e., married filing jointly with an income between \$131,450 and \$200,300). This means that a gift to charity for \$1,000 will lead to a reduction in tax paid of $28\% \times \$1,000 = \$280$.

Fudd's actual tax paid is based on a taxable income of \$160,000 is

$$(16,050 \times 0.10)+$$
$$(65,100 - 16,050) \times 0.15+$$
$$(131,450 - 65,100) \times 0.25+$$
$$(\$160,000 - \$131,450) \times 0.28 = \$33,529$$

His average tax rate is \$33,529/160,000 or 20.96%

The average tax rate will always be lower than is marginal rate, since the average includes the fraction of the income taxed at the lower tax rate of the low income brackets.

His effective tax rate is based upon a gross income of \$200,000.
Effective Tax Rate = \$33,529/\$200,000 = 16.76%

The effective tax rate results from dividing the amount of paid taxes by the income without considering deductions or exemptions.

The spread between average and effective income taxes (20.96 versus 16.76%) is low in this case, compared to many large corporations with large tax breaks. With corporations the difference between average and effective rates are significantly influenced by the marginal rates since the corporate rates do not monotonically increase as income increases. The rates have a bimodal or camel back shape. As can be seen from Table 6.5, the corporate rate increases to 39% and then drops back to 34% for incomes between \$335,000 and \$10 million before increasing to 35% above \$10 million. The rates move between 39 and 34% for incomes above \$100,000. This convoluted rate structure and aggressive tax sheltering strategies yield significant reductions in taxable income and result in large differentials between average and effective rates. The IRS reported that in 1999 the effective rate paid (on average) by the top 10,000 largest corporations was 20%, considerably less than their average tax rate of near 34%.

6.15 NET OPERATING LOSSES

A company has a net operating loss (NOL) when it reports a higher amount of allowable deductions than revenue. NOL is generally close, but not equal, to the loss (if any) reported in

---

[3] Gross income is the income prior to consideration of exemptions and deductions.

the company's income statement. As previously discussed, there are deductions for tax purposes that may be computed differently in the company's accounting system (e.g., depreciation) and may not appear in the income statement (e.g., incentives to locate in certain areas), or vice versa, may not be acceptable for tax purposes (e.g., some public relations and travel expenses). Individuals and all forms of business organizations can report a NOL.

No income taxes are levied in the year that an individual or company has an NOL. Furthermore, the loss can be used to revise the taxable income of previous years and to lower the taxable income of future years. The NOL applied to past years is called a carryback, and the amount applied to future years is a carryforward.

The rules applied to carryback and carryforward have changed with every major revision of the tax code, and depend on the type of loss, income, and even the type of institution (banks, for example, have their set of rules in this area). Being aware of the concepts of carryback and carryover should be enough for the intended level of this book. Carryback has been limited to the previous 3 years and carryforward to the next 15 years for most situations applicable to construction companies. The following example helps in the understanding of the process.

Suppose that a company reports a NOL of $1 million for 2009. It had a taxable income of $200,000 in each of the previous 4 years. How would this company apply this NOL to revise previous tax computations and as a deduction for coming years?

The carryback process is usually performed before the carryforward because it results in immediate tax relief (a refund) from the IRS. An exception happens when a company finds that the marginal tax rate of previous years is lower than the expected marginal rate for coming years. In such (uncommon) case, the difference in rates may justify the wait. In this case, the company chooses to apply the carryback first. The taxable income of each of the 3 previous years is revised to zero:

- In 2008, the amount available to carry back is $1 million. After subtracting the $200,000 originally reported as taxable income, the revised taxable income is $0, and the amount available to further carryback is $800,000.

- In 2007, the revised taxable income is also zero, and the remaining carryback is $600,000.

- In 2006, the same process is applied. The revised taxable income is zero, and the remaining carryback is $400,000.

- The taxable income for 2005 cannot be revised, despite the $400,000 of NOL still available, since this year is more than 3 years before the NOL.

The company will send the IRS revised tax returns showing zero taxable income for 2008, 2007, and 2006. It will get a refund for all taxes paid in those years (not $200,000 for each year, but the tax amount paid in each year).

The unapplied portion of the NOL is saved as carryforward for the coming years. If the company has a taxable income before applying the NOL of $100,000 in 2010, it can set to $0 the taxable income and still have $300,000 available. If it has a taxable income of $450,000 in 2011, it would apply all the remaining funds from the NOL to pay taxes on a taxable income of $150,000.

6.16 SUMMARY

This chapter has discussed issues related to taxation which influence the bottom line of a construction company and impact financial decisions. Clearly, minimizing the amount of taxes paid increases the profit and retained earnings controlled by the company and ultimately distributed to the owners and stock holders. One area, which is often discussed in conjunction with taxation, is contributions made by companies to ensure the welfare of employees. Companies are involved in paying Social Security contributions required by federal law, unemployment and Workman's Compensation insurance required by state governments, and so-called "fringe" benefits required of union contractors by union contracts. Although these are not strictly taxes, they do have a similar impact on the profitability of the company and must be carefully analyzed. The company's top management and particularly the chief financial officer (CFO) must continuously study and analyze the impact of taxation on all financial and organizational decisions. Taxes are business costs that lead to a reduction in before tax income and must be considered seriously. These costs require at least the same level of professional and financial expertise that is required from a technical perspective to build innovative and successful construction projects.

REVIEW QUESTIONS AND EXERCISES

6.1 Explain in your own words the concept of marginal tax rate.

6.2 Does a single individual making $500,000 pay the same marginal tax rate as another making $1 million? Explain.

6.3 You decide to sell a house that you bought 5 years ago with the purpose of renting it. Its purchase price was $300,000, and now its market value is about $500,000. How much should you expect to pay in taxes for the sale of this house?

6.4 What is the average tax rate paid by a single person with a $60,000 adjusted gross income for 2008?

6.5 What is the name of the amount of money that can be deducted directly from the computed tax, instead of from the taxable income?

6.6 Which business enterprises pay untaxed profits to owners/partners who then pay the taxes on their individual returns?

6.7 Peachtree Construction, Inc., which is incorporated as a corporation, reported a NOL of $500,000 in 2008. Given the following data, how much of this NOL is likely to be available as a deduction from its profit in 2009?

- The company was created in 2006.
- It had a NOL of $100,000 in 2006.
- It reported a profit before taxes of $400,000 in 2007.

6.8 A corporation had a net income before taxes of $1.2 million for 2008. Find the tax liability for this company.

6.9 Pegg Construction, a partnership of three brothers, had a net income before taxes of $300,000. The brothers have agreed that all profit will be shared in equal parts. John Pegg is single and has an income of $500,000 from other sources. How much will John pay in taxes for the $100,000 from the partnership with his siblings? Use 2008 tables.

6.10 Let us assume that we own our personal residence. One question of interest is, "If we depreciate real property (e.g., land, buildings, etc.), can we depreciate the house in which we live?" Depreciation allows for the recovery of the cost of doing business. Since, in most cases, we do not "do business" in our own home, our residence is not a depreciable asset. You can, however, think of some instances in which a person conducts business at home. Can such a person working at home, depreciate the cost of his house?

6.11 Why are governments willing to agree to accept the concept of depreciation, which leads in many cases to a considerably lower level of taxation for business entities? The answer to this question provides an insight into how modern economies operate. How are economies impacted by depreciation?

6.12 Verify the 5- and 7-year property class percentages given in Table 6.2 by applying the 200% depreciable basis (DB) approach to a piece of equipment with a nominal value of $1,000. For the 7-year property class, in what year is the switch from %200 DB to straight line made based on the percentages in the Table 6.2. Only half of the first and last years are considered in the computations. This means that for a 5-year class property, instead of deducting $100\% \times (2/5) = 40\%$ in the first year, only half ($40\%/2 = 20\%$) is deducted.

6.13 Why do you think the 150% DB method is used as the basis for depreciation for 15- and 20-year classes of equipment as given in Table 6.2?

6.14 Find the yearly depreciation of a $100,000 5-year class concrete mixer, first assuming that it is new when purchased, and then that it is purchased as used equipment.

Chapter 7

Project Planning

CONTENTS

Building Information Modeling

A construction project model is a stand-in for the actual structure to be built. A model can show the geometry of the project, as is the case of its blueprints, or can depict a more conceptual aspect, such as the sequence of its building operations or the cost of its components. Although the term *model* may seem uncommon, every participant in the many aspects involved in a construction project requires of one that can answer to their particular needs. As a result, a single project has dozens of independent, consecutively developed models, each one with possible mistakes that can be propagated into the others. A drawing error in the

Building Information Systems

architectural drawings may result in more mistakes in its structural drawings, which in turn may affect the shop drawings made by the mechanical subcontractor, and so on. The mistake could be caught at some point in this chain of graphical and conceptual models. However, it is cumbersome for, say, the mechanical contractor to warn the structural engineer of the mistake so that the structural can in turn advise the architect to fix it. An error is frequently fixed in some of these project models, but not in all. Every stakeholder is affected in time, cost, and even safety by the multiple, independent models typical of current construction practices.

Building information modeling (BIM) attempts to build a single common model of a project that can then be used and updated as changes are introduced or mistakes are detected and fixed. The Associated General Contractors of America has developed a useful booklet introducing BIM. This booklet defines BIM as follows:

> Building Information Modeling *is the development and use of a computer software model to simulate the construction and operation of a facility. The resulting model, a* Building Information Model, *is a data-rich, object-oriented, intelligent and parametric digital representation of the facility, from which views and data appropriate to various users' needs can be extracted and analyzed to generate information that can be used to make decisions and improve the process of delivering the facility.*[1]

To reinforce an important aspect mentioned, BIM is fundamentally computer based. The software model universally includes a three-dimensional (3-D) representation of the project's geometry, from which blueprint views and sections can be easily created. This 3-D model allows a virtual walkthrough of the facility and can identify any geometrical conflicts, such as a mechanical duct running into a structural girder. The visualization capabilities of BIM are useful, but by no means are the only intended content or use for a BIM model. A fully developed BIM model should include other information such as the specifications for each construction element (e.g., the hardware specified for each door or for each roof-top unit), its cost, and even its timing in the construction schedule.

BIM has been vigorously advanced in recent years, and the capabilities of new software and hardware make possible its widespread adoption by the construction industry. There are also hurdles to overcome. To whom does the model belong? What happens if the model, after having been reviewed by multiple stake holders, has a costly or fatal mistake? How much information should be kept private by the various stake holders? These are significant questions that will likely need years of legal testing to determine.

7.1 INTRODUCTION

The planning of a project involves the concept of an objective or facility, and a *scope of work* defining the work product or deliverable. The bid package consisting of the plans and specifications establishes the scope of work to be performed. To be properly managed, the scope of work must be broken into components that define work elements or building blocks, which need to be accomplished to realize the end objective. The assumption is that the project is the summation of its subelements.

Definition of the subelements is important because it determines how the project is to be realized in the field. The subelements are often referred to as work packages. The

---

[1] Ernstrom, R., Hanson, D., Hill, D., Jarboe, J., *et al.* (2006). "The Contractors' Guide to BIM." Arlington, VA: The Associated General Contractors of America.

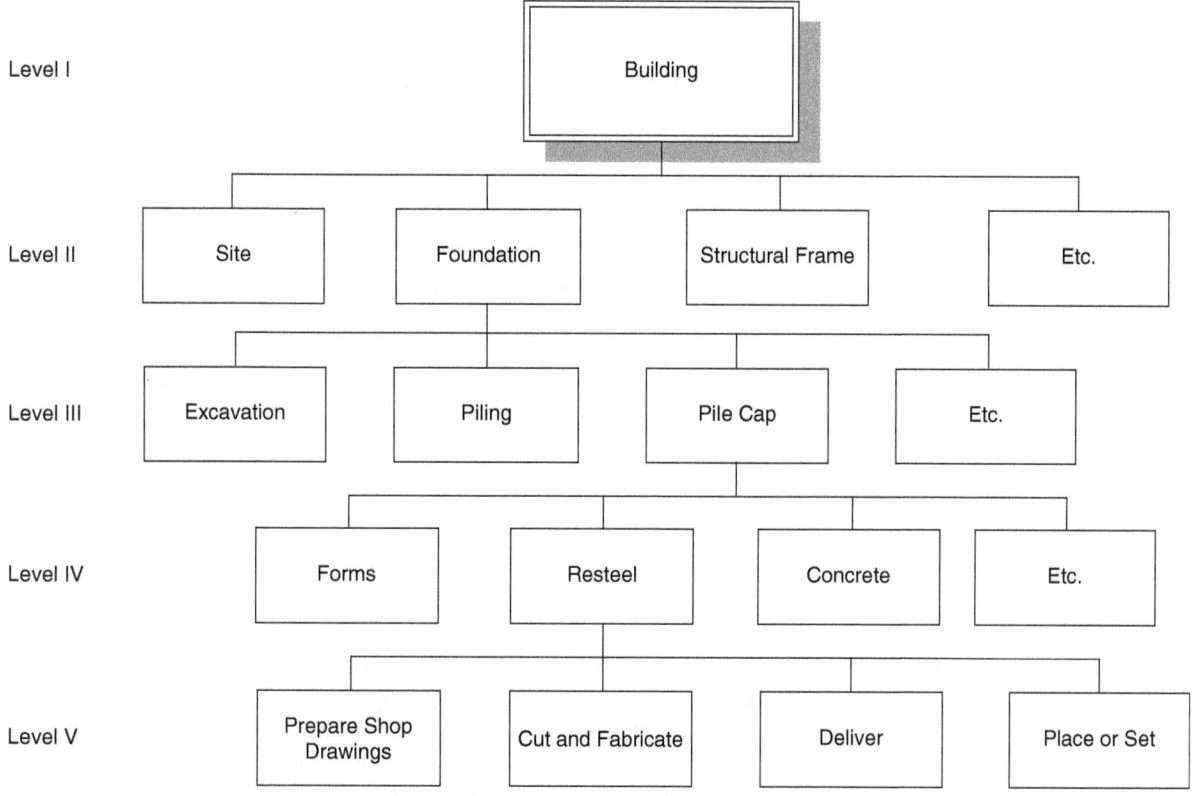

Figure 7.1 Work Breakdown Structure (WBS) Example.

summation of the work packages can be shown in a hierarchical format called a work breakdown structure (WBS). Figure 7.1 is an example of a WBS for a small building.

The *building* defined at Level I is subdivided into major subsystems at Level II. The *foundation* work is again broken into major work activities at Level III. Similarly, the *pile cap* consists of work tasks such as forming, installation of reinforcement, etc.

Development of a WBS requires a thorough understanding of the project scope of work. Experience in building is key to establishing a functional WBS. The WBS and the hierarchy of work packages of which it is composed are used to determine the status of a project and to manage the project from a time, cost, and quality perspective. Mentally, building a WBS structures the work that must be physically accomplished to realize the project and its end objective.

Planning can be thought of as the definition and sequencing of the work packages within a given project. That is:

$$Planning = Work\ Breakdown + Work\ Sequencing$$

Planning leads to a refinement of the scope of work as established in the contract documents. A good plan reduces uncertainty and improves efficiency.

The WBS also assists in determining the amount of planning needed. That is, it defines the level of planning required. For instance, if we are traveling to Washington, D.C., certain major elements of the travel should be planned. If we are traveling by air we need to book an airline ticket. We need to determine what to pack. Typically, we will need a place to stay, so we probably will reserve a hotel room unless we are staying with friends. At a lower level in our planning hierarchy, we must determine how to get from the airport to the hotel. If time is critical, we may reserve a limousine to meet us at the airport.

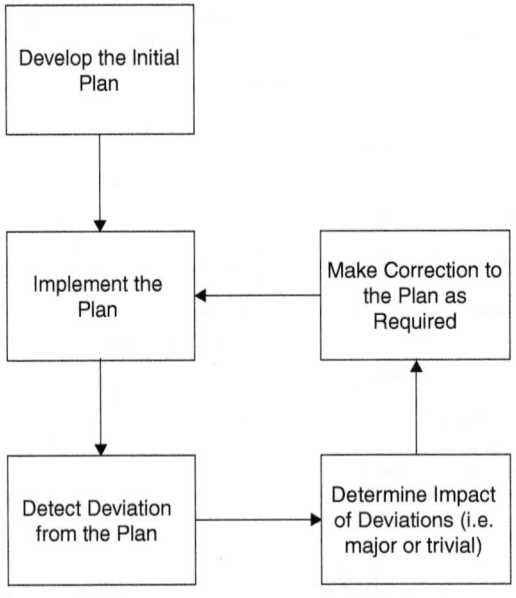

Figure 7.2 The Planning/Management Cycle.

On the other hand, we may decide to make a decision on arrival as to whether to take a cab, a shuttle bus, or the Washington Metro Mass Transit system. In effect, the development of the WBS and the definition of work packages is an exercise in "thinking things out ahead of time." It provides an orderly mechanism that facilitates planning. How detailed and comprehensive the WBS must be will vary with the situation and the complexity of the project.

Planning allows us to develop a framework for project execution, monitoring, and control. This minimizes uncertainty, clarifies subobjectives within the overall objective, assists in establishing sequencing of activity, and helps to avoid crisis management.

Planning is, however, an ongoing task and continues throughout the life of the project. As General Dwight D. Eisenhower once said: "Plans are nothing, Planning is everything." Initial planning is inevitably impacted by events that cause changes to the plan. The success of a project is tied to the manager's agility in identifying deviations from the plan and solving the challenges precipitated by these deviations. Figure 7.2 reflects this cycle of planning in terms of a simple flowchart. Once the WBS has been developed, it provides the framework within which planning can proceed throughout the life of the project. It becomes the vehicle for identifying deviations, assessing their impact, and making appropriate corrections to the existing plan. Therefore, "planning is everything," and it is an ongoing activity until the project is completed.

7.2   DEVELOPING THE WORK BREAKDOWN STRUCTURE

Work packages are the building blocks of the WBS. They should be defined to assist the manager in determining project status or level of completion of the project. One definition of the WBS is "the progressive hierarchical breakdown of the project into smaller pieces to the lowest practical level to which cost is applied." When monitoring and controlling a project, cost and time are areas of primary interest. The WBS is extremely useful in developing both cost and time (schedule) plans.

In establishing the WBS, the following guidelines need to be considered:

1. Work packages must be clearly distinguishable from other work packages.

2. Each work package must have a unique starting and ending date.

3. Each work package should have its own unique budget.

4. Work packages should be small enough that precise measurement of work progress is possible.

For example, in Figure 7.1 one work package at Level IV defines work associated with the installation of reinforcing steel in a pile cap. This work package is (a) clearly defined and separate from other work packages, (b) has a starting and ending date, and (c) has a cost budget that is unique and is small enough for accurate progress measurement. The work packages at Level V become more generic and more difficult to distinguish as unique. For instance, tasks such as cutting and setting steel are short and assigning a unique budget to this level of work becomes difficult. Therefore, the Level V work packages in Figure 7.1 can be viewed as subtasks that are prorated to the more unique Level IV work packages.

7.3 A WORK BREAKDOWN EXAMPLE

To better understand the concept of WBS, consider the small gas station project for which simplified plans and concept drawings are shown in Appendix H.

In construction, the various aspects of the work that contribute to breakdown of the project into packages relate to:

1. Methods used to place work.

2. Skills needed for the work.

3. Craft workers involved.

4. Critical resources (e.g., cranes, crews, etc.)

The definition of work packages can be facilitated by using categories which help in establishing a level of uniqueness. For this example, the following four categories have been defined:

1. *Location or area* within the project (e.g., foundation–pile cap).

2. *Material type* (e.g., concrete, resteel, etc.)

3. *Method of installation* (e.g., excavation).

4. *Types of resources* required (e.g., labor and equipment needed).

In the gas station project, the construction requires a *foundation* for the load bearing walls. The foundation can be thought of as a location (as well as a structural support system). The *location* or *area* of the work is a physical part of the construction. That is, one can walk up and touch the location of the package. A work package defining the floor slab in Section A on the third floor of an academic building is something we can locate and physically touch in the completed facility. The fact that the slab is concrete (i.e., rather than wood or metal) is another parameter that is important. It has implications from a procurement, as well as from a placement and work content, point of view.

Location and *material type* will influence the method of placement or installation. The *method of installation* and the material type will determine the human skills and equipment resources needed for installation. The method of placement or installation dictates the *types of resources* required, thus differentiating one package from another. For instance, in one case we may be placing concrete using a concrete pump, whereas in another situation the concrete may be transported using concrete buggies. In each case, labor and equipment resources, the budget, and the productivity of the concrete placement will be different.

Table 7.1 Work Packages for the Gas Station Project

(1) Earthwork for Parking and Service (P&S) Area	(10) Interior Built-ins (e.g., Cabinets, etc.)
(2) Asphalt Paving for P&S Area	(11) Interior Painting
(3) Concrete Hardstands in P&S Area	(12) Interior Drywall
(4) Concrete Foundations	(13) Interior Doors, Frames, Hardware, etc.
(5) Walls–Masonry Bearing Walls	(14) Interior Floor Coverings (if required)
(6) Walls–Prefab Metal Sandwich Panels	(15) Exterior Brick Façade
(7) Interior Concrete Floors	(16) Exterior Glazing
(8) Built-up Roof	(17) Exterior Doors and Signage
(9) Roof Gutters/Drainage	(18) Mechanical Systems
	(19) Electrical Systems

7.4   WORK PACKAGES FOR THE GAS STATION PROJECT

Let's develop a set of work packages and a WBS for the gas station construction. First, locations that are work package related will be determined. As noted previously, the building foundation can be considered a location. It would be important to know whether the scope of work includes the parking and service area surrounding the station. For the purposes of this exercise, we will assume it is within the scope of work. *Location* work packages would be as follows:

1. Parking and service area.

2. Foundation.

3. Building walls/structural panels.

4. Building roof.

5. Interior floors/slabs (separate from the foundation).

6. Interior finishes.

7. Exterior finishes.

8. Electrical systems.

9. Mechanical systems.

Adding the category of *material type* expands the number of work packages as shown in Table 7.1. Although this listing is, by no means, complete, it indicates the level of detail which must be considered even in a relatively small project (when only two levels of hierarchy are defined).

If mechanical work is expanded to cover location, material type, methods, and resources, the following partial list of work packages would be added to the hierarchy of the WBS.

1. Excavation of waste water system.

2. Drainage tile installation–waste water.

3. Septic tank installation.

4. Fresh water lines (piping).

5. Sinks, basins, and toilets installation.

6. Hot water system installation.

7. Pneumatic air system installation.

Figure 7.3 shows a WBS based on this partial development of work packages.

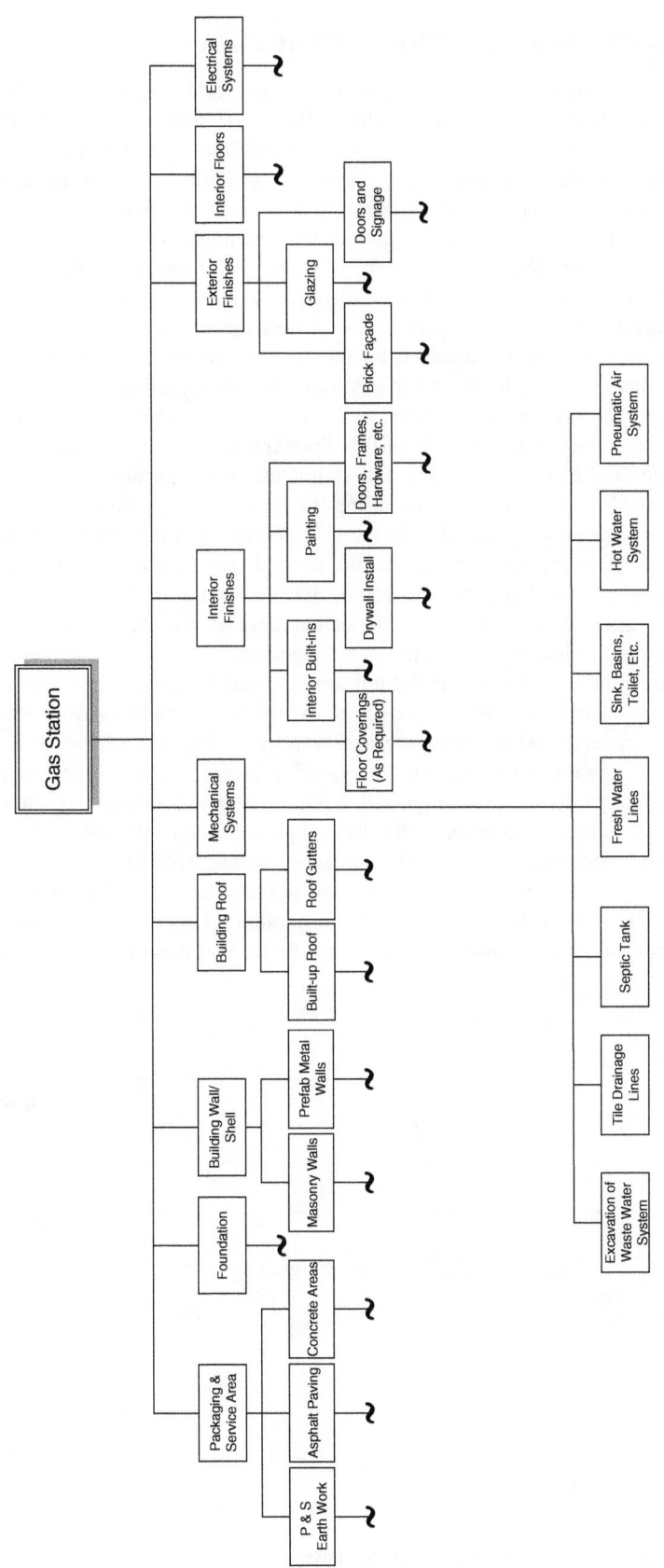

Figure 7.3 WBS for Gas Station Project.

7.5   DETERMINING SEQUENCE OF WORK PACKAGES

Having broken the work in to work packages, activities which facilitate time management and control can be defined and logically placed in sequence. The word *activity* is generally used when discussing time control or scheduling to refer to the work elements that appear in the schedule in their expected sequence or logical order. As mentioned in Chapter 1, the word *technology* implies that operations have a logic or sequence. An understanding of this logic is a key to successful project management.

In arranging the work package sequence for time control, the criteria of (a) *location*, (b) *material*, (c) *method*, and (d) *required resources* developed in Section 7.3 must be reconsidered from the perspective of how these criteria impact the order or sequence of work activities. For instance, location can determine sequence. It is normal to complete the structural frame for floor 1 of a building before beginning work on the structural frame for floor 2. This could be considered a *physical* constraint because the second floor frame cannot be supported until the first floor frame is completed. Such physical constraints or *physical logic* are common and characteristic of construction operations (e.g., the floor must be complete before installing the floor covering, etc.). Locational aspects of a work package may, therefore, determine its sequence in the overall project.

In some cases, physical requirements do not dictate order or sequence. For instance, in finish work relating to a restroom, fixtures such as sinks, toilets, partitions, etc. must be installed. Wall and floor finishes such as ceramic tiles must also be completed. Since there is no physical constraint, it is a management decision as to whether the walls and floors are completed first or the fixtures are installed first. In this case, the situation is controlled by a management decision (e.g., the fixture subcontractor is available first and is instructed to proceed) and the sequence is driven by *management logic*.

Again consider the small gas station project. A preliminary sequencing of the work packages is shown in Figure 7.4. One activity is defined to address mobilizing men, material, and machines to the site and preparation of the site. Then the locational or area work packages are ordered in sequence starting with the foundation and completing with the interior and exterior finishes and the mechanical and electrical systems that can be worked on at the same time (i.e., in parallel). This preliminary sequencing provides the framework for a more detailed schedule development to be presented in Chapter 8.

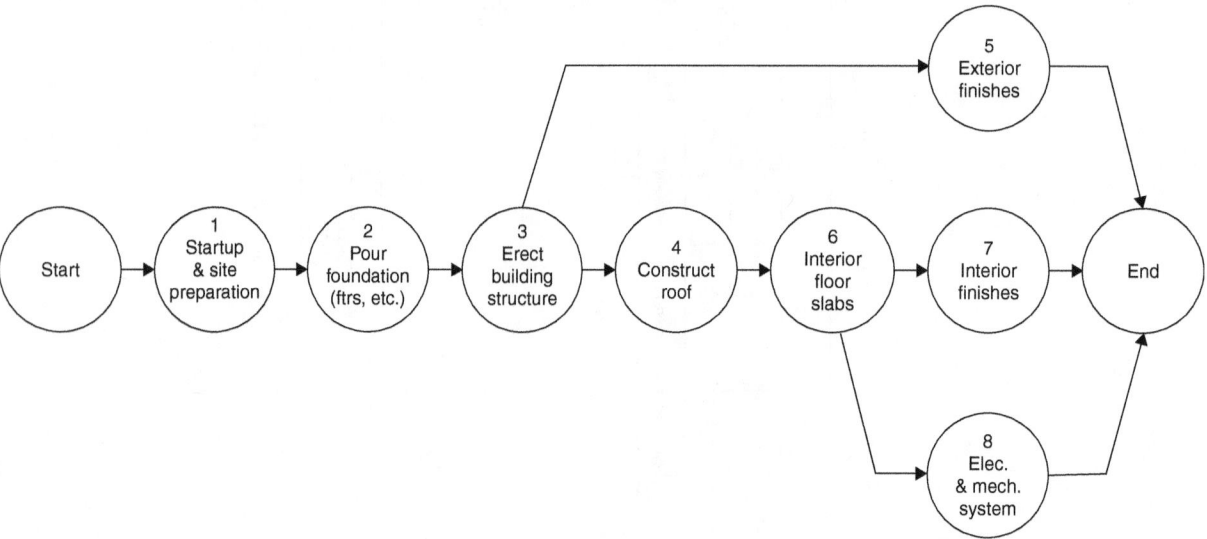

Figure 7.4 Preliminary project breakdown.

Following the preparation of the site, the footers and pier foundations are to be poured. After the footers and pier foundations have reached sufficient strength, the building structure is erected. It should be noted that, in this case, the floor slabs of the service bays, the show room, and the office areas as well as the utility room and toilets are not poured until the building structure is erected. Since the foundation consists of a *ring footing*[2] and individual piers to support walls and columns, etc., the building shell and roof can be completed prior to casting the various floor slabs throughout the building. When the roof is completed and the building is "closed in," work inside the building can proceed without concern for the weather.

As a more detailed time plan (i.e., schedule) is developed, consideration must be given to other time-consuming activities that are not necessarily identified using the location, material, method, and resource criteria.

Administrative actions, such as inspections, permit issuance, noise constraints, etc. must be considered in developing the time schedule logic.

Deliveries of materials and similar logistical issues must also be factored into the schedule.

Finally, certain special activities tied to the physical properties of the materials or procedures required (e.g., curing of concrete, moisture content measures for soil compaction, etc.) must be included in the time schedule.

A well-defined WBS facilitates the development of both preliminary and detailed schedules.

7.6   ESTIMATE DEVELOPMENT AND COST CONTROL RELATED TO THE WORK BREAKDOWN STRUCTURE

A good WBS facilitates cost control during the life of the project.[3] Work packages are defined so that they have their own unique budgets. When referring to work packages in the context of cost control, the terminology *cost accounts* or *control accounts* is often used. During the bidding process, the contractor prepares an estimate of cost that becomes the basis for the bid price submitted for the proposed work. If the bid is accepted, the detail estimate used for bid submittal is converted to a *budget* that serves as a cost baseline to control spending during the life of the project. The concept of cost control is shown schematically in Figure 7.5.

Based on a refinement of the bid estimate, a control budget is prepared. The budget structure is tied to the breakdown of the project into major cost elements. For small and simple projects, such as the paving of a residential driveway the cost breakdown of the budget may consist of relatively few elements (e.g., labor, materials, equipment, special item, etc.). For larger and more complex projects however, the structure and level of detail of the cost breakdown is key to effective control of project spending. In the case of the gas station project, for instance, budgets for each of the work packages shown in Figure 7.3 would be developed. The summation on these individual work package budgets can then be used to track total project costs and determine an overall cost status for the project at any time during construction.

In large and complex projects (e.g. large buildings, manufacturing plants, etc.), a comprehensive WBS is required. Literally thousands of work packages must be defined and a consistent and reliable system of referring to these elements of the work breakdown is essential. To provide consistency and structure to the management of large cost control

---

[2] A ring footing is a footing that supports the periphery of the building.
[3] If a WBS is used, the major items are work packages and the control accounts are subelements of the work packages.

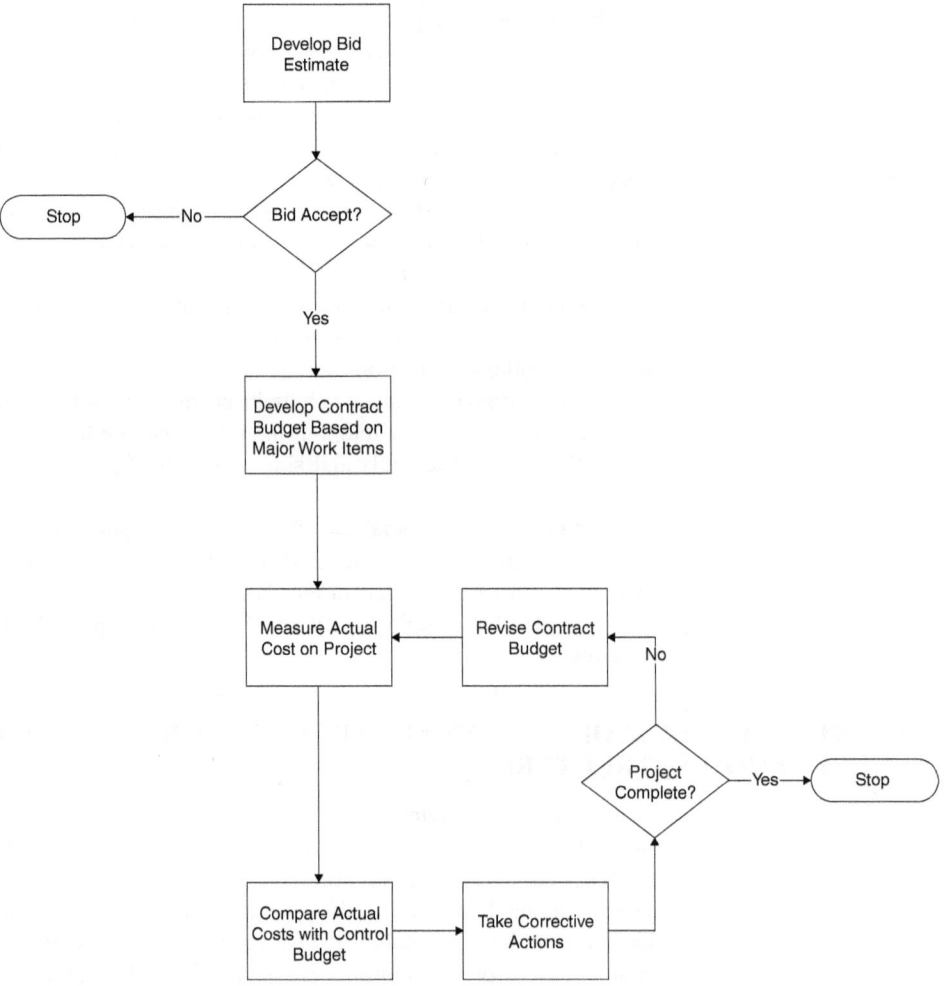

Figure 7.5 Cost Control Cycle.

systems, a *code of accounts* is used as a template or guide in defining and cataloging the cost centers within a project. Several cost coding systems are presented in Chapter 17.

To be consistent with the WBS, such *cost coding systems* use a numerical or alpha-numerical labeling that is multilevel or hierarchical in nature. This numerical labeling must be (a) suitable for computerization, (b) compatible with the company's financial account-ing system, and (c) facilitate gathering and reporting of information for broad cost categories (e.g., cost of all slab concrete) and for highly detailed information (e.g., cost of labor for pump installation-pump 101 in the cooling tower). The cost accounting system should facilitate the aggregation or "rolling up" of costs. The hierarchical or multilevel nature of the WBS provides an excellent format for the collection and review of cost throughout the life of a complex project.

7.7   ROLE OF CODE OF ACCOUNTS

The development of cost coding systems is discussed in some detail in Chapter 17. In general these labeling or coding systems focus on the multiple or hierarchical levels of cost spending typical of complex construction projects. In defining work packages, location of the work, material type, method of installation, and resources required help the various levels of the

Table 7.2 Cost Code Structure (Example)

Level	Project	Area	Discipline	Trade
1	21300			
2		804		
3			724	
4				112

Cost Code = 21300 − 804 − 724 − 112

breakdown. In cost code development in these identifiers can be characterized in many ways. A typical cost code might define (a) the project designator, (b) area of work, (c) work discipline (e.g., civil, mechanical, electrical, etc.), and (d) the trade specialty required. Table 7.2 gives an example of how such a code might be developed. The code in Table 7.2 is 14 digits long. Therefore, computer databases that are used to store and retrieve cost information must be able to handle long numerical designators or labels such as the one shown here.

On complex or unique projects, preparation of work package budget control sheets may be appropriate. Such a sheet allows for collection of data regarding actual costs versus estimated cost for each work element or package. A typical sheet for the interior slab construction on the small gas station project is shown in Figure 7.6.

The sheet acts as documentation of the control budget for each work package. Locations for the base cost calculations for each resource type (e.g., materials, installed equipment, labor, and equipment) are provided. As work packages are completed, the "as-built" cost and productivity achieved are recorded. This is backup material for the cost estimating database and can be used as a reference when estimating the cost of similar work in future. It also supports the comparison activity (e.g., compare actual costs with control budget) in the flowchart of Figure 7.5.

Work Package Identification 06-123 Description: Concrete Placement Interior Floor Slabs						
Materials						**Actual Productivity**
Resource Code	Description	Unit	Qty	Unit Cost	Extension	
101	Concrete, 2500psi	CY	30	40.00	1200.00	
Installed Equipment						Notes
	Crew Labor	NR	Hours	Cost/Hr	Extension	Cost Summary
020	Foreman	1	8	30.00	240.00	Actual Cost
029	Laborer	4	8	15.00	480.00	Labor =
022	Finisher	1	8	20.00	160.00	Materials =
063	Pump Operator	1	8	25.00	200.00	Equipment =
				Total	1080.00	Variation from
Equipment Not Charged As Indirects						Budget
		NR	Hours	Cost/Hr	Extension	
505	Vibrator	1	8	10.00	80.00	
517	Finisher	1	8	15.00	120.00	
308	Concrete Pump	1	8	150.00	1200.00	
				Total	1400.00	

Figure 7.6 Work Package Control Account Sheet.

Source: This sheet is based upon a system presented by J. M. Neil in the text *Construction Cost Estimating for Project Control*, published by Prentice-Hall, 1982.

7.8   SUMMARY

Construction project planning focuses heavily on time and cost control. Planning is a continuous task. Deviations from the original plan are the norm rather than the exception. Therefore, an organized approach to identifying change from the original plan is critical. It has been noted that:

> "*In order to* **manage**, *one should be able to* **control**.
> *In order to* **control**, *one should be able to* **measure**.
> *In order to* **measure**, *one should be able to* **define**.
> *In order to* **define**, *one should be able to* **quantify**."
> —D. Burchfield, 1970

The WBS approach provides a rigorous way to quantify, define, measure, and control the elements or work packages of a given scope of work. Breaking a construction project into work elements to be managed is essential for both time and cost planning. In addition to time and cost planning, a number of other planning efforts are needed when constructing a facility. Decisions and supporting plans must be developed to address many dimensions of the project. The following is a partial list of other plans which must be developed:

- Procurement plan.
- Safety plan.
- Subcontracting plan.
- Quality plan.
- Communication plan.
- Organizational plan.
- Completion and start-up plan.

The effective manager is constantly involved in developing and updating plans. Planning is a never-ending task. Therefore, one is continuously improving and perfecting the plan. Planning is everything.

REVIEW QUESTIONS AND EXERCISES

7.1 Develop cost codes for each element of the WBS shown in Figure 6.3. *Hint*: If you assign the account number 01-00-00 for the overall project, then a good account for the *packaging and service area* could be 01-01-00, and *asphalt paving* could be the account 01-01-02. Cost control is discussed in detail in Chapter 17.

7.2 Develop a two-level WBS consisting of at least 16 work packages for the building of the Brooklyn Bridge as described in Chapter 1.

7.3 Develop a preliminary WBS for a small one-story commercial building to be constructed on the site of an existing small-frame structure. It is 30 by 60 feet in plan (see illustration). The exterior and interior walls are of concrete block. The roof is constructed of bar joists covered with a steel roof deck, rigid insulation, and built-up roofing. The ceiling is suspended acoustical tile. The floor is a concrete slab on grade with an asphalt tile finish. Interior finish on all walls is paint.

  a. Show the first level of this structure for the total project (WBS is developed from top to the bottom).

  b. Select one work package (or building system) of the first level and develop the second-level structure for this work package.

7.4 Describe how you would develop a subcontracting plan for the construction of a small child-care facility

Chapter 8

Project Scheduling

CONTENTS

Four-Dimensional Modeling

Traditional design and construction planning tools, such as two-dimensional (2-D) drawings and network diagrams, do not support the timely and integrated decision making necessary to move projects forward quickly. They do not provide the information modeling, visualization, and analysis environment necessary to support the rapid and integrated design and construction of facilities. Synthesis of construction schedules from design descriptions and integrated evaluation of design and schedule alternatives are still mainly manual tasks. Furthermore, the underlying representations of a design and a construction schedule are too abstract to allow the multiple stakeholders to visualize and understand the cross-disciplinary impacts of design and construction decisions.

Four-dimensional (4-D) technologies are now being used by planners, designers, and, engineers to analyze and visualize many aspects of a construction project, from the three-dimensional (3-D) design of a project to the sequence of construction to the relationships among schedule, cost, and resource availability data. These intelligent 4-D models support computer-based analysis of schedules with respect to cost, interference, safety, etc., and improve communication of design and schedule information.

Extending the traditional planning tools, visual 4-D models combine 3-D CAD models with construction activities to display the progression of construction over

3d wireframe building project with white background.
© mustafa deliormanli/iStockphoto

time. However, 4-D models are time consuming to generate manually and do not currently support analysis programs. The difficulty and cost of creating and using such models are currently blocking their widespread adoption. The construction knowledge necessary to build 4-D models has been formalized and developed by a methodology that guides project planners in generating 4-D models from 3-D product models. This formalized knowledge enables project managers to create and update realistic schedules rapidly and to integrate the temporal and spatial aspects of a schedule as intelligent 4-D models.

8.1 INTRODUCTION

As discussed in the previous chapter, a project's work breakdown structure (WBS) defines work packages that can be sequenced (scheduled) to coordinate the project construction. The resulting project schedule is of paramount importance for many management functions, such as the proper coordination of material procurement and subcontractor mobilization. The schedule provides a concise means to communicate the intended construction strategy and can reveal any inconsistencies in its design. This chapter examines the steps required to develop a project schedule. They involve:

- Breaking down the scope of the work into manageable portions, namely its activities. This first step was discussed in the previous chapter.
- Estimating each activity's duration.
- Establishing sequence relationships among activities. If one activity is called *fabricate formwork for deck A* and another is called *pour concrete for deck A*, then finishing the former is a necessary condition for executing the latter.
- Finding the earliest and latest possible times for the execution of each activity. This is determined by the duration of the activities and their mutual relationships.

- Looking up the project's critical sequence of activities and the leeway that activities not included in this sequence can have. This step is an outcome from the scheduling of the previous step.
- Reviewing, optimizing, communicating, updating, and in general, using the schedule.

8.2 ESTIMATING ACTIVITY DURATIONS

After a list of activities is developed, the duration of each one must be estimated. The duration of any activity is largely determined by:

- *The nature of the work that the activity entails.* All other factors being equal, a concrete pour in the 50th floor of a building is likely to take longer than one in the 2nd floor.
- *The quantities of the work.* Quite evidently, an activity involving 6,000 square foot of formwork will take longer than another consisting of 3,000 square foot.
- *The technique used for the task.* A concrete pour usually takes longer if it is done using a tower and buggies than if it is done with a concrete pump.
- *The resources used to perform the work.* Twenty masons are likely to finish the same amount of work in a shorter time than 15 masons. A larger fleet of scrapers will finish an earthmoving operation faster than a smaller one, if the conditions discussed are met.
- *The working hours for the resources.* An activity will be finished in fewer days if it is performed using 10-hour shifts instead of 8-hour shifts.

These factors are extremely important in shaping the duration of an activity, but other factors must be also taken into consideration. An earthmoving operation, for example, requires that the number of trucks carrying the material is balanced with the equipment used to load the trucks. If the operation is *balanced* with the use of five trucks, the increase in productivity achieved by using seven units is minimal. Chapter 14 discusses equipment productivity in more detail. Similarly, an excessive number of crews working in a small space dramatically decreases the productivity of each one. This and other factors affecting productivity will be discussed in Chapter 16.

With all the factors considered here, it would appear that estimating durations is a colossal endeavor. In fact, there are many ways to simplify this part of the scheduling process. Figure 8.1 shows an excerpt from *Building Construction Costs*, published annually by R. S. Means, Inc. It shows the "normal" productivity for a crew fabricating and placing footing formwork. Similar data are available in other sources.

03 11 Concrete Forming

03 11 13 – Structural Cast-In-Place Concrete Forming

03 11 13.45 Forms In Place, Footings		Crew	Daily Output	Labor-Hours	Unit	Material	2008 Bare Costs		Total	Total Incl O&P
							Labor	Equipment		
0010	**FORMS IN PLACE, FOOTINGS**	R031113-40								
0020	Continuous wall, plywood, 1 use		C-1	375	.085	SFCA	5.25	3.08		8.33	10.60
0050	2 use	R031113-60		440	.073		2.89	2.63		5.52	7.25
0100	3 use			470	.068		2.10	2.46		4.56	6.15
0150	4 use			485	.066		1.71	2.38		4.09	5.60

Figure 8.1 Excerpt from R.S. Means Construction Cost Data showing productivity data.

8.2.1 Example: Estimating the Duration of an Activity

A project schedule has an activity called *Form Continuous Wall Footings*. There are 6,000 square feet of contact area (SFCA) of continuous wall footing formwork. Estimate this activity's duration.

As mentioned previously, R. S. Mean's *Building Construction Costs* and similar publications provide productivity data that greatly simplify the estimating of activity durations. Figure 8.1 indicates that, assuming three uses for each plywood section, one carpentry crew can fabricate 470 SFCA per day. Using one crew, this output results in 6,000/470 = 12.76 days. If two crews are used, then the total duration is halved to 12.76/2 = 6.39 days. Rounding this result to the next highest whole number, the final result for this example is 7 days.

The result of this example should be checked by the superintendent, or even better, the foreman assigned to perform the work. In fact, the duration of many (if not most) activities is determined in practice by experience instead of a mathematical analysis.

The effect of individual deviations in the estimate of activity durations may not be significant. Errors in individual activity duration estimates often cancel each other, as long as there is no bias toward excessive conservatism or risk and the activities are not repetitive. If the project requires the repetition of similar activities, then it is important to accurately calculate their duration.

Activity durations are always expressed in working days, as opposed to calendar days. An activity with duration of 5 days would take a full week to be completed if the project has a Monday-to-Friday working week. It is customary in the construction industry to use round numbers for durations. If more precision is required (and can be realistically estimated), then durations should be expressed in working hours.

8.3   BAR CHARTS

The basic modeling concept of the bar chart is the representation of a project work item or activity as a time scaled bar whose length represents the planned duration of the activity. Figure 8.2(a) shows a bar representation for a work item requiring four project time units (e.g., weeks). The bar is located on a time line to indicate the schedule for planned start, execution, and completion of the project work activity.

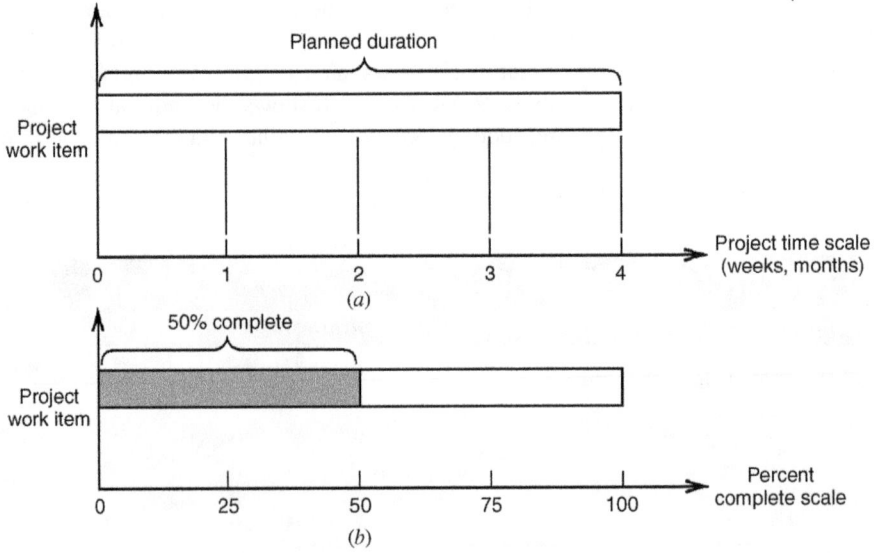

Figure 8.2 Bar chart model: (a) plan focus and (b) work focus.

In practice the scaled length of the bar is also used as a graphical base on which to plot actual performance toward completion of the project work item see Figure 8.2(b). In this way the bar chart acts both as a planning-scheduling model and as a reporting-control model. In this use of the bar chart, the length of the bar has two different meanings:

- The physical length of the bar represents the planned duration of the work item.
- It also provides a proportionally scaled baseline on which to plot at successive intervals of time, the correct percentage complete.

Figure 8.2(b) shows a bar for a project work item that has been half completed. In a situation in which the work rate is constant and field conditions permit, this would occur in half the planned duration. If, however, actual work rates vary from time to time according to resource use and field conditions, then the work will be half-completed sooner or later than half the planned duration. In this modeling concept, actual work progress is modeled independently of the actual time scale of the bar chart.

Figure 8.3(a) shows a schedule for a project consisting of three activities. Activity A is to be carried out in the first 4 months, activity B in the last 4 months, and activity C in the third month. Actual progress in the project can be plotted from time to time on these bars as shown in Figure 8.3(b).

In this manner, project status contours can be superimposed on the bar chart as an aid to management control of the project. By using different shading patterns, the bar chart can indicate monthly progress toward physical completion of the activities.

Project bar chart models are developed by breaking down the project into a number of components. In practice, the breakdown rarely exceeds 50 to 100 work activities and generally focuses on physical components of the project. If a project time frame is

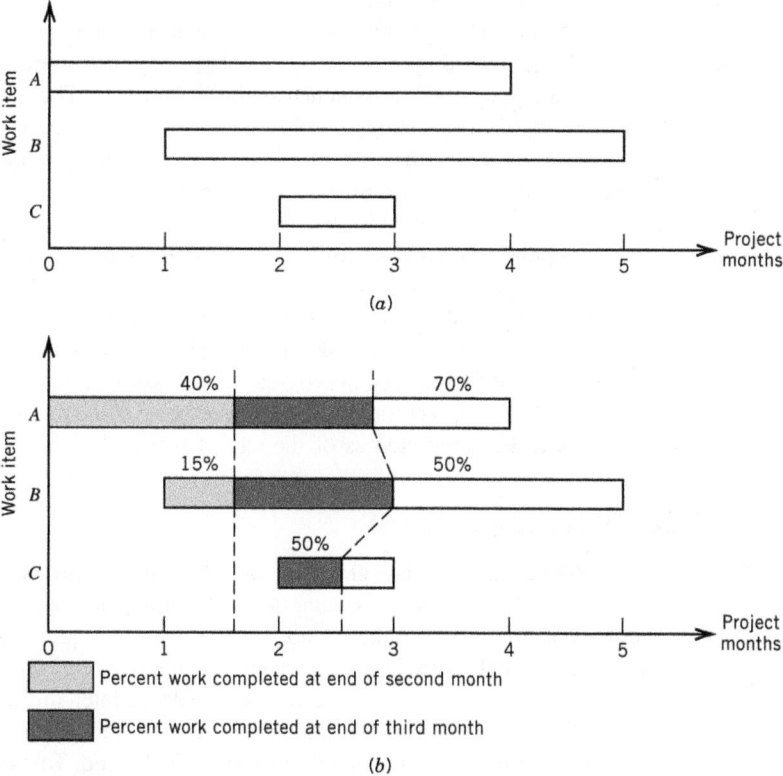

Figure 8.3 Bar chart project models: (a) bar chart schedule (plan focus) and (b) bar chart updating (control focus).

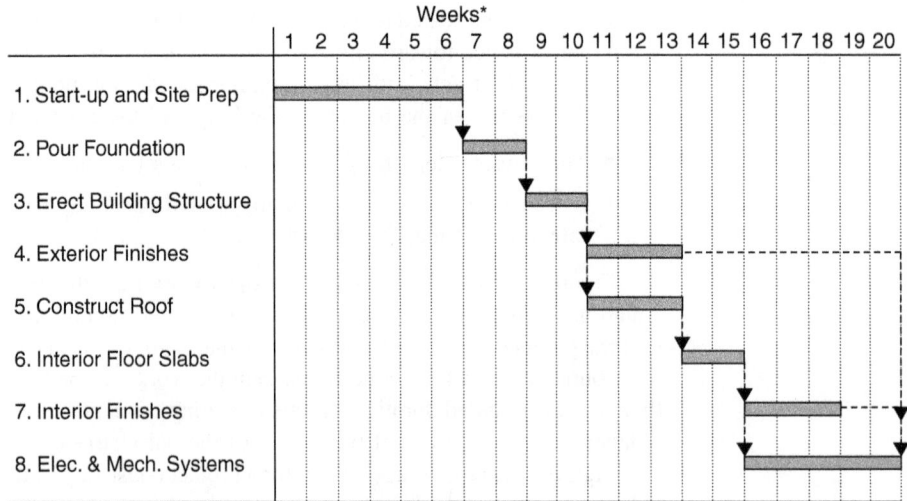

*Weeks are assumed to be working weeks consisting of 5 working days.

Figure 8.4   Preliminary Bar-net schedule for the small gas station.

established, the relative positioning of the project work activities indicates the planned project schedule and the approximate sequence of the work. One disadvantage of the traditional bar chart is the lack of precision in establishing the exact sequence between activities. This problem can be addressed by using directional links or arrows connecting the bars to give a precise indication of logical order between activities. This connected bar chart is sometimes referred to as a bar-net.

A connected bar chart (bar-net) showing the major activities defined in the preliminary project breakdown diagram for the small gas station (Figure 7.4) is shown in Figure 8.4. The bars are positioned in sequence against a time line. The sequence or logic between the bars is formalized by connecting the end of the preceding bar to the start of the following bar, as for instance, the end of bar 3. Erect building structure is connected using a directional link or arrow to the two activities that follow it (Activities 5 and 4). Positioning the eight activities as bars in their logical sequence using the arrow connectors against a time line plotted in weeks allows us to visually determine that the duration of the entire project is roughly 20 weeks.

This diagram also allows one to determine the expected progress on the project as of any given week. For example, as of week 11, Activities 1, 2, and 3 should be completed. Activities 4 and 5 should be in progress. If we assume a linear rate of production (i.e., half of a 2-week activity is completed after 1 week), we could assume that one-third of activities 4 and 5 will be completed as of the end of week 11.

8.4   ACTIVITY PRECEDENCE DIAGRAMS

A precedence diagram is a graphical tool that shows a project's activities as rectangles or circles joined by arrows indicating their mutual dependency or *relationship*. The resulting network clearly indicates the intended sequence of activities, as shown in Figure 8.5, and is the fundamental graphical component of the *Critical Path Method (CPM)*, whose scheduling mechanism will be discussed in detail later in this chapter.

To build a precedence diagram, the logical sequence or scheduling logic that relates the various activities to one another must be developed. To gain a better understanding of the role played by sequencing in developing a schedule, consider a simple pier made up of two lines of piles with connecting headers and simply supported deck slabs.

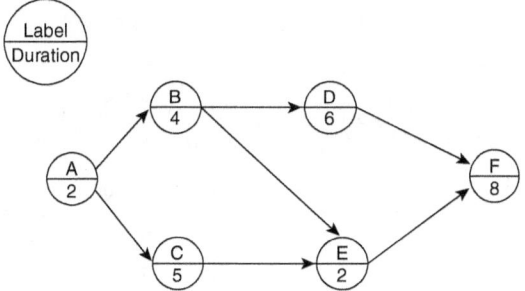

Figure 8.5 Precedence diagram.

A schematic view of a portion of the pier is shown Figure 8.6(a). The various physical components of the pier have been identified and labeled. An exploded view of the pier is shown in Figure 8.6(b) and shows each physical component individually separated but in the same relative positions. Notice that abbreviated labels have now been introduced. Clearly, these figures are schematic models (i.e., not physical models), but they have rather simple conceptual rules so that the physical relationship between the components of the structure is clear.

Figure 8.6 Simple schematic models (a) Schematic view of pier (b) Exploded view of pier (Antill and Woodhead, 1982).

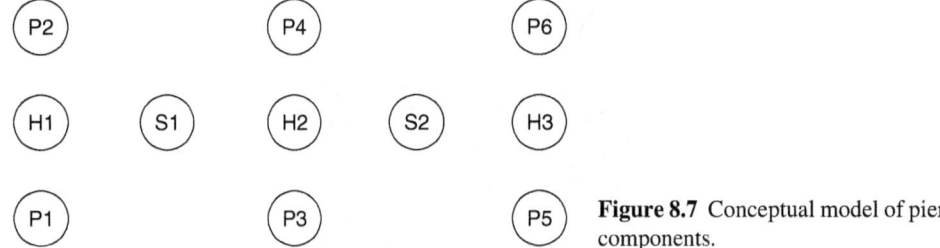

Figure 8.7 Conceptual model of pier components.

Now suppose that each component or element is represented by a labeled circle (or node). Figure 8.7 gives a "plan" view of the pier components shown in Figure 8.6. Such an abstraction or model can be used as the basis for portraying information about the physical makeup of the pier or about the order in which the physical components will actually appear on the site.

For example, an indication of the adjacency of physical components or the relational contact of physical components may be required. A model to portray these properties requires a modeling element (say a line) to indicate that the property exists. Assuming the modeling rationale of Figure 8.8(a), the various nodes of Figure 8.7 can be joined by a series of lines to develop a graph structure portraying the physical component adjacency or contact nature of the pier. If the idea of contact is expanded to indicate the order in which elements appear and physical contact is established, a directed line modeling rationale may be used, as shown in Figure 8.7(b). Using this conceptual modeling rule, Figure 8.9 can be developed. This figure shows, for example, that header 1 (H1) can only appear (i.e., be built) after piles 1 and 2 (i.e., P1, P2) appear; in fact, H1 is built around, on top of, and therefore in contact with P1 and P2.

Finally, if the order of appearance of physical elements is to be modeled for all elements, whether or not in contact, a directional arrow such as that shown in Figure 8.8(c) may be necessary.

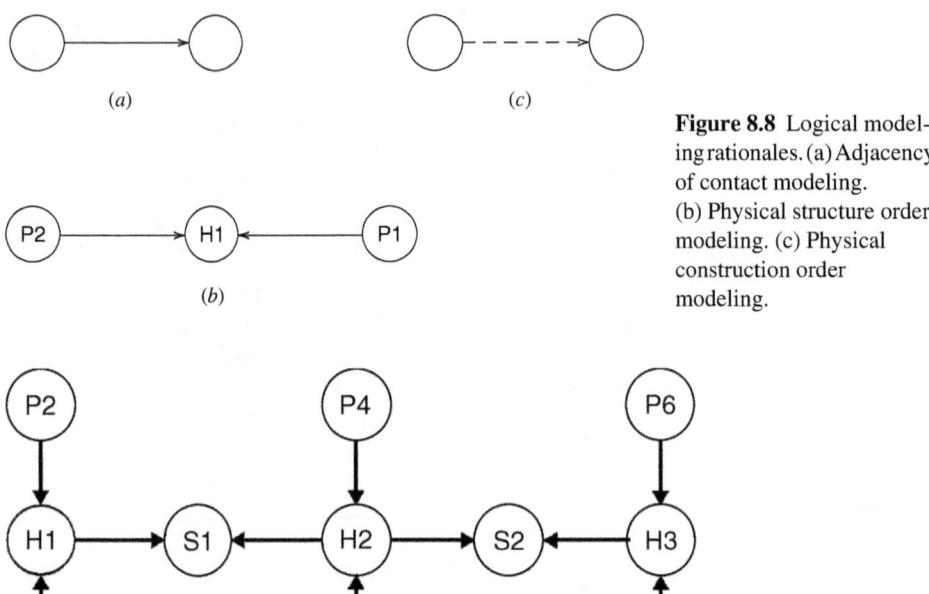

Figure 8.8 Logical modeling rationales. (a) Adjacency of contact modeling. (b) Physical structure order modeling. (c) Physical construction order modeling.

Figure 8.9 Conceptual model of pier components relationships.

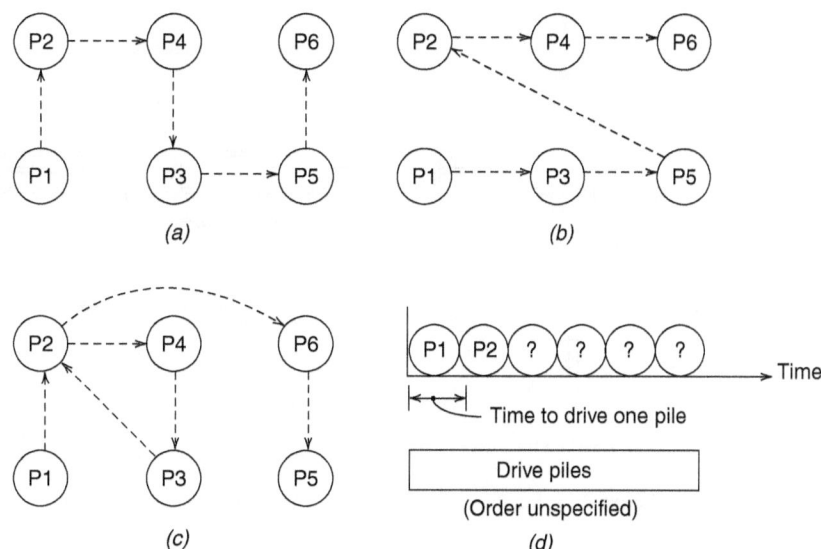

Figure 8.10 Construction sequence and activity modeling. (a) Alternate row pile driving. (b) Sequential row pile driving. (c) Field mishap alteration to pile driving sequence. (d) Bar chart model of pile driving operation (Antill and Woodhead, 1982).

As an example of these modeling techniques, consider the pier pile-driving operation. A number of possible pile-driving sequences are shown in Figure 8.10(a), (b), and (c). In Figure 8.10(a) it is assumed that the pile-driving rig is swung by its mooring cables to drive the piles alternatively from one line to the next (i.e., P1 and P2) before being relocated for the next set of piles (i.e., P3 and P4), and so forth. In Figure 8.10(b) the pile-driving sequence is along one line first (i.e., P1, P3, and P5) and then along the other line (i.e., P2, P4, P6, and so on). Figure 8.10(c) shows a situation that may result if field events interrupt the planned sequence. In this case the figure indicates a situation in which, for example, P2 is broken or lost during pile-driving operations, so that to conserve time the pile-driving rig moves on to drive P4 and P3 and then returns to redrive a new P2 before resuming normal pile-driving sequences. This situation is common in practice and indicates the major difficulty with scheduling models in relation to what actually happens in the field.

Figure 8.10(d) indicates the basic modeling rationale of bar charts wherein specific identification with individual piles is hidden. First, it implies the concept that each pile requires a certain time to appear in its driven position on site; second, it implies that the actual sequence of driving piles on the site is not absolutely fixed or essential to field management.

8.4.1 More About Relationships

Any relationship implies that there is a reason for sequencing the activities the way that the relationship stipulates. Two activities may be performed at different times without having a mutual relationship. Consider Figure 8.12, showing that driving the piles of the east side of the small bridge shown in Figure 8.11 will be completed before other operations can begin. In this case the piles of the east side must be driven before: (a) the piles of the west side are driven; (b) the east piles are capped, and then (c) the east abutment wall can begin to be built. It could well be that the duration of capping the piles results in the west piles being finished before the abutment wall can begin. However, there is no relationship between the driving of the west piles and the beginning of the east abutment wall. If a relationship is drawn, the resulting network, and therefore the project schedule, could be illogical. For

Figure 8.11 Small bridge section view (Courtesy Sears and Sears).

instance, the driving the west piles could be delayed by an unforeseen circumstance. A schedule linking the completion of the piles on the west abutment wall to the start of the east abutment wall would indicate that the start of the east abutment must wait until west the piles are completed, which is not true. Perhaps the most common mistake made by beginning schedulers is to link two activities that happen to normally be in sequence but have no real impediment to be performed simultaneously or even in inverse order.

A relationship between two activities can be the result of physical restrictions or of a deliberate management decision.

- *Physical relationships:* The relations among the activities in each of the two legs (E and W) of Figure 8.12 are determined by physical necessity. Piles must be driven to be capped, and abutment walls are on top of the pile caps. There is no alternative for this sequence at execution time.

- *Strategic relationships:* In Figure 8.12, there is no physical reason to start with the east side of the structure. The relationship reflects a management choice, which could be changed if, for example, the west side is available before the east side. Relationships introduced for strategic reasons are weaker than those reflecting physical constraints, in the sense that while the strategy can be changed, it is difficult or impossible to change an activity sequence resulting from physical conditions.

- *Safety:* Some relationships result from the need to guarantee the safety of the construction personnel. It would be a bad idea to build a sidewalk around a building while bricks are being placed in the upper floors. In this case, a finish-to-start relationship would be drawn between the activities "Brick façade," as predecessor, and "Build sidewalks" as successor.

Figure 8.12 Segment of precedence diagram for small bridge.

- *Other considerations:* A traditional rule in establishing relationships is that all required resources will be available, or in other words, that activities should not be constrained by resource limitations. The rationale is that limitations in resource availability at a given time could be overcome in the future, and therefore, such relationships limit the strategies available to the project manager. In many cases in which the number of the resources involved in a particular activity is unlikely to change, this rule is not followed. The bridge pile-driving example previously discussed could be approached by assuming that there are no limitations in the number of available pile rigs, and therefore, that both sides can be completed simultaneously. Most schedulers would know whether there is any realistic possibility of using two pile rigs, and if such possibility is remote, the schedule would be developed with the assumption that only one would be used.

8.4.2 Example: Developing the Gas Station Precedence Diagram[1]

Using the bar chart (which includes the relationships among activities) shown in Figure 8.4 and information from the WBS given in Figure 7.3 for the small gas station, the original 8 activities can be expanded to a set of 22 activities providing a more comprehensive scheduling framework for this project. The next step in scheduling is to assign durations to each of the activities. The scheduler must consult with field personnel and get the best estimate of the duration for each activity based on the methods selected for accomplishing the work, the resources available, and the experience the field management has in estimating productivity and time durations with these methods and resources.

For instance, based on the crew size, the equipment available, and the method of placement, the field superintendent can establish how long it will take to cast the concrete footers. The superintendent will know whether the concrete is to be mixed on site or brought in using a transit mix truck. He will also know the type of forms to be used, the nature of the reinforcement to be installed, the nature of any embedment and penetration required, and the placement and number of anchor bolts required for the building structure. The estimate of duration for a given activity is given in working days. Table 8.1 gives a listing of the expanded list of activities together with estimated durations for each activity. An expanded connected bar chart based on Table 8.1 and including the logical sequencing of the 22 activities is shown in Figure 8.13. The network for this project is shown along with its scheduling results in Figure 8.25.

8.5   GENERALIZED RELATIONSHIPS

The relationships discussed so far describe situations in which the end of an activity is a necessary condition to begin a successor. Since the relationship links the finish of the predecessor to the start of the successor, all relationships discussed so far are of the finish-to-start (FS) type. This type of relationship is by far the most commonly used in any precedence diagram. In fact, many schedulers use only these *traditional relationships* to build their networks.

The FS relationship is only one of four possible ways in which two activities can be related. It is possible that the start of an activity is a condition for the start of a follower, in what is called a start-to-start (SS) relationship. Similarly, the finish of an activity can constrain the finish of another in a finish-to-finish (FF) relationship. Even another possibility is that the start of an activity constrains the finish of another in a start-to-finish (SF) relationship. This section discusses these more generalized ways to relate the activities in a precedence diagram.

---

[1] A discussion of the small gas station schedule using an alternative method, called arrow notation, is presented in Appendix E.

Table 8.1 Durations of Activities for the Small Gas Station

Activity	Title	Duration (Days)
1	Mobilize	10
2	Obtain permits	15
3	Site work	8
4	Exterior utilities	12
5	Excavate catch basin	2
6	Excavate footers	5
7	Excavate foundation piers	6
8	Pour footers, etc.	8
9	Erect bldg. frame	10
10	Exterior brick facade	14
11	Exterior fascia panels	4
12	Roof construction	15
13	Landscaping	12
14	Pour interior slabs	10
15	Glazing and doors	6
16	Interior walls	10
17	Elec. & mech. Systems	25
18	Shelves	3
19	Floor coverings	6
20	Interior finishes	8
21	Final inspection	1
22	Demobilization	3

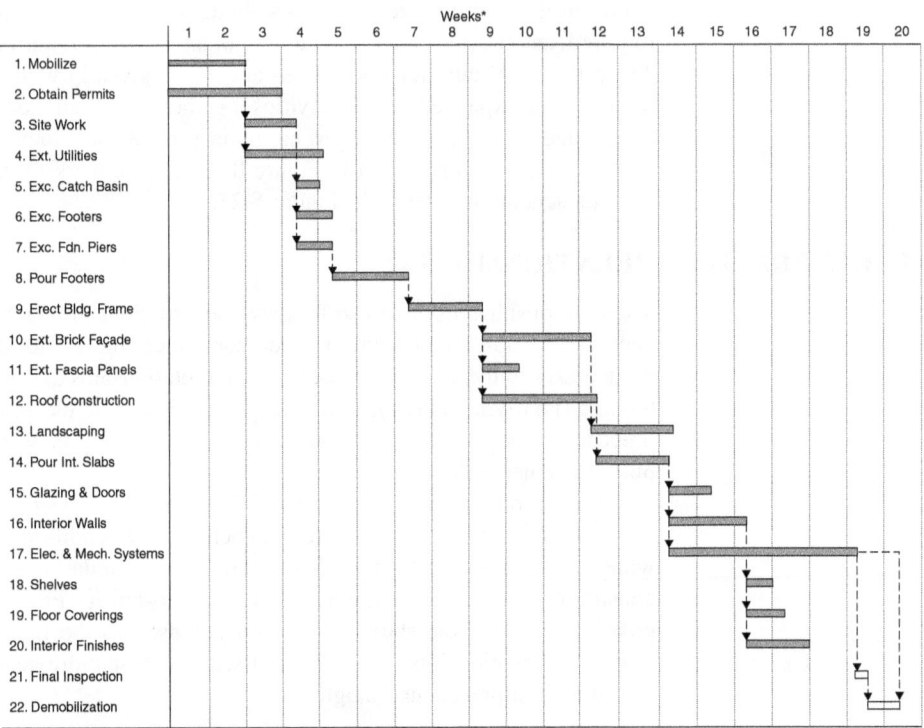

*Weeks are assumed to be work weeks consisting of 5 working days.

Figure 8.13 Expanded bar-net schedule for the small gas station.

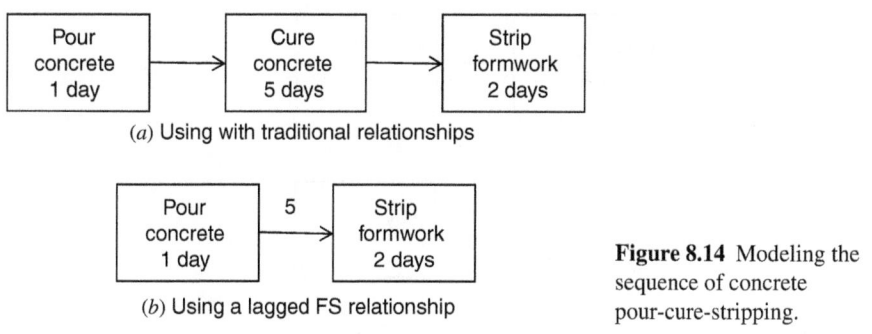

(a) Using with traditional relationships

(b) Using a lagged FS relationship

Figure 8.14 Modeling the sequence of concrete pour-cure-stripping.

8.5.1 Finish-to-Start Relationships

Sometimes, traditional FS relationships are inefficient to model the relation between two activities, even when the finish of the predecessor is a condition for the start of the successor. For example, in any project involving cast-in-place concrete, the concrete is poured and then allowed to cure for a number of days before its formwork can be stripped. The traditional solution to model the time between the end of the pouring operation and the start of the form-stripping operation is to include a third activity between them to account for the curing period, as shown in Figure 8.14(a).

Although this traditional solution is effective in projects with a few concrete pours, it is impractical for project with many pours. The number of activities in the precedence network for a multistory building with hundreds of pours would be increased by these *concrete curing time* activities that are included just to delay the start of the *form-stripping* activities. An alternative to this solution is to include a lag time (or simply, LAG) in the arrow relating pouring and stripping, replacing the concrete curing time activity. As shown in Figure 8.14(b), the start of the stripping activity is the same in the traditional and the lagged alternatives. The relationship used is still of type FS, but it cannot be categorized as traditional anymore.

8.5.2 Start-to-Start Relationships

SS relationships are used when a successor activity must begin after at least a LAG number of days after the start of the predecessor activity. The first segment of a long pipe can be laid before its trench is completely finished. All that is required is that the laying begins when an appropriate length of trench is available. This could be modeled using traditional FS relationships, as shown in Figure 8.15(a). The excavation is broken in two segments so that pipe laying can begin midway into the activity. This solution is not only wasteful by creating an additional activity, but it is also confusing because it implies that the excavation has two stages. In fact, the excavation is totally independent of the succeeding pipe laying. Compare this traditional solution with Figure 8.15(b) showing the same scenario with an SS relationship with lag time. The solution is more elegant and intuitive than its traditional counterpart.

8.5.3 Finish-to-Finish Relationships

This type of relationship signifies that the successor activity must be finished after at least the number of days indicated in the LAG between the two activities. The previous example of the trenching and pipe laying operation is also applicable here. Suppose that the trenching operation takes 7 days and the pipe must finish at least 3 days after the excavation has been completed. Traditional relationships can depict this condition, as shown in

(*a*) Using traditional relationships

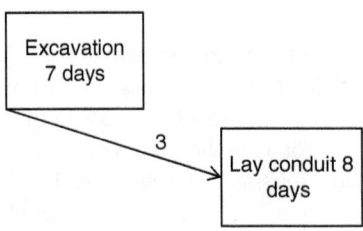

(*b*) Using an SS relationship   **Figure 8.15** Start-to-Start (SS) relationships.

Figure 8.16(a). As in the previous cases, this solution is confusing and inefficient. The FF relationship shown in Figure 8.16(b) would be more efficient and elegant. An FF relationship requires the finish of the successor to occur at least LAG days after the finish of the predecessor. The relationship does not indicate that pipe laying *must* end 3 days after the trench has been finished. It indicates that the pipe laying must be finished *at least* 3 days after the trench is available.

(*a*) Using traditional relationships

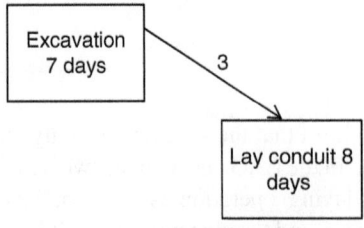

(*b*) Using an FF relationship   **Figure 8.16** Finish-to-Finish relationships.

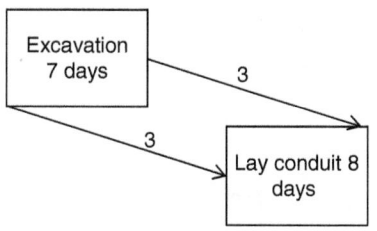

Figure 8.17 Double relationship between trenching and pipe laying.

8.5.4 Activities with Multiple Mutual Relationships

Two activities can have more than one mutual relationship, as long as the relationships specify different types of conditions, and the predecessor–successor relation is not swapped. The same two activities can be related by, say, an SS and an FF relationship, as previously discussed in the example of the pipe laying and trenching activities. But if a relationship indicates that activity A is the predecessor of Activity B, then another cannot indicate that B precedes A. Figure 8.17 shows the double relationships between trenching and pipe laying for the example previously discussed. Notice that in this case, a double relationship is the only way to express the two conditions desired for the previous SS and FF examples. The FF alone implies that the pipe could be laid after at least 1 day of beginning the excavation; the SS alone does satisfy the condition of ending at least 3 days after the excavation is completed. However, field conditions could extend the planned duration of the excavation to, say, 11 days. In such case, the SS relationship would lead to the nonsensical conclusion that the pipe can finish before the excavation. The start and finish dates of the successor are always determined by the relationship that "pushes" it farther in the timeline. This means that only one of the relationships will drive the scheduling of the successor. It is tempting but unwise to remove all nondriving relationships to simplify the precedence diagram. As the project gets underway, the assumptions about durations and lags may be changed. An originally nondriving relationship may end up being crucial for the project.

8.6   OVERVIEW OF THE CRITICAL PATH METHOD SCHEDULING PROCEDURE

When two activities are linked by a traditional relationship, the predecessor must finish before the successor can begin. The earliest possible dates to execute each activity in the network can be found when this rationale is successively applied to all activities in the diagram. This process is known as the CPM *forward pass* and also determines the minimum possible duration of the project. A similar procedure is then performed backward, from the last to the first activities in the diagram. This second step is called the CPM *backward pass* and computes the latest dates by which each activity can be performed without increasing the project's minimum duration. The difference between the earliest and latest possible dates to execute each activity is its *total float*, which as the name implies, is the time that an activity can "drift" or be "floated" between its two extreme dates. A chain of critical activities will not have any available float. The timely completion of this chain of activities is critical for the scheduled completion of the entire project. This activity chain is appropriately called the *critical path* of the project and is so central to the schedule that the entire scheduling technique is called the *critical path method*. The small precedence diagram shown in Figure 8.18 will be used to illustrate CPM's scheduling computations. It contains seven activities linked by traditional relationships (FS with LAG = 0). The second item in each box is the activity duration in working days. The most common way to perform the computations by hand (without computer software) is by marking the network, as done in the discussion that follows.

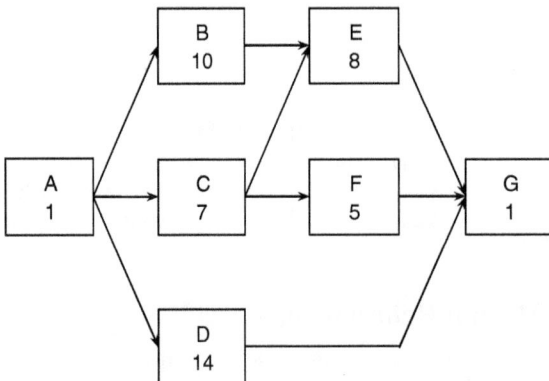

Figure 8.18 Example Diagram with traditional relationships.

8.7   FORWARD PASS (TRADITIONAL RELATIONSHIPS)

The forward pass is the first phase of the scheduling computations. It computes the minimum date at which each activity can be performed while observing the constraints imposed by the relationships among activities.

The initial activity, A in the case of Figure 8.18, is assigned an early start date (ESD) of zero. The early finish date (EFD) for this initial activity is equal to its ESD plus its duration, that is $0 + 1 = 1$.

All of A's immediate successors B, C, and D are assigned an ESD equal to the EFD of A (using a more formal notation: $ESD_B = ESD_C = ESD_D = 1$). Their EFDs are computed similarly to A, by adding their respective durations to their ESDs. Figure 8.19 shows the computations for the first four activities. The broken-line arrows show the order of computations for activity C.

The ESD for each activity must be interpreted with care. It shows the number of full working days that must elapse before the activity can begin. For example, 1 working day must elapse before Activity B can begin. Activity B will start at the beginning of the second working day. This level of detail will be important to find the calendar dates over which each activity will be executed, as discussed later on.[2]

The ESD and EFD for the successors to B, C, and D (i.e., E and F) are computed similarly. Because the EFD of C is 8, this number is repeated as ESD for F. But E poses a

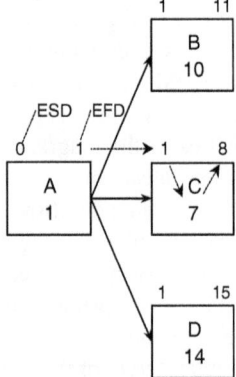

Figure 8.19 ESD and EFD computation for first four activities.

---

[2] Some authors use the so-called start plus-one convention instead of the method used here to find ESDs. Each activity's ESD appears shifted by one. Activity B would show an ESD of 2 and an EFD of $2 + 10 - 1 = 11$ (the same EFD shown in Figure 8.19). The start-plus-one convention has the advantage of showing directly the working dates for the start and finish of each activity, but it is more prone to error in manual computations.

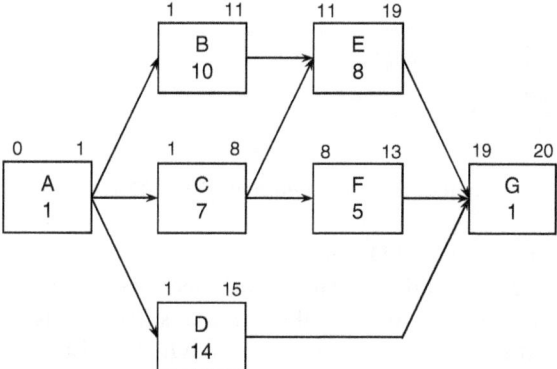

Figure 8.20 Forward pass results.

new situation. It is an immediate successor to both B and C. Its ESD will be the larger of its predecessors' EFD, that is,

$$ESD_E = Max(EFD_B, EFD_C) = Max(11, 8) = 11$$

An FS relationship means that the earliest that the successor can be started is when the predecessor is finished. Any date earlier than 11 would violate the relationship between B and E, but the ESD of 11 does keep the FS relationship between C and F.

The ESD of G is, therefore, the maximum of the EFDs of its predecessors:

$$ESD_G = Max(EFD_E, EFD_F, EFD_D) = Max(19, 13, 15) = 19$$

The EFD of the last activity G $(19 + 1 = 20)$ is also the minimum duration that this project can have. It is impossible to complete this project in less than 20 working days while complying with the relationships and durations shown in its precedence diagram. The complete forward pass is shown in Figure 8.20.

8.8   BACKWARD PASS (TRADITIONAL RELATIONSHIPS)

The second phase of CPM computations is the backward pass, whose progression through the network is opposite to that of the forward pass, as the name suggests. The backward pass computes the latest date by which each activity must finish to avoid extending the project duration. Take, for example, Activity D in the example. It could end as late as by working day 19 without affecting the early start of Activity G. But, if Activity D ends even 1 day after working day 19, it would "push" the start of Activity G and would result in a project duration of more than 20 working days.

The latest possible start date for each activity calculated by the backward pass is called the activity's *late start date* (LSD). Similarly, the latest finish date for each activity is its *late finish date* (LFD).

The backward pass begins by assigning an LFD to the last activity. In most cases, this date is the same as the EFD already found for this last activity. For our example, Activity G would have an LFD of 20.[3] The LSD for G is computed by subtracting its duration T_G:

$$LSD_G = EFD_G - T_G = 20 - 1 = 19$$

---

[3] It is advisable to clearly identify the last activity in a project's precedence diagram. Although the CPM algorithm can handle, in principle, diagrams with multiple ending activities by assigning the same LFD to all activities with no successors. For example, Activity E in the example network could have another successor H. Activities G and H would have no successors, resulting in a diagram with two ending activities. The LFD for both activities would be set at 20. However, to avoid confusion and even mistakes, in it is advisable to add a single final milestone (a type of activity discussed later here) set as the successor to all ending activities.

Figure 8.21 Assigning an LFD = EFD to the last activity.

The last activity's LFD is the only point at which the forward pass provides any data relevant to the backward pass. The forward and backward passes can even be performed in separate drawings, to avoid visual clutter. This first step is shown in Figure 8.21.

The next step in the backward pass is to find the LFD of all predecessors to the last activity. This date is set as the last activity's LSD. In the example, G has three predecessors: D, E, and F. Their LFDs are set as follows:

$$LFD_D = LSD_G = 19$$
$$LFD_E = LSD_G = 19$$
$$LFD_F = LSD_G = 19$$

Their respective LSDs are found by subtracting their duration from their LFD.

$$LSD_D = LFD_D - T_D = 19 - 14 = 5$$
$$LSD_E = LFD_E - T_E = 19 - 8 = 11$$
$$LSD_F = LFD_F - T_F = 19 - 5 = 14$$

Figure 8.22 shows the backward pass computations, including the activities discussed that follows. To avoid confusion with the forward pass, each activity's LFD and LSD are located below the activity box.

When an activity has a single successor, its LFD is simply the LSD of the successor. For example, the LFD for B is 11 because it is the LSD of its successor, E. When an activity has more than one successor, as is the case with C, its LFD is the minimum of the LFDs of its followers. Therefore, using the data shown in Figure 8.22, the LFD for C is computed as follows:

$$LFD_C = Min(LSD_E, LSD_F) = Min(11, 14) = 11$$

It should be clear that LFD_C could not be set at working day 14 without extending the project's duration. Since C precedes E, this would mean that E would have to begin after 14 working days of project execution. Yet, its own late start has been set at 11. If it would start after 14 days, its finish would be $14 + 8 = 22$. Because E precedes G, the minimum start date of the latter would be now set at 22, and its finish date at $22 + 1 = 23$. The project would have a scheduled duration that is $23 - 20 = 3$ working days over the minimum duration found in the forward pass.

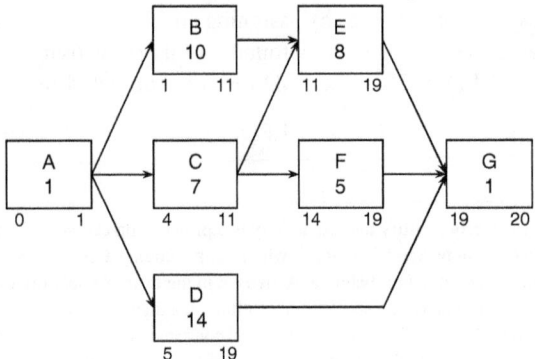

Figure 8.22 Backward pass results.

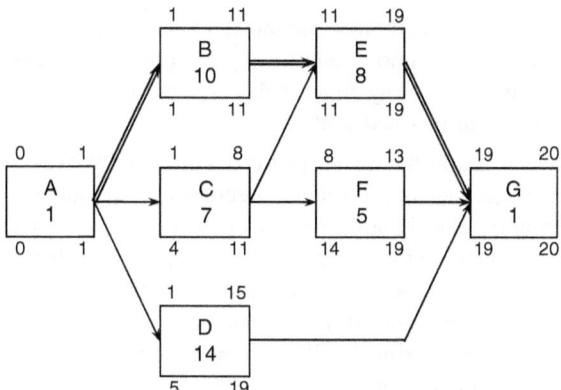

Figure 8.23 Combined forward and backward passes.

The combined forward and backward passes are shown in Figure 8.23. As previously mentioned, the forward pass computations are shown at the top of each activity box, and the backward pass dates are shown below the box in the figure.

8.9   CRITICAL PATH

It can be noticed in Figure 8.18 that the longest sequence of activities that must be performed to complete the example project is A-B-E-G. The total duration of this chain of activities is $1 + 10 + 8 + 1 = 20$ working days. Any other activity sequence from start to finish yields a smaller total duration. It is possible to check this by listing all possible paths for the example diagram.

Activity path A-C-F-G: $1 + 7 + 5 + 1 = 14$ days

Activity path A-C-E-G: $1 + 7 + 8 + 1 = 17$ days

Activity path A-D-G: $1 + 14 + 1 = 16$ days

The forward pass provides a straightforward way to find the duration of the project. Enumerating all possible activity paths is possible for the small example network, but it would be quite impractical for any project with even a few dozen activities. The backward pass provides additional information by allowing the identification of this longest path. Any activity with the same ESD and LSD (or EFD and LFD) is part of the longest path of activities spanning from the beginning to the end of the network. Notice that only the activities that we already identified as belonging to the longest path, namely A, B, E and G, have this property.

The activity path with the maximum total duration is called the *critical path* of the project and its activities are called *critical activities*. The path is marked with double-line relationships, as shown in Figure 8.23. The critical path is of paramount importance for the control of the project. Understanding its implications is essential for rational management decisions. The following list is a good starting point to get a perspective of the properties and implications of a project's critical path.

- All projects will have at least one critical path. There could be a tie in the total duration of two or more paths, resulting in multiple critical paths. If D had a duration of 18 days, the path A-D-G would become critical, in addition to the original critical path A-B-G-E.

- All critical activities belong to a critical path. This is a biunivocal correspondence. An activity cannot be critical without belonging to a critical path.

- A delay in any critical activity, even by a single day, will extend the duration of the project.

- All noncritical activities (anyone not belonging to a critical path) can be delayed by at least 1 day without lengthening the project duration. The maximum number of days for this leeway in start date is determined by each activity's total float, as discussed in the next section.

- Since the duration of a project is determined by its critical path, saving time in any noncritical activity does not decrease at all the project's duration. Moreover, shortening the duration of any critical activity also shortens the duration of the critical path, and therefore of the whole project. If there are several critical paths running in parallel, each one must be shortened by the same amount of days. Eventually, as more days are cut back, another sequence of activities that was originally non-critical will become critical, and this new path will have to be considered in any additional reductions. For example, if Activity B in the example project were shortened to 7 days, a new critical path consisting of activities A-C-E-G would be created, in addition to the original path of A-B-E-G. Notice that some activities may belong to more than one critical path.

The list suggests the importance of the critical path for the planning and control of any project. For example, if two activities are competing for the use of a piece of equipment and one of them is critical, it makes sense to give priority to the critical activity over the non-critical.[4] If a project must be expedited to recover some lost time, it does not make sense to require all subcontractors to show up to work over the weekend. Only those subcontractors performing critical activities should increase their working days, since all others would just increase the project cost by paying overtime, without reducing the project duration by a single day.

8.10   ACTIVITY FLOATS

8.10.1 Total Float

The difference between the early start ESD and the late start LSD of an activity is its total float (TF). For any Activity I:

$$TF_i = LSD_i - ESD_i$$

The same numerical result can be obtained by using finish dates. TF_i is also equal to $LFD_i - EFD_i$.

An activity's TF is, therefore, the leeway between the earliest date at which it can start and the latest date at which it can start without resulting in a delay for the entire project. The early start is physically determined by the activity's predecessors. Its late start is determined by the goal of not extending the project's duration.

Critical activities have a TF of zero, and noncritical activities have a TF greater than zero.[5] If any activity "uses up" its TF by starting at its LSD, it becomes critical (as well as all its succeeding activities on the longest path from it to the last activity in the network). If any activity, critical or not, is delayed from its ESD by more than its TF, the whole project will be delayed in the amount of time that the TF is exceeded.

TFs for the example project are shown in Figure 8.24 as the uppermost of the two figures shown vertically at the bottom center of each activity box.

---

[4] There are circumstances in which giving preference to the noncritical activity would be granted. Such a case may arise, for example, when the noncritical activity precedes others with a constrained schedule for the availability of their own equipment or other resources.
[5] This is true when the LFD of the last activity is set to be equal to its EFD. There are circumstances when these two dates must differ, and the resulting TF has a slightly different interpretation.

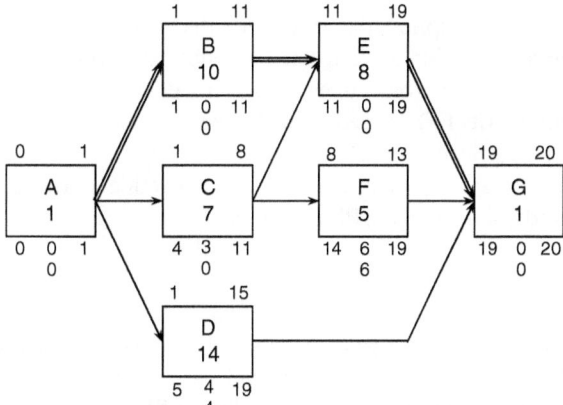

Figure 8.24 Complete scheduling results.

The previous section discusses how the critical path can be identified by looking for those activities whose ESD is the same as their LSD (and therefore, their EFD is equal to their LFD).

8.10.2 Free Float

The free float (FF) of an activity is the difference between the minimum early start ESD of its successors and its own early finish EFD. For an Activity I with several successors j_1, j_2, j_3 . . . , this can be expressed as follows:

$$FF_i = MIN(ESD_j) - EFD_i \text{ for } j = 1, 2, 3, \ldots$$

In the example project network, the FF for activity D is computed as the ESD of its only successor G (19) minus the EFD of activity D (14).

$$FF_D = ESD_G - EFD_D = 19 - 14 = 5$$

Activity C has two successors, E and F. In this case, the minimum of the two ESD must be selected:

$$FF_C = MIN(ESD_E, ESD_F) - EFD_C = MIN(11, 8) - 8 = 8 - 8 = 0$$

The FFs computed for the example project are the numbers in the bottommost row below each activity box in Figure 8.24.

The FF of an activity assumes that all its predecessors and successors begin as early as possible and that the activity itself ends as early as possible. This assumption is much more restrictive than the one used for an activity's TF. The TF assumes that all predecessors to the activity begin at their ESD, and all successors begin as late as possible, that is, at their LSD. Because the assumptions for an activity's FF are more restrictive than those for its TF, the TF of any activity is always equal or greater than its FF. Because critical activities have TF = 0 and TF >= FF, it follows that the FF of all critical activities is zero. (In fact, *all* types of floats for critical activities are zero.). Noncritical activities can have FF of zero, as can be seen in the example project.

8.10.3 Interfering Float

The interfering float (IF) of an activity is the difference between its TF and its FF. For any Activity I:

$$IF_i = TF_i - FF_i$$

As the name implies, the IF of an activity is the portion of its TF that "pushes" succeeding activities up to their latest possible start times. Because these successors are now at their LSD, the project does not suffer a delay. For example, the IF of Activity C is 3 (its TF) minus 0 (its FF). Indeed, it can be seen that each day that C delays its early start of 1 increases the earliest possible start of Activity F.

Since the computation of IF is trivial for practical purposes, it is rarely included in the data explicitly shown in a CPM schedule.

8.10.4 Independent Float

The final and most restrictive of all types of activity floats is the independent float (IndF). This float assumes that the activity ends as late as possible (at its LSD), and all its successors begin as early as possible, at their ESD. From this assumption, its follows that the IndF of an Activity I whose followers are $j_1, j_2, j_3 \ldots$ can be computed as follows:

$$\text{IndF}_i = \text{MIN}\left(\text{ESD}_j\right) - \text{LFD}_i \text{ for } j = 1, 2, 3, \ldots$$

Using this formula to compute the IndF of Activity C in the example project,

$$\text{IndF}_C = \text{MIN}(\text{ESD}_E, \text{ESD}_F) - \text{LFD}_C = \text{MIN}(11, 8) - 11 = 8 - 11 = -3. \text{ Use } 0.$$

The computations for all the floats previously discussed yield a result of zero as a minimum. But as shown in the case of Activity C, the IndF formula can result in a negative number. Since time cannot be negative, this result simply means that there is no way for this activity to satisfy the conditions assumed for this float. IndF is rarely shown in a CPM schedule report, although it can be argued that it is as relevant as the others. An activity with IndF greater than 0 can start at any point between its ESD and LSD without affecting its successors and with the worst-case scenario of having all predecessors starting at their LSD. Therefore, delaying the start of this activity in case of conflict in the use of resources is the most conservative alternative possible, even if the other activities competing for the resource are also non-critical.

8.11 SCHEDULING THE SMALL GARAGE PROJECT

The small garage project provides a more realistic example of the scheduling process than the minimum precedence diagram used so far.

The forward-pass algorithm is applied repetitively starting at the source node labeled *START*. All events related to the START node are given the value 0. Two activities (e.g., 1 and 2) follow the START node. Calculations for all of the activities are shown in Table 8.2. Calculated ESD and LSD values are shown on each activity in Figure 8.25. Based on the calculations in Table 8.2, it can be seen that the duration of the longest and therefore critical path in the network is 96 working days.

The minimum project duration based on this critical path analysis is 96 working days or slightly over 19 working weeks. If we are plotting the project duration in calendar days, this will equate to $(19 \times 7) + 1$ or 134 calendar days.[6] If the project is to begin on March 1, the estimated project completion date will be July 12.

The forward-pass calculations establish the minimum project duration (based on no delays to the critical activities). It does not, however, identify the critical path. To identify the critical set of activities that constrain the project to a minimum duration of 96 days, we must apply the backward-pass algorithm. To start the backward-pass algorithm, the LFD (END) is set to 96 days. This is equivalent to setting the LFD for Activity 22 to 96 days. As we have just seen, based on our forward-pass calculations, the LFD of Activity 22 cannot

---

[6] Working days are differentiated from calendar days. There are typically 5 working days in a week as opposed to 7 calendar days.

Table 8.2 Forward-Pass Calculations for the Small Gas Station Project

Activity	Calculations	
Start	ESD(START) = 0	EFD(START) = 0
1	ESD(1) = max[EFD(START)] = 0	EFD(1) = 0 + 10 = 10
2	ESD(2) = max[EFD(START)] = 0	EFD(2) = 0 + 15 = 15
3	ESD(3) = max[EFD(1)] = 10	EFD(3) = 10 + 8 = 18
4	ESD(4) = max[EFD(1)] = 10	EFD(4) = 10 + 12 = 22
5	ESD(5) = max[EFD(2), EFD(3)] = 18	EFD(5) = 18 + 2 = 20
6	ESD(6) = max[EFD(3)] = 18	EFD(6) = 18 + 5 = 23
7	ESD(7) = max[EFD(3)] = 18	EFD(7) = 18 + 6 = 24
8	ESD(8) = max[EFD(4), EFD(5), EFD(6), EFD(7)] = 24	EFD(8) = 24 + 8 = 32
9	ESD(9) = max[EFD(8)] = 32	EFD(9) = 32 + 10 = 42
10	ESD(10) = max[EFD(9)] = 42	EFD(10) = 42 + 14 = 56
11	ESD(11) = max[EFD(9)] = 42	EFD(11) = 42 + 4 = 46
12	ESD(12) = max[EFD(9)] = 42	EFD(12) = 42 + 15 = 57
13	ESD(13) = max[EFD(10), EFD(11)] = 56	EFD(13) = 56 + 12 = 68
14	ESD(14) = max[EFD(12)] = 57	EFD(14) = 57 + 10 = 67
15	ESD(15) = max[EFD(10), EFD(11), EFD(14)] = 67	EFD(15) = 67 + 6 = 73
16	ESD(16) = max[EFD(10), EFD(11), EFD(14)] = 67	EFD(16) = 67 + 10 = 77
17	ESD(17) = max[EFD(10), EFD(11), EFD(14)] = 67	EFD(17) = 67 + 25 = 92
18	ESD(18) = max[EFD(15), EFD(16)] = 77	EFD(18) = 77 + 3 = 80
19	ESD(19) = max[EFD(15), EFD(16)] = 77	EFD(19) = 77 + 6 = 83
20	ESD(20) = max[EFD(15), EFD(16)] = 77	EFD(20) = 77 + 8 = 85
21	ESD(21) = max[EFD(13), EFD(17), EFD (18), EFD(19), EFD(20)] = 92	EFD(21) = 92 + 1 = 93
22	ESD(22) = max[EFD(21)] = 93	EFD(22) = 93 + 3 = 96

be less than 96 days. If we set LFD (22) to a duration greater than 96 days, the finish date for the project will be extended.

The calculations for the backward-pass are given in Table 8.3, and the LSD and LFD values for each activity are shown in Figure 8.25. Float calculations are shown in Table 8.4.

Table 8.3 Backward-Pass Calculations for the Small Gas Station Project

Activity	Calculations	
END	LFD(END) = 96	LSD(END) = 96
22	LFD(22) = min[LSD(END)] = 96	LSD(22) = 96–3 = 93
21	LFD(21) = min[LSD(22)] = 93	LSD(21) = 93–1 = 92
20	LFD(20) = min[LSD(21)] = 92	LSD(20) = 92–8 = 84
19	LFD(19) = min[LSD(21)] = 92	LSD(19) = 92–6 = 86
18	LFD(18) = min[LSD(21)] = 92	LSD(18) = 92–3 = 89
17	LFD(17) = min[LSD(21)] = 92	LSD(17) = 92–25 = 67
16	LFD(16) = min[LSD(18), LSD(19), LSD(20)] = 84	LSD(16) = 84–10 = 74
15	LFD(15) = min[LSD(18), LSD(19), LSD(20)] = 84	LSD(15) = 84–6 = 78
14	LFD(14) = min[LSD(15), LSD(16), LSD(17)] = 67	LSD(14) = 67–10 = 57
13	LFD(13) = min[LSD(21)] = 92	LSD(13) = 92–12 = 80
12	LFD(12) = min[LSD(14)] = 57	LSD(12) = 57–15 = 42
11	LFD(11) = min[LSD(13), LSD(15), LSD(16), LSD(17) = 67	LSD(11) = 67–4 = 63
10	LFD(10) = min[LSD(13), LSD(15), LSD(16), LSD(17)] = 67	LSD(10) = 67–14 = 53
9	LFD(9) = min[LSD(10), LSD(11), LSD(12)] = 42	LSD(9) = 42–10 = 32
8	LFD(8) = min[LSD(9)] = 32	LSD(8) = 32–8 = 24
7	LFD(7) = min[LSD(8)] = 24	LSD(7) = 24–6 = 18
6	LFD(6) = min[LSD(8)] = 24	LSD(6) = 24–5 = 19
5	LFD(5) = min[LSD(8)] = 24	LSD(5) = 24–2 = 22
4	LFD(4) = min[LSD(8)] = 24	LSD(4) = 24–12 = 12
3	LFD(3) = min[LSD(5), LSD(6), LSD(7)] = 18	LSD(3) = 18–8 = 10
2	LFD(2) = min[LSD(5)] = 22	LSD(2) = 22–15 = 7
1	LFD(1) = min[LSD(3), LSD(4)] = 10	LSD(1) = 10–10 = 0

Figure 8.25 Expanded network schedule for the small gas station project.

Table 8.4 Float Calculations for the Small Gas Station Project

Activity	Total Float	Free Float	Interfering Float
* 1	TF(1) = 10 − 10 = 0	FF(1) = 10 − 10 = 0	IF(1) = 0 − 0 = 0
2	TF(2) = 22 − 15 = 7	FF(2) = 18 − 15 = 3	IF(2) = 7 − 3 = 4
* 3	TF(3) = 18 − 18 = 0	FF(3) =18 − 18 = 0	IF(3) = 0 − 0 = 0
4	TF(4) = 24 − 22 = 2	FF(4) = 24 − 22 = 2	IF(4) = 2 − 2 = 0
5	TF(5) = 24 − 20 = 4	FF(5) = 24 − 20 = 4	IF(5) = 4 − 4 = 0
6	TF(6) = 24 − 23 = 1	FF(6) = 24 − 23 = 1	IF(6) = 1 − 1 = 0
* 7	TF(7) = 24 − 24 = 0	FF(7) = 24 − 24 = 0	IF(7) = 0 − 0 = 0
* 8	TF(8) = 32 − 32 = 0	FF(8) = 32 − 32 = 0	IF(8) = 0 − 0 = 0
* 9	TF(9) = 42 − 42 = 0	FF(9) = 42 − 42 = 0	IF(9) = 0 − 0 = 0
10	TF(10) = 67 − 56 = 11	FF(10) = 56 − 56 = 0	IF(10)= 11 − 0 = 11
11	TF(11) = 67 − 46 = 21	FF(11) = 56 − 46 = 10	IF(11) = 21 − 10 = 11
* 12	TF(12) = 57 − 57 = 0	FF(12) = 57 − 57 = 0	IF(12) = 0 − 0 = 0
13	TF(13) = 92 − 68 = 24	FF(13) = 92 − 68 = 24	IF(13) = 24 − 24 = 0
* 14	TF(14) = 67 − 67 = 0	FF(14) = 67 − 67 = 0	IF(14) = 0 − 0 = 0
15	TF(15) = 84 − 73 = 11	FF(15) = 77 − 73 = 4	IF(15) = 11 − 4 = 7
16	TF(16) = 84 − 77 = 7	FF(16) = 77 − 77 = 0	IF(16) = 7 − 0 = 7
* 17	TF(17) = 92 − 92 = 0	FF(17) = 0 − 0 = 0	IF(17) = 0 − 0 = 0
18	TF(18) = 92 − 80 = 12	FF(18) = 92 − 80 = 12	IF(18)= 12 − 12 = 0
19	TF(19) = 92 − 83 = 9	FF(19) = 92 − 83 = 9	IF(19) = 9 − 9 = 0
20	TF(20) = 92 − 85 = 7	FF(20) = 92 − 85 = 7	IF(20) = 7 − 7 = 0
* 21	TF(21) = 93 − 93 = 0	FF(21) = 0 − 0 = 0	IF(21) = 0 − 0 = 0
* 22	TF(22) = 96 − 96 = 0	FF(22) = 0 − 0 = 0	IF(22) = 0 − 0 = 0

8.12 GENERALIZED RELATIONSHIP SCHEDULING COMPUTATIONS

The scheduling computations for precedence diagrams with generalized relationships follow the same rationale discussed for traditional relationships. First, a forward pass establishes the earliest dates that each activity can be performed while keeping the logic of the diagram. A backward pass computes the latest possible dates to execute each activity without extending the duration found by the forward pass. Finally, the critical path and activity floats are computed.

The small example used to discuss traditional computations has been modified as shown in Figure 8.26 by introducing several generalized relationships. Although the activities have kept their relative position and duration, almost all of them are related by

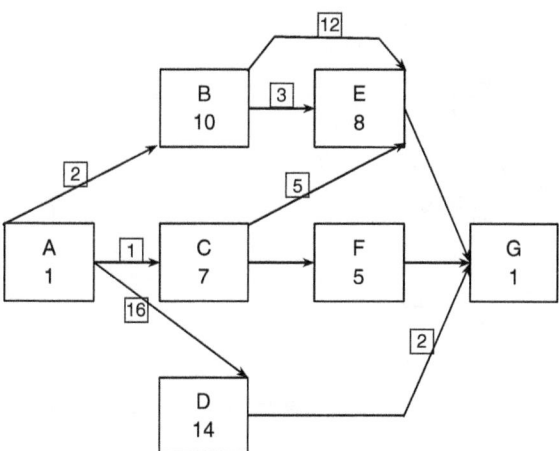

Figure 8.26 Small precedence diagram with generalized relationships.

generalized relationships differing from the original example. This diagram is an extreme case of the use of generalized relationships, done for the purposes of explanation. Most CPM diagrams will have a much smaller proportion of nontraditional relationships.

8.12.1 Forward Pass (Generalized Relationships)

As with traditional relationships, an ESD of 0 is assigned to the initial Activity A. Because this activity includes the three types of generalized relationships usually used in construction project diagrams, it can reveal the scheduling procedure for each one. Figure 8.27 shows a detail of the computations that follow.

The relationship between A and B, which can be called R_{AB} for simplicity, is SS. Therefore, the start of A (0) plus the lag of R_{AB} (2) will be used in the computation of the early start ESD of Activity B. Its EFD will be its ESD plus its duration T.

$$ESD_B = ESD_A + LAG_{AB} = 0 + 2 = 2$$
$$EFD_B = ESD_B + T_B = 2 + 10 = 12$$

The relation between A and C, R_{AC}, is a traditional FS relationship to which a lag of 1 has been added. Therefore, the ESD of C is determined by adding this lag to the EFD of A. The EFD is computed by adding the activity's duration.

$$ESD_C = EFD_A + LAG_{AC} = 1 + 1 = 2$$
$$EFD_C = ESD_C + T_C = 2 + 7 = 9$$

The relationship between A and D, R_{AD}, is FF. Therefore, the EFD of D is determined as $EFD_A + LAG_{AD}$, and then ESD_D results from *subtracting* T_D from its EFD.

$$EFD_D = EFD_A + LAG_{AD} = 1 + 16 = 17$$
$$ESD_D = EFD_D - T_D = 17 - 14 = 3$$

The three activities considered so far have a single predecessor, namely Activity A, since the latter is the initial activity in the precedence diagram. Activity E, in contrast, is preceded by B and C. Each one of the two relationships between B and E must be evaluated separately. Similarly to the computations already discussed for traditional relationships, the ESD and EFD of E will be dictated by the one of its two predecessors that "pushes" it the most into the future. That is, considering the FF relationship between B and E (since it is an FF relationship, EFD is computed first):

$$EFD_E = EFD_B + LAGB_{BE(FF)} = 12 + 12 = 24$$
$$ESD_E = EFD_E - T_E = 24 - 8 = 16$$

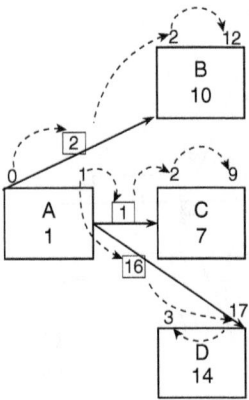

Figure 8.27 Forward pass for first four activities of generalized network.

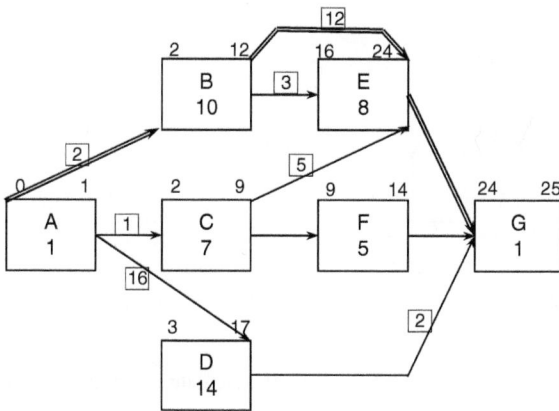

Figure 8.28 Complete forward pass for small generalized network.

Considering the FS relationship between B and E:

$$\text{ESD}_E = \text{EFD}_B + \text{LAGB}_{BE(FS)} = 12 + 3 = 15$$
$$\text{EFD}_E = \text{ESD}_E + T_E = 15 + 8 = 23$$

Finally, considering the FF relationship between C and E:

$$\text{EFD}_E = \text{EFD}_C + \text{LAG}_{CE} = 9 + 5 = 14$$
$$\text{ESD}_E = \text{EFD}_E - T_E = 14 - 8 = 6$$

The controlling relationship in this case is the FF between B and E, since it results in the maximum of the three possible early dates:

$$\text{ESD}_E = \text{MAX}(16, 15, 6) = 16$$
$$\text{EFD}_E = \text{MAX}(24, 23, 14) = 24 \ (\text{also, EFD}_E = \text{ESD}_E + T_E = 16 + 8 = 24)$$

Activity F has a traditional relationship with its sole predecessor C:

$$\text{ESD}_F = \text{EFD}_C = 9$$
$$\text{EFD}_F = \text{ESD}_F + T_F = 9 + 5 = 14$$

The last activity G has traditional relationships with E and F, and an FS with lag relates it to D. Therefore,

$$\text{ESD}_G = \text{MAX}(\text{EFD}_E, \text{EFD}_F, [\text{EFD}_D + \text{LAG}_{DG}]) = \text{MAX}(24, 14, [17 + 2]) = 24$$
$$\text{EFD}_G = \text{ESD}_G + T_G = 24 + 1 = 25$$

The complete forward pass is shown in Figure 8.28.

8.12.2 Backward Pass (Generalized Relationships)

The backward pass follows the logic discussed for traditional relationships. As in traditional relationships, it determines the latest possible point in time to start each activity without extending the project's duration. The difference with traditional relationships is that the lag between activities must be subtracted instead of added in the process of examining activities from last to first. Moreover, the computations must be adjusted to include FF and SS relationships, as will be discussed.

The first step in the backward pass is the assignment of an LFD to the last activity, normally the same as its EFD. Its LSD is its LFD minus its duration. In the example,

$$\text{LFD}_G = \text{EFD}_G = 25$$
$$\text{LSD}_G = \text{LFD}_G - T_G = 25 - 1 = 24$$

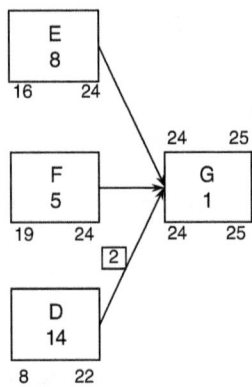

Figure 8.29 Late date computations for activities D, E, F and G.

Activity G's predecessors are scheduled next. Since activities E and F are linked to G by traditional relationships, their LFD is equal to the LSD of G. LSDs are found by subtracting their respective duration from their LFD:

$$LFD_F = LFD_E = LSD_G = 24$$
$$LSD_F = LFD_F - T_F = 24 - 5 = 19$$
$$LSD_E = LFD_E - T_E = 24 - 8 = 16$$

Activity D includes a lag of 2 in its FS relationship to G. The LFD and LSD computations are similar to the ones for E and F, but now the lag between D and G is subtracted in the computations:

$$LFD_D = LSD_G - LAGD_G = 24 - 2 = 22$$
$$LSD_D = LFD_D - T_D = 22 - 14 = 8$$

The late dates of the activities considered so far are shown in Figure 8.29.

The next activity in the last-to-first sequence of the backward pass is C. It has a traditional relationship to F, and an FF relationship to E. The rationale used so far applies here, with the caveat that since the FF relationship links the respective finishes of C and E, it is LFD_E that must be considered:

$$LFD_C = MIN(LSD_F, [LFD_E - LAG_{CF}]) = MIN(19, [24 - 5]) = 19$$
$$LSD_C = LFD_C - T_C = 19 - 7 = 12$$

Notice that in this case there is a tie between the two possible relationships, since $24 - 5 = 19$. Such ties are allowed by the algorithm and in fact are common in projects with a tight execution schedule.

Activity B is related to its successor by two relationships, and therefore, both must be considered in the computations.

$$LFD_B = MIN([LFD_E - LAG_{BE(FF)}], [LSD_E - LAG_{BE(FS)}])$$
$$= MIN([24 - 12], [16 - 3]) = 12$$
$$LSD_B = LFD_B - T_B = 12 - 10 = 2$$

The initial Activity A has the three types of generalized relationships with its successor. The computations are best understood by analyzing separately each relationship. Notice that, as expected, SS_{AB} relates directly the LSDs of A and B. LFD_A is computed *after* LSD_A is found, inverting the computation order used so far for these two dates.

LFD_A and LSD_A will be equal to the minimum LSD and LFD imposed by each of the three relationships.

Considering the SS relationship to B:

$$LSD_A = LSD_B - LAG_{AB} = 2 - 2 = 0; \quad LFD_A = LSD_A + T_A = 0 + 1 = 1$$

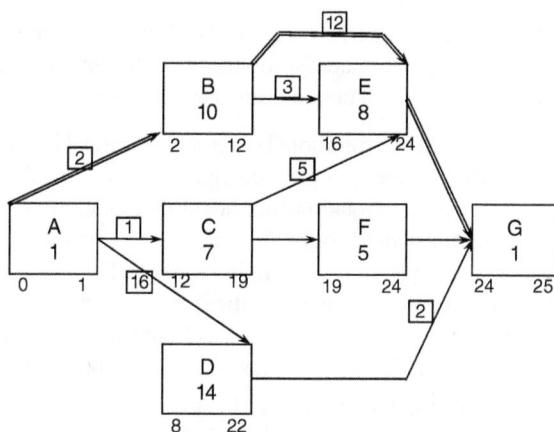

Figure 8.30 Complete backward pass for generalized relationships example.

Considering the FS relationship to C (See Fig. 8.30):

$$LFD_A = LSD_C - LAG_{AC} = 12 - 1 = 11; \quad LSD_A = LFD_A - T_A = 11 - 1 = 10$$

Considering the FF relationship to D:

$$LFD_A = LFD_D - LAG_{AD} = 22 - 16 = 6; \quad LSD_A = LFD_A - T_A = 6 - 1 = 5$$

Now the LSD and LFD can be determined as follows:

$$LFD_A = MIN(LFD_A \text{ computed using each relationship}) = MIN(1, 11, 6) = 1$$
$$LSD_A = MIN(LSD_A \text{ computed using each relationship}) = MIN(0, 10, 5) = 0$$
(Or, $LSD_A = LFD_A - T_A = 1 - 1 = 0$)

8.12.3 Total Float and Critical Path (Generalized Relationships)

The TF of each activity is the difference between its LFD and its EFD (or its LSD and its ESD). TFs are shown as the uppermost numbers of the two stacked at the bottom center of each activity box in Figure 8.31, which shows all the scheduling results for the example network.

The critical path is the chain of activities with TF = 0. For the example network, it consists of the chain of activities A-B-E-G. The relationships linking the critical activities are drawn with double lines in Figure 8.31.

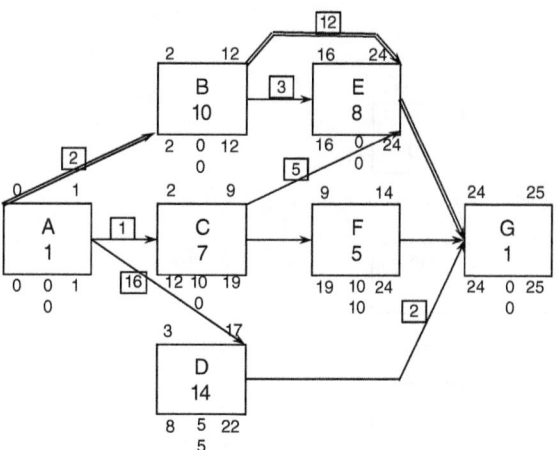

Figure 8.31 Complete computations for generalized relationships example.

All the critical path properties previously discussed for precedence diagrams with only traditional relationships also apply for those with generalized relationships. Moreover, there are a few additional issues to consider for generalized relationships:

- Critical activities in a traditional diagram are critical in their start, finish, and duration. If any of these three elements change for any critical activity, the project will be delayed. Generalized relationships are more subtle in defining the critical path because in some circumstances, only the start or the finish of an activity is truly critical. Consider Activity E in the example network. The FF relationship with Activity B is critical, and therefore, what is critical for E is to *finish* on time. Its duration could perfectly change to, for example, 10 without extending the project execution time, if it starts by day 14. This type of flexibility is not possible in a traditional network.

- Two activities may be critical and yet one or more relationships linking them may not be part of the critical path. This is the case for the FS relationship between activities B and E.

8.12.4 Free Float (Generalized Relationships)

The FF for activities in a generalized precedence diagrams is conceptually identical to the FF for activities in a traditional network already discussed. As before, it is the leeway, if any, that an activity has before it "pushes" the ESD of any successor. The FF of the activity is the minimum leeway between it and its successors. The computation of this leeway depends on the type of relationship. For an Activity I followed by another Activity J, the following formulas apply.

$$FS \text{ relationships}: FF \text{ between } j \text{ and } i = ESD_j - LAG_{ij} - EFD_i$$
$$SS \text{ relationships}: FF \text{ between } j \text{ and } i = ESD_j - LAG_{ij} - ESD_i$$
$$FF \text{ relationships}: FF \text{ between } j \text{ and } i = EFD_j - LAG_{ij} - EFD_i$$

Consider the case for Activity A of the example network, whose FF computation is shown graphically in Figure 8.32.

$$FF_A = MIN([2 - 2 - 0], [2 - 1 - 1], [17 - 16 - 1]) = 0$$

Notice that each of the three possibilities for FF_A result in FF = 0. This is expected. Activity A is critical, and therefore all its floats are 0. Moreover, Activity A drives the ESD of all its three successors, and therefore, there can be no leeway with any of them. Lastly, remember that for any Activity J, $TF_j >= FF_j$. $TF_A = 0$, and its FF cannot be greater. The FF of each activity is shown in Figure 8.31 as the lower number of the two stacked at the center and below its box.

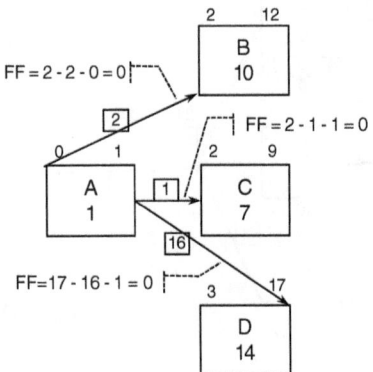

Figure 8.32 Free Float computation for Activity A.

8.12.5 Interfering and Independent Floats (Generalized Relationships)

As in traditional diagrams, the IF of an activity is the difference between its total float TF and its FF. This computation is so trivial that the IF is rarely explicitly included in the scheduling reports.

The IndF of an activity is, as before, the leeway that may exist when all the activity's successors are located in their ESDs, and it is located in its LSD. The leeway formula for each type of generalized relationship is identical to the one for its FF, with the only difference of substituting the ESD/EFD of successors with their respective LSD/LFD. Formulas for TF and FF never result in negative numbers. This is not the case with IndF. Use 0 when a formula yields IndF < 0.

$$\text{FS relationships}: \text{IndF between j and i} = \text{ESD}_j - \text{LAG}_{ij} - \text{EFD}_i$$
$$\text{SS relationships}: \text{IndF between j and i} = \text{ESD}_j - \text{LAG}_{ij} - \text{ESD}_i$$
$$\text{FF relationships}: \text{IndF between j and i} = \text{EFD}_j - \text{LAG}_{ij} - \text{EFD}_i$$

IndF is greater than 0 for relatively few activities in a typical network. Applying the formulas, it can be verified that all activities in the example network have IndF $= 0$. As mentioned in the discussion of traditional networks, it is uncommon to include this float in a schedule report.

8.13   WORKING TO CALENDAR DATES

An activity schedule must include the calendar dates intended for each start and finish. Contractual terms almost universally specify calendar days for a project's duration, time to mobilize after the Notice to Proceed, finish deadline, and similar time-sensitive issues. This practice is hardly surprising, since many project owners may not fully understand the meaning and implications of working days versus calendar days. Moreover, calendar dates are simpler to enforce from a contractual viewpoint: a contractor could allege, perhaps truly, that their company does not work on Flag Day, and therefore it could not be included in the working day count. Vice versa, an owner could claim that July 4th could be included in the count. Such gray areas are avoided by specifying durations and deadlines in calendar dates.

The key step to relate a project's working days to their corresponding calendar dates is to count the number of working days in a calendar in which the first calendar date of project execution is marked as working day 1. Each following day is successively marked in the calendar. Nonworking days such as weekend and holidays are skipped in the count. This simple process is illustrated in Figure 8.33, showing the calendar for a project with a start date of December 23. Its contractor has chosen not to work on Saturdays and Sundays, as well as on December 25 and January 1. The workday to calendar date conversion depends on each particular contractor. Another contractor could have a different work calendar.

Sun	Mon	Tue	Wed	Thu	Fri	Sat
20-Dec	21	22	23 WD1	24 WD2	25	26
27	28 WD3	29 WD4	30 WD5	31 WD6	1-Jan	2
3	4 WD7	5 WD8	6 WD9	7 WD10	8 WD11	9
10	11 WD12	12 WD13	13	14	15	16

Figure 8.33 Project calendar with working days.

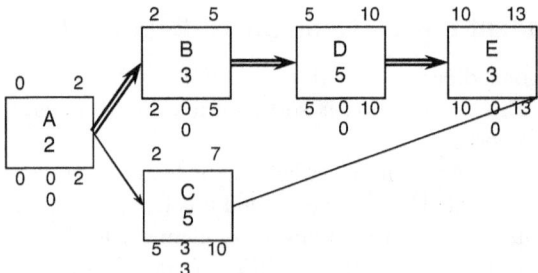

Figure 8.34 Example network for working to calendar dates example.

Table 8.5 Working to calendar dates conversion for example network

| Activity | EARLY DATES | | | | LATE DATES | | | |
| | ESD | | EFD | | LSD | | LFD | |
	Working Days	Calendar Date	Working Days	Calendar Date	Working Days	Calendar Date	Working Days	Calendar Date
A	0	23-Dec	2	24-Dec	0	23-Dec	2	24-Dec
B	2	28-Dec	5	30-Dec	2	28-Dec	5	30-Dec
C	2	28-Dec	7	4-Jan	5	31-Dec	10	7-Jan
D	5	31-Dec	10	7-Jan	5	31-Dec	10	7-Jan
E	10	8-Jan	13	12-Jan	10	8-Jan	13	12-Jan

To illustrate this procedure, let us assume that the calendar of Figure 8.33 applies to the small project whose schedule is shown in Figure 8.34.

As previously mentioned when discussing the forward pass computation process, the ESDs shown in Figure 8.34 indicate the number of working days that must elapse before each activity can start. The ESD of 5 for Activity D indicates that 5 working days (in this case, the sum of the durations of activities A and B) must have elapsed before Activity D can begin. Therefore, the early calendar date for activity D is $5 + 1 = 6$. In the calendar shown in Figure 8.33, this means that Activity D will begin on December 31. This one-day shift applies as well to all other activities, as can be seen in Table 8.5. Late start dates (LSDs) are also shifted by one day, as can be seen in Table 8.5.

The 1-day shift just discussed does not apply to early and late finish dates (EFDs and LFDs) shown in the diagram. Their corresponding calendar dates can be read directly in the conversion calendar. The calendar EFD for Activity D is, therefore, January 7. All finish calendar dates are also shown in Table 8.5.

There are no differences between traditional and generalized relationships in the conversion from working days to calendar dates.

8.14   SUMMARY

A project activity schedule developed using the techniques discussed in the last two chapters can be of great help to the project manager entrusted with its execution. Summarizing the discussion to this point, such schedule will provide:

- A clear definition of the work to performed, divided into the assignable segments constituted by the project activities.

- A precedence diagram showing assumptions and strategic decisions about the sequence of activities.

- Start and finish dates for each activity.

- The number of days that each activity can be delayed before resulting in a delay in the overall project.

Originally, CPM utilized a network convention called *arrow notation* as discussed in Appendix E. This chapter has emphasized *precedence notation* since it is more widely used today. CPM is used for the procurement and control of suppliers, subcontractors, and the contractor's own forces. Furthermore, it can be used to optimize decisions that otherwise result in unnecessary effort and waste. For example, as introduced in the Critical Path section, it makes no sense to expedite a project duration by requiring all ongoing activities to work overtime. A CPM-based schedule will show which activities are critical, and therefore, should be expedited. Expediting noncritical activities would only add to their float, without shortening the project duration at all. With all its positive aspects, a project schedule can be detrimental to the management of a project when its role and limitations are not well understood. A schedule should be a management resource for rational decision making in the field, instead of the control mechanism to which field management must adjust their decisions. Any project manager approaching project management as an effort to conform to the schedule instead of a continuous process to conform to the *objectives* of the schedule will soon find that, regardless of its initial merits, any schedule must be updated to reflect the changing conditions of the field.

REVIEW QUESTIONS AND EXERCISES

8.1 Develop the precedence diagram network for the following project and then calculate the TF and FF for each activity.

Activity	Duration	Immediately Following Activities
a	22	dj
b	10	cf
c	13	dj
d	8	—
e	15	cfg
f	17	hik
g	15	hik
h	6	dj
I	11	j
j	12	—
k	20	—

H	7	F
I	5	G
J	3	G
K	8	C,G
L	9	H,I
M	4	F,J,K
N	7	D,K
0	8	E,K
P	6	M,N
Q	10	N,0
R	5	L,O,P
S	7	Q,R

8.3 From the following network data, determine the critical path, starting and finishing times, and TFs and FFs.

Activity	Description	Duration	Immediate Predecessors
ST	Start milestone	-	-
A	Excavate stage 1	4	ST
B	Order and deliver steelwork	7	ST
C	Formwork stage 1	4	A
D	Excavate stage 2	5	A
E	Concrete stage 1	8	C
F	Formwork stage 2	2	C, D
G	Backfill stage 1	3	E
H	Concrete stage 2	8	E, F
I	Erect steelwork	10	B, H
J	Backfill stage 2	5	G, H
END	End milestone	-	I, J

8.2 Make a clear and neat sketch of the network specified below using precedence notation. On the precedence diagram calculate and show the early start (EST), early finish (EFT), late start (LST), late finish (LFT), and TF in each activity. Start calculations with day zero 0. Show the critical paths with colored pencil.

Label	Duration	Must Follow Operations
A	2	—
B	4	A
C	7	A
D	3	A
E	5	A
F	7	B
G	6	B

8.4 Using the information given in figures shown, develop a precedence notation CPM network for the bridge project described. Certain logical relationships are implied by the bar chart that has been supplied to you by the field superintendent who has been chosen to run the job. According to contract specifications your company must submit a network scheduling (CPM) to the owner. Knowing that you area CPM expert, your boss has given you the job of setting up the network. He also has asked you to calculate the project duration in working days. The duration of each activity can be developed from the dates given on the bar chart.

Mobilization and procurement activities are as follows:

Act	Duration	Description	Followers
1	10	Shop drawings, abutment, and deck steel	11
2	5	Shop drawings, foot steel	6
3	3	Move in	7,8
4	15	Deliver piles	9
5	10	Shop drawings, girders	26
1	15	Deliver abutment and deck steel	16
6	7	Deliver footer steel	12
6	25	Deliver girders	28

Hints and Restrictions

- There is only one excavation crew.
- There is one set of formwork material for footers and abutments.
- Project duration should be approximately 65 days long.
- General sequence of activities:
  - Excavation, piles driven, footer, abutment, deck
  - Forming, pouring, curing, stripping

Example project, bridge profile.

Example project, bridge transverse section for problem 7.4.

Activity	No.	Schedule (June → Sept)
Exc. abut. no. 1	8	17–21 (June)
Make abut. forms	7	17–21 (June)
Exc. abut. no. 2	10	22–23 (June)
Drive piles abut. no. 1	9	5–8 (July)
Forms & steel foot. no. 1	12	9–12 (July)
Drive piles abut. no. 2	13	9–13 (July)
Pour foot. no. 1	14	13 (July)
Strip foot. no. 1	15	14 (July)
Form & steel foot. no. 2	17	15–16 (July)
Pour foot. no. 2	19	19 (July)
Form & steel abut. no. 1	16	20–23 (July)
Strip foot. no. 2	21	20 (July)
Pour abut. no. 1	18	26–27 (July)
Strip & cure abut. no. 1	20	28–30 (July/Aug)
Back abut. no. 1	22	2–4 (Aug)
Rub. conc. abut. no. 1	30	2–4 (Aug)
Form & steel abut. no. 2	23	2–5 (Aug)
Pour abut. no. 2	24	6–9 (Aug)
Strip & cure abut. no. 2	25	10–12 (Aug)
Rub. conc. abut. no. 2	32	13–17 (Aug)
Back abut. no. 2	27	13–17 (Aug)
Set girders	28	13–16 (Aug)
Deck forms & steel	29	17–20 (Aug)
Pour & cure deck	31	23–25 (Aug)
Strip deck forms	33	26–30 (Aug)
Saw contraction joints	34	26 (Aug)
Painting	35	31–7 (Aug/Sept)
Guardrail	36	31–2 (Aug/Sept)
Clean up	37	8–10 (Sept)
Inspection	38	13 (Sept)

Bar chart schedule for bridge project for problem 8.4. (Based on an example in Clough and Sears, *Construction Project Management*).

To calculate the work days from the bar chart, assume June 17 is a Thursday. Workdays are Monday through Friday.

8.5 A new road section with concrete pavement, shown in longitudinal section, is 11,600 feet long. It is to be constructed in accordance with the following conditions:

- The balanced earthworks from station 00 to 58(00) may be done at the same time as the balanced earthworks from station 58(00) to 116(00) using two separate independent crews.
- The double-box culvert will be built by one crew, and another crew will build the two small culverts. Concrete may be supplied either from the paving batch plant or from small independent mixers at the culvert sites, whichever is expedient.
- One small slip-form paver will do all the concrete paving work, and all the shouldering will then follow with one crew after the concrete pavement is cured.
- Seeding the embankments with grass must be left as late as possible.

Prepare a network diagram and determine the minimum possible project duration.

If independent concrete mixers are used for the culverts, what is the latest day for delivery of the paving batch plant to the site, so that the paving crew may have continuity of work (no idle time at all)?

Activity Description	Duration
Deliver rebars—double-box culvert	10
Move in equipment	3
Deliver rebars—small culverts	10
Set up paving batch plant	8
Order and deliver paving mesh	10
Build and cure double-box culvert, station 38	40
Clear and grub, station 00–58	10
Clear and grab, station 58–116	8
Build small culvert, station 85	14
Move dirt, station 00–58	27
Move part dirt, station 58–116	16
Build small culvert, station 96	14
Cure small culvert, station 85	10
Cure small culvert, station 96	10
Move balance dirt, station 58–116	5
Place subbase, station 00–58	4
Place subbase, station 58–116	4
Order and stockpile paving materials	7
Pave, station 58–116	5
Cure pavement, station 58–116	10
Pave, station 00–58	5
Cure pavement, station 00–58	10
Shoulders, station 00–58	2
Shoulders, station 58–116	2
Guardrail on curves	3
Seeding embankments with grass	4
Move out and open road	3

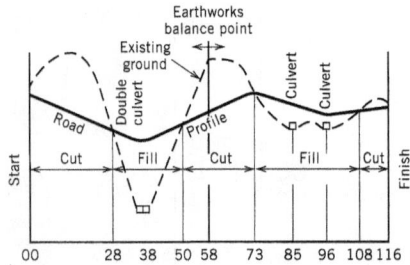

8.6 From the following network data, determine the critical path, early start, early finish, late start, late finish, TF, and FF for each activity. Highlight the critical path.

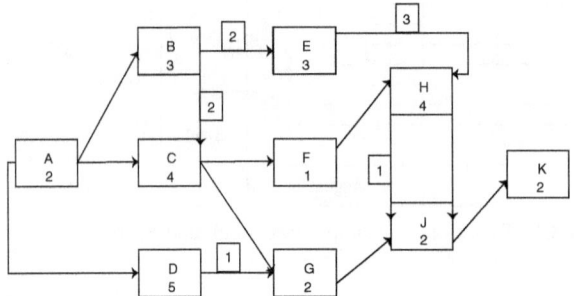

8.7 You need to provide a project's owner with a 3-week lookahead schedule. Identify one benefit and one limitation of using a bar chart for the look-ahead schedule.

8.8 Develop an activity diagram for the following activity list. Label each activity and minimize crossing relationship lines.

Activity	Immediate Succeeding Activities
A	N
B	F
C	K, J
D	G
E	A, M (SS)
F	E
G	M (SS), P
H	None
I	None
J	D, L
K	D, L
L	G
M	H, I
N	I (FF)
O	P (FF)
P	H

Chapter 9

Scheduling: Program Evaluation and Review Technique Networks and Linear Operations

CONTENTS

Lean Construction

Lean Construction

Lean construction has been called a project delivery system, a collection of management techniques, and even a management philosophy. It is based on a successful set of ideas and techniques used in the manufacturing industry, generally called "lean production" or " the Toyota production system." This management approach was created and first implemented by Taiichi Ohno at Toyota Motors in Japan. Ohno installed a system designed to minimize the amount of inventory in the plant and even stop the assembly line at any time a defective part was detected.

Building a construction project, of course, is different from assembling a car. The adaptation of lean techniques for construction has been a gradual process that began in the mid-1990s, with techniques that attempt to mimic the essence of the Toyota system. Much of the current thrust of lean construction lies in its approach to project planning and control. Traditional planning assumes that a detailed plan can be created for the entire duration of a project. Control consists of identifying deviations from the plan, and steering the project execution back into compliance. This philosophy is not too different from a traditional car production line approach. The "production line" is the plan, to which workers must yield even if it leads to suboptimal decisions.

In contrast, lean construction asserts that it is virtually impossible to develop a detailed plan of events that are months, or even weeks, in the future. Instead of the top-down approach of traditional management, top management provides deadlines and other production goals, and implementation details are largely left to the project field personnel. Subcontractors and foremen are supposed to stop the production line (i.e., do an immediate hold and review) in the case of a questionable request. Weekly meetings have protocols to investigate the root causes of execution problems. Middle managers, such as the project's superintendent, are the enablers of the detailed short-term plans developed at the work face. For example, a key duty of middle management is procuring the resources required by the short- and long-term plans.

Implementing lean construction can be difficult, in part because all these roles and procedures seem so logical that many contractors allege that they already use them. However, the orderly, cascading discipline for planning and control advocated by lean construction is rarely in place without a concerted effort. The percent plan complete (PPC) is a popular metric used in lean construction, consisting of the proportion of tasks executed as planned in a given week. A 50% PPC means that only half of the tasks that were planned to be done in a week were actually carried out. It does not imply a lack of activity. As an automobile manufacturing worker in a traditional car production line keeps adding parts to the inventory, each construction crew finds ways to keep busy, even at the expense of what was deemed to be a rational plan just a week before. A 50% PPC is typical for many contractors, revealing a lack of planning reliability, process transparency and communication, among other negative implications.

9.1 INTRODUCTION

Bar charts and critical path method (CPM) networks assume that all activity durations are constant or deterministic.[1] An estimate is made of the duration of each activity prior to commencement of a project, and the activity duration is assumed to remain the same (e.g., a nonvariable value) throughout the life of the project. In fact, this assumption is not realistic. As soon as work begins, due to actual working conditions, the assumed durations for each activity begin to vary. The variability of project activities is addressed in a method developed by the U.S. Navy at approximately the same time as CPM. This method was called the Program Evaluation and Review Technique (PERT).

PERT incorporates means to consider uncertainty in the project planning by assuming that the duration of some or all of the project activities can be variable. The variability is defined in terms of three estimates of the duration of each activity as follows:

1. Most pessimistic duration.
2. Most optimistic duration.
3. Most likely duration.

Let's assume that a 20,000-square-foot slab on grade is to be cast in place. For scheduling purposes, the project superintendent is asked for estimates of the three durations above (i.e., most pessimistic, optimistic and likely) rather than for a single constant duration. The three estimates are used to calculate an *expected* activity duration. The calculations are loosely based on concepts from mathematical probability. The expected duration, t_e, is assumed to be the average value of a probability distribution defined by the

---

[1] Logic is also considered to be constant or invariable throughout the life of the project for each activity.

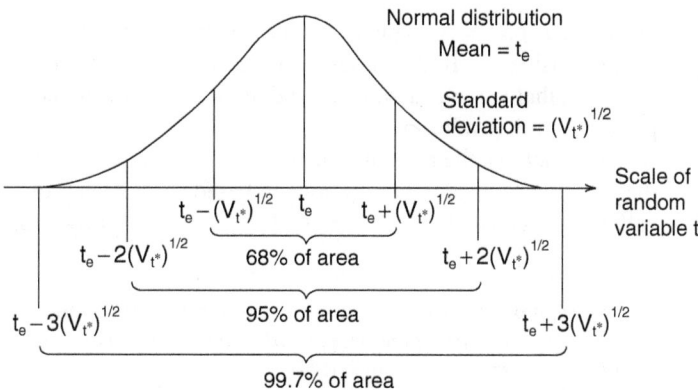

Figure 9.1 Selected areas under the normal distribution curve.

three-estimate set. The expected duration, t_e, of each activity with variable characteristics is given as:

$$t_e = [t_a + 4t_m + t_b]/6$$

Where

t_a is the most optimistic duration estimate

t_m is the most likely duration estimate

t_b is the most pessimistic duration estimate

For instance, if for the slab pour, the three estimates from the superintendent are:

$t_a = 5$ days

$t_m = 8$ days

$t_b = 12$ days

The expected activity duration is calculated as:

$$t_e = (5 + 4(8) + 12)/6 = 49/6 = 8.17 \text{ days, say } 9 \text{ working days}$$

The expected value for each activity with a constant value is t_e = constant.

Once the t_e values for each variable duration activity have been calculated, the longest path and project duration are determined using the same methods used by CPM. The probability of completing the project within a predetermined time duration is calculated by assuming that the probability distribution of the total project duration is normally distributed with the longest path of t_e values as a mean value of the normal distribution.

The normal distribution is defined by its mean value x (i.e., in this case the value of the longest path through the network) and the value, σ, which is the so-called *standard deviation* (σ) of the distribution. The standard deviation is a measure of how widely about the mean value the actual observed values are spread or distributed. Another parameter called the *variance* is the square of the standard deviation, i.e., σ^2. It can be shown mathematically that 99.7% of the values of normally distributed variables will lie in a range defined by three standard deviations below the mean and three standard deviations above the mean (see Figure 9.1).

In PERT, the standard deviation σ of the normal distribution for the total project duration is calculated using the variance of each activity on the critical path. The variance[2] for each PERT activity is defined as:

$$\sigma^2 = [(t_b - t_a)/6]^2$$

---

[2]The variance is the standard deviation squared.

where the expression inside the brackets approximates the standard deviation of the activity being considered. If the variance of each activity on the longest (i.e., critical) path is summed, that value is assumed to be the variance of the normal distribution of the entire project duration values.

The fact that the normal distribution is used to represent the probability distribution of the possible total project durations is based on a basic concept from probability theory called the *Central Limit Theorem*. This is explained by Moder and Phillips as follows:

> *Theorem*
>
> *Suppose m independent tasks are to be performed in order; (one might think of these as the m tasks that lie on the critical path of a network). Let $t_1^*, t_2^*, \ldots tm^*$ be the times at which these tasks are actually completed.*
>
> *Note that these are random variables with true means $t_1, t_2, \ldots, t_m$, and true variances $V_{t1}^*, V_{t2}^*, \ldots, V_{tm}^*$, and ... actual times are unknown until these specific tasks are actually performed. Now define T^* to be the sum:*
>
> $$T^* = t_1^* + t_2^* + \cdots + t_m^*$$
>
> *And note that T^* is also a random variable and thus has a distribution. The Central Limit Theorem states that if m is large, say four or more, the distribution of T^* is approximately normal with mean T and variance VT^* given by*
>
> $$T = t_1 + t_2 + \cdots + t_m$$
>
> $$V_T^* = V_{t1}^* + V_{t2}^* + \cdots + V_{tm}^*$$
>
> *That is, the mean of the sum, is the sum of the means; the variance of the sum is the sum of the variances; and the distribution of the sum of activity times will be normal regardless of the shape of the distribution of actual activity performance times (Moder and Phillips, 1964).*

9.2   AN EXAMPLE PROGRAM EVALUATION AND REVIEW TECHNIQUE NETWORK

To demonstrate the use of the PERT approach, consider the small arrow notation network[3] shown in Figure 9.2. Three estimate durations for each activity in this activity network are given in Table 9.1.

The t_e values shown for each activity are calculated using the formula:

$$t_e = (t_a + t_m + t_b)/6$$

For instance, t_e for Activity 7 is:

$$t_e = (4 + 4(7) + 13)/6 = 45/6 = 7.5$$

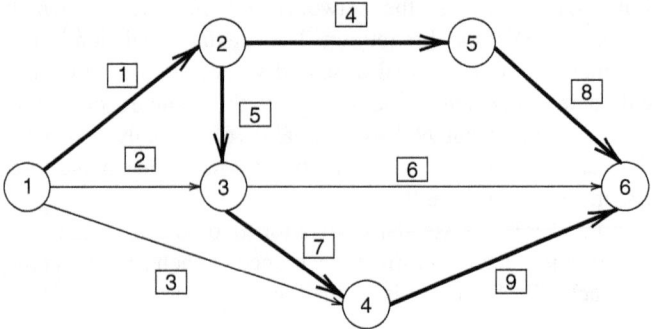

Figure 9.2 Small PERT network.

---

[3] See Appendix E describing the use of arrow notation.

Table 9.1 Three Estimate Values and Calculated Values for Each Activity

Activity	t_m	t_a	t_b	t_e	Var
1	3	1	5	3	0.44
2	6	3	9	6	1.00
3	13	10	19	13.5	2.25
4	9	3	12	8.5	2.25
5	3	1	8	3.5	1.36
6	9	8	16	10	1.23
7	7	4	13	7.5	2.25
8	6	3	9	6	1.00
9	3	1	8	3.5	1.36

The variance for each activity is approximated by the equation:

$$V = [(t_b - t_a)/6]^2$$

For Activity 7, the variance is:

$$V_7 = [(13 - 4)/6]^2 = 81/36 = 2.25$$

Using the forward and backward pass methods described in Chapter 8, two paths have an expected duration of 17.5 days. These paths are shown below.

	DURATION	**VARIANCE**
Path 1	$3 + 8.5 + 6 = 17.5$	$Var = .444 + 2.25 + 1.0 = 3.694$
(1-4-8)		
Path 2	$3 + 3.5 + 7.5 + 3.5 = 17.5$	$Var = .444 + 1.361 + 2.25 + 1.361 = 5.416$
(1-5-7-9)		

The mean of the normal distribution is therefore assumed to be 17.5 days. The variances of the two longest paths are calculated by adding variances of the individual activities in each path. The variance of path two (5.416) is greater than that of path one (3.694). Because this means a greater spread of the probable total project durations, the variance of path two is selected as the variance to be used for further PERT calculations. The PERT normal distribution for total project duration is shown in Figure 9.3.

The normal distribution is symmetrical about the mean. The standard deviation will be

$$\sigma = \text{SQRT}(5.416) = 2.327.$$

PERT answers the question: "What is the probability (given the variable durations of the activities) that the project can be completed in N days?" The probability of

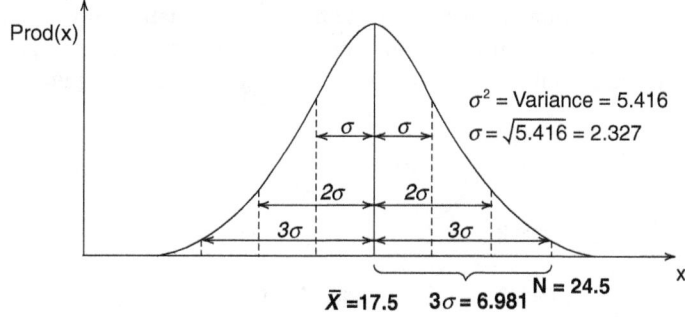

Figure 9.3 Normal distribution of total project durations for small PERT network.

completing the project is given by the area under the normal distribution to the left of the value N selected for investigation. Because we know that 99.7% of the area (representing probability) under the normal distribution is in the range of 3σ below the mean and 3σ above the mean, we can say that there is a better than 99.7% chance that the project can be completed in $[x + 3\sigma]$ or $[17.5 + 3(2.327)] = 24.5$ days or less. That is, at least 99.7% of the area under the normal curve is to the left of the value 24.5 days in Figure 9.3. In other words, we can be almost 100% sure that the project can be completed in 25 days or less.

What if we want to know the probability of completing in 19 days? Given the values of the mean and the variance, we can use a cumulative normal distribution function table such as that shown in Appendix K to calculate the area under the curve left of the value 19. First, we must calculate the Z value which for a given value x (e.g., 19):

$$Z = \frac{\text{Mean} - x}{\sqrt{\text{Variance}}} \quad \text{or} \quad Z = \frac{(\bar{X} - x)}{\sigma}$$

where σ is the standard deviation of the cumulative normal distribution. In our case:

$$Z = \left|\frac{17.5 - 19}{\sqrt{5.416}}\right| = \frac{1.5}{2.327} = 0.644$$

Consulting the cumulative normal distribution function table given in Appendix J with a Z value of 0.644, yields a value of .7389 or 73.89% probability of completing the project in 19 days. What would be the probability of completing the project in 16 days?

9.3   PROGRAM EVALUATION AND REVIEW TECHNIQUE SHORTCOMINGS

In fact, the PERT results are too optimistic. The method of using the t_e value to determine the longest path through the project network and assuming that the duration of this path is the most probable value for the total project duration is not totally accurate. Although PERT introduces elements of probability into the calculation of the project duration, it consistently underestimates the duration. The principal cause of this underestimation is a condition known as *merge event bias*. Briefly, merge event bias occurs when several paths converge on a single node. Figure 9.4 is a simplified depiction of how several paths in a schedule network might converge on a single node.

PERT calculations give the early expected finish time of this node as the summation of times on the longest path leading to the node. This path then becomes part of the longest path through the network that determines expected project duration. However, because the duration of the activities on the paths are random variables, it is possible that some other path converging on the node could have an activity with a random duration longer than its expected (mean) duration. Thus, this longer path would determine the early finish time of the node. That this potential longer path is *not* taken into account in the PERT calculation leads to an underestimation of project duration.

Additionally, the PERT method assumes statistical independence between activities. This assumption allows the variance of activities along a path to be added, giving the variance of the duration of the project. The assumption of independence, however, may not

Figure 9.4 Merge event bias.

always be appropriate. For instance, weather can create a positive correlation between activities, and a delay in one activity may create a negative correlation between activities.

One solution to the difficulties noted above is computer simulation. Because Monte Carlo simulation of schedule networks does not use a single number to represent activity durations, it avoids the merge event bias described previously.

9.4   LINEAR CONSTRUCTION OPERATIONS

Often construction sites have linear properties that influence the production sequence. Road construction, for instance, is worked in sections that require that a set of work processes be completed in a particular sequence before the section is completed. The individual sections can be thought of as "processing through" a series of workstations.

For example, a road job may be subdivided into 14 sections that must be completed (see Figure 9.5[a]). This type of breakdown is typically established based on centerline stationing (e.g., section 1 is defined as running from station 100 to station 254.3). Each of the 14 stations must undergo the following work activities: (a) rough grading, (b) finish grading, (c) aggregate base installation, (d) 5-inch concrete pavement, (e) 9-inch concrete pavement, and (f) curb installation.

Each of the 14 sections can be thought of as being processed by crews and equipment representing each of the six work processes. Because the site is linear, the normal way for the work to proceed would be to start with section 1, then go on to the second section and so forth. This implies that the sections will first be rough graded, then finish -graded, then aggregate base will be placed, and so on. A bar chart indicating this sequence of activity is shown in Figure 9.5(b).

The bar chart indicates that work activity overlaps such that several operations are in progress simultaneously during the middle of the job. The required sequential nature leads

(a)

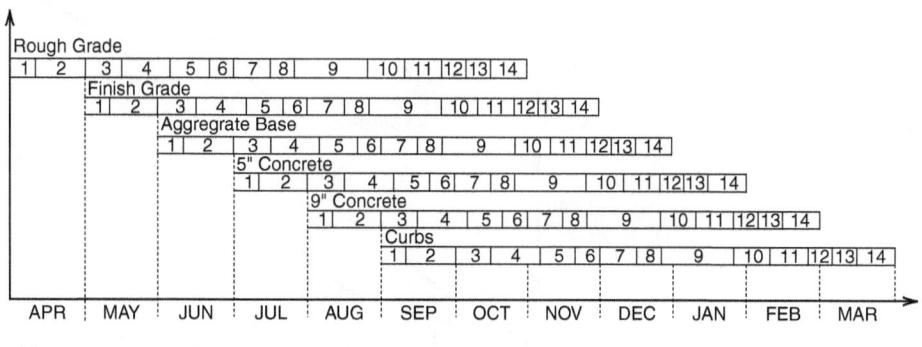

(b)

Figure 9.5 Road project divided into 14 sections.

to a "boxcar" effect. That is, a section must complete 5-inch concrete before proceeding to 9-inch concrete. Therefore, the sections can be thought of as a "train" or "parade" of work that must pass each of the stations represented by the six construction processes.

Many types of projects exhibit this kind of rigid work sequence. A high-rise building, for instance, requires each floor to pass through a set of operations. Each floor can be thought of as a "car" in the "train" of work to be completed. Construction processes such as erect formwork, install reinforcing steel and embedments, and pour concrete can be viewed as workstations through which each floor must pass.

Tunnels are worked in sections in a fashion similar to road or pipeline work. Each section must be processed through work processes such as drill, blast, remove muck, and advance drilling shield. This again leads to a repetitive sequence that is rigidly sequenced.

9.5   PRODUCTION CURVES

Bar charts and network schedules provide only limited information when modeling linear operations and projects. They typically do not readily reflect the production rate or speed with which sections or units are being processed. Because the rate of production will vary across time, this has a major impact on the release of work for following work processes. Delays in achieving the first units of production occur as a result of mobilization requirements. As the operation nears completion, the rate of production typically declines because of demobilization or closeout considerations. The period of maximum production is during the mid-period of the process duration. This leads to a production curve with the shape of a "lazy S," as shown in Figure 9.6. The slope of the curve is flat at the beginning and the end, but steep in the midsection. The slope of the curve is the production rate.

These curves are also called time-distance, time-quantity, or velocity diagrams because they relate units of production (i.e., quantities or distance) on the y axis (vertical, ordinate) with time plotted on the x axis (horizontal, abscissa). The slope of the curve relates the increase in production units on the y axis with the increment of time as shown on the x axis. The slope of the curve, therefore, represents the number of units produced over a given time increment. This is the rate of production.

The production curves for a typical road job are shown in Figure 9.7. The curves indicate the beginning and ending points in time for each of the processes. The slope of each curve is

Figure 9.6 Production curve.

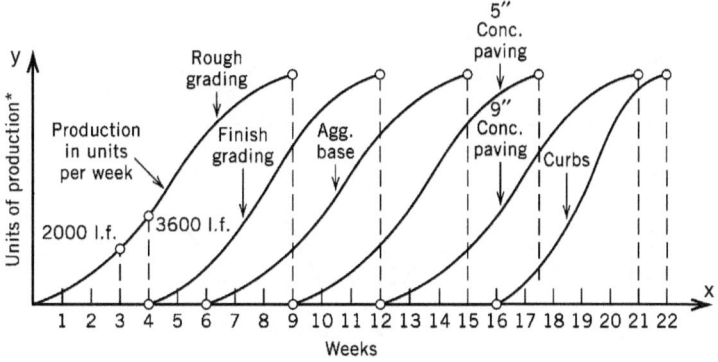

Figure 9.7 Velocity diagrams for a road construction project.

* Units of production (e.g., l.f. for rough and finish grading, tons of aggregate base, cu. yd. or sq. ft. of paving, etc.)

the production rate for each process. The distance between the beginning points of each process establishes the lag between processes. The aggregate base operation begins in week 6 and lags the finish-grading operation by 2 weeks. This means that 2 weeks of work (i.e., completed finish-grade sections) are built up before the aggregate base operation is started.

Leading processes generate work area or availability so that follow-on processes have a "reservoir" of work from which to operate. Reservoirs of work are cascaded so that units of work must be available from an "upstream" process reservoir before work is available at a following process reservoir. This illustrates that work flow moves from leading to following processes.

In addition to indicating the rate of production, production curves or velocity diagrams are helpful in establishing the project status. The planned status of the job as of week 12 can be determined by simply drawing a vertical line at week 12 on the x axis of Figure 9.7. This will intersect the aggregate base and 5-inch concrete curves. It also represents the beginning of work on the 9-inch concrete pavement (overlaying the 5-inch base concrete). It can be readily determined that:

1. Both rough and finish grading should be completed.
2. Approximately 80% of the aggregate base has been placed.
3. Placement of the 5-inch concrete base is approximately 30% complete.
4. Placement of 9-inch concrete is just commencing.

The planned status of construction as of week 12 is shown in Figure 9.8.

There is a definite advantage in balancing the production rates between processes. Balancing rates means ensuring that the slopes of the production curves that interact are roughly parallel and do not intersect. If rates are not balanced, the situation shown in

Figure 9.8 Planned status of construction as of week 12.

Figure 9.9 Unbalanced process production rates.

Figure 9.9 can develop. In this example, the slope (production rate) of process B is so steep that it catches or intersects the process A curve at time M. This requires a shutdown of process B until more work units can be made available from A. Again, at time L, process B overtakes the production in process A, resulting in a work stoppage. This is clearly inefficient since it requires the demobilization and restarting of process B. The stoppage of process B at M also causes a "ripple" effect since this causes a shutdown of process C at time N.

It should be clear that these stoppages are undesirable. Thus, processes should be coordinated so as to avoid intersections of production curves (e.g., times M, N, and L). Obviously, one way to avoid this is to control production in each process so that the slopes of the curves are parallel. This implies the need to design each process so that the resources used result in production rates that are roughly the same for all interacting construction processes. Because the six curves for the road job are roughly parallel, we can assume that the production rates have been coordinated to avoid one process overtaking its leading or preceding process.

9.6   LINE OF BALANCE CONCEPTS

Line of balance (LOB) is a graphical method for production control integrating bar-charting and production curve concepts. It focuses on the planned versus actual progress for individual activities and provides a visual display depicting differences between the two. Indication of these discrepancies enables management to provide accurate control in determining priorities for reallocation of labor resources. Those activities indicated ahead of schedule can be slowed by directing part or all of their labor crews to individual activities that lag behind schedule. This obviously assumes that resources are interchangeable. This can present a limitation to the application of this procedure in construction.

The LOB method serves two fundamental purposes. The first is to control production and the second is to act as a project management aid. Each of these objectives is interrelated through development and analysis of four LOB elements. These elements provide the basis for progress study on critical operations throughout the project duration. The four elements are:

1. The objective chart.
2. The program chart.

Figure 9.10 Objective chart.

3. The progress chart.

4. The comparison.

The *objective chart* is a straight-line curve showing cumulative end products to be produced over a calendar time period. The number of end products may be specified in the contract. Assume that the units being considered in this example are precast panels for the exterior of a high-rise building. A typical objective chart is shown in Figure 9.10.

This example indicates a total of 30 units to be delivered or completed by June 1; 60 units to be delivered and completed by July 1; 180 units to be delivered and completed by November 1; and a total of 240 units to be delivered and completed by February 28. The contract start date is shown as March 1.

The *program chart* is the basic unit of the LOB system. It is a flow process chart of all major activities, illustrating their planned, sequenced interrelationships on a "lead-time" basis. Three aspects to consider in development of the program chart are determination of (a) operations to be performed, (b) the sequence of operations, and (c) processing and assembly lead time.

The program chart indicated in Figure 9.11 describes the production process for the 240 units mentioned in the objective chart. Each activity (A through E) has associated with it a lead time (latest start time) signified by an event starting symbol (☐) and an event coordination symbol (△) signifying its end or completion. These event coordination

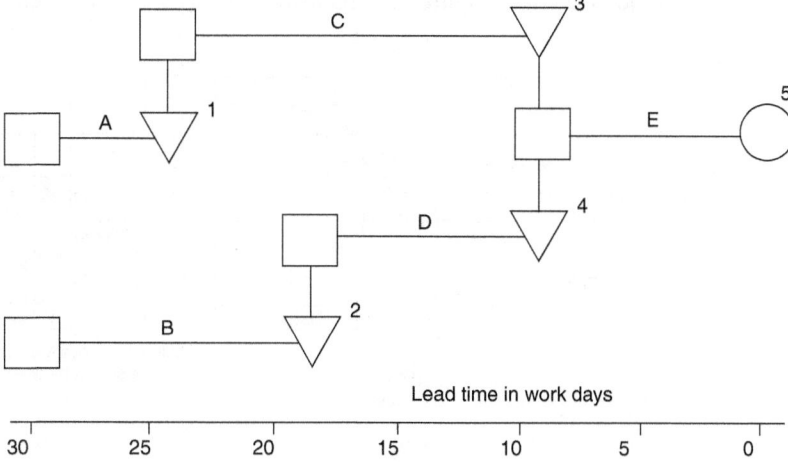

Figure 9.11 Program chart with lead time in workdays.

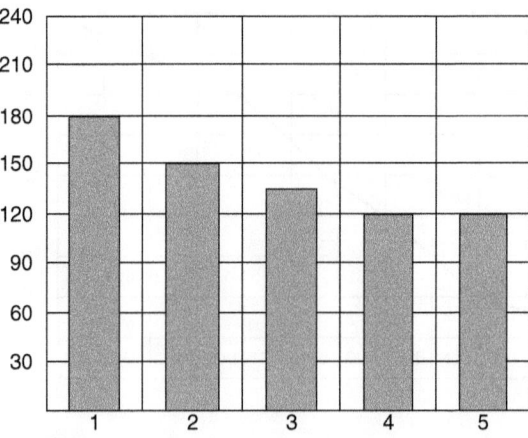

Figure 9.12 Progress Chart.

symbols, referred to as *progress monitoring points*, are labeled from top to bottom and from left to right. All five activities must be completed before one unit can be ready for delivery. This takes 30 working days as shown on the program chart's lead-time scale.

The *progress chart* is drawn to the same vertical scale as the objective chart and has a horizontal axis corresponding to the progress monitoring points labeled in chronologic order. Vertical bars represent the cumulative progress or status of actual performance at each monitoring point, usually based on visiting the site and measuring actual progress (e.g., assessing status of completion).

The progress chart of Figure 9.12 indicates that on a given day when inventory was taken, 120 units had passed through monitoring point 5. In other words, the vertical height of bar 5 is equal to the number of units actually having completed station 5. This corresponds to activity E in the program chart, which is the last activity in the production process. Similarly, activity D (bar 4) had completed 120 units and activity C (bar 3) had completed 130 units; activity B (bar 2) had completed 150 units; and activity A (bar 1) had completed 180 units.

In the comparison, actual progress is compared to expected progress. The objective, program, and progress charts are then used to draw the LOB by projecting certain points from the objective chart to the progress chart. This results in a step-down line graph indicating the number of units that must be available at each monitoring point for actual progress to remain consistent with the expected progress as given by the objective chart. Figure 9.13 indicates the LOB and the method used to project it from

Figure 9.13 Progress Chart with Line of Balance.

the objective chart to the progress chart. The procedure for striking the LOB is as follows:

1. Plot the balance quantity for each control point.
    a. Starting with the study date (e.g., 1 Sept) on the horizontal axis of the objective chart, mark off to the right the number of working days (or weeks or months, as appropriate) of lead time for that control point. This information is obtained from the program chart.
    b. Draw a vertical line from that point on the horizontal axis to the cumulative objective curve.
    c. From that point draw a horizontal line to the corresponding bar on the progress chart. This is the balance quantity for that bar.
2. Join the balance quantities to form one stair-step-type line across the progress chart.

Analysis of the LOB reveals that activities 2 and 5 are right on schedule, whereas activities 3 and 4 show values below the number of units required. Activity 1 shows surplus. This surplus is the difference between the 180 units actually completed by activity 1 and the 157 units indicated as necessary by the LOB. On the other hand, activities 3 and 4 are lagging by 5 and 15 units, respectively. The LOB display enables management to begin corrective action on activities 3 and 4 to ensure that they do not impede the progress rate of the remaining units.

9.7   LINE OF BALANCE APPLIED TO CONSTRUCTION

To illustrate the use of LOB in a construction context, consider a high-rise building in which repetitive activity sequences are a part of the floor-to-floor operation. To ensure a smooth flow of production, a schedule is needed that accounts for the interrelationships between different activities. This becomes even more obvious when an additional constraint such as limited formwork is involved. Each floor consists of four sections (A, B, C, and D). These sections can be viewed as processed units.

Each floor section must be processed through the following work activities:

1. Erect forms.
2. Place reinforcing steel.
3. Place concrete.
4. Dismantle forms.
5. Place curtain wall (exterior façade).
6. Place windows.

Figure 9.14 shows a schematic of the status of activities at a given point in time. At the time illustrated, work is proceeding as follows:

1. Erect forms section A, floor $N + 5$.
2. Place reinforcing steel, section D, floor $N + 4$.
3. Place concrete section C, floor $N + 4$.
4. Dismantle forms sections B, floor $N + 1$.
5. Place curtain wall section D, floor N.
6. Place windows section D, floor $N - 1$.

Crews proceed from section A to B to C to D.

The diagram in Figure 9.15 shows the LOB objective chart for a 10-story building. The program chart for a typical section is shown above the objective chart. During the first

Figure 9.14 Schematic of floor cycle work tasks.

2 weeks the floor cycle required is one floor (four sections) per week. For weeks 2 through 6 the rate of floor production is 1.5 floors (six sections) per week. That is, six floors must be completed in the 4-week period from week 2 to week 6. In the last 2 weeks, the rate is reduced to one floor per week. The lead times required for various activities are shown on the bar program chart above the objective. To strike a LOB for the beginning of week 5, the lead times are projected as described in Section 9.5. A diagram of this projection is shown in Figure 9.16. The LOB values can be calculated by determining the slope relating horizontal distance (lead time) to vertical distance (required sections). The slope of the objective during weeks 5 to 6 is six sections (1.5 floors) per 40 hours (1 week) or 6/40 sections per hour.

Figure 9.15 Program chart and objective.

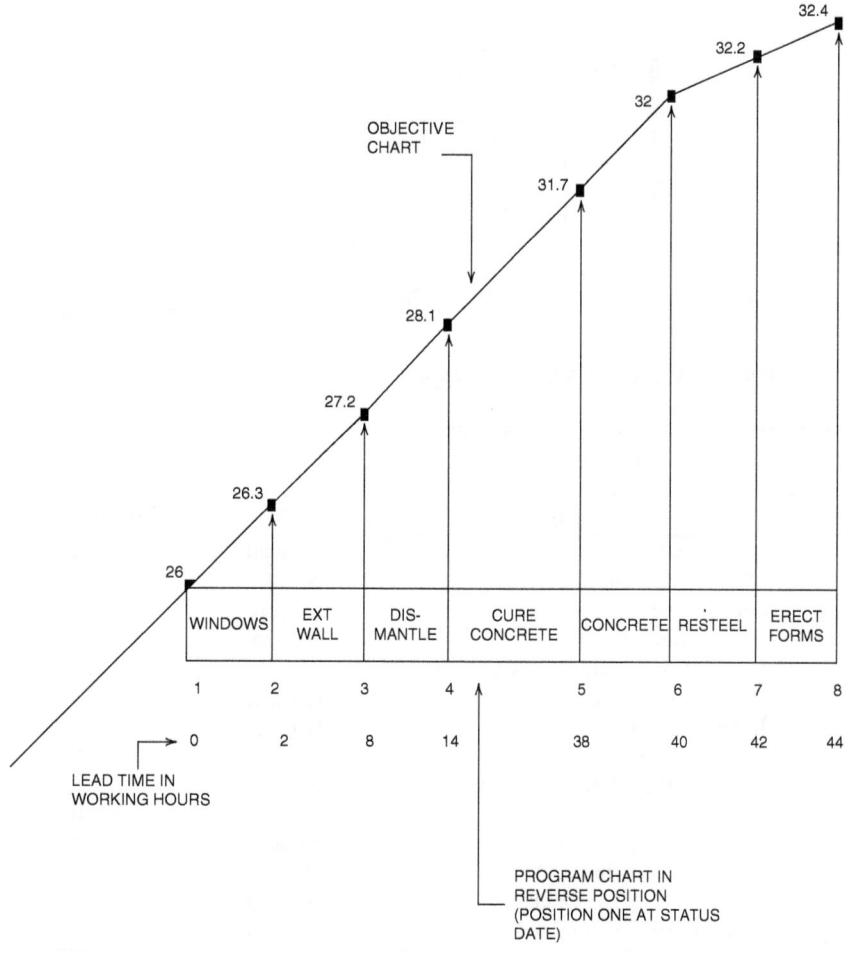

Figure 9.16 Enlarged projection of program chart onto objective chart.

During the remaining weeks, the slope is four sections (one floor) per 40 hours or 1/10 section per hour. The LOB for control point I is given as:

LOB(I) = Section completed as for week 5 + [(slope) × (lead time of control point I)]

The number of sections to be completed as of week 5 is 6.5 or 26 sections (6.5 × 4). Therefore,

LOB(1) = 26
LOB(2) = 26 + (6/40)2 = 26.3
LOB(3) = 26 + (6/40)8 = 27.2
LOB(4) = 26 + (6/40)14 = 28.1
LOB(5) = 26 + (6/40)38 = 31.7
LOB(6) = 26 + (6/40)40 = 32

Control points 7 and 8 plot to the flatter portion of the objective:

$$LOB(7) = 32 + (1/10)(42 - 40) = 32.2$$
$$LOB(8) = 32 + (1/10)(44 - 40) = 32.4$$

The LOB for week 5 is shown in Figure 9.17. Field reports would be used to establish actual progress, and a comparison will determine whether actual progress is consistent with expected progress.

Figure 9.17 Line of balance for week 5.

REVIEW QUESTIONS AND EXERCISES

9.1   (a) Using PERT, calculate expected project duration and determine the critical path in the network defined by the data below. (b) What is the probability of completing this project in 32 weeks or less?

Activity #	Activity	Type	Duration (Weeks)	Followed by Act #
10	Prefab Wall Forms	constant	2	40
20	Excavate Cols and Walls	constant	3	50, 60, 70
30	Let Elec and Mech Subcontract	t_a, t_m, t_b	3, 4, 8	60, 70
40	Deliver wall Forms	constant	4	80, 90, 100
50	Forms, Pour & Cure Wall & Col Fig	t_a, t_m, t_b	6, 7, 8	80, 90, 100
60	Rough-in Plumbing	t_a, t_m, t_b	5, 7, 10	110
70	Install Conduit	$t_a, t_m\ t_b$	9, 11, 15	110
80	Erect Wall Forms & Steel	constant	9	110
90	Fabricate & Set Interior Column Forms	constant	6	120
100	Erect Temporary Roof	t_a, t_m, t_b	12, 16, 18	140
110	Pour, Cure & Strip Walls	constant	10	130
120	Pour, Cure & Strip Int. Walls	constant	6	140
130	Backfill for Slab on Grade	constant	1	140
140	Grade & Pour Floor Slab	constant	5	END

9.2   (a) Given the data below for a small pipeline project, based on a PERT analysis what is the expected project duration? (b) What is the probability of completing this project in 120 days or less?

Activity	Description	t_a	t_m	t_b	Followed by Activity
1	Start	0	0	0	2
2	Lead Time	10	10	10	3, 4, 5
3	Move to Site	18	20	22	6
4	Obtain Pipes	20	30	100	8, 9, 10
5	Obtain Valves	18	20	70	11
6	Lay Out Pipeline	6	7	14	7
7	Dig Trench	20	25	60	8, 10
8	Prepare Valve Chambers	17	18	31	11
9	Cut Specials	7	9	17	11
10	Lay Pipes	18	20	46	12
11	Fit Valves	8	10	12	13, 14
12	Concrete Anchors	11	12	13	13, 14, 15
13	Finish Valve Chambers	8	8	8	17
14	Test Pipeline	5	6	7	16
15	Backfill	8	10	20	16
16	Clean Up	2	3	10	17
17	Leave Site	3	4	5	18
18	End				

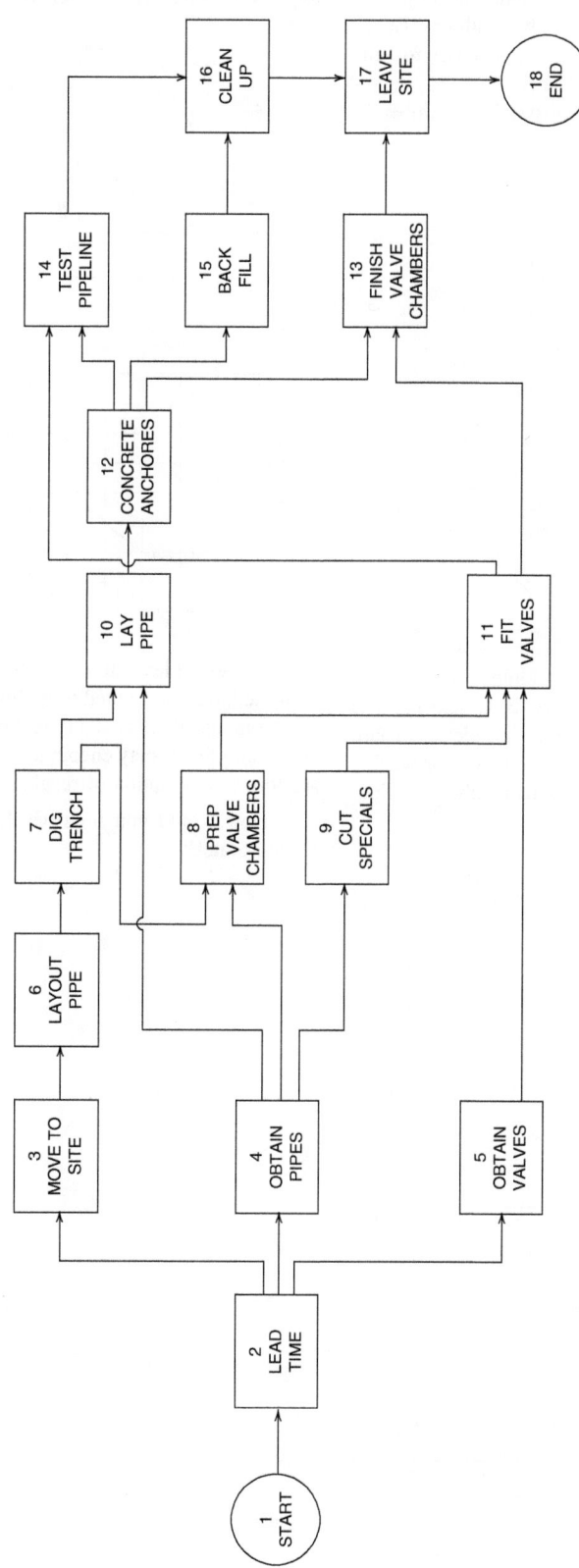

Network For Problem 8.2

169

9.3 Consider the road job described in Figure 9.5, which consists of 14 road sections to be completed. The objective chart for this job is given. Assume that each month consists of 20 working days on the average. The program chart for this process is also shown. Calculate the LOB for the study representing the beginning of month 9 (i.e., day to right of number 8).

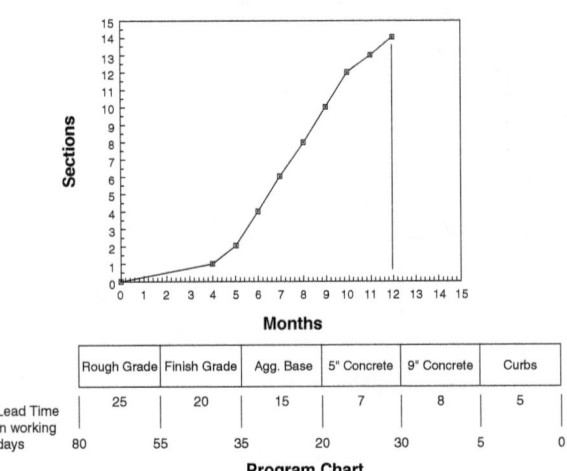

Months

	Rough Grade	Finish Grade	Agg. Base	5" Concrete	9" Concrete	Curbs
	25	20	15	7	8	5
Lead Time in working days	80	55	35	20	30	5	0

Program Chart

9.4 The objective chart for a 10-story building is show below. Each floor is divided into four sections (A, B, C, and D). The production for a typical section is shown in the objective chart. Calculate the LOB values for control points 1 to 8 for week 5 (200 hours). Determine the LOB values in numbers of floor sections.

Typical floor w/ 4 sections

Control Pts	1	2	3	4	5	6	7	8
	ERECT FORMS	RESTEEL	CONCRETEL	CURE CONCRETE	DISMANTLE FORMS	BRICK	WINDOWS	
	120	112	102	96	24	10	4	0
Lead Time in hours	2 hrs	2 hrs	2 hrs	72 hrs	6 hrs	6 hrs		

9.5 Given the following charts, calculate the LOB quantities for July 1. What are these charts called?

Lead time in Work Days

9.6 Discuss at least three disadvantages of PERT that, in your opinion, have contributed to its limited use. Do not include the shortcomings discussed in Section 8.3. Instead, discuss the challenges that it may encounter in its data collection, software support, interpretation of results, and similar practical issues.

9.7 How would you improve the popularity of LOB in the United States?

Chapter 10

The Mathematics of Money

CONTENTS

Using Life Cycle Cost to Assess Sustainable Construction Costs

Sustainable or "green" construction centers on the efficient use of resources that create and support the built environment. The efficient use of energy, such as the electricity required to run a home, is a key aspect of sustainable construction especially during the occupancy phase of a building's life cycle. Residential construction accounts for about one-third of total expenditures in the U.S. construction industry. Unfortunately, the rate of electricity losses in U.S. homes have increased at a higher rate than the consumption of electricity, as shown in the figure (Source: U.S. Department of Energy's Annual Energy Review, 2008). This means that despite the increased public awareness of sustainability issues, homes have become less energy efficient over time.

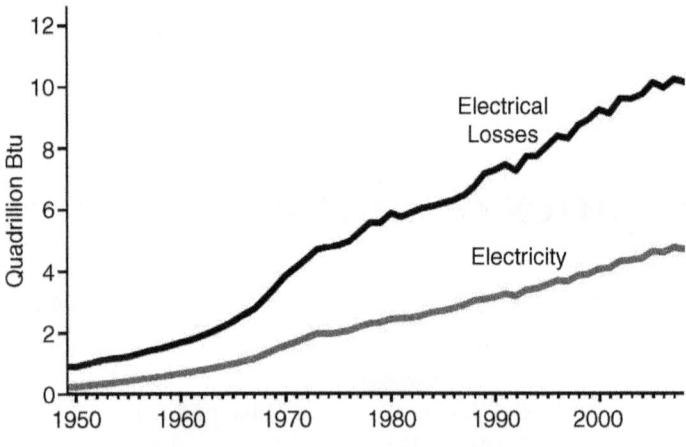

(a) Electrical losses in US residences, 1950–2008

(b) Electric Demand Water Heater

Life Cycle Cost (LCC) is a financial technique increasingly used in sustainable construction analyses to compare high-efficiency construction materials, systems, and designs with traditional alternatives that initially may be less expensive but have higher operational and maintenance costs. As its name implies, LCC examines varying alternatives that impact the overall costs and savings of a process, product, or process over a given time frame. In addition to considering an appropriate time frame, one must include an estimate of the cost of capital and in some cases an inflation rate. LLC analysis depends on a thorough understanding of the time value of money, which will be discussed in detail in this chapter.

A common dilemma confronted by homeowners considering newer green sustainable building systems is whether the use of these technologies can be financially justified. For example, a water heater system which does not use a tank (shown schematically in the figure) is more energy efficient than a traditional water heater, and therefore, preferable from a sustainability perspective. But, is it a good idea to install one? The initial cost of this new technology is considerably higher than the initial cost of a traditional water heater, and many owners would be hesitant to adopt it. For example, the owner of a new residence in Fort Collins, Colorado had this dilemma and decided to analyze the system using LCC. He found that for his particular circumstances, the LCC of the new system was lower than the cost of the traditional hot water tank alternative and installed a high-efficiency unit. In this type of analysis, LCC can be an important tool for the objective assessment of sustainable technologies.

10.1   INTRODUCTION

Management of money is critical to the success of any company. Therefore, a knowledge of the role money plays in the day to day operations of a construction company is of the utmost importance to a construction manager. Issues such as the borrowing of money, expenditure of money for expansion of company operations, and similar topics play a major role in the success of any construction enterprise. The role of money will be examined in this and the following two chapters.

Construction firms acquire funds, as most individuals do, by borrowing money from banks and similar lending sources. It is typical, for instance, to borrow money to finance operations in the field pending receipt of progress payments from the client. As will be discussed in the next chapter on project cash flow, this type of "inventory financing" allows a contractor to cover overdrafts that occur during a project when the progress payments from the client lag behind the costs of construction while work is in progress.

A more long-term variety of borrowing occurs when purchasing new equipment or expanding staff. A construction company may decide to move into a new geographical location or industry sector (e.g., waste water treatment plant construction, bridge construction, and so on) to pursue work. To do this, new equipment and a new staff of professionals will be needed. This can be financed out of funds available within the firm (i.e., the company's net worth). In many cases, however, it may be decided to borrow this money (much as a farmer borrows money for seed at the beginning of the planting season) and to retire the debt based on revenues generated by the new company office or construction sector activity.

When money is borrowed, the lender will require a fee for the use of the borrowed funds.

> The practice of charging a fee for borrowing money goes back to the dawn of recorded history: indeed much of the earliest mathematical literature known to us deals with questions related to interest.[1]

Anyone who has used a credit card is familiar with the concept of interest. The payment of interest over time for use of money has generated a branch of mathematics, which quantitatively analyzes the impact of the time value of money. This chapter addresses the mathematics of borrowing and investment.

10.2   INTEREST IN ACTION

Interest and its application over time impact the cost of any transaction involving the lending, borrowing, and investment of money. In fact, the cost of money and the time during which money is tied up in any business decision process are crucial financial management factors. To illustrate this point, consider the following cases.

1. Assume you are buying a new car. The purchase price is $26,050 and you plan to pay this amount over a 60-month time period. One option involves a $1,000 cash discount and a loan at the rate of 6.94%. A second option is to finance the car with a 5.9% interest loan but no discount. Which option is better? *(Answer: the two are the same; you'd pay around $500 per month taking either one)*

2. What will be the balance of your mutual funds account when you retire at 65 if you are 25 now and begin depositing $100 per month right away? Assume a 12% annual interest. *(Answer: Over a million dollars)*

3. What is the down payment for a house worth $250,000, if you can get a 30-year fixed interest rate of 7.50% and want to cap the payments to $1,500 per month? *(Answer: Exactly $35,473.56)*

---

[1] Maor, E. (1994) *The story of a number*. Princeton: Princeton University Press.

Similar decisions are frequently faced by any business enterprise. Typical questions relating to financial management are:

1. Should my company continue leasing this building, or should it build a new one?
2. Should the firm rent or purchase a certain piece of equipment?
3. Is this 3-year project with a profit of $1 million better than another that can be finished in a year with a $300,000 profit?

The discipline called *engineering economy* considers the time and cost of borrowing as they relate to these and many other questions involving money as a construction resource. The management of expensive, long-term construction projects definitely requires a thorough understanding of the principles and techniques presented here. Money management on complex long-term projects is too important to be dealt with using "horseback questimates." The mathematical approach of engineering economy is increasingly useful even for small, personal decisions like those described above. Calculators, software, and thousands of websites provide support in dealing with the topics covered here. Construction managers need to have a good working understanding of the concepts discussed in this chapter.

10.3   TIME VALUE OF MONEY

Let us consider how interest and time influences many everyday monetary decisions. If someone asked you to decide between receiving $1,000 right now or receiving it 1 year from now, wouldn't you choose to get the money right away? You could invest the money, pay off a debt, or simply keep it until there is a good reason to spend it.

If you were lending $1,000 for 1 year to a perfect stranger, wouldn't you expect to receive more than $1,000 back? After all, you could have invested this money in a savings account at the bank or in your own business. The borrower would have had the potential to invest this money, whereas you would be deprived from this possibility. In short, it is reasonable that anyone lending money asks for "rent," and anyone borrowing money must expect to pay "rent" for its use.

10.4   INTEREST

As noted previously, *interest* is the fee that a lender charges for the use of their money. From the borrower's viewpoint, it is the "rent" paid for the use of someone else's money. Normally, it is directly proportional to the amount of money loaned or borrowed. That is, a $2,000 loan almost always earns twice the interest amount that a $1,000 loan earns. Interest is also proportional to the time that money is loaned or borrowed. A 3-year loan earns more interest than a 1-year loan for the same dollar amount.

The money that $1 earns for each unit of time that it is loaned is called its *interest rate*. This rate is expressed as a percentage: If $20.00 earns $2.40 over a year, then the interest rate in this case is $2.40/$20.00 = 0.12 or 12% per year. It is so common to deal with interest rates that often it is referred to simply as "interest," with the "rate" being implied by the context of the situation. To avoid confusion, the money earned by charging interest is frequently called *total interest*. It should be easy to notice which of these terms is being used in a specific context. If the interest refers to a percentage, then the complete term should be *interest rate*.

The interest rate established for any business transaction is of paramount importance. Let us suppose that you take a $200,000 mortgage loan to purchase a house. If you agree to pay back the loan over 30 years at a 15% interest per year, your monthly payment toward repaying the loan will be $2,528.89. If the interest is 7% per year, your payment would be $1,330.60, approximately half the previous amount. Whereas the total interest paid over the 30 years would be $710,399.70 at 15%, it would be $279,017.80 at 7%.

In both cases, you end up paying more for interest than the original borrowed amount. But why would anyone agree to pay 15% instead of 7%? The degree of investment risk, the inflation rate, and the stability of the market where the money will be used are important factors determining the interest rate of a particular loan. For instance, lenders charge higher interest rates to borrowers with a higher risk of defaulting on the loan. Other factors can also influence rates, such as the level of competition among lenders and the cost of the money to them (since often lenders, in turn, borrow their money). The setting of the interest rate is discussed in the context of a construction project in Chapter 12.

10.5   SIMPLE AND COMPOUND INTEREST

There are two fundamentally different ways to compute interest. Let us suppose that you need $10,000, which you will pay back at the end of 3 years, plus all the interest accumulated over the 3 years. If you have to pay a 10% interest rate, how much money will you have to pay back? The original borrowed amount, or loan *principal*, is $10,000. The total interest, as previously mentioned, depends on the principal, the time that the money was borrowed, and the interest rate. Also, it depends on the type of interest: whether it is simple or compound.

10.5.1 Simple Interest

You ask Uncle Fudd to lend you the $10,000. He agrees to allow you to pay simple interest, as a concession to one of his favorite relatives. Paying simple interest means that you pay interest on the principal, but *not* on the interest accumulated from previous periods. Therefore, your debt at the end of each year is computed as follows:

Year (n)	Base for applying i	Interest i for period	Total owed (F)
0 (now)			10,000
1	10,000	$10,000 \times 0.10 = 1,000$	11,000
2	10,000	$10,000 \times 0.10 = 1,000$	12,000
3	10,000	$10,000 \times 0.10 = 1,000$	13,000

For simple interest, the future value F_n of a principal P borrowed over n periods at an interest rate i can be computed as $F_n = P + P \times i \times n$, or more directly,

$$F_n = P \times (1 + i \times n)$$

In this case, we can compute $F_3 = 10,000 \times (1 + 0.10 \times 3) = 10,300$ without developing the table.

10.5.2 Compound Interest

You decide to use a bank to borrow the $10,000 instead of asking your uncle. Banks never use simple interest to compute your debt. Instead, the interest owed after each time period is computed by adding the total interest accumulated from previous periods to the loan principal, and then multiplying this total by the interest rate. The computations below help to illustrate how this process works

Year (*n*)	Base for applying (*i*)	Interest for period	Total owed (*F*)
0 (now)			10,000
1	10,000	$10,000 \times 0.10 = 1,000$	11,000
2	11,000	$11,000 \times 0.10 = 1,100$	12,100
3	12,100	$12,100 \times 0.10 = 1,210$	13,310

Using the same notation as for simple interest above, you can see that

$$F_n = P \times (1 + i)^n$$

Using this formula, we can find F_3 directly:

$$F_3 = 10,000 \times (1 + 0.10)^3 = 13,310$$

The formula also works in the opposite direction. One can find the principal that would result in a future value, *F*, after *n* periods at an interest rate of *i* using the formula

$$P = F_n / (1 + i)^n.$$

The *P* found here is usually called the *present value* (*P*) of the future amount, *F*.

Compound interest is more logical and equitable than simple interest. Why would the interest accumulated to date be exempt of consideration, as in simple interest? In fact, as you can appreciate in the mortgage loan example in the previous section, many times the earned interest is higher than a loan's principal. Simple interest is never used for business transactions.

10.6   NOMINAL AND EFFECTIVE RATE

A lender could advertise in big letters a 1% interest rate and then add in small letters that it refers to the monthly rate, not the annual rate as many people would assume. Technically, the lender would be telling the truth: an interest rate can refer to any time unit, from seconds to decades, but many people would be deceived by such advertising strategy. To protect consumers, Congress passed in 1968 the *Truth in Lending Act*, which among other issues, requires that lenders clearly indicate interest rates as an annual percentage of each received dollar. This *annual percentage rate (APR)* also includes upfront commissions and fees converted to equivalent annual percentages.

Requiring that lenders disclose their APR was an important step for consumer protection. However, an APR can still be misleading. The APR simply adds the interest rates charged in 1 year, instead of compounding these intermediate periods. If your credit card charges 1% per month, then its APR is $(1\%/\text{month}) \times (12\,\text{months/year}) = 12\%$. This is the *nominal interest rate* of your card, but it is not its *effective interest rate*. If you have $10,000 in credit card charges, the APR suggests that you will owe $10,000 \times (1.12) = \$11,200$ a year from now (assuming that you skip payments during the year). However, since the card company computes financial charges every month, you would owe $10,000 \times (1 + 0.01)^{12} = \$11,268.25$ at the end of the year, using the compound formula for *F* that we examined in the previous section. The $68.25 difference is substantial and perfectly legal for the credit card company to charge. Almost all credit cards charge interest by the day, which means that total the annual interest that you would owe almost certainly would be

$$\$10,000 \times \left(1 + {}^{0.12}/_{365}\right)^{365} = \$11,274.75$$

It is possible to find the effective annual rate that you pay on a loan (usually called its *annual percentage yield* or APY) if you know its APR and the number of times per year that the interest is compounded. The following formula finds a loan's APY when its APR and the number of times per year that the APR is compounded (the C in the formula) are known. If the interest is compounded monthly, then $C = 12$, the number of months in a year.

$$APY = \left(1 + {APR}/{C}\right)^{C} - 1$$

In the previous example, APR $= 12\%$ compounded monthly. Its APY is:

$$APY = \left(1 + {0.12}/{12}\right)^{12} - 1 = 12.6825\%$$

An APY is not too different from its APR compounded C times, but a seemingly small difference can be significant. If you take \$100,000 to repay after 15 years at $i = 12.0000\%$ and at 12.6825\%, the future amounts owed are:

> For $i = 12.0000\%$: F $= \$100,000 \times (1 + 0.120000)^{15} = \$547,356.57$
> For $i = 12.6825\%$: F $= \$100,000 \times (1 + 0.126825)^{15} = \$599,580.20$

The difference of \$52,223.63 is due exclusively to the difference between APR and APY.

Once a loan's APY is computed, n must be expressed in years for all calculations. Alternatively, the APR can be divided by the number of compounding periods per year (e.g., 12 for monthly compounding), and the number of periods n expressed in the same units. For a 30-year loan with an APR of 12% compounded monthly, one could use $i = 12.6825\%$/year (the APY previously calculated) and $n = 30$ years, or $i = 1\%$/month (i.e. 12%year/12 months/year) and $n = 360$ months.

10.7   EQUIVALENCE AND MINIMUM ATTRACTIVE RATE OF RETURN

If you have an offer to receive \$100 now or \$101 a year from now, which option would you choose? Assume that you have enough resources to think about this offer in terms of investing the money, as opposed to spending it on some "spur of the moment" purchase. Even with enough money to subsist, many people will choose the \$100 now, for the reasons previously discussed in the section on the time value of money. But what happens if instead of \$101, you are offered \$500 a year from now? What if the future amount offered is \$400? Or \$300? Or \$150? Deciding which offer to take gets more tricky. It is quite possible that you would choose the \$500 a year from now, but some of the lower amounts would probably be attractive enough to prefer them over the \$100 offered now in this hypothetical example.

Let us take this mental exercise to the point that you are offered an amount that is just high enough to make you indifferent to having it now or in the future. The two amounts are said to be *equivalent* to you. The "you" in the previous sentence is important. Someone else could have a different idea of what is equivalent to \$100 now. Moreover, you would certainly require a higher amount 2 years from now than what you would get a year from now to declare equivalence in preference.

Equivalence is usually considered in terms of interest rates instead of total interest. If your equivalent amount for \$100 is \$150 received a year from now, then the total interest would be \$150 − \$100 = \$50. The interest rate would be \$50/\$100 = 50%. This particular interest rate is called your *minimum attractive rate of return (MARR)*. In an actual business environment, MARR is not found as arbitrarily as previously shown. If you are in heavy construction and in your experience a reasonable return on equity is 20%, then you would not aspire to a 50% MARR. Or, if you are paying 10% interest on your borrowed money,

---

then your MARR cannot be less than 10%. If your MARR is less than 10%, you would lose money when borrowing at 10%; you would pay more borrowing costs than you would make on the return from your MARR. If your Uncle Fudd makes 25% on his money and is willing to let you invest in his company, then you could establish this rate as your MARR. If Uncle Fudd can consistently offer you 25%, why should you accept less?

Computing MARR is a sensitive and sophisticated issue for any company and will not be covered in detail here. However, keep in mind that when an interest rate i is specified in this chapter, it is assumed to be the MARR for at least one of the stake holders in the problem.

10.8   DISCOUNT RATE

What we have called so far *interest rate* is sometimes called *discount rate*. When we discussed the concept of interest, it was presented from the perspective of someone beginning with an amount P (at time present) and ending in the future with an amount F. When we know the value in the future of a given sum and want to know its equivalent value now, one speaks of *discounting* the future amount, F, to the present at some i value. We want to find P given F. P is equivalent to the known value F when F is discounted using the formula $P = F_n/(1+i)^n$. This formula is the inverse of $F_n = P \times (1+i)^n$.

The equivalence of P and F can be extended to include several Fs. Suppose that we are offered a payment of $55.00 a year from now and a second payment of $60.50 2 years from now. If our MARR is 10%, we can find that the first future amount is equivalent to a present amount of $55.00/1.10 = $50, and the second amount is equivalent to $60.50/1.10^2 = $50 now. We will be willing to accept $50 + $50 = $100 now, is equivalent to the two amounts in the future.

10.9   IMPORTANCE OF EQUIVALENCE

Equivalence is a handy way of moving values such as P and F from point to point in time. For instance we can use this principle to compare payment now (cash and carry) versus purchase of merchandise on credit. If a seller offers 10% credit, we can purchase a $100 music player and arrange to pay for it in two payments: one payment of $55.00 a year from now and another payment of $60.50 2 years from now. As noted, these two payments are equivalent to $100 today at 10%. The seller may be willing to let us take home the merchandise with our written promise to pay back these two amounts at the specified points in time. (More likely, we would prefer to pay two equal payments of $57.62; such equal amounts are called *annuities*, as will be discussed later.) The concept of equivalence allows us to compare various payment options.

10.10   SUNK COSTS

Any cost that has already been expended is called a *sunk cost*. As such, it cannot be revoked and therefore, is irrelevant to future decisions. The following example illustrates the concept of sunk costs.

You purchased new laptop computer 2 months ago for $2,000. However, a new model with the same features is available now for $1,000. Cousin Elmer offers you $1,200 for the old computer. Should you sell the old computer and purchase the new model?

The answer, from a financial viewpoint, is that you should sell the old computer and buy the new model. The cost of the original laptop is irrelevant to the decision of keeping or selling it. Many people have difficulty in selling a possession at a loss compared to the price they paid for it. In this case, keeping the computer would avoid the sense of having taken a bad decision by purchasing it in the first place. But that is a faulty logic. The key point to consider should be whether you are better off by your decision. What is relevant is that selling the old computer would give you an extra $200, and keeping it would prevent you from realizing this profit.

10.11   CASH FLOW DIAGRAMS

So far, we have used only words to describe our examples. In most real situations, this is a cumbersome approach at best, especially when compared to the power and simplicity of graphics to convey information. *Cash flow diagrams* provide a graphical means for describing situations in which interest and time interact. Developing the appropriate cash flow diagram is often the critical step in solving an engineering economy analysis. After the proper diagram is developed, the numerical solution of the problem is straightforward.

A cash flow diagram has the structure shown in Figure 10.1.

Specifically, it consists of the following:

1. A timeline consisting of a horizontal line, which begins at present time, usually represented as time 0. The end of each future period is drawn as a tick mark on the line. The timeline normally ends at period n, that is, the last time period in the analysis. Since n can be relatively large, the line is frequently broken as shown.

2. Arrows placed on the tick marks show each sum of money. An inflow of money is shown as an arrow pointing up at the point the money or value of an asset was earned; similarly, the arrow for a money outflow is drawn pointing down. A person's inflow is another person's outflow (e.g., lender/borrower; seller/buyer), and therefore it is important to keep track of which side of a transaction applies to the situation at hand. If a person buys a car and pays it in two payments of $15,000, then the value of the car appears as an arrow pointing up (it is a new asset for the person), and the two payments are shown as arrows pointing down. From the seller's perspective, the arrows would point exactly opposite to the buyer's.

3. Comments and additional data are written near the main timeline and arrows. The interest used is usually included in this way.

It is tempting to skip the diagram and jump into the numerical solution just by reading the problem. This is a relatively safe way to approach a simple problem, including many of the examples discussed in this chapter. However, real-life situations tend to be complicated, with multiple amounts of money flowing in and out of the scenario. A cash flow diagram is indispensable in these real-life analyses because it is quite easy to forget or misplace some of the amounts that must be included. A cash flow diagram is the easiest way of making sure that the whole cash flow sequence is in view and each individual amount is included in the computations.

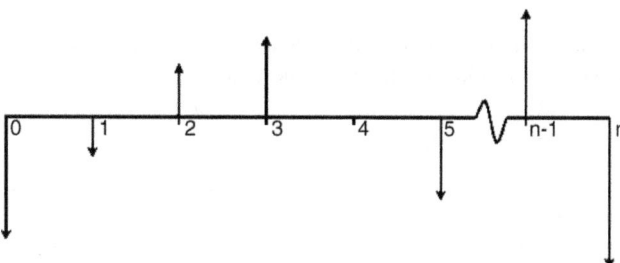

Figure 10.1 Structure of a Cash Flow Diagram.

10.12   ANNUITIES

An *annuity* is a series of equal payments (or receipts) paid out or received in a sequence over a period of time. Our examples so far have consisted of single amounts of money in the present or the future. We have discussed situations such as a $100 article purchased at a 20% interest, which can be paid as $120 a year from now or $144 in 2 years. But one-time payments are the exception, not the norm in a real business environment. Many common business transactions consist of purchasing an article or service by paying a number of equal amounts of money (*installments*) over a number of time periods. For example, a person rarely purchases a $30,000 car on credit by promising to make a single payment of $40,446 in 36 months from date of purchase, based on an APR of 10% compounded monthly. Much more frequently, the arrangement will be that the person pays $968 per month for 36 months, assuming the same principal of $30,000, as well as the same APR and compounding frequency.

You know by now that the $44,446 can be found with the formula $F_n = P \times (1+i)^n$ so that $F = \$30{,}000 \times (1.008333)^{36} = \$40{,}446$. But how do we know that the 36 payments of $968 are equivalent to $30,000 in the present with the lending terms presented?[2] We could check that this is the case by finding the present value, P, of each payment, which we will call A_n, and then adding all these individual A_ns. Since the APR is 10% compounded monthly, the effective monthly interest is $i = 10\%/12 = 0.83333\%$ or 0.008333.

$$P_1 = \$968/(1+0.008333)^1 = \$960.00$$
$$P_2 = \$968/(1+0.008333)^2 = \$952.07$$
$$P_3 = \$968/(1+0.008333)^3 = \$944.20$$
$$P_4 = \$968/(1+0.008333)^4 = \$936.39$$
$$[\ldots]$$
$$P_{36} = \$968/(1+0.008333)^{36} = \$718.00$$

The total present value, P, is:

$$P_{\text{total}} = P_1 + P_2 + P_3 + P_4 + [\ldots] + P_{36}$$
$$P_{\text{total}} = \$30{,}000$$

This present value P is the amount of merchandise that can be purchased by n installments A at an interest i. In this case, the computations check that 36 installments of $968 and an APR of 10% compounded monthly can purchase *exactly* $30,000 of merchandise.

Finding the present value of a series of equal payments this way is totally impractical. Instead of a number of individual F to P conversions, we can find the present value of a series of equal payments (for a given interest rate and a specified number of payment periods) using a simple, single formula.

To develop this formula, we need to consider once more the previous example.

You can see that what we did can be expressed as the following mathematical series:

$$P = A/(1+i)^1 + A/(1+i)^2 + A/(1+i)^3 + [\ldots] + A/(1+i)^n$$

You may remember from Algebra that a series like this can be converted into a single equivalent formula. In this case, the resulting formula is:

$$P = A \times [(1+i)^n - 1]/[i \times (1+i)^n]$$

---

[2] $F = 30{,}000(1+0.1/12)^{36} = 30{,}000 \times (1.008333)^{36} = 30{,}000\,(1.34820) = 40{,}446$. It is important to remember when using i in these formulas, that it has to be expressed as a decimal value (e.g., 10% is 0.1) and 0.1 divided by 12 (used to convert APR to a monthly interest rate) is 0.008333.

For the previous example,

$$P_{total} = 968 \times [(1 + 0.008333\%)^{36} - 1]/[0.008333\%^*(1 + 0.8333\%)^{36}]$$
$$P_{total} = 968 \times 30.99142 = \$30,000$$

This formula is a bit more complicated than the basic P from F conversion, but it is much more straightforward than the individual amount conversion of the example, especially when a high number of Fs need to be converted.

10.12.1 Finding *A* Given *P*

The inverse of the formula can be used to find A given P, i and n. The most common business situation is to know the price of a merchandise and then to compute the monthly (or more generically, the *periodic*) amount to be paid for the merchandise. For $i = 0.008333\%$ and $n = 36$, the A to P formula for $P = \$1$ will always result in the value 30.99142. The P to A computation consists of using the inverse of this factor. Therefore, to find the monthly payment of a \$30,000 car paid with 36 equal payments over 36 months at an APR of 10% compounded monthly, the following computation applies:

$$A = \$30,000/30.99142 = \$968$$

More generally, the formula to convert P to A can be written as follows, which is simply the inverse of the formula for A to P:

$$A = P \times [i \times (1 + i)^n]/[(1 + i)^n - 1]$$

10.13   CONDITIONS FOR ANNUITY CALCULATIONS

Let us recapitulate the conditions that a series of payments or receipts must fulfill to be considered an annuity.

1. All payments must be equal. Figure 10.2a is an annuity, but Figure 10.2b is not one because not all payments are equal.
2. Payments must take place without interruption throughout a number of periods and the periods must be of the same length or duration (e.g., week, month, and so on). For example, it is not acceptable to have one period of 2 weeks followed by a period of 2 days. Figure 10.2c is not an annuity since one payment is missing. This causes the payment periods to vary in duration.
3. Payments normally begin at the end of the first period. Figure 10.2d begins after the first period, and therefore, cannot be directly considered an annuity.
4. All payments must carry the same interest.

Annuities do not necessarily refer to annual payments. Although the semantics are contradictory (think of *annual*), any series of payments fulfilling the aforementioned conditions is an annuity regardless of its time unit. An alternative expression for an annuity sequence is *uniform payment series*.

10.14   CALCULATING THE FUTURE VALUE OF A SERIES OF PAYMENTS

A similar mathematical process can be used to find the future value of an annuity. Consider the following situation. Assume a person saves \$1,000 per year over 20 years at

a. Standard Annuity

b. Unequal Payments

c. Missing Payment

d. First payment not at Period 1

Figure 10.2 Standard and Defective Annuity Schemes.

10% annual interest. What would be the balance of this savings account at the end of the 20 years?

In this case, we are interested in the amount of the account at the end of the n periods instead of at present. The first $1,000, deposited at the end of year 1, would have accumulated interest over 19 years. The last one, at the end of the twentieth year, would have no interest gain (in this example, depositing $1,000 at the end of the 20 years to retire them right away does not make much sense, but in many other situations this is reasonable). This situation is shown in Figure 10.3.

The balance of the savings account would be:

$$F_1 = \$1,000 \times (1 + 0.10)^{19} = \$6,115.91$$
$$F_2 = \$1,000 \times (1 + 0.10)^{18} = \$5,559.92$$
$$F_3 = \$1,000 \times (1 + 0.10)^{17} = \$5,054.47$$
$$[\ldots]$$
$$F_{19} = \$1,000 \times (1 + 0.10)^{1} = \$1,100.00$$
$$F_{20} = \$1,000 \times (1 + 0.10)^{0} = \$1,000.00$$

The balance would be the sum of all the individual deposit future values:

$$F_{total} = F_1 + F_2 + F_3 + [\ldots] + F_{19} + F_{20}$$
$$F_{total} = \$57,275.00$$

Figure 10.3 Future Value of an Annuity.

Following a line of reasoning parallel to the one discussed for the present value of an annuity, we find that a single formula can take care of the 21 individual computations. The formula for the future value of an annuity (e.g., to convert A to F) is:

$$F = A \times \{[(1+i)^n - 1]/i\}$$

We would find the same result for F_{total} by directly applying the given formula:

$$F = \$1{,}000 \times \{[(1+0.10)^{20} - 1]/0.10\} = \$1{,}000 \times 57.2749995$$
$$F = \$57{,}275.00$$

As with present value, we can use the inverse of the formula to compute an annuity given a goal future value, that is, to convert F to A. To find the annual payment that would result in a balance of \$50,000 after 20 years at 10%,

$$A = \$50{,}000/57.2749995 = \$872.98$$

The direct formula to find A given F is the inverse of the previous formula to find F given A:

$$A = F \times \{i/[(1+i)^n - 1]\}$$

Annuity calculations address many practical situations such as:

- Given an assumed MARR, how much should you save each month to accumulate \$400,000 over 20 years? This is a sinking fund type of calculation. Assuming an MARR of 10% APR, the answer is \$526.75.

- What is the amount of the monthly interest payment you should expect to pay for a 15-year mortgage in the amount of \$100,000 at a fixed borrowing rate of 6.0%? The answer is \$843.85. In a typical mortgage agreement, taxes and insurance would be added to this interest amount. If taxes are \$2,500 per year (\$2,500/12 monthly) and insurance is \$1,000 annually (\$1,000/12 monthly), your mortgage payment would be around \$1136 per month.

Real estate and financial professionals are involved with calculations similar to these examples on a daily basis.

10.15   SUMMARY OF EQUIVALENCE FORMULAS

Table 10.1 gives a summary of the formulas and factors we have discussed linking P to F, F to P, A to F, F to A, A to P, and P to A. The factors relating the principal variables (P, F, and A) have the standard "full" names shown. Moreover, they are frequently referred to by abbreviations in parentheses at the right of their full names. The abbreviations are expressed as "fractions" and are borrowed from nomenclature used in statistics. An abbreviation such as F/P is read "Find F given P." P/A is read "Find P given A."

1. Compound amount factor (F/P)
2. Present worth factor (P/F)

Table 10.1 Interest Factors

Type of Cash Flow Name	Single Payment		Uniform Series (Annuities)			
	Compound Amount Factor	Present Worth Factor	Compound Amount Factor	Sinking Fund Factor	Present Worth Factor	Capital Recovery Factor
Description	Find F given P (F/P)	Find P given F (P/F)	Find F given A (F/A)	Find A given F (A/F)	Find P given A (P/A)	Find A given P (A/P)
Formula	$(1+i)^n$	$\dfrac{1}{(1+i)^n}$	$\dfrac{[(1+i)^n - 1]}{i}$	$\dfrac{i}{[(1+i)^n - 1]}$	$\dfrac{[(1+i)^n - 1]}{i^*(1+i)^n}$	$\dfrac{i^*(1+i)^n}{[(1+i)^n - 1]}$

3. Compound amount factor uniform series (*F/A*)

4. Sinking fund factor (*A/F*)

5. Present worth factor uniform series (*P/A*)

6. Capital recovery factor (*A/P*)

These six conversion factors along, with the concept of equivalence, are the building blocks that make complex engineering economy calculations possible. They are sufficient for most financial situations involving time value of money. Given an interest rate (e.g., a MARR), we can find the amount of a future payment *F* based on a *P* and interest accruing at the rate, *i* over some number, *n*, of time periods (in shorthand, we can convert *P* to *F*). Conversely, given *F*, *i*, and *n*, we can calculate *P*. We have discussed that annuities (*A*) are so common in business that formulas have been developed to allow the conversion of these sets of payments to an equivalent present value (that is, converting *A* to *P*) or to a point in the future (converting *A* to *F*). We also found that the formulas for these conversions work in either direction. That is, the formula converting *P* to *A* is the inverse of the formula converting *A* to *P*.

The formulas we have discussed consist of an amount of money times an expression involving *i* and *n*. For example $F = P \times (1+i)^n$. The expression $(1+i)^n$ will always result in the same number for a given *i* and *n* (e.g., for $i = 10\%$ and $n = 2$, the result is always $(1 + 0.10)^2 = 1.21$). If we make a table listing the result of this expression for many combinations of *i* and *n*, we can simply look up its result for a given *i* and *n*. Because the formula relating *F* and *P* is relatively simple, the advantage of using tables is not so obvious. However, in cases where more complicated situations must be analyzed, such as the one for taking *A* to *P* or combinations of calculations are required in the same problem (e.g., *F* to *P* plus *A* to *P*, and the result converted to *F*), the advantage of using tables is more apparent. Tables for the various interest factors shown in Table 10.1 are given in Appendix G.

The tables are easy to use. Let us repeat the example used in the section where we calculated the future value of a series of payments by the use of a formula. In the example, a person saves $1,000 per year over 20 years at 10% annual interest. We want to find the balance of this savings account at the end of the 20 years.

Using the tables in Appendix G, we look for the page for interest = 10.00% (the 10.00% is shown at the top of the page). Then, find the line for $n = 20$. Finally, choose the appropriate column to find the factor. In this case, we are looking for the compound amount factor, which is described on the table as "Find F given A." Looking up the number, it is 57.27500.

This factor is the future amount of $1 after 20 years of depositing $1000/year at an interest of 10%/year. To find the amount for $1,000/year, we multiply the annual deposit by the factor:

$$F = \$1000 \times 57.27500 = \$57,275.00$$

This, of course, is the same result that we previously obtained using the formula to find F given A.

10.16   WORTH ANALYSIS TECHNIQUES: AN OVERVIEW

Financial options can be compared by calculating the Worth of each option and determining which is most attractive. Consider the following situation, which requires comparison of options.

Cousin Thaddeus offers to pay you $3,000 every year for the next 10 years if you give him $20,000 now. Let us suppose that your MARR is 10%, and that a local bank is offering this rate. How can you determine whether Cousin Thaddeus' offer is better than putting the money in the bank?

There are three techniques that can be used to compare investments involving time and money. This offer would be attractive:

1. If the present value of a $3,000 annuity over 10 years at your MARR is greater than the $20,000 that Thaddeus will borrow from you; or

2. If the $3,000 per year is greater than the annuity resulting from investing $20,000 over 10 years at your MARR; or

3. If the interest that yields 10 payments of $3,000 from an initial investment of $20,000 is greater than your MARR.

The three options should result in the same decision to accept or reject your cousin's offer.

(1) The Present Worth (PW) of an investment consists of the present value of all the amounts involved in the investment. If this sum, which is the net present worth of the entire investment, is negative (i.e., less than $20,000), then you would better off if you placed the $20,000 in a bank for 10 years at your MARR.
In our case,

$$PW = -20,000 + 3,000(P/A, 10\%, 10) = -20,000 + 3,000 \times 6.1446$$
$$= -1,566.30$$

Because the present worth of the entire investment is negative, we should reject Thaddeus's proposition.

(2) We can determine the equivalent annual worth (EAW) of Thaddeus' proposal. All the money is converted into equivalent annuities and these annuities are added up. If the resulting net annuity is a negative number, you would be better off if you placed your money in a bank for 10 years at your MARR and withdrew $3,000 every year. For our example,

$$EAW = -20,000(A/P, 10\%, 10) + 3,000 = -20,000 \times 0.1627 + 3,000$$
$$= -254.91$$

Consistent with the previous finding, and because the annual equivalent is negative, we conclude that the proposed arrangement is not attractive.

(3) The method of determining the effective rate of interest given n, and two of the values P, F, or A (i.e., P and F, or P and A, or F and A) is called *internal rate of return (IRR)*. The concept of IRR is discussed briefly in the following paragraphs.

Based on Uncle Thaddeus' proposal, you are offered a 10-year annuity of $3,000 per year based on a payment to him of $20,000. Using this technique, you can use an interpolation approach to determine which rate of interest you will be receiving. If this interest rate is less than your MARR, then by definition the investment opportunity yields less than you require. In such case, you should consider placing your money with an investment firm or bank offering an interest rate which meets of exceeds your MARR.

To find the investment IRR, we use either of the equations given previously, with two differences: We begin with the assumption that PW or EAW is 0; and second, the interest rate we are calculating is the (yet unknown) IRR. In other words, we could use *any* of the following equations:

$$PW = 0 = -20{,}000 + 3{,}000(P/A, IRR, 10)$$
$$EAW = 0 = -20{,}000(A/P, IRR, 10) + 3{,}000$$

If we use PW: $20{,}000 = 3{,}000 \, (P/A, IRR, 10)$

$$(P/A, IRR, 10) = 20{,}000/3{,}000 = 6.6667$$

Knowing that (P/A, IRR, 10 year) is 6.6667, we can proceed by looking up in a compound interest table (see Appendix G) the two values between which we find 6.6667. We can then interpolate to find our unknown value of i.

The IRR in this case turns out to be 8.15%.

Because this interest rate is lower than our MARR of 10%, we also conclude using the IRR approach that the offer made by Thaddeus is not attractive for us.

The following sections discuss the use of PW, EAW and IRR in greater detail.

10.17   PRESENT WORTH ANALYSIS

There are many situations in which it is important to know whether we are paying too much upfront for something that will yield future profit, or how much money should we ask for something that will be paid for in the future. In such cases, PW is an excellent way to proceed, as we will see in the following examples. This technique does, however, have its limitations. First, we need to estimate our MARR, which is a relative quantity, and can change depending on many circumstances. Second, this is not the simplest approach when we need to choose between alternatives with different lifespan: A 10-year project with a total profit whose present worth of $100,000 is obviously less attractive than a 1-year project whose profit's present worth is $90,000. Finally, of all four analysis techniques, it probably is the one whose results are most difficult to explain to a layperson.

10.17.1 Example: Down Payment on a Car

How much is the down payment for a $30,000 car, if you want to limit your payments to $450 per month for 60 months? You will pay a 12.00% APR, compounded monthly (therefore, $i = 1\%$/month)

The car dealer doesn't care whether the car is paid for with a $30,000 check right away or by a check from the bank financing your loan and another check from you, as long as the two checks add up to $30,000. The bank will pay the dealer the PW of your 60 installments of $450 each compounded at 12%. If you can figure this PW, then you can find you must pay upfront to complete the $30,000. In other words:

(Down payment) + (PW of installments)   = Value of your purchase
(Down payment) + $450 \times (P/A, 1\%, 60)$ = $30,000
Down payment = $30{,}000 - \$450 \times 44.9550 = \$9{,}770.23$

This example is useful. It applies to any similar situation involving installment payments, from a bulldozer to your mortgage loan to your living room sound component.

10.17.2 Example: Small Excavator

Fudd and Nephews, Inc., purchased a small excavator for $100,000 two years ago. The excavator was supposed to generate a net benefit of $30,000 per year, but it has generated only $5,000 per year. It is expected to generate the same amount of net benefit of $5,000 for the next 4 years, with a negligible market value after these 4 years. Another contractor offers to purchase the machine for $15,000 now. Should Fudd sell the machine?

In this case, we have to decide the better of two undesirable options. Regardless of whether or not Fudd sells the machine, it will end up losing money. Because the machine has not lived up to expectations, the first impulse for many people would be to sell the machine and move on. But the original plans are water under the bridge. The chief consideration is whether or not the $15,000 offered for the excavator is more than the PW of the annuity of $5,000 over 4 years. This PW, of course, is not $5,000 × 4 = $20,000 because the money is received over a period of time. Using the formula for the PW of an annuity that we previously discussed, we find that:

$$\text{PW at a MARR of } 13.0\% = \$14{,}872$$
$$\text{PW at a MARR of } 12.6\% = \$15{,}000$$
$$\text{PW at a MARR of } 12.0\% = \$15{,}187$$

If Fudd's MARR is less than 12.6%, then the best deal is to keep the excavator because the PW of the money stream that it will generate over its remaining economic life is higher than the $15,000 offered for it now.

In this case, as before, there could be other factors that are not readily visible that may influence the decision. If Uncle Elmer's ulcer acts up each time he receives a repair bill from another breakdown of this excavator, it may be better to get rid of it. Notice, however, that even this rationale considers only what will happen from now on.

10.18 EQUIVALENT ANNUAL WORTH

The EAW of an investment is its value expressed as a uniform dollar amount over a time period. More concisely, it is the equivalent annuity of an investment. When the cost of an investment is the main issue under consideration, then the term *equivalent annual cost (EAC)* instead of EAW is frequently used. Both terms refer to the same concept.

If the PW of an investment's cost at, say 8%, is $15 million and the PW of its revenue is $16 million, then the investment is desirable. It will result in more income than the minimum required by its investors at their MARR: $16 million − $15 million = $1 million. If you "spread out" $1 million present value dollars over 10 years at an interest rate of 8% (8.14%, to be more accurate), the resulting annuity is $150,000. Any PW value can be expressed as an EAW by just converting the PW of the whole investment into an equivalent annuity. The formula for this conversion, as previously discussed, is

$$\text{EAW} = P \times [i \times (i+1)^n]/[(i+1)^n - 1]$$

Using factors from a compound interest factor table (CRF column in Appendix G)

$$\text{EAW} = P \times (A/P, i, n)$$

(*A/P*, *i*, *n*) is the *capital recovery* factor.

A future value F can also be converted into equivalent annuities, using a similar procedure:

$$\text{EAW} = F \times [i]/[(i+1)^n - 1]$$

Or, using factors from a compound interest factor table,

$$EAW = F \times (A/F, i, n)$$

$(A/F, i, n)$ is called the *sinking fund* factor, as discussed in Section 10.15.

A given annuity does not need to be converted at all. It is already in the desired format.

The EAW of an investment involving initial costs, annuities, and future values is the sum of the EAW of each component.

10.18.1 Example: Best Job Offer

Greg is undecided between two job offers. Fudd Associates, Inc., offers a $10,000 sign-up bonus and $50,000/year, which will be revised after 3 years. ABC Contractors, Inc., offers an annual salary of $54,000, which will also be revised after 3 years. Greg has made 12% on the money invested in the stock market and believes that he can keep getting the same return if he invests the entire $10,000 bonus. Both companies offer comparable benefits and Greg likes both of them. Based on the salary offer, which offer is better?

Solution: The timeframe for the comparison is 3 years and the interest is 12%. The question is whether the extra $4,000/year are more attractive than the initial $10,000 bonus. The EAW for the $10,000 is:

$$EAW = 10,000 \times (A/P, 12\%, 3) = \$10,000 \times 0.41635 = \$4,163.50$$

Based exclusively on salary considerations, Fudd Associates, Inc.'s offer is slightly better ($163.50/year) than the offer made by ABC Contractors, Inc. Since this is such a small difference, other considerations will play a role (e.g., office location, job assignments, etc.) because the two alternatives are practically the same.

10.19 INTERNAL RATE OF RETURN

The Internal Rate of Return (IRR) of an investment is the interest rate at which the investment has a PW of 0. The PW of any investment depends on the interest used for the computations. More specifically, the higher the interest, the lower the investment's PW. You can recall that an investment with a positive PW is desirable and one with a negative PW is not. An investment with a PW of 0 is exactly breaking even at the interest rate used in the PW calculations. (When an investment's PW is 0, its EAW is also 0. The IRR is usually defined as the interest resulting in PW = 0, but it is also perfectly correct to define it as the interest at which the investment's EAW = 0)

The concept of IRR is best understood in the context of an example. Suppose that an investment consists of the following:

Initial cost: $100,000

Net benefits: $27,057/year

Investment span: 5 years

The PW of this investment can be expressed as:

$$PW = -100,000 + 27,057 \times (P/A, i, 5)$$

If we look this P/A for several interest rates, we can develop the following table:

$(P/A, 12\%, 5) = 3.60478$ $PW12\% = -\$100,000 + \$27,057 \times 3.79079 = -\$2,465$
$(P/A, 11\%, 5) = 3.69590$ $PW11\% = -\$100,000 + \$27,057 \times 3.69590 = \$0$
$(P/A, 10\%, 5) = 3.79079$ $PW10\% = -\$100,000 + \$27,057 \times 3.79079 = \$2,567$

The results indicate that if we want to get a return of 12% on our money, this investment is not desirable; whereas if we want a 10% return, the project is generating money above and beyond the desired return. Eleven percent is the IRR for this project because it the interest rate that leaves no shortfall or surplus money.

The IRR of simple investment scenarios can be computed by manipulating the time value of money formulas that we have discussed. For example, Cousin Wolfgang wants to borrow $1,000 and pay you back $1,500 five years from now. What is the IRR of this investment? Because $F = P \times (1 + i)^n$, we can solve for i:

$$\$1,500 = \$1,000 \times (1 + i)^5$$
$$i = (\$1,500/\$1,000)1/5 - 1 = 8.45\%$$

Unfortunately, most cases involve the use of factors for $(P/A, i, n)$ and $(P/F, i, n)$. The mathematical formulas for each of these factors involve i multiple times. It is practically impossible to find a neat solution for i as in the previous example for even one of these complex formulas, let alone a combination of them. The practical ways to find the IRR of most investments is by using financial calculators, financial functions in a computer spreadsheet, or by trial and error. The trial and error process will be demonstrated in Chapter 11 by calculating IRR values comparing project cash flow situations. An example of IRR applied to owner financing using bonds will be presented in Chapter 12.

10.20   LIMITATIONS OF THE INTERNAL RATE OF RETURN METHOD

The IRR method has some limitations. If we need to choose among two investments A and B, it could happen that A had the higher IRR, indicating that it is the most desirable. However, B could have the larger PW of the two when computed using MARR as the interest rate. This is frequently the case when there are large differences in the size of the investments. For example, investment A may involve $100,000 and have an IRR of 17%, whereas investment B can have an IRR of 15% but involve $1 million in approximately the same time frame. If the company's MARR is 12% and the PW of both projects is computed using this interest, B will have the largest PW. The results of the two methods (PW and IRR) are contradictory in such case. There are other more obscure shortcomings, such as the capital reinvestments assumptions of the method, which are not appropriate for discussion at this introductory level. In general, IRR is not well suited for selecting mutually exclusive options unless their size and time frame is similar.

The IRR, in summary, is intuitive and proceeds from having the expected amounts in an investment to finding which interest rate is implicit in the investment. In contrast, the PW and EAW methods check these amounts against a goal interest rate (normally the company's MARR), and their results are more difficult to interpret by a non-specialist. The IRR, as discussed previously, has its share of drawbacks, and must be part of the analysis arsenal instead of outright replacing the other two methods.

10.21   A PRACTICAL EXAMPLE USING PRESENT WORTH ANALYSIS

In the purchase of equipment, contractors are confronted with how to recover the cost of purchase and operation over a given life span. The following example takes a engineering economy approach to solving this using a MARR of 10%.

Assume we buy an earth hauler for $80,000. The following data regarding anticipated costs for the operation of the equipment piece are available.

1. Initial cost $80,000

2. Annual operational costs (operator, fuel, oil, etc.) $20,000

3. Tire replacement at the end of years 2 and 4 $11,000

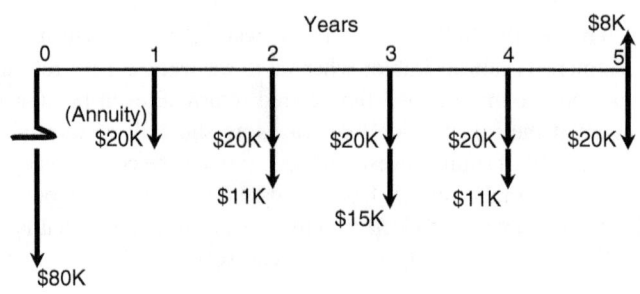

Figure 10.4 Cash Flow Diagram for Equipment Problem.

4. Major overhaul at the end of year 3	$15,000
5. Sale at the end of year 5 (salvage value)	$8,000

The cash flow diagram for this situation is shown in Figure 10.4.

The first thing we must do is to refer all costs to present value. That is, we calculate the PW for all cost and the resale value. The following present values are calculated:

1. The cost of the unit – $80,000 – is at Present Value.

2. The operational costs each year of $20,000 are considered as a uniform series of payments. Therefore, A is $20,000 and P is calculated. Find P, given A.

$$P = A(P/A, 10\%, 5) = \$20,000(3.7905) = \$75,820.$$

3. The tire replacement costs of $11,000 are assumed to occur at the end of year 2 and 4. Since the payments are not a uniform series, we find P, given F for each expenditure. (Why can't we use find P, given A?)

$$P = F((P/F, 10\%, 2) + (P/F, 10\%, 4)) = \$11,000(0.8264 + 0.6830)) = \$16,603$$

4. The major overhaul occurs at the end of year 3. Therefore,

$$P = F(P/F, 10\%, 3) = 0.7512(\$15,000) = \$11,200$$

5. Because the resale or salvage value represents revenue, it is considered to have a negative value in this calculation (i.e., costs are plus and revenues are negative). Then,

$$P = -8,000(P/F, 10\%, 5) = -\$8,000(0.6209) = -\$4,967$$

Summation of the values calculated in (1) through (5) yields

$$P = \$80,000 + \$75,820 + \$16,603 + \$11,270 - \$4,967 = \$178,726.$$

To find the amount which must be recovered each year from clients, we spread the present value calculated across the 5-year period as an annuity or uniform series using a capital recovery factor (A/P)—see Figure 10.5. Using Appendix G to find A given P, we can see that the factor is 0.26380

$$A = P \times (\text{capital recovery factor}) = \$178,726 \times (0.26380) \approx \$41,100$$

10.22   COMPARISON USING EQUIVALENT ANNUAL WORTH

Let us consider a situation comparing two alternatives using an MARR of 7%. A new reservoir has been constructed providing water to a mid-size town called Littleville (See Figure 10.6).

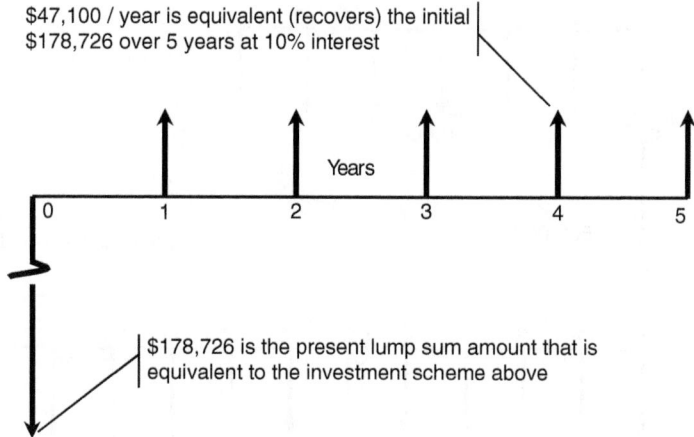

Figure 10.5 Distribution of Costs/Revenue Using A/P.

Two options are available for transporting water from the reservoir to Littleville town. Option A involves building a 10-mile long gravity pipeline at an initial cost of $2.8 million. Option B envisions using a pumping station to lift water over some barrier high ground. This option reduces the length of the pipeline to a length of 2 miles. A 40-year service life is used to compare these two alternatives.

	Costs for pipeline A	Costs for pipeline B
Initial investment pipeline	$2.8 million	$1.5 million
Cost of pumping station	$0	$500,000
Annual operation and maintenance (O & M) costs	$30,000	$60,000
Annual power costs during first 10 years	$0	$40,000
Annual power costs after 10 years	$0	$120,000

To compare these two options, we refer all cost to present value.
For Plan A, the cash flow diagram is shown in Figure 10.7.

$$P(1) = \$2,800,000$$
$$P(2) = \$30,000 \ (P/A, 7\%, 40) \ \text{Using the Tables in Appendix G}$$
$$\text{Therefore}, P(2) = \$30,000 \ (13.332) = \$399,960.$$
$$P(\text{total}) = \$2,800,000 + \$399,960 = \$3,199,960$$

The EAW is

$$\text{EAW} = \$3,199,960 \ (A/P, 7\%, 40) = \$3,199,960(.07501) = \$240,029.00$$

Plan A – Gravity Line Plan; Plan B – Line Using Pumping Station

Figure 10.6 Alternatives for Littleville water supply.

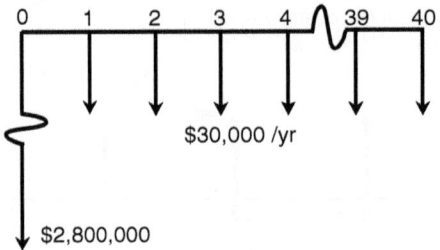

Figure 10.7 Cash Flow for Plan A.

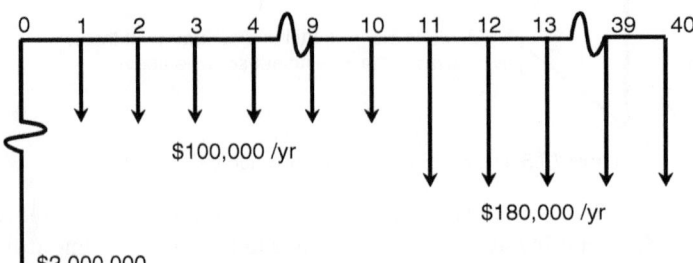

Figure 10.8 Cash Flow for Plan B.

For Plan B, the cost of operating the pumping station must be included. The diagram describes this option as a cash flow diagram (See Figure 10.8)

The present value of the cost of building the pipeline with pumping state is $1,500,000 + $500,000 or $2,000,000. Therefore,

$$P(1) = \$2,000,000$$

The cost of O & M plus power for the first 10 years is $60,000 + $40,000 or $100,000/year. The present value of this as a uniform series over 10 years is

$$P(2) = \$100,000 \, (P/A, 7\%, 10) = \$100,000 \, (7.024) = \$702,400.00$$

Cost of O & M and power for the last 30 years is $60,000 + $120,000 or $180,000/year. The present value of this alternative is calculated as follows:

First we calculate P(3) by referring the uniform series of payments from years 11 through 40 to year 10. This yields:

$$P(3) = \$180,000 \, (P/A, 7\%, 30) = \$180,000 \, (12.409) = \$2,233,620$$

This gives the present value of the last 30 years of expense consolidate as a single payment at year 10. This value must now be referred from year 10 to year 0 (i.e., PW at time 0). Therefore, we have a given F, find P situation. Using the tables in Appendix G,

$$P(4) = \$2,233,620 \, (P/F, 7\%, 10) = \$2,233,620 \, (0.5084) = \$1,135,572.$$

$$P(\text{total}) = P(1) + P(2) + P(4) = \$2,000,000 + \$702,400 + \$1,135,572 = 3,837,972$$

Converting this to EAW over the 40-year life span (i.e., find A given P),

$$EAW = \$3,837,972 \, (A/P, 7\%, 40) = \$3,837,972(0.07501) = \$287,886.31.$$

Since $240,029 is < $287,886, Option A is less expensive and should be selected.

10.23 SUMMING UP

No enterprise can survive the modern business environment without a good grasp of the concepts and techniques about the time value of money. At the most obvious level, the company must be able to estimate the payments that it will make to repay borrowed money.

But many other more subtle issues can be equally important. Determining the attractiveness of a business scheme, comparing several alternatives, finding the true cost of a business proposition when interest is considered, are examples of the immediate and critical usefulness of these techniques This chapter has introduced some of the basic concepts which comprise the foundations of time value of money and the mathematical techniques available for considering value in various borrowing and repayment sequences.

REVIEW QUESTIONS AND EXERCISES

10.1 If your credit card's APR is 24% compounded daily, what is the effective annual interest rate that you are paying?

10.2 Explain why is it that sunk costs are not considered in the analysis of investment alternatives.

10.3 You want to accumulate $500,000 in a savings account in 20 years. If the bank pays 6% compounded annually, how much should you deposit in the account?

10.4 What is the future value 8 years from, of $ 2,000 invested today at a periodic interest rate of 12% compounded annually?

10.5 What would be the result of the previous problem, if simple interest were used?

10.6 You must decide between job offers from CPM Construction and Fudd Associates. Both companies concentrate on sustainable commercial construction, have similar fringe benefits and offer comparable salary increases. CPM Construction offers you a $20,000 sign-up bonus, whereas Fudd Associates will give you $30,000 after 3 years in the job. Which offer would you take? Why? Consider engineering economy principles in your explanation.

10.7 You want to purchase a car with a sticker price of $25,000. The car dealer offers you a $2,000 discount and a 48-month, 8.5% APR compounded monthly, or no discount with a 4.0% APR on a 48-month loan. Which offer is better?

10.8 Repeat problem 10.7, considering an inflation of 4.0%. How does inflation affect the results of the previous problem?

10.9 Develop the table below comparing the monthly payments for a 30-year $200,000 mortgage loan with its APR compounded monthly.

APR	Monthly Payment	Difference
0%		
4%		
8%		
12%		
16%		

10.10 The total monthly payment for a home mortgage loan is affected by the property tax and insurance. This amount is around 15% of the principal plus interest payment that you calculated in the previous problem. Moreover, as a rule of thumb, a family should not devote more than 28% of its gross income for home loan payments. With this information, find the amount that a family must earn to afford a $200,000 mortgage loan using the APRs of the previous problem. The table below should help you in your computations.

A	B	C	D	E
APR	Monthly Payment (From previous problem)	Total payment (1.15 × Column B)	Annual Payment (12 × Column C)	Minimum Annual Income (Column D/0.28)
0%				
4%				
8%				
12%				
16%				

10.11 Develop the table comparing the monthly payments for a $200,000 mortgage loan with an 8.00% APR compounded monthly

Years	Monthly Payment	Difference
100		
30		
15		
10		
5		

10.12   Find the minimum amount that a family should earn to afford the loan scenarios in problem 10.11. Use the same assumptions of the previous similar problem, and the table given.

A	B	C	D	E
Years	Monthly Payment (From previous problem)	Total payment (1.15 × Column B)	Annual Payment (12 × Column C)	Minimum Annual Income (Column D/0.28)
100				
30				
15				
10				
5				

10.13   CPM Construction plans to buy a truck for $150,000 and sell it for $15,000 at the end of 5 years. The annual operating cost of the vehicle is $60,000. What is the equivalent annual cost of this truck, if the company uses a MARR of 12%?

10.14   What is the present value of the truck described in problem 13?

10.15   CPM Construction plans to charge $100,000/year for the truck described in problem 10.13 (because the annual operating cost is estimated at $60,000, this means that the annual income will be $40,000). What is the IRR of this investment? Given CPM's MARR, is this an attractive investment?

Chapter 11

Project Cash Flow

CONTENTS

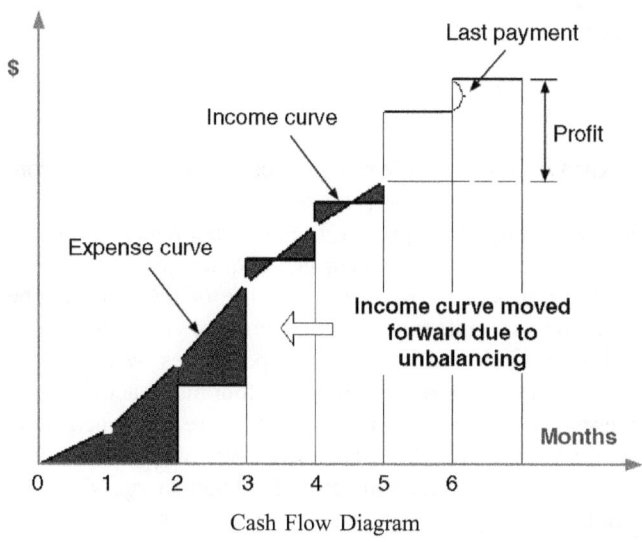

Cash Flow Diagram

Scenario Testing

Construction company operations are project based. Cash flows can be estimated by attempting to assess flows from (a) projects in progress, (b) projects under contract but not yet begun, and (c) potential projects which will start during the coming financial accounting period. These sources of income can be viewed as (a) "birds in the hand," (b) "birds in the bush," and (c) "birds flying in the sky." In other words, cash flows can be projected from projects in progress and projects which may, with some probability, start in the coming period for which forecasts are being made. The advent of spreadsheet analysis and high speed computing has led to "scenario testing" of future cash flow expectations.

Spreadsheets allow managers to run probabilistic cash flow projections that take into account the factors previously noted. More advanced analysis can also factor in historical evidence of payment trends and the potential impact of macroeconomic factors. These techniques go beyond the typical best-, expected-, and worst-case scenario modeling and may rely on Monte Carlo simulation, Markov modeling, or the use of *fuzzy data* sets to build up statistically valid outcomes. At the most advanced level, when future cash flows are tied to a multitude of unknowns, probabilistic techniques may be employed in combination with real-options theory to gain an improved view of the impact of a financial decision (e.g., accepting or declining a project or changing market strategy) on value creation for a company. This level of analysis used to be in the economist's realm, but it is now commonplace in the finance and business development groups of corporations.

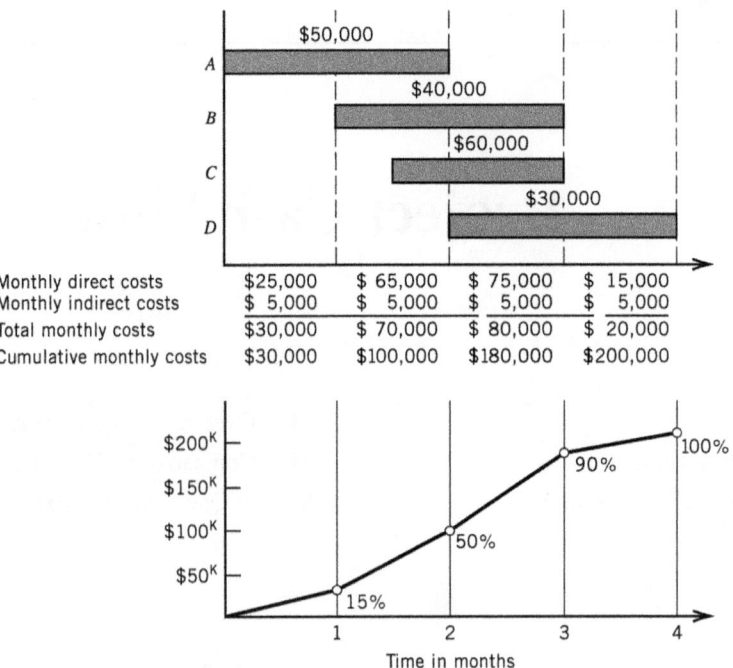

Figure 11.1 Development of the S-curve.

11.1   CASH FLOW PROJECTION

The projection of income and expense during the life of a project can be developed from several time-scheduling aids used by the contractor. The sophistication of the method adopted usually depends on the complexity of the project. In many contracts (e.g., public contracts such as those used by state agencies), the owner may require the contractor to provide an S-curve of estimated progress and costs across the life of the project. The contractor develops this by constructing a simple bar chart of the project, assigning costs to the bars, and smoothly connecting the projected amounts of expenditures over time.

Consider the highly simplified project (Figure 11.1) in which four major activities are scheduled across a 4-month time span. Bars representing the activities are positioned along a time scale indicating start and finish times. The direct costs associated with each activity are shown above each bar. It is assumed that the monthly cost of indirect charges (i.e., site office costs, telephone, heat, light, and supervisory salaries, which cannot be charged directly to an activity) is $5,000. Assuming for simplicity that the direct costs are evenly distributed across the duration of the activity, the monthly direct costs can be readily calculated and are shown below the time line. The direct charges in the second month, for example, derive from activities A, B, and C, all of which have a portion in the period.

The direct charge is simply calculated based on the portion of the activity scheduled in the second month, as:

$$\begin{array}{lll}
\text{Activity } A: & 1/2 \times 50{,}000 = \$25{,}000 \\
\text{Activity } B: & 1/2 \times 40{,}000 = \$20{,}000 \\
\text{Activity } C: & 1/3 \times 60{,}000 = \underline{\$20{,}000} \\
& & \$65{,}000
\end{array}$$

The figure shows the total monthly and cumulative monthly expenditures across the life of the project. The S-curve is nothing more than a graphical presentation of the cumulative expenditures over time. A curve is plotted below the time-scaled bars through the points of cumulative expenditure. As activities come online, the level of expenditures increases and the curve has a steeper middle section. Toward the end of a project, activities are winding down and expenditures flatten again. The points are connected by a smooth curve because the assumption is that the expenditures are relatively evenly distributed over each time period. This curve is essentially a graphical portrayal of the outflow of monies (i.e., expense flow) for both direct and indirect costs.

11.2   CASH FLOW TO THE CONTRACTOR

The flow of money from the owner to the contractor is in the form of progress payments. As already noted, estimates of work completed are made by the contractor periodically (usually monthly) and are verified by the owner's representative. Depending on the type of contract (e.g., lump sum, unit price, etc.), the estimates are based on evaluations of the percentage of total contract completion or actual field measurements of quantities placed. This process is best demonstrated by further consideration of the four-activity example just described. Assume that the, contractor originally included a profit or markup in his bid of $50,000 (i.e., 25%) so that the total bid price was $250,000. The owner retains 10% of all validated progress payment claims until one-half of the contract value (i.e., $125,000) has been built and approved as an incentive for the contractor to complete the contract. The retainage will be deducted from the progress payments on the first $125,000 and eventually paid to the contractor on satisfactory completion of the contract. The progress payments will be billed at the end of the month, and the owner will transfer the billed amount minus any retainage to the contractor's account 30 days later. The amount of each progress payment can be calculated as:

$$\text{Pay} = 1.25(\text{indirect expense} + \text{direct expense})$$
$$-0.10[1.25(\text{indirect expense} + \text{direct expense})]$$

The minus term for retainage drops out of the equation when 50% of the contract has been completed. Because of the delay in payment of billings by the owner and the retainage withheld, the revenue profile lags behind the expense S-curve as shown in Figure 11.2.

The revenue profile has a stair-step appearance because the progress payments are transferred in discrete amounts based on the preceding equation. The shaded area in Figure 11.2 between the revenue and expense profiles indicates the need on the part of the contractor to finance part of the construction until such time as he is reimbursed by the owner. This difference between revenue and expense makes it necessary for the contractor to obtain temporary financing. Usually, a bank extends a line of credit against which the contractor can draw to buy materials, make payments, and pay other expenses while waiting for reimbursement. This is similar to the procedure used by major credit card companies in which they allow credit card holders to charge expenses and carry an outstanding balance for payment. Interest is charged by the bank (or credit card company)

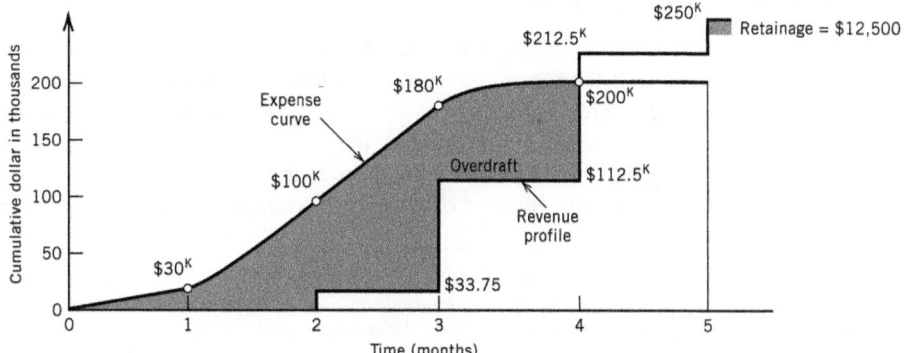

Figure 11.2 Expenses and income profiles.

Figure 11.3 Influence of front, or mobilization, payment on expense and income profiles.

on the amount of the outstanding balance or overdraft.[1] It is, of course, good policy to try to minimize the amount of the overdraft and, therefore, the interest payments. The amount of the overdraft is influenced by a number of factors, including the amount of markup or profit the contractor has in his bid, the amount of retainage withheld by the owner, and the delay between billing and payment by the owner.

Interest on this type of financing is usually quoted in relationship to the prime rate. The *prime rate* is the interest rate charged preferred customers who are rated as reliable and who represent an extremely small risk of default (e.g., Proctor and Gamble, Exxon, etc.). The amount of interest is quoted in the number of points (i.e., the number of percentage points) above the prime rate. The higher-risk customers must pay more points than more risky borrowers. Construction contractors are normally considered high-risk borrowers; if they default, the loan is secured only by some materials inventories and partially completed construction. In the event that a manufacturer of household appliances defaults, the inventory of appliances is available to cover part of the loss to the lender. Additionally, since construction contractors have a historically high rate of bankruptcy, they are more liable to be charged higher interest rates in most of their financial borrowings.

Some contractors offset the overdraft borrowing requirement by requesting front, or mobilization, money from the owner. This shifts the position of the revenue profile so that a reduced, or zero, overdraft occurs (Figure 11.3). Since the owner is normally considered less of a risk than the contractor, the owner or client can borrow short-term money at a lower interest rate. If the owner agrees to this approach, he or she essentially takes on the interim financing requirement normally carried by the contractor. This can occur on cost-reimbursable contracts in which the owner has great confidence in the contractor's ability to complete the project. In such cases it represents an overall cost savings to the owner because otherwise the contractor will charge for the higher financing rate he or she must pay.

11.3   OVERDRAFT REQUIREMENTS

To know how much credit must be made available at the bank, the contractor needs to know what the maximum overdraft will be during the life of the project. With the information given regarding the four-activity project, the overdraft profile can be calculated and plotted. For purposes of illustration, the interest rate applied to the overdraft will be assumed to be 1% per month. That is, the contractor must pay the bank 1% per month for the amount of the overdraft at the end of the month. More commonly, daily interest factors may be employed

---

[1] Similar examples of this type of inventory financing can be found in many cyclic commercial undertakings. Automobile dealers, for instance, typically borrow money to finance the purchase of inventories of new car models and then repay the lender as cars are sold. Clothing stores buy large inventories of spring or fall fashions with borrowed money and then repay the lender as sales are made.

for the purpose of calculating this interest service charge. Month-end balances might otherwise be manipulated by profitable short-term borrowings at the end of the month. The calculations required to define the overdraft profile are summarized in Table 11.1.

The table indicates that the payment by the owner occurs at the end of a month based on the billing at the end of the previous month. It is assumed that the interest is calculated on the overdraft and added to obtain the amount financed. This amount is then reduced by the amount received from the owner for previous billings. To illustrate: The overdraft at the bank at the end of the second month is $100,300. The interest on this amount is $1,003 and is added to the overdraft to obtain the total amount financed ($101,303). To obtain the overdraft at the end of the third month, the progress payment of $33,750 is applied to reduce the overdraft at the beginning of the third month to $67,553. The overdraft at the end of the period is, then, $67,553 plus the costs for the period. Therefore, the overdraft is $67,553 plus $80,000, or $147,553. The information in the table is plotted in Figure 11.4. The overdraft profile appears as a sawtooth curve plotted below the baseline. This profile shows that the maximum required is $149,029. Therefore, for this project the contractor must have a line of credit that will provide at least $150,000 at the bank plus a margin for safety, say $175,000 overall to cover expenses.

Requirements for other projects are added to the overdraft for this project to get a total overdraft or cash commitment profile. The timing of all projects presently under construction by the contractor leads to overlapping overdraft profiles that must be considered to find the maximum overdraft envelope for a given period of time. Bids submitted that may be accepted must also be considered in the projection of total overdraft requirement. The plot of total overdraft requirements for a set of projects is shown in Figure 11.5.

Cash flow management involves the techniques described in this chapter—and much more. Cash flows are affected by a significant degree of uncertainty. A cash flow management model of a relatively simple kind involves making provision for a set of at least 50 variables and requires a computer program to secure sufficient, timely, and usable cash management decision-making information.

11.4 COMPARISON OF PAYMENT SCHEMES

Analysis using internal rate of return (IRR) as described in Chapter 10 is helpful in comparing the economic value to a contractor of varying payment schemes. This comparison is made by determining the IRR of competing payment sequences. It provides a vehicle for examining the economic impact of (a) varying retainage policies, (b) delay in payment strategies, and (c) the payment of a mobilization item to the contractor.

Consider the small four-activity project of Figure 11.1. The owner will consider the payment of a mobilization item at the end of the first period. This will be deducted from the final payment to the contractor. The amount of the payment will be $20,000. To determine the impact of this payment, the rate of return on the original payment sequence will be compared with the rate of return given the mobilization payment. Figure 11.6 shows a diagram of the original payment and expenditure sequence. Expenditures as taken from Table 11.1 are shown above the baseline in the figure. Revenues are shown below the line.

The IRR for a given sequence of payments and expenditures must be found by determining which value of i satisfies the following relationship.

$$\sum^{all\ I} PW[REV(I)] - \sum^{all\ I} PW[EXP(I)] = 0$$

where

$\quad\quad$ Rev(i) = revenue for period i

$\quad\quad$ Exp(i) = expenditure for period i

$\quad\quad\quad$ PW = present worth of these values

Table 11.1 Overdraft Calculations

	Month					
	1	2	3	4	5	6
Direct cost	$25,000	$65,000	$75,000	$15,000		
Indirect cost	5,000	5,000	5,000	5,000		
Subtotal	30,000	70,000	80,000	20,000		
Markup (25%)	7,500	17,500	20,000	5,000		
Total billed	37,500	87,500	100,000	25,000		
Retainage withheld (10%)	3,750	8,750	0	0		
Payment received			$33,750	$78,750	$100,000	$37,500
Total cost to date	30,000	100,000	180,000	200,000		
Total amount billed to date	37,500	125,000	225,000	250,000		
Total paid to date			$33,750	112,500	212,500	250,000
Overdraft end of month	30,000	100,300	147,553	90,279	(8,818)[b]	(46,318)[b]
Interest on overdraft balance[a]	300	1,003	1,476	903	0	0
Total amount financed	30,300	101,303	149,029	91,182	(8,818)	0

[a] A simple illustration only. Most lenders would calculate interest charges more precisely on the amount/time involved employing daily interest factors.
[b] Parentheses indicate a positive balance in this case.

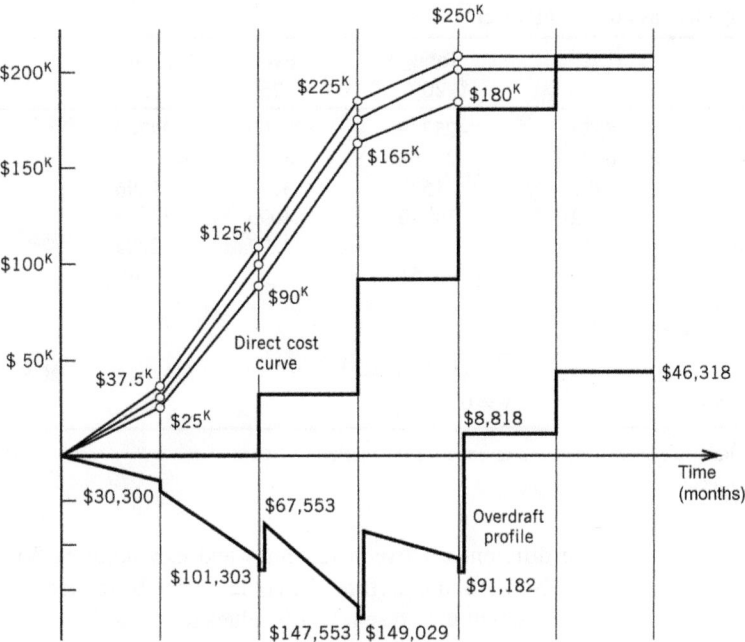

Figure 11.4 Plot of maximum overdraft.

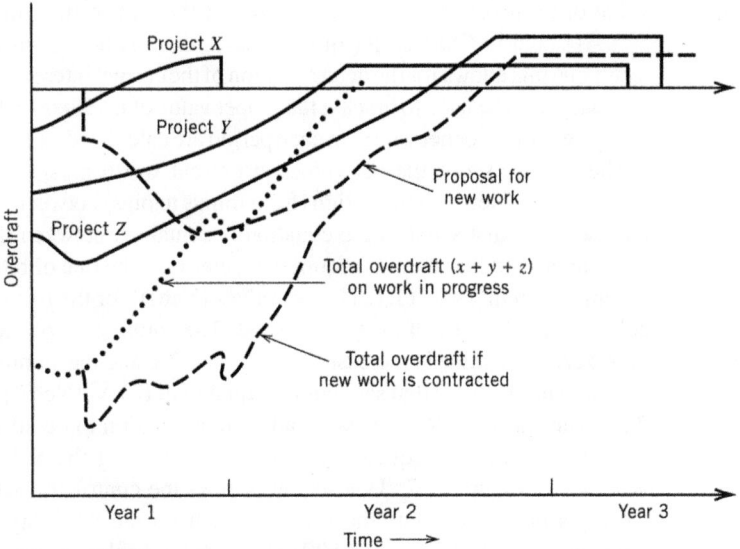

Figure 11.5 Composite overdraft profiles.

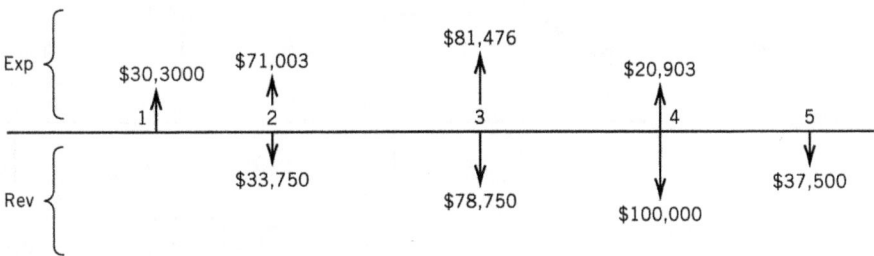

Figure 11.6 ROR for small bar chart problem.

Table 11.2 ROR Calculations for Small Project

N	NET[a]	PWF[b] @ 20%	Total @ 20%	PWF @ 25%	Total @ 25%	PWF @ 22%	Total @ 22%
1	−30300	.8333	−25249	.8000	−24240	.8196	−24834
2	−37253	.6944	−25868	.6400	−23842	.6719	−25030
3	−2726	.5787	−1577	.5120	−1396	.5507	−1501
4	79097	.4822	38140	.4096	32398	.4514	35704
5	37500	.4019	15071	.3277	12289	.3700	13875
			$\Sigma = +517$		$\Sigma = -4971$		$\Sigma = -1786$

$$\frac{X}{2\%} = \frac{517}{(1786 + 517)}$$

$$X = 0.45\%$$

$$ROR = 20\% + 0.45\%$$
$$= 20.45\%$$

[a]A negative net value indicates expenses exceed revenue for this period.
[b]PWF = Present Worth Factor.

The difference between revenues and expenditures for a given period is Rev(i) − Exp(i) = Net(i), and therefore, the equation can be reduced to $\sum$PW[Net(i)] = 0. In other words, the sum of the present worth values for all period NET values must equal to 0. This assumes that all expenditures and all revenues are recognized at the end of each period i. The IRR will be the value of interest that satisfies this equation or, in effect, causes the present value of all expenditures (recognized at the end of each month) to equal the present value of all revenues (again recognized at the end of the month).

As noted in Chapter 10, in most cases there is no closed-form mathematical solution technique that allows for the determination of the correct interest value i. Therefore, an iterative approach must be used to bracket the proper value of i. Values for the present value of revenues and expenditures or net values in each period are calculated using an assumed value of interest, i. The summation of the net values at present worth must equal zero. If the value of the summation of Net(i) changes sign (from minus to plus) between two different values of i, then the value of i that satisfies the equation is contained between those two values. Table 11.2 summarizes the calculations required to determine the rate of return for the original payment scheme given in Table 11.1. The net values [Net(i)] for the five periods are shown in the first column. A value of i of 20% is selected. The summation of the net values is calculated to be +517. The i value is increased from 20 to 22% and the summation of net values becomes −1786. The value of i that satisfies the equality of PW$\sum$[Net(i)] = 0 must be between 20 and 22%. The correct IRR value is found by interpolation to be 20.45%.

The alternative sequence of payment including the $20,000 mobilization payment is shown in Figure 11.7. This sequence pays the contractor $20,000 at the end of the first work period and deducts the prepayment from the final payment. The final payment is reduced from $37,500 to $17,500. The cash flow calculations for this sequence are shown in Table 11.3. It can be seen that the mobilization payment causes the revenue profile to

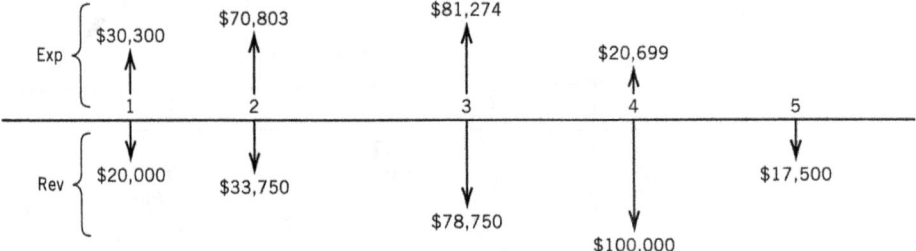

Figure 11.7 ROR for small bar chart problem with mobilization payment.

Table 11.3 Overdraft Calculations with Mobilization Payment

	Month					
	1	2	3	4	5	6
Direct cost	$25,000	$65,000	$75,000	$15,000		
Indirect cost	5,000	5,000	5,000	5,000		
Subtotal	30,000	70,000	80,000	20,000		
Markup (25%)	7,500	17,500	20,000	5,000		
Total billed	37,500	87,500	100,000	25,000		
Retainage withheld (10%)	3,750	8,750	0	0		
Payment received		$20,000	$33,750	$78,750	$100,000	$17,500
Total cost to date	30,000	100,000	180,000	200,000	200,000	
Total amount billed to date	37,500	125,000	225,000	250,000	250,000	
Total paid to date		$20,000	$53,750	$132,000	$232,500	$250,000
Overdraft end of month	30,000	80,300	127,353	69,877	(29,424)[a]	(46,924)[a]
Interest on overdraft balance	300	803	1,274	699	0	0
Total amount financed	$30,300	$81,103	$128,627	$70,576	(29,424)	0

[a]Parentheses indicate a positive balance.

Table 11.4 ROR Calculations to Include Mobilization Payment

N	Net[a]	PWF[b] 30%	Total @ 30%	PWF 32%	Total @ 32%	PWF 34%	Total @ 34%
1	−10300	.7692	−7923	.7575	−7802	.7463	−7687
2	−37053	.5917	−21925	.5739	−21265	.5569	−20635
3	−2524	.4552	−1149	.4348	−1097	.4156	−1049
4	79301	.3501	27765	.3294	26122	.3101	24591
5	17500	.2693	4713	.2495	4366	.2315	4051
			$\Sigma = 1482$		$\Sigma = 324$		$\Sigma = -729$

$$\frac{X}{2\%} = \frac{324}{(324 + 729)}$$

$$X = 0.62$$

$$ROR = [32 + .62]\%$$

$$= 32.62\%$$

[a] A negative net value indicates expenses exceed revenue for this period.
[b] PWF = Present Worth Factor.

move closer to the expense curve, thus reducing the area between the two. This also reduces the overdraft and peak financial requirement.

The calculations to determine the IRR of this payment sequence are given in Table 11.4. The correct value is bracketed between 32 and 34%. The final rate of return (ROR) is 32.6%. This indicates that payment of the mobilization payment at the end of the first period increases the rate of return on this project to the contractor in the amount of approximately 12%. This is due in part to the reduction in the amount of inventory financing that must be carried by the contractor. What would be the impact of paying a $30,000 mobilization payment immediately on commencement of the job? This is left as an exercise for the reader to determine the change in the ROR.

REVIEW QUESTIONS AND EXERCISES

11.1 Given the following cost expenditures for a small warehouse project (to include direct and indirect charges), calculate the peak financial requirement, the average overdraft, and the ROR on invested money. Sketch a diagram of the overdraft profile.

Assume 12% markup
Retainage 10% throughout project
Finance charge = 1.5% month
Payments are billed at end of month and received one month later

Month	1	2	3	4
Indirect + Direct Cost ($)	$69,000	$21,800	$17,800	$40,900

11.2 The table and graph that follows represent a contractor's overdraft requirements for a project. Complete the table shown for costs, markup, total worth, retainage, and pay received. Retainage is 10%, markup is 10%, and interest is 1% per month.

The client is billed at the end of the month. Payment is received the end of the next month, to be deposited in the bank the first of the following month.

Overdraft	−50,000	−120,500	−82,205	−13,727	+10,336
Interest	−500	−1,205	−822	−137	—
Cumulative	−50,500	−121,705	−83,027	−13,864	+10,336

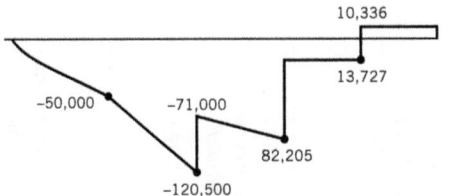

	1	2	3	4	5
Direct cost					
Indirect cost	10,000	10,000	5,000		
Total cost					
Markup					
Total worth					
Retainage					
Pay received					

11.3 Given the following data with the direct costs for each activity as shown, calculate the ROR of the contractor. Assume that (a) the markup is 15%; (b) retainage is 5% on the first 50% of worth, and 0% thereafter; (c) payment requests are submitted at the end of each month, and payments are received one month later; and (d) the finance charge is 1% per month of the amount of the overdraft at the end of the month.

Timing and allocation	$25,000	$65,000	$75,000	$15,000	
			Total direct costs		$180,000
Indirect costs $5000/month	5,000	5,000	5,000	5,000	
			Total indirect costs		$20,000
	$30,000	$70,000	$80,000	$20,000	$200,000

11.4 A contractor is preparing to bid for a project. He has made his cost estimate together with the schedule of work. His expected expenses and their time occurrence are as shown in the following table. For simplicity of analysis he assumed that all expenses are recognized at the end of the month in which they occur.

(a) The contractor is planning to add 10% to his estimated expenses to cover profits and office expenses. The total will be his bid price. He is also planning to submit for this progress payment at the end of each month. Upon approval, the owner will subtract 5% for retainage and pay the contractor one month later. The accumulated retainage will be paid to the contractor with the last payment (i.e., end of month 13).

(i) Develop the cash flow diagram.

(ii) What is the peak financial requirement and when does it occur?

(b) Assume the same as in part (a), except that the owner will retain 10% instead of 5%. Plot the cash flow diagram and calculate the peak financial requirement.

Month	Mobilization Demobilization	Subcontractors	Materials	Payroll	Equipment	Field Overhead
0	$40,000	$0	$0	$0	$0	$0
1	0	10,000	10,000	10,000	20,000	1,000
2	0	30,000	20,000	15,000	10,000	5,000
3	0	30,000	30,000	20,000	20,000	6,000
4	0	40,000	30,000	20,000	30,000	6,000
5	0	50,000	40,000	40,000	20,000	6,000
6	0	50,000	40,000	40,000	15,000	6,000
7	0	40,000	30,000	40,000	10,000	6,000
8	0	40,000	10,000	20,000	10,000	6,000
9	0	70,000	10,000	10,000	10,000	6,000
10	0	30,000	5,000	5,000	10,000	6,000
11	0	30,000	5,000	5,000	5,000	6,000
12	20,000	50,000	0	5,000	5,000	5,000
Total	$60,000	$470,000	$230,000	$230,000	$165,000	$65,000

Total cost = $60,000 + $470,000 + $230,000 + $230,000 + $165,000 + $65,000 + $1,220,000
Profits + overhead 10% = $122,000
Bid price = $1,342,000

11.5   What is the cash position (overdraft) amount for the 1-Aug column in the table?

Date	1-Jun	1-Jul	1-Aug	1-Sep	1-Oct	1-Sep
Cash Inflow ($ received)	-	-	97	3,483	20,975	107,235
Cash Outflow ($ expended)	100	3,584	11,579	83,324	27,908	288
Net cash flow for month	(100)		(11,482)	(79,841)	(6,933)	106,947
Cash Position (Overdraft)						

Chapter 12

Project Funding

CONTENTS

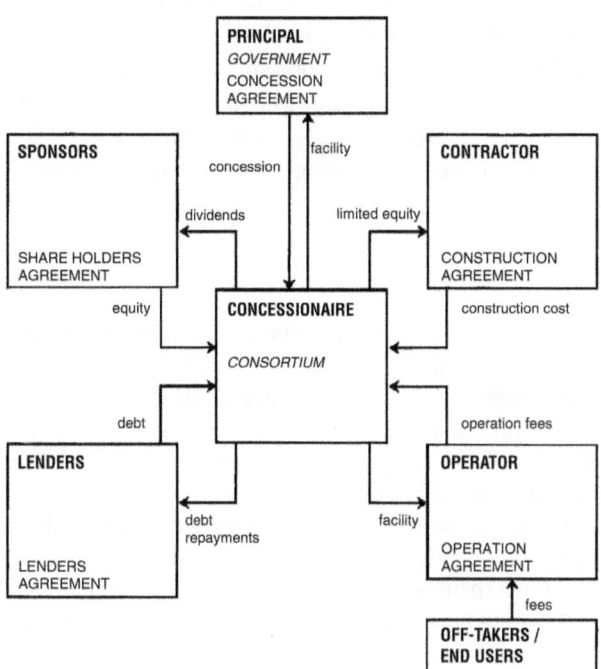

Stakeholders in BOT Funding

Build Operate and Transfer (BOT)[1]

Pollalis (1996) defines build operate and transfer (BOT) as follows:

> *In the BOT approach, a private party or concessionaire retains a concession for a fixed period from a public party, called principal (client), for the development and operation of a public facility. The development consists of the financing, design, and construction of the facility, managing and maintaining the facility adequately, and making it sufficiently profitable. The concessionaire secures return of investment by operating the facility and, during the concession period, the concessionaire acts as owner. At the end of the concession period, the concessionaire transfers the ownership of the facility free of liens to the principals at no cost.*

The modern implementation of BOT concepts is generally accredited to the Turkish government under the leadership of Prime Minister Turgut Ozal. This method of funding and constructing a large infrastructure program was initiated in Turkey starting in 1984. In fact, the financing and construction of the Suez Canal by the French in the 19th century was done using a system that would be considered BOT by today's standards (Levy, 1996).

Traditionally, highways, dams, public buildings (e.g., jails), tunnels, etc. have been constructed using funds that were generated from taxes levied by public entities (e.g., federal, state, or municipal government). In many cases, tax payers have rebelled and

---

[1] BOT is also referred to as Public Private Partnership (PPP).

Confederation Bridge Crossing, Prince Edwards Island, Canada

failed to support the issuance of bonds and similar borrowing instruments to allow construction of critically needed public facilities. In developing countries with relatively weak economies, modest tax revenues have led to a delay in developing infrastructure to support national development.

The concept of privatization as defined by the BOT approach became popular in the early 1980s. Privatization addresses the problems of developing infrastructure projects by using private funds to finance and construct public projects. For instance, if a bridge is needed to connect two political entities separated by a river or a strait, a private consortium can raise the funds and construct the bridge recovering the cost and effort involved by charging tolls to users of the bridge. The Confederation Bridge linking New Brunswick with Prince Edward Island in Canada is an example of a BOT infrastructure project.

12.1   MONEY: A BASIC RESOURCE

The essential resource ingredients that must be considered in the construction of a project are usually referred to as the *four M's*. These basic construction resources are (a) money, (b) machines, (c) manpower, and (d) materials. They will be examined in the next few chapters in this same order. Here, the first of these resources to be encountered in the construction process, money, is considered. Money (i.e., actual cash or its equivalent in monetary or financial transactions) is a cascading resource that is encountered at various levels within the project structure. The owner or developer must have money available to initiate construction. The contractor must have cash reserves available to maintain continuity of operations during the time spent awaiting payment from the owner. The major agents involved in the flow of cash in the construction process are shown in simple schematic format in Figure 12.1.

Rising construction costs have increased the pressure on the construction industry to carefully monitor and control the flow of money at all levels. As a result, more emphasis is being placed on cash flow and cost control functions in construction management than ever before. In the planning phases, more thorough investigations and more accurate cost

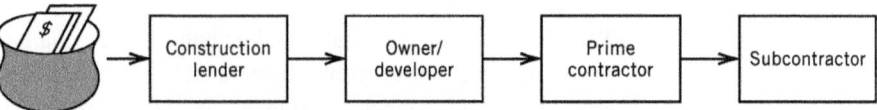

Figure 12.1 Project money flow.

estimates are being required for those seeking financial backing. To remain competitive, contractors are being forced to monitor their cost accounts more closely and to know where losses are occurring. In this chapter, the methods by which the owner/entrepreneur acquires project funding will be considered. The relationship between the flow of money from owner to contractor and its impact on the contractor's project financing was discussed in Chapter 11.

12.2   CONSTRUCTION FINANCING PROCESS

The owner's financing of any significant undertaking typically requires two types of funding: *short-term* (construction) funding and *long-term* (mortgage) funding. The short-term funding is usually in the form of a construction loan, whereas the long-term financing involves a mortgage loan over a term ranging from 10 to 30 years.

The short-term loans provide funds for items such as facility construction, land purchases, or land development. Typically these short-term loans extend over the construction period of the project. For large and complex projects, this can be a period of 6 to 8 years as in the case of utility power plants. A short-term loan is provided by a lending institution, based on the assurance that it will be repaid with interest, by some other loan. This subsequent mortgage loan constitutes the long-term financing. Therefore, the first objective of any entrepreneur is to seek a commitment for long-term, or permanent, financing from a mortgage lender. Regardless of the type of project, this commitment will permit the construction loan, and any other funding required, to be obtained with relative ease or, at least, more easily.

Unless he or she is in a position to raise the funds required directly by the issue of personal securities, the entrepreneur will seek to obtain a commitment from one of several alternate sources, including real estate investment trusts (REITs), investment or merchant banks, commercial banks, savings and loan associations, insurance companies, governmental agencies (e.g., Veterans Administration, Federal Housing Administration), or, in special cases, from one of the international development banks. Public institutions often raise project construction funds by the sale of bonds. The choice of lender depends on the type and size of project. The choice of security employed depends on a number of factors such as relative cost, the time period for which the funds will be available, and the degree of flexibility involved (the freedom to pay out or refinance) as to whether there are any restrictions involved and whether there is any sacrifice of control to the lender. The funding of some larger projects may be handled by a consortium of international bankers (e.g., the Channel Tunnel).

Lending institutions are cautious; they are not interested in financing failures or in owning partially completed projects. Therefore, they normally undertake a great deal of research and evaluation prior to providing a commitment for funding. At a minimum, an entrepreneur will be expected to provide the following as part of the loan application:

1. A set of financial statements for the firm.
2. Personal financial statements from the principals of the firm.
3. Proof of clear title to the land for the project and documentation that it has an appropriate zoning.
4. Preliminary floor plans and elevations for the project.
5. Preliminary cost estimates.
6. A market research study to verify expected income.
7. A detailed pro forma indicating projected income and expense throughout the life of the mortgage loan.

12.3   LONG-TERM PRO FORMA EXAMPLE

An example of a long-term finance pro forma for a venture involving the construction and leasing of a 75-unit apartment complex is shown in Figure 12.2a. This document indicates that the annual income from the proposed apartment complex project will be $306,830. The requested loan is $2,422,000, and the annual debt service (i.e., interest) on this amount is $236,145, realizing an income after debt service of approximately $70,000. The ratio between income and debt service is 1.3. Lenders normally wish this ratio to be below 1.3. The basis for the loan amount is given in Figure 12.2b. Items 1 to 34 are construction-related items and are developed from standard references [e.g., *R. S. Means Building Construction Cost Data* (published annually)] based on unit measures such as square footage. The lender normally has a unit-price guide for use in verifying these figures. Items 35 to 46 in Figure 12.2b cover nonconstruction costs that are incurred by the entrepreneur. It should be noticed that the interest for the construction loan is included in the costs carried forward to the long-term financing.

The method used to calculate the actual dollar amount of the loan is of great interest to the entrepreneur. The interest the developer pays for the use of the borrowed money is an expense, and it is generally considered prudent business policy to minimize expenses. One way to minimize the interest expense would be to borrow as little as possible. This is not, however, the way a developer moves toward the objective of financing the project. The developer seeks primarily to protect personal assets (or those of the company) in efforts to complete the project. The more the developer invests, the more he stands to lose if the project fails. With this consideration in mind, the developer may seek to minimize his or her own investments. That is, the developer tries to expand a small initial asset input into a large amount of usable money. This is called *leverage*. He takes a small amount and levers, or amplifies, it into a large amount.

Figure 12.2a Pro forma for 75 apartment units.

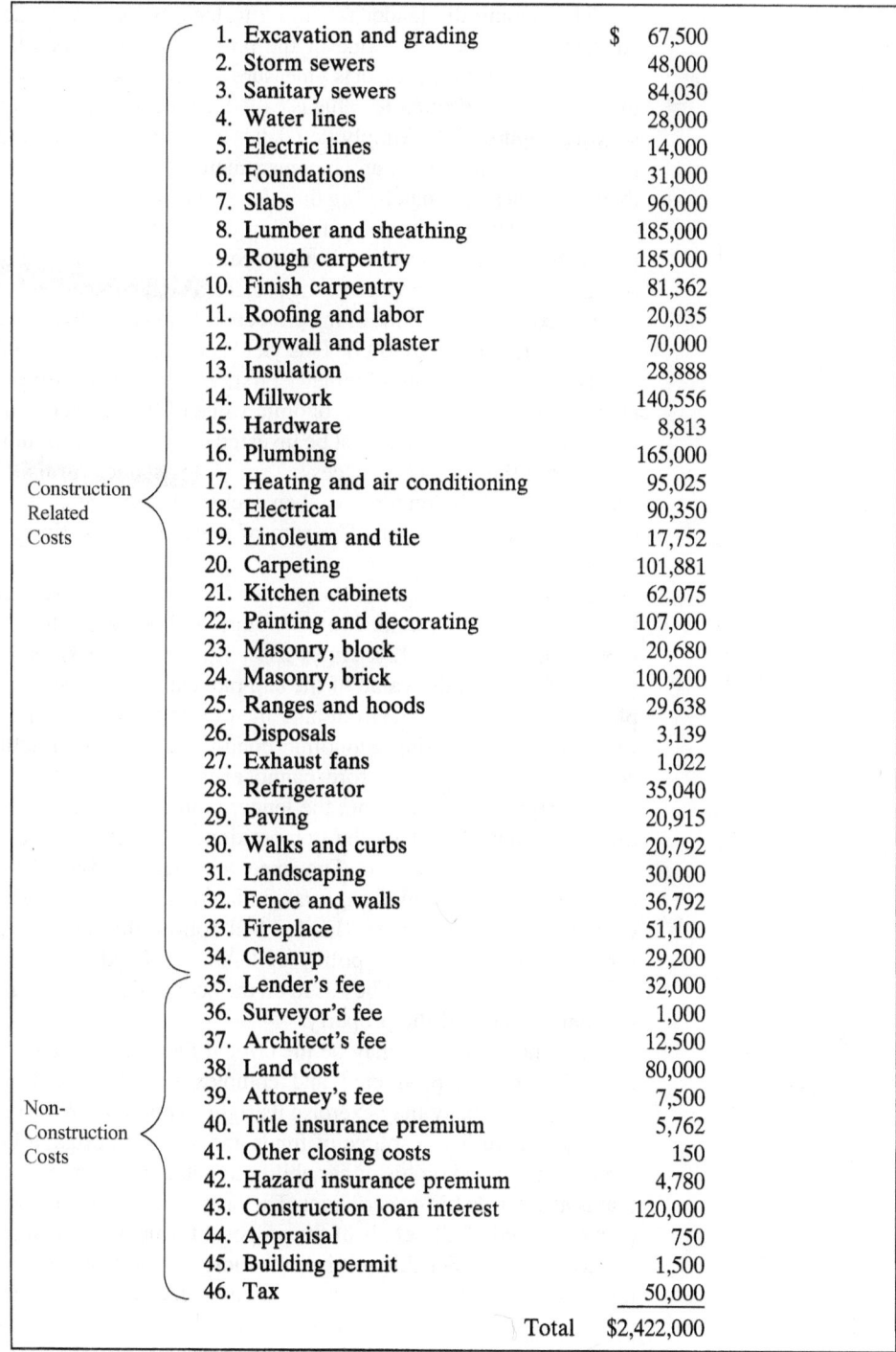

	1. Excavation and grading	$ 67,500
	2. Storm sewers	48,000
	3. Sanitary sewers	84,030
	4. Water lines	28,000
	5. Electric lines	14,000
	6. Foundations	31,000
	7. Slabs	96,000
	8. Lumber and sheathing	185,000
	9. Rough carpentry	185,000
	10. Finish carpentry	81,362
	11. Roofing and labor	20,035
	12. Drywall and plaster	70,000
	13. Insulation	28,888
	14. Millwork	140,556
	15. Hardware	8,813
	16. Plumbing	165,000
Construction	17. Heating and air conditioning	95,025
Related	18. Electrical	90,350
Costs	19. Linoleum and tile	17,752
	20. Carpeting	101,881
	21. Kitchen cabinets	62,075
	22. Painting and decorating	107,000
	23. Masonry, block	20,680
	24. Masonry, brick	100,200
	25. Ranges and hoods	29,638
	26. Disposals	3,139
	27. Exhaust fans	1,022
	28. Refrigerator	35,040
	29. Paving	20,915
	30. Walks and curbs	20,792
	31. Landscaping	30,000
	32. Fence and walls	36,792
	33. Fireplace	51,100
	34. Cleanup	29,200
	35. Lender's fee	32,000
	36. Surveyor's fee	1,000
	37. Architect's fee	12,500
	38. Land cost	80,000
Non-	39. Attorney's fee	7,500
Construction	40. Title insurance premium	5,762
Costs	41. Other closing costs	150
	42. Hazard insurance premium	4,780
	43. Construction loan interest	120,000
	44. Appraisal	750
	45. Building permit	1,500
	46. Tax	50,000
	Total	$2,422,000

Figure 12.2b Construction cost breakdown for 75 apartment units.

The amount of the mortgage loan should be a happy medium between too much and too little. If the mortgage is too small, there will not be enough to cover the project. On the other hand, if the mortgage is too large, the developer will find that the individual mortgage payments will exceed available revenue and may be unable to meet all financial obligations.

The amount the lender is willing to lend as long-term funding is derived from two concepts: the economic value of the project and the capitalization rate (cap rate). The economic value of the project is a measure of the project's ability to earn money. One method of predicting the economic value is called the income approach to value and is the method shown in Figure 12.2a. Simply stated, it is the result of an estimated income statement of the project in operation. Like any income statement, it shows the various types of revenue and their sum. These are matched against the predicted sums of the different expenses. Although the predicted net income is a function of many estimated numbers, commonly a fairly reasonable degree of accuracy is achieved. The expected net income divided by the cap rate produces the economic value of the project. The cap rate used in Figure 12.2a is 9.5%. The capitalized economic value of the project is obtained by dividing the net income ($306,830) by the cap rate factor (0.095). This yields an economic value of $3,229,789.

How is the cap rate obtained? First, a lender generally provides a mortgage that is about 75% of the estimated economic value of the project. This is done because 25% of the value, or thereabouts, must be invested by the developer and will serve as an incentive for making the project a success. That is, the lender furnishes 75% and the developer furnishes 25%. The lender must then decide what the interest rate will be and takes up the developer's rate of return. The sum of these numbers, times their respective portions, gives the cap rate.

As an example, suppose that the lender decides that the interest rate will be 8.5% and that the developer's planned rate of return will be 12%. Then, the cap rate is obtained as 8.5% times 75% plus 12% times 25%, which gives 9.375% or 0.09375 as the cap rate factor. Obviously, the value of the cap rate can be adjusted by the values that the lender places on the own interest rate and the developer's rate of return. These numbers are a function of the existing economic conditions and thus fluctuate with the state of the economy. The lender, therefore, cannot exert as much influence on their values as might at first be expected. In addition, the lender is in business to lend and wisely will not price himself or herself out of the competition. He will attempt to establish a rate that is conservative but attractive. The expected income divided by the cap rate yields the economic value. The mortgage value may then be on the order of 75% of the calculated economic value. Not every lender will follow this type of formula approach; some, for example, may have a policy of lending a fixed proportion of their own assessed valuation, which may not be based on the economic value but instead on their estimate of the market value of the property.

The mortgage loan may be the critical financial foundation of the entire project and may also involve protracted and complex negotiations. For this reason, the project developing company may exercise its right to hire a professional mortgage broker whose business it is to find a source of funds and service mortgage loan dealings. The broker's reputation is based on his or her ability to obtain the correct size mortgage at the best rate that is also fair to his or her client. The broker acts as an advisor to the client, keeping the client apprised of all details of the proposal financing in advance of actually entering into the commitment. For this service, the mortgage broker receives a fee of about 2% of the mortgage loan, although the rate and amount will vary with the size of the loan.

12.4   MORTGAGE LOAN COMMITMENT

Once the lending institution has reviewed the venture and the loan committee of the lender has approved the loan, a preliminary commitment is issued. Most institutions reserve their final commitment approval until they have reviewed and approved the final construction plans and specifications.

The commitment issued is later embodied in a formal contract between the lender and borrower, with the borrower pledging to construct the project following the approved plans, and the lender agreeing that upon construction completion, and the achievement of target occupancy, the agreed upon amount of funds will be provided at the stated interest rate for the stated period of time. As noted previously, the actual amount of funds provided generally is less than the entire amount needed for the venture. This difference, called owner's equity, must be furnished from the entrepreneur's own funds or from some other source. The formal commitment will define the floor and ceiling amounts of the long-term loan.

During the construction period, no money flows from the long-term lender to the borrower. Funds necessary for construction must be provided by the entrepreneur or obtained from a short-term construction lender. Typically the lender of the long-term financing will pay off the short-term loan in full, at the time of construction completion, thereby canceling the construction loan and leaving the borrower with a long-term debt to the mortgage lender.

12.5   CONSTRUCTION LOAN

Once the long-term financing commitment has been obtained, the negotiation of a construction loan is possible. Commercial banks often make construction loans because they have some guarantee that the loan will be repaid from the long-term financing. However, even in these situations, there are definite risks involved for the short-term lender. These risks relate to the possibility that the entrepreneur or contractor may, during construction, find themselves in financial difficulties. If this occurs, it may not be possible for the entrepreneur/contractor to complete the project, in which case the construction leader may have to take over the job and initiate action for its completion. This risk is offset by a discount (1–2%), which is deducted from the loan before any money is disbursed. For example, if the amount of construction money desired is $1,060,000, the borrower signs a note that he will pay back $1,020,000. The borrower, in effect, pays immediately an interest of $20,000. This is referred to as a discount and may be viewed as an additional interest rate for the construction loan. One way to minimize these risks is to require the borrower to designate his intended contractor and design architect. The lender may also require that all contactors involved in the construction be bonded as well. Some commercial banks evaluate and seek to approve the owner's intended contractor, his prime subcontractors, and the owner's architect, as a prerequisite to approving the construction loan. This evaluation extends to an evaluation of their financial positions, technical capabilities, and current workloads.

To minimize the risks involved, the banks will also base their construction loans on the floor of the mortgage loan, and only 75 to 80% of this floor will be lent. Of course, the developer may need additional funds to cover construction costs. One way to ensure this is to finance the gap between the floor and ceiling of the long-term mortgage loan. The entrepreneur goes to a lender specializing in this type of financing and obtains a standby commitment to cover the difference or gap between what the long-term lender provides and the ceiling of the long-term mortgage. Then, if the entrepreneur fails to achieve the breakeven rent roll, he still is ensured of the ceiling amount. In this situation, the construction lender will provide 75 to 80% of the ceiling rather than the floor. If the floor of the loan is $2.7 million and the ceiling is $3 million, the financing of the gap can lead to an additional $240,000 for construction (i.e., 80% of $300,000). Financing of the gap is usually expensive, requiring a prepaid amount of as much as 5% to the gap lender. In the preceding example, this would be $15,000 paid for money that may not be required if the rent roll is achieved. Nevertheless, the additional $240,000 of construction funding may be critical to completion of the project and, therefore, the $15,000 is well spent in ensuring that the construction loan will include this gap funding.

Once the construction loan has been approved, the lender sets up a draw schedule for the builder or contractor. This draw schedule allows the release of funds in a defined

Figure 12.3 (*a*) Profile of project financing by the entrepreneur and (*b*) draw schedule.

pattern, depending on the site and length of the project. Smaller projects, such as single unit residential housing, will be set up for partial payments based on completion of various stages of construction (i.e., foundation, framing, roofing, and interior), corresponding to the work of the various subcontractors who must be paid (see Figure 12.3). For larger projects, the draw schedule is based on monthly payments. The contractor will invoice the owner each month for the work put in place that month. This request for funds is usually sent to the owner's representative or architect who certifies the quantities and value of work in place. Once approved by the architect and representative, the bank will issue payment for the invoice, less an owner's retainage (see Chapter 3).

The owner's retainage is a provision written into the contract as an incentive for the contractor to complete the project, as well as a reserve fund to cover defective work that must be made good by the contractor before the retainage is released. Typically retainage is 10%, although various decreasing formulas are also used. When the project is completed, approved, cleared, and taken over by the owner, these retainages are released to the builder.

In addition to the funds mentioned, the developer should be aware that some front money is usually required. These funds are needed to make a good-faith deposit on the loan to cover architectural, legal, and surveying fees and for the typical closing costs.

12.6 OWNER FINANCING USING BONDS

Large corporations and public institutions commonly use the procedure of issuing bonds[2] to raise money for construction projects. A bond is a kind of formal IOU issued by the borrower promising to pay back a sum of money at a future point in time. Sometimes this proviso is supported by the pledging of some form of property by way of security in case of default by the borrower. A series of bonds or debentures, issued on the basis of a prospectus, are the general type of security issued by corporations, cities, or other institutions, but not by individual owner-borrowers. In this discussion, owner financing means financing arrangements made by those corporations or institutions that are the owners of the project property. In the illustrative material that follows, "Joe" stands as a surrogate for "any borrower" ("Joan" would have served as well). During the period in which he has use of the money, the borrower promises to pay an amount of interest at regular intervals. For instance, Joe borrows $1,000 and agrees to pay back the $1,000 (referred to as the principal) in full at the end of 10 years. He pays an annual interest of 8% (at the end of each year). That is, he, in effect, pays a rent of $80.00 per year on the principal sum of $1,000 for 10 years and then pays back the amount borrowed. The rent is payable at the end of each year. The sequence of payments for this situation would be as appears in Figure 12.4. When a series of bonds is issued, there may be a commitment to pay the interest due in quarterly installments rather than in one amount at the end of the year. A bond, as a long-term promissory note, may take any one of a variety of forms depending on the circumstances; mortgage bonds involve the pledging of real property, such as land and buildings; debentures do not involve the pledging of specific property. Apart from the security offered, there is the question of interest rates and the arrangements to be made for the repayment of the principal sum. Sometimes a sinking fund may be set up to provide for the separate investment, at interest, of capital installments that will provide for the orderly retirement of the bond issue. Investors find this type of arrangement an attractive condition in a bond issue.

In preparing for a bond or debenture issue, financial statements must be drawn up and sometimes a special audit may be required. A prospectus for the issue may need to be drawn up, and this will involve settling the terms of issue and of repayment, the interest rates payable, and the series of promises or conditions related to the issue, such as its relative status in terms of priority of repayment, limitations on borrowing, the relative value of the security, and the nomination of a trustee to watch the interests of bond or debenture holders. These details are usually settled with the aid of specialists such as a certified public accountant (CPA) firm or the mortgage broker.

Public bodies may need the approval of some local regulatory authority, and corporations may have to file and have approved a prospectus for the proposed bond issue. Charters or other constitutional documents must, of course, confer on the public body or corporation

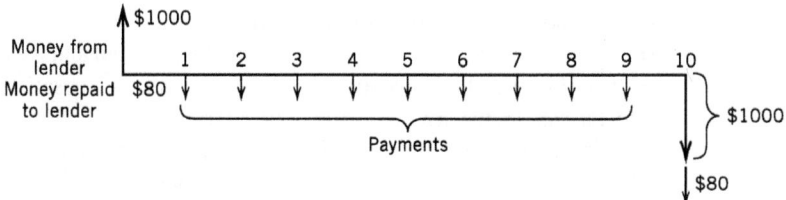

Figure 12.4 Sequence of payments for a bond.

---

[2] In this case, bonds refer to financial borrowing instruments.

Figure 12.5 Profile of revenue-expense for a bond issue.

the power to borrow money in this way; this power is exercised by the council or by the board of directors or governors. For public offerings that are particularly attractive, banks bid for the opportunity to handle the placement of the bonds. The banks recover their expense and profit by offering to provide a sum of money slightly less than the amount to be repaid. As noted previously, this is called *discounting* the loan. The fact that more will be repaid by the borrower than is lent by the lender leads to a change in the actual interest rate. This is established through competitive bidding by the banks wishing to provide the amount of the bond issue. The bank that offers the lowest effective rate is normally selected; this represents the basic cost incurred for the use of the money.

Consider the following situation in which a city that has just received a baseball franchise decides to build a multipurpose sports stadium. The design has been completed, and the architect's estimate of cost is $40.5 million. The stadium building authority has been authorized to issue $42 million in bonds to fund the construction and ancillary costs. The bonds will be redeemable at the end of 50 years with annual interest paid at 5% of the bond principal. Neither the term nor its rate purported to be representative of current market conditions. At this time the term for any bond issue would tend to be shorter and its rate higher. In some commercial dealings, "index number" escalation clauses are also occasionally seen. The banks bid the amounts for which they are willing to secure payment support. Suppose the highest bid received is $41 million.

To determine the effective rate of interest, the internal rate of return (IRR) can be calculated as discussed in Chapter 10. The profile of income and expense is shown in Figure 12.5. The effective rate of interest is that rate for which the present worth of the expenses is equal to the present worth (PW) of the revenue (in this case, $41 million). That is,

$$PW \text{ (revenue)} = PW \text{ (expenses)}$$

The annual interest paid is $2.1 million. This in an annuity A (Uniform Series) with $n = 50$ years. The $42 million must be repaid as a single payment at the end of the 50 years (therefore, it is a future amount F).

Using this information, this expression for the bond issue problem becomes:

$$\$41,000,000 = \$2,100,000 \, (P/A, i, 50 \text{ yr}) + \$42,000,000 \, (P/F, i, 50 \text{ yr})$$

In taking this approach to a solution, a trial-and-error method must be employed to solve the equation. That is, values of i must be assumed and the equation solved to see if the relationship [i.e., PW(revenue) − PW(expenses) = 0] is satisfied. In this case, two initial interest candidates are chosen for consideration $i = 5.00\%$ and $i = 6.00\%$.

The present worth factors P/A and P/F can be found consulting compound interest tables or using the formulas discussed in Chapter 10. The right side of the equation is evaluated as follows:

For $i = 5.00\%, P/A = 18.256$ and $P/F = 0.0872$
PW $= \$2,100,000 \times (18.256) + \$42,000,000 \times (0.0872) = \$42,000,000$
Difference: $\$42,000,000 - \$41,000,000 = +\$1,000,000$

For $i = 6.00\%$, $P/A = 15.762$ and $P/F = 0.0543$
$$PW = \$2,100,000 \times (15.762) + \$42,000,000 \times (0.0543) = \$35,380,000$$
$$\text{Difference} = \$35,380,000 - \$41,000,000 = -\$5,620,000$$

Since the equation balance goes from plus to minus, the value satisfying the relationship is between 5.00 and 6.00%. Using linear interpolation (i expressed in decimal form, 0.05 and 0.06), the effective interest rate is found to be

$$\dot{z} = 0.05 + \left\{ (0.06 - 0.05) \times \left[\frac{1,000,000}{1,000,000 + 5,620,000} \right] \right\}$$
$$= 0.05 + \left\{ 0.01 \times \left[\frac{1,000,000}{6,620,000} \right] \right\} = 0.0515$$

Therefore, the effective rate of interest is $i = 5.15\%$.

REVIEW QUESTIONS AND EXERCISES

12.1 What is the present level of the prime rate? How does this rate relate to the current financing and overdraft charges for new building construction in your locality? (How many points above the prime is this rate?) Does this overdraft rate vary with the magnitude of the monies involved?

12.2 Referring to the example in Figure 12.2a, suppose the market rent for a two-bedroom unit is $550 per month and for a three-bedroom unit is $650 per month. If the going cap rate is 10%, rework the pro forma calculations for the apartment project of Figure 8.2. Then determine the lender's interest rate. What is the new breakeven vacancy factor?

12.3 What determines the number of draws a builder can make in completing a facility? What is the existing policy regarding number of draws in your locality?

12.4 Suppose that for the multipurpose sports stadium example considered in the text, the bond issue was for 40 years and the annual interest rate is 9% of the bond principal. Using the architect's estimate of cost of $41 million, determine the effective interest rate.

12.5 Suppose in the preceding problem that the bonds are financed by a bank that discounts the bond issue to $40 million. What is the new effective interest rate?

12.6 What were the reasons that originated the real estate bubble of the early to mid-2000s? Do some research to justify your reasoning.

12.7 What should be done to avoid a repetition of the real estate bubble of the early to mid-2000s? Support your analysis with pertinent facts.

Chapter 13

Equipment Ownership

CONTENTS

Radio Frequency Identification Application in Construction

Electronic identification tags can be used for tracing and identifying construction equipment on construction sites. With radio frequency identification (RFID) technology, no line of sight or direct contact is required between the reader and the tag. Because RFID does not rely on optics, it is ideal for dirty, oily, wet, or harsh environments. RFID is an automatic identification technology, similar to bar code technology, with positive identification and automatic data transfer between a tagged object and a reader. Because the RFID tags are read by low wattage radio waves, instead of light waves (as with bar codes), they will communicate through non-metallic materials such as paint, plastic, grease, and dirt and are impervious to vibration, light, water, and heat up to 100°C in most cases.

A RFID system consists of two major components, called *reader* and *tag*, that work together to provide the user with a noncontact solution to uniquely identify people, assets, and locations. The reader performs several functions, one of which is to produce a low-level radio frequency magnetic field. This radio frequency magnetic field serves as a carrier of power from the reader to the passive (no battery required) RFID tag. When a tag is brought into the magnetic field produced by the reader, the recovered energy powers the integrated circuit in the tag and the memory contents are transmitted back to the reader. Once the reader has checked for errors and validated the received data, the data are

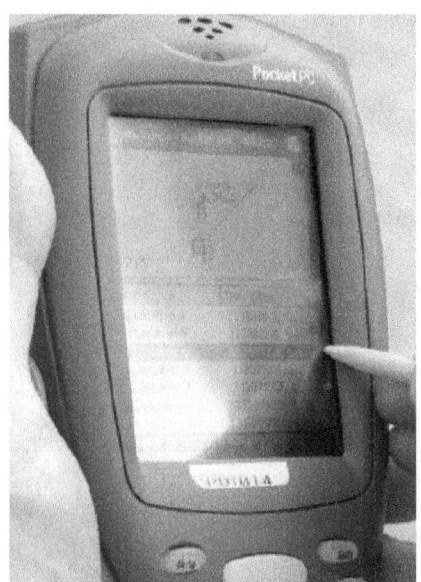

Hand held RFID device

decoded and restructured for transmission to a user in the format required by the host computer system. The RFID tags used are both readable and writable. This capability enables information to be written back to the tag for enhanced asset management. RFID tags do not require a line of sight for identification, and readability is not affected by bright lighting situations.

13.1   GENERAL

Equipment resources play a major role in any construction activity. Decisions regarding equipment type and combination can have a major impact on the profitability of a job. In this respect, the manager's goal is to select the equipment combination that yields the maximum production at the best or most reasonable price. Quite obviously, the manager must have a basic understanding of the costs associated with a particular piece of equipment. This person must also be capable of calculating the rate of production of the piece or combination of equipment. The cost and the rate of production combine to yield the cost per unit of production. For example, if it is estimated that the cost of a particular fleet of haulers and loaders is $500 per hour and the production rate is 750 cubic yards per hour, the unit price can be easily calculated as $0.66 per cubic yard.

Construction equipment can be divided into two major categories. Productive equipment describes units that alone or in combination lead to an end product that is recognized as a unit for payment. Support equipment is required for operations related to the placement of construction such as movement of personnel and materials and activities that influence the placement environment. Typical production units are pavers, haulers, loaders, rollers, and entrenchers. Hoists, lighting sets, vibrators, scaffolds, and heaters represent typical classes of support equipment. In most cases, equipment units are involved either in handling construction materials at some point in the process of placing a definable piece of construction (e.g., crane lifting a boiler, pavers spreading concrete or asphalt into lifts on a base course) or in controlling the environment in which a piece of construction is realized (e.g., heaters controlling ambient temperature, prefabricated forms controlling the location of concrete in a frame or floor slab).

In heavy construction, large quantities of fluid or semifluid materials such as earth, concrete, and asphalt are handled and placed, leading to the use of machines. The equipment mix in such cases has a major impact on production, and the labor component controls production rates only in terms of the skill required to operate machines. Therefore, heavy construction operations are referred to as being equipment intensive. Heavy construction contractors normally have a considerable amount of money tied up in fixed equipment assets because capitalizing a heavy construction firm is a relatively expensive operation.

Building and industrial construction require handwork on the part of skilled labor at the point of placement and are therefore normally not as equipment intensive.

Equipment is required to move materials and manpower to the point of installation and to support the assembly process. Emphasis is on hand tools; and, although heavy equipment pieces are important, the building and industrial contractors tend to have less of their capital tied up in equipment. Also because of the variability of equipment needs from project to project, the building contractor relies heavily on the renting of equipment. The heavy construction contractor, because of the repetitive use of many major equipment units, often finds it more cost effective to own this equipment.

13.2   EQUIPMENT OWNING AND OPERATING COSTS

The costs associated with construction equipment can be broken down into two major categories. Certain costs (e.g., depreciation, insurance, and interest charges) accrue whether the piece of equipment is in a productive state or not. These costs are fixed and directly related to the length of time the equipment is owned. Therefore, these costs are called fixed, or ownership, costs. The term fixed indicates that these costs are time dependent and can be calculated based on a fixed formula or a constant rate basis. On the other hand, the operation of a machine leads to operating costs that occur only during the period of operation. Some of these costs accrue because of the consumption of supplies, such as tires, gas, and oil, and the widespread practice of including the operator's wages in the operating costs. Other costs occur as a result of the need to set aside moneys for both routine and unscheduled maintenance. Thus, operating costs are variable costs.

The total of owning and operating costs for items of equipment such as tractors, shovels, scrapers, dozers, loaders, and backhoes is typically expressed on an hourly basis. These two categories of cost accrue in different ways. Ownership costs are usually arrived at by relating the estimated total service life in hours to the total of those costs. If the equipment is idle for some of those hours, the relevant costs would be taken up as part of general operating overhead; when the equipment is in use, the hourly costs are charged to the job or project.

Operating costs are variable in total amount, being a function of the number of operating hours, but these hourly costs are found to be relatively constant.

The hourly charge for a piece of equipment is made up of four elements. An allowance for estimated hourly overhead costs is added to the ownership and operating costs. The fourth element is an amount for income or profit. A schematic illustration of this breakdown of the hourly charge for a piece of equipment is shown in Figure 13.1.

Ownership costs are composed of two elements: first, an estimate for depreciation on the cost of using the equipment itself. Each piece of equipment represents an estimated number of hours of useful service life and the depreciable value, the major part of its original cost, is divided by the total hours to yield a charging rate for this element of equipment costs. The second component of ownership costs consists of estimates of allowance for interest, insurance, and taxes.

Operating costs cover a broader range of items, the principal elements being: fuel, oils and lubricants, hydraulics fluid, grease, filters, and other supplies; maintenance, general

Figure 13.1 Cost components in a production unit.

overhauls, and repairs; and parts replacement (e.g., cutting edges, blades, and buckets), tire replacements, and the like. Also included here are the direct labor costs—the operator's wages—including all of the expense loadings for holidays, sick leave, and insurance.

To the direct operating costs just enumerated are added allowances for general overhead expenses and the indirect costs of supervisory labor. This total establishes the total hourly cost of owning and operating a unit of equipment. A percentage markup is added to provide for an income or profit element.

Some of these costs are incurred and paid for concurrently with the operation of the equipment, but the allowances or estimates included for items such as repairs and maintenance are provisions for costs that will have to be paid at some future time.

General administrative costs (e.g., overhead) are an allowance that forms part of the hourly charging rate. It includes items such as telephones, stationery, postage, heat, light and power, and the costs of idle equipment in general, are aggregated together as general overhead expense.

13.3   DEPRECIATION OF EQUIPMENT

The method by which depreciation is calculated for tax purposes must conform to standards established by the Internal Revenue Service (IRS). Federal law has introduced the use of fixed percentages as given in published tables to calculate the amount of depreciation for various classes of equipment and depreciable property. The tables have replaced accelerated methods referred to as the declining balance and sum-of-years-digits (SOYD) methods, which were used prior to 1981. The methods used under pre-1981 legislation are still relevant in understanding the tables presently used and are still required in some situations, and therefore, they will be described briefly.

The four most commonly used methods of calculating depreciation on equipment prior to 1981 are:

1. Straight line

2. Declining balance (DB)

DB and sum of years are referred to as *accelerated* methods because they allow larger amounts of depreciation to be taken in the early years of the life of the asset. (Only the DB method will be described in this chapter.) The contractor usually selects a method that offsets or reduces the reported profit for tax purposes as much as possible. In effect, for companies paying taxes at the corporate rate (assume 34%), each dollar of depreciation reduces the amount of tax paid by 34 cents (assuming that the depreciation does not reduce revenue below zero).

Most heavy construction contractors assume that each machine in the fleet is a small "profit center" and will attempt to apply any depreciation associated with a piece of equipment to offset the profit generated by that machine. The major factors to be considered in calculating the depreciation of an asset are shown in Figure 13.2. The three major factors form the three sides of the depreciation "box" that are linked by the method of depreciation selected. They are:

1. Initial cost or basis in dollars

2. Service life in years or hours

3. Salvage value in dollars

The amount that can be depreciated or claimed by way of a tax deduction is the difference between the initial net value of the asset and its residual or salvage value. This is referred to as the depreciable amount and establishes the maximum number of depreciation dollars available in the asset during its service life.

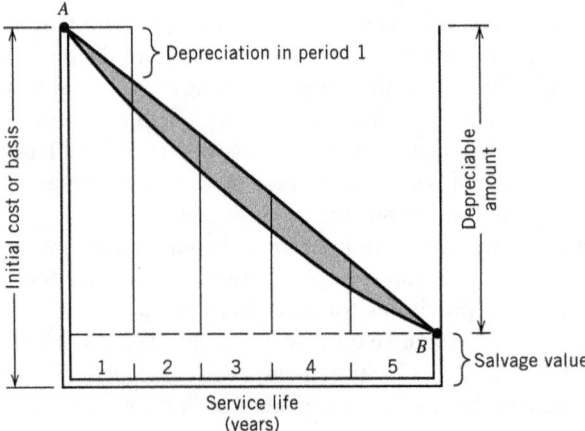

Figure 13.2 Factors in depreciation.

The declared initial cost of the asset must be acceptable in terms of the IRS definition of depreciable cost. For instance, suppose a $75,000 scraper is purchased. The tires on the scraper cost $15,000. These tires are considered a current period expense and therefore are not depreciable. That is, they are not part of the capital asset for purpose of depreciation. The tires are considered consumable and have a service life different from that of the asset. In this case, the initial value of the scraper for depreciation purposes is $60,000.

The initial depreciable cost or basis is often referred to as the net first cost. In addition to the purchase price minus major expenses, items such as tires, freight costs, and taxes are included in the net first cost and are part of the amount depreciable. If we have purchased a rubber-tired wheeled tractor, the net first cost for purposes of depreciation would be arrived at as follows:[1]

Purchase price	$84,000 (Price at factory, FOB)[1]
Less tires	$ 4,000
	$80,000
Plus tax at 5%	$ 4,000
Plus freight	$ 2,800
Net first cost	$86,800

The depreciable basis for the calculation of depreciation allowances is this first cost of $86,800.

The concept of salvage value implies that there is some residual value in the piece of equipment (i.e., scrap value) at the end of its life. Unless this value exceeds 10% of the first cost of the equipment, this value is neglected and the entire first cost is considered to be available for depreciation. In the case cited, if the salvage value is less than $8,680, the entire first cost will be considered as depreciable and the piece of equipment will yield tax payment reductions in the amount of $29,512 for a contractor paying 34% of marginal tax (i.e., $86,800 × 0.34) across its service life.

The IRS publishes tables indicating the appropriate service life values. Most construction equipment items fall into the 3-, 5-, or 7-year service life categories. Manufacturers typically publish tables such as that shown in Table 13.1 indicating a variable service life based on operating conditions. Service life is defined by the IRS tables, and the only question has to do with the category or class of property to which an equipment type is to be assigned.

Given the present highly defined system of depreciation based on fixed tabular percentages, decisions regarding depreciation are simplified as to whether an accelerated

---

[1] FOB is discussed in Section 16.2 of Chapter 16. In this case it indicates the cost of the equipment at the factory prior to shipment.

Table 13.1 Estimated Service Life Table (Caterpillar Tractor Co.)

Type of Equipment	Excellent Conditions: Hours	Average Conditions: Hours	Severe Conditions: Hours
Track-type tractors Traxcavators Wheeled loaders Wheeled tractors Scrapers	12,000	10,000	8,000
Motor graders	15,000	12,000	10,000

To determine the cost per hour due to depreciation, the above information may be used as follows:

$$\text{Depreciation cost per hour} = \frac{\text{Purchase price } - \text{ Tire value}}{\text{Estimated service life in hours}}$$

or linearly prorated system of depreciation is to be used. To better understand the concepts behind the tables and the prorated system, two of the basic methods of calculating depreciation will be discussed in the following sections.

13.4   STRAIGHT-LINE METHOD

An accountant (and the IRS) would describe the straight-line method of calculating allowable depreciation as being based on the assumption that the depreciation, or the loss in value through use, is uniform during the useful life of the property. In other words, the net first cost or other basis for the calculation, less the estimated salvage value, is deductible in equal annual amounts over the estimated useful life of the equipment. An engineer would call this a linear method. This simply means that the depreciable amount is linearly prorated or distributed over the service life of the asset. Let us assume that we have a piece of equipment that has an initial cost or base value of $16,000 and a salvage value of $1,000. The service life is 5 years and the depreciable amount is $15,000 (initial cost minus salvage value). If we linearly distribute the $15,000 over the 5-year service life (i.e., take equal amounts each year), we are using the straight-line method of depreciation. The amount of depreciation claimed each year is $3000. This is illustrated in Figure 13.3.

Figure 13.3 Straight-line depreciation.

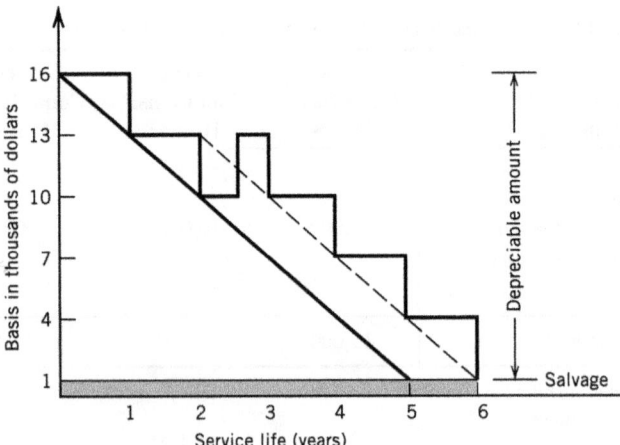

Figure 13.4 Adjustment of basis.

The remaining value of the piece of equipment for depreciation purposes can be determined by consulting the stepwise curve of declining value. During the third year of the asset's service life, for example, the remaining base value, or book value, of the asset is $10,000. If we connect the points representing the book value at the end of each year (following subtraction of the depreciation), we have the "straight line."

The concept of the base value, or book value, has further tax implications. For instance, if we sell this asset in the third year for $13,000, we are receiving more from the buyer than the book value of $10,000. We are gaining $3,000 more than the depreciated book value of the asset. The $3,000 constitutes a capital gain. The reasoning is that we have claimed depreciation up to this point of $6,000, and we have declared that as part of the cost of doing business. Now the market has allowed us to sell at $3,000 over the previously declared value, demonstrating that the depreciation was actually less than was claimed. We have profited and, therefore, have received taxable income. Prior to the 1986 tax law, a capital gain was not taxed at the full rate but at approximately half of the tax rate for normal income. Presently, capital gains are taxed as normal income (i.e., 34%). Business entities have been pressing for the reinstatement of the alternate capital gains tax rate.

The base value for depreciation is affected if we modify substantially the piece of equipment. Assume in the preceding example, that in the third year we perform a major modification on the engine of the machine at a total cost of $3,000. Since this is a capital improvement, the term basis is used to refer to the depreciation base. The modification increases the base value of the unit by $3,000 as shown in Figure 13.4 It also may extend the service life of the asset. Something similar occurs if we make some improvements to a building. The value is increased and this added value can be depreciated.

If we can depreciate real property, can we depreciate the house in which we live? Depreciation represents a cost of doing business. In most cases we do not "do business" in our own home, and therefore our home is not a depreciable asset. You can, however, think of some instances in which a person conducts some business at home. Special depreciation rules apply to that situation.

13.5 DECLINING BALANCE

One of the accelerated methods previously (prior to 1981) used is the declining balance (DB) method. When applied to new equipment with a useful life of at least 3 years, the effective rate at which the balance is reduced may be twice the straight-line rate. For this reason, the expression double-declining balance (DDB) is used when this IRS option is applied to new assets. For assets that are not purchased new but are secondhand, the

Table 13.2 Double-Declining-Balance Method

SLY	Rate Applied to Balance (%)	Book Value End of Previous Year ($)	Depreciation for this Year ($)	Book value End of this Year ($)
1	40	16,000.00	6,400.00	9,600.00
2	40	9,600.00	3,840.00	5,760.00
3	40	5,760.00	2,304.00	3,456.00
4	40	3,456.00	1,382.40	2,073.60
5	40	2,073.60	829.44	1,244.16
			TOTAL $14,755.84	

optional rate is 150% of the straight-line rate. In this method, it is the rate that is important because it remains constant throughout the calculations.

Formally stated, in the DB method, the amount of depreciation claimed in the previous year is subtracted from the book value (base value) at the beginning of the previous year before computing the next year's depreciation. That is, a constant rate is applied to a balance, which is declined each year by the amount claimed in the previous year. For new equipment the rate is calculated by dividing 200% by the number of service life years (SLY) (i.e., 200/SLY). For used equipment the rate is 150% divided by the SLY.

To illustrate, consider the $16,000 piece of equipment used in discussing the straight-line method. We will assume the piece is purchased new at this price. Because the service life of the unit is 5 years, the constant rate to be applied will be 200%/5 = 40%. The calculations for this example are summarized in Table 13.2.

A repetitive process of calculation can be detected. The constant rate of 40% (column 2) is applied to the book value at the end of the previous year (column 3) to obtain the depreciation (column 4). The reduced value of the property is column 3 minus column 4, as shown in column 5. The "Book Value End of This Year" for year N is the "Book Value End of Previous Year" for year $N + 1$. It follows that the value in column 3 for year 2 will be the same as the value in column 5 for year 1.

Another interesting fact is noted. The amount of depreciation taken over the 5-year service life is less than the depreciable amount. The book value at the end of 5 years is $1,244.16 and the salvage is $1,000. Therefore, $244.16 has not been recovered. Typically, the method is changed to the straight-line approach in the fourth or fifth year to ensure closure on the salvage value. This underlines the fact that the only role played by the depreciable value in the DB method is to set an upper limit on the amount of depreciation that can be recovered. That is, an asset may not be depreciated below a reasonable salvage value. A common mistake is to apply the rate to the depreciable value in the first year. The rate is always applied to the total remaining book value, which in this example during the first year is $16,000.

If the piece of equipment had been purchased used, for $16,000, the procedure would be the same but the rate would be reduced. In this case, the rate would be 150%/15 or 30%.

The 150% calculations are summarized in Table 13.3. In this situation, because $1,689.12 in unclaimed depreciation would remain at the end of year 5, the method could be changed to the straight-line approach in the fourth or fifth year with some advantage. A comparison of the DDB methods and the straight-line method is shown in Figure 13.5.

The proportionately higher rate of recovery in the early service life years is revealed by this figure. More depreciation is available in the first year using the DDB method ($6,400) than in the first 2 years using the straight-line method ($6,000). Equipment rental firms that intend to sell the equipment after the first 2 years of ownership are in a good position to capitalize on this feature of the accelerated methods. Of course, if they sell at a price well above the book value, they must consider the impact of the capital gains tax.

Table 13.3 150 Declining-Balance Method

SLY	Rate (%)	Book Value End of Previous Year ($)	Depreciation for this Year ($)	Book Value End of this Year ($)
1	30	16,000.00	4,800.00	11,200.00
2	30	11,200.00	3,360.00	7,840.00
3	30	7,840.00	2,352.00	5,488.00
4	30	5,488.00	1,646.40	3,841.60
5	30	3,841.60	1,152.48	2,689.12
			TOTAL $13,310.88	

Figure 13.5 Comparison of double-declining-balance and straight-line methods.

13.6 PRODUCTION METHOD

It was stated previously that the contractor tries to claim depreciation on a given unit of equipment at the same time the equipment is generating profit to reduce the tax that might otherwise be payable. The production method allows this because the depreciation is taken based on the number of hours the unit was in production or use for a given year. The asset's cost is prorated and recovered on a per-unit-of-output basis. If the $16,000 equipment unit we have been discussing has a 10,000-hour operation time, the $15,000 depreciable amount is prorated over 10,000 hours of productive service life. This method is popular with smaller contractors since it is easy to calculate and ensures that the depreciation available from the asset will be recovered at the same time the unit is generating profit. A reasonable estimate of the total operating hours for a piece of equipment may be obtained by referring to the odometer on the unit together with the logbook or job cards.

In some cases, unless this method is used, the units may be depreciated during a period when they are not generating income and, consequently, the full benefit of the depreciation deduction may be lost. The objective of the contractor is to have a depreciation deduction available in years in which it can be more effectively applied to reduce taxable income. It may not be possible to defer depreciation and to take it in years in which it can be applied with more advantage. Therefore, the strategy should be to have it available in the years in which profits are likely to be high. The production method ensures that depreciation deduction is available when the machine is productive and theoretically profitable or income producing.

At the time of construction of the Alaska pipeline, contractors with contracts for the access road to parallel the pipeline purchased large equipment fleets in anticipation of project start-up. Then, environmental groups delayed the project several years during which time the contractors were forced to put their equipment fleets in storage. Because these units were not in use and were not productive and profitable during this delay, the contractors claimed no depreciation.

Nevertheless, the production method allowed them to apply the depreciation at the proper time when the job mobilized and the units were put into production.

In some situations, the production method might be less desirable. If we own an entrenching machine (service life of 10,000 hours) but only operate it 500 hours per year, using the production method would stretch the period of recovery out over 20 years. If the machine is sold after 5 years, we would have claimed only one-quarter of the available depreciation. One advantage offsetting this apparent disadvantage is that we might have a smaller adjustment to make by way of capital gain on the sale. In such a case, clearly a method other than the production method would be more appropriate and a more balanced way of dealing with the matter.

13.7   DEPRECIATION BASED ON CURRENT LAW

For equipment placed in service prior to 1981, depreciation was calculated using methods such as the straight-line or DDB described previously. For equipment placed in service during the period 1981 to 1986, contractors were required to depreciate using either the Accelerated Cost Recovery System (ACRS) method or the alternate ACRS system, which was equivalent to a straight line prorating of the cost across the life of the asset (e.g., the straight-line or production method).

A set of tables defines accelerated depreciation amounts for equipment placed in service after 1986, and has been only minimally revised in the years since. The depreciation system defined by these tables is referred to as the Modified ACRS or MACRS.

The amount depreciated is calculated using prescribed depreciation methods for each class of equipment. For example, 3-, 5-, 7-, and 10-year property are depreciated using the 200% declining-balance method with a switch to the straight-line method at a time that maximizes the deduction. In addition, the "half-year convention" is used to calculate the first-year depreciation.

The alternate MACRS method uses the straight-line method as did its predecessor, the alternate ACRS.

The accelerated MACRS values are given in Table 13.4. To better understand the basis of the values in the table, consider the following situation. A $100,000 piece of equipment is to be depreciated using the accelerated MACRS method. It is assumed that the equipment has a 5-year property life. The MACRS table is based on using the 200% DB method. The rate of depreciation will be 40% (200% divided by 5 years). However, due to the half-year convention, only half of the 200 DB depreciation is taken in the first year. Therefore, the effective percent is 20 and $20,000 can be depreciated in the first year. The remaining value is $80,000. The second-year depreciation is 40% of $80.000, or $32,000. This amounts to 32% of the original $100,000 basis. The balance is now declined to $80,000—$32,000, or $48,000. Again for the third-year depreciation the 200 DB

Table 13.4 MACRS Table for Accelerated Depreciation

Recovery year	Annual Recovery (Percent of Original Depreciable Basis)					
	3-Year Class (200% d.b.)	5-Year Class (200% d.b.)	7-Year Class (200% d.b.)	10-Year Class (200% d.b.)	15-Year Class (150% d.b.)	20-Year Class (150% d.b.)
1	33.00	20.00	14.28	10.00	5.00	3.75
2	45.00	32.00	24.49	18.00	9.50	7.22
3	15.00	19.20	17.49	14.40	8.55	6.68
4	7.00	11.52	12.49	11.52	7.69	6.18
5		11.52	8.93	9.22	6.93	5.71
6		5.76	8.93	7.37	6.23	5.28
7			8.93	6.55	5.90	4.89
8			4.46	6.55	5.90	4.52
9				6.55	5.90	4.46
10				6.55	5.90	4.46
11				3.29	5.90	4.46
12					5.90	4.46
13					5.90	4.46
14					5.90	4.46
15					5.90	4.46
16					3.00	4.46
17						4.46
18						4.46
19						4.46
20						4.46
21						2.25

method yields 40% of $48,000, or $19,200. The depreciation table for this equipment is as follows:

Year	Depreciation	Book Value
1	$20,000	$80,000
2	$32,000	$48,000
3	$19,200	$28,000
4	$11,520	$17,280
5	$11,520	$ 5,760
6	$ 5,760	$ 0

It will be noted that although this is a 5-year property class equipment, it is depreciated out across a 6-year period. Also, following the third year, a switch from 200 DB to straight-line method is made to close on a residual or salvage value of zero. By dividing the depreciation amounts for each year by the original basis of $100,000, it will be seen that the percentage of depreciation in each year is the same as the values given in Table 13.4. To confirm that this method is used to determine the percentages in Table 13.4, the reader should try to calculate the annual depreciation amounts for a $100,000 piece of equipment with a 10-year service life.

13.8 DEPRECIATION VERSUS AMORTIZATION

Depreciation is a legitimate cost of business that recognizes the loss in value of equipment over time. As such, it is an expense and can be deducted from revenues, resulting in a lowering of taxes (e.g., 34 cents per dollar of depreciation as noted previously). This yields the contractor a tax savings that can be used to replace the equipment. However, this savings would represent only 34% of the original value of the equipment. To provide for the replacement of the equipment at some point in the future, contractors charge the client an amount that provides a fund to purchase new equipment. This practice of charging clients an amount to be used to purchase replacement equipment is referred to as amortizing the equipment. This is a protocol throughout the industry and allows the contractor to accumulate (i.e., escrow) funds for renewing the equipment fleet over time.

For instance, the contractor will charge clients an annual amount of $20,000 for a $100,000 equipment with a service life of 5 years. This provides $100,000 at the end of 5 years to purchase replacement equipment. Of course, due to inflation and escalating prices, a piece of replacement equipment may cost $120,000. This would indicate that the contractor should recover $24,000 per year to escrow the needed $120,000 for a new machine. Part of this amount will be recovered through depreciation due to reduced taxes: $34,000 will be available through the reduction of taxes. The contractor may consider this in calculating the amount of back charge to the client.

Since the amortization charge leads to larger revenue and the possibility of incurring income, the contractor may end up paying some tax on the amount charged to the client to amortize equipment. There is a complex interaction between depreciation and amortization, and this must be studied in the context of each equipment piece and the tax structure of each company.

13.9   INTEREST, INSURANCE, AND TAX (IIT) COSTS

In addition to the amortization/depreciation component, the ownership costs include a charge for other fixed costs that must be recovered by the equipment owner. Throughout the life of the unit, the owner must pay for insurance, applicable taxes, and either pay interest on the note used to purchase the equipment or lose interest on the money invested in equipment if the unit was paid for in cash. These costs are considered together as what can be called the interest, insurance, and tax (IIT) costs. Recovery of these charges is based on percentages developed from accounting records that indicate the proper levels that must be provided during the year to offset these costs. The percentages for each cost are applied to the average annual value of the machine to determine the amount to be recovered each hour or year with respect of these cost items.

The average annual value is defined as:

$$AAV = C(n+1)/2n$$

where AAV is the average annual value, C is the initial new value of the asset, and n is the number of service life years. This expression assumes that the salvage value is zero. What the formula does is level the declining value of the asset over its service life so that a constant average value on an annual basis is achieved. This is indicated graphically in Figure 13.6.

Applying this formula to a machine with initial capital value of $16,000 and zero salvage value, the average annual value is calculated as:

$$AAV = \$16,000 \times (6)/10 = \$9,600$$

The area under the rectangle in Figure 13.6 representing the AAV equals the area under the plot representing the straight-line decline in value. Using this fact, the formula can be derived. If we consider the salvage value, the area under the stepped curve is increased by the area of the shaded segment in Figure 13.6b. Therefore, the AAV is increased somewhat. The appropriate expression for AAV including the salvage value is:

$$AAV = [C(n+1) + S(n-1)]/2n$$

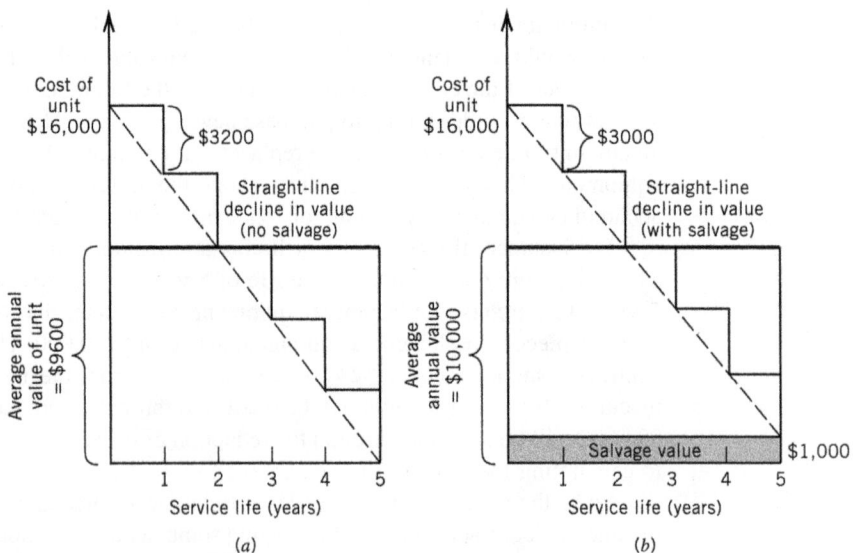

Figure 13.6 Interpretation of average annual value: (*a*) average annual value without salvage value included and (*b*) average annual value considering salvage value.

For the $16,000 piece of equipment considered, this yields

$$AAV = [\$16,000\ (6) + \$1,000\ (4)]/10 = \$10,000$$

Verification of this expression is left as an exercise for the reader.

Assume that the proper levels of the annual provision to cover IIT costs for the unit are as follows:

$$Interest = 8\% \text{ of AAV}$$
$$Insurance = 3\% \text{ of AAV}$$
$$Taxes = \underline{2\% \text{ of AAV}}$$
$$Total = 13\% \text{ of AAV}$$

The amount to cover these ownership costs must be recovered on an hourly basis by backcharging the owner. Therefore, an estimate of the number of hours the unit will be operational each year must be made. Assume the number of hours of operation for the unit is 2,000 hours per year. Then, the IIT cost per hour would be:

$$IIT = 0.13(AAV)/2000 = 0.13(9600)/2000 = \$0.624 \text{ or } \$0.62 \text{ per hour}$$

Manufacturers provide charts that simplify this calculation.

The interest component may be a nominal rate or an actual rate or, again, it may reflect some value of the cost of capital to the company. Some contractors also include here a charge for the protective housing or storage of the unit when it is not in use. These adjustments may raise the annual provision to cover IIT costs by from 1 to 5% with the following effect (in this case, each 1% charge may represent a 5 cents per hour increase in the charging rate).

Percentage of AAV (%)	General Provision for IIT Costs ($)	Hourly Rate (Base 2,000 Hours, $)
13	1248	0.62
14	1344	0.67
15	1440	0.72
16	1536	0.77
17	1632	0.82
18	1728	0.87

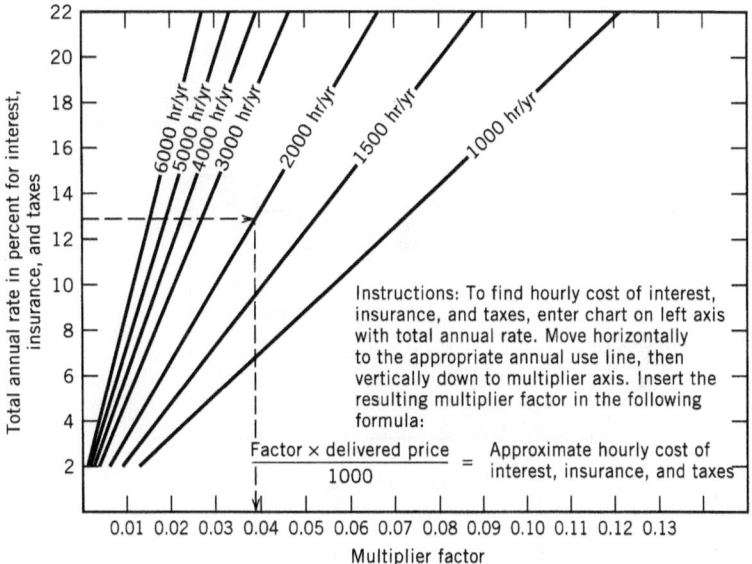

Figure 13.7 Guide for estimating hourly cost of interest, insurance, and taxes (Caterpillar Tractor Co.).

Ultimately it is the competitive situation that sets practical limits to what may be recovered. That is why the tax limitation strategies discussed previously are of such importance.

Figure 13.7 shows a chart for calculating the hourly cost of IIT. To use the chart the total percent of AAV and the estimated number of annual operating hours are required. Entering the y axis with the percent (use 13% from the example computing IIT) and reading down from the intersection of the 13% line with the 2,000-hour slant line, the multiplier factor is 0.039. The hourly charge for IIT is calculated as:

$$\text{IIT/hour} = \text{factor} \times \text{delivery price}/1000$$

In the example discussed,

$$\text{IIT/hour} = 0.039 \times 16,000/1000 = \$0.624 \quad \text{or} \quad \$0.62 \text{ per hour}$$

If the amortization/depreciation costs using the straight-line method for the $16,000 unit in the example are $1.50 per hour, the owner must recover $1.50 plus $0.62 or $2.12 per hour for fixed costs.

13.10 OPERATING COSTS

The major components contributing to the operating or variable costs are fuel, oil, and grease (FOG), tire replacement (on rubber-wheeled vehicles), and normal repairs. Normally, historical records (e.g., purchase vouchers) are available that help in establishing the rate of use of consumables such as fuel, oil, and tires. Maintenance records indicate the frequency of repair. The function that best represents the repair costs to be anticipated on a unit starts low and increases over the life of the equipment. Because repairs come in discrete amounts, the function has a stepwise appearance (see Figure 13.8).

The following guidelines for establishing the amount to set aside for repairs are taken from Caterpillar Tractor material. To estimate hourly repair costs, select the appropriate

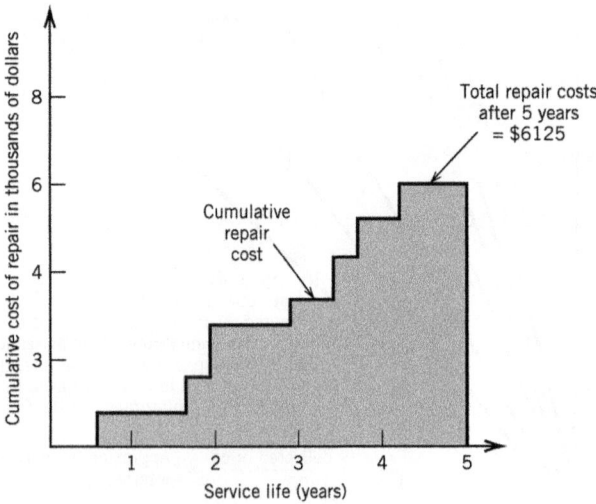

Figure 13.8 Repair cost profile.

multiplier factor from the table below and apply it in the following formula:

$$\text{Repair factor} \times (\text{delivered price} - \text{tires})/1000 = \text{estimated hourly repair reserve}$$

	Operating Conditions		
	Excellent	**Average**	**Severe**
Track-type tractors	0.07	0.09	0.13
Wheel tractor scrapers	0.07	0.09	0.13
Off-highway trucks	0.06	0.08	0.11
Wheel-type tractors	0.04	0.06	0.09
Track-type loaders	0.07	0.09	0.13
Wheel loaders	0.04	0.06	0.09
Motor graders	0.03	0.05	0.07

The cost of tires on rubber-wheeled vehicles is prorated over a service life expressed in years or hours. Therefore, if a set of tires has an initial cost of $15,000 and a service life of 5,000 hours, the hourly cost of tires set aside for replacements is:

$$\text{Hourly cost of tires} = \$15,000/5000 = \$3.00$$

13.11 OVERHEAD AND MARKUP

In addition to the direct costs of ownership and operation, general overhead costs must be considered in recovering costs associated with equipment ownership and operation. Overhead charges include items such as the costs of operating the maintenance force and facility including: (a) wages of the mechanics and supervisory personnel, (b) clerical and records support, and (c) rental or amortization of the maintenance facility (i.e., maintenance bays, lifts, machinery, and instruments). The industry practice is to prorate the total charge to each unit in the equipment fleet based on the number of hours it operates as a fraction of the total number of hours logged by the fleet. For instance, if the total number of hours logged by all units in the fleet was 20,000 and a particular unit operated 500 hours, its proportion of the total overhead would be 500/20,000 × 100, or 2.5%. If the total cost of overhead for the year is $100,000, the unit must recover $2,500 in backcharge to the client to cover its portion of the overhead. Overhead rates are updated annually from operating records to ensure adequate coverage. If overhead costs overrun projections, the coverage will be inadequate and the overrun will reduce profits.

The last component of the total charge associated with a unit of production is the profit expressed as a percentage of total hourly operating costs, which, in turn, may be expressed in cubic yards of material moved or in some other bid-relevant measure. The amount of profit per cubic yard, square foot, or linear foot is a judgment that contractors must make based on their desire to win the contract and the nature of the competition. In a tight market in which competition is strong, the allowable margin of profit that still allows the bidder to be competitive maybe only 1 or 2%. In a "fat" market, in which a lot of jobs are available, the demand is greater and the client is ready to pay a higher markup to get the work under way. Competition is bidding higher profit so the amount of profit can be adjusted upward.

Bidding strategy will include attention to the concept of marginal costs, which may permit the acceptance of jobs yielding less than the desired rate of return. In general, bidding, based on margins as low as 1 or 2% is uncomfortably close to what one might call the disaster area; the area of operating losses.

REVIEW QUESTIONS AND EXERCISES

13.1 What are the major cost components that must be considered when pricing out a piece of equipment? How can a contractor manipulate amortization for a piece of equipment to increase or reduce direct costs charged per unit of production? Why are tires on a rubber-tired vehicle not considered for depreciation?

13.2 What is the cost of a new CAT Standard Scraper 631G? (You may be surprised by its cost!)

13.3 You have just bought a new pusher dozer for your equipment fleet. Its cost is $100,000. It has an estimated service life of 4 years. Its salvage value is $12,000.

  a. Calculate the depreciation for the first and second year using the straight-line and DDB methods.

  b. The IIT components of ownership cost based on average annual value are:

Tax:	2%
Insurance:	2%
Interest:	7%

What cost per hour of operation would you charge to cover IIT?

13.4 You have just bought a used track-type tractor to add to your production fleet. The initial capitalized value of the tractor is $110,000. The estimated service life remaining on the tractor is 10,000 hours and the anticipated operating conditions across

the remainder of its life are normal. The salvage value of the tractor is $12,000. The tractor was purchased on July 1, 1997.

  a. What amount of depreciation will you claim for each calendar year between 2007 and 2010?

  b. What percent of the total depreciable amount is taken in the first year?

  c. The IIT components of ownership cost based on average annual value are:

Tax:	3%
Insurance:	2%
Interest:	8%

What cost per hour of operation would you charge to cover IIT?

  d. If the total average operating cost for the tractor is $23.50 per hour and the amount of overhead cost prorated to this tractor for the year is $4,000, what would be your total hourly cost for the operation of the tractor (during the first year of its service life)?

13.5 Verify the 5- and 7-year property class percentage given in Table 12.4 by applying the 200% DB approach to a piece of equipment with a nominal value of $1,000. For the 7-year property class, in what year is the switch from 200% DB to straight line made based on the percentages given in the table?

Chapter 14

Equipment Productivity

CONTENTS

Grader with TopCon 3D-MC

Computer and Total station

Receiver

Laser-Based Machine Control

Construction equipment using laser control technology can achieve higher levels of productivity. The guiding of road construction equipment in curving contours requires references such as hubs, staking, or elevated string lines. These benchmarks limit productivity because their installation is slow, subject to human error, and requires skilled operators to accurately steer the machine using rudimentary control methods. Attempts to guide equipment in curves using radio communication have been tried but this solution is still slow and unreliable.

New systems use three modules to control the piece of equipment:

1. Survey plans are uploaded in a total station using a laptop computer. The total station converts the digital information into an infrared laser beam.
2. A receiver, mounted on the blade of the equipment, intercepts the laser beam emitted by the total station and continuously determines (20 upgrades per second) the blade's current position and grade with respect the theoretical ones defined by the designer plans.
3. The interface between the positioning information and the actual steering of the equipment is performed through the use of a control system device, which converts the digital data into machine hydraulic valve pulses.

The main benefit of these systems is the obvious gain of productivity generated by this innovation. According to some research carried out by manufacturers of such guide systems, the laser devices can triple the productivity of equipment on highway projects as well as drastically increase their levels of precision and performance. Laser-based systems represent the next generation of equipment controlling devices bringing an alternative to the existing slower and unreliable manual systems.

14.1   PRODUCTIVITY CONCEPTS

Now that a basis for charging each unit of production has been established, the rate of production, or the number of productive units that can be generated per hour, per day, or other period of time must be considered. Our discussions here will be limited primarily to heavy construction units such as haulers, graders, and dozers. The concepts developed, however, are applicable to all types of construction equipment performing basically repetitive or cyclic operations. The cycle of an equipment piece is the sequence of tasks, which is repeated to produce a unit of output (e.g., a cubic yard, a trip load, etc.).

There are two characteristics of the machine and the cycle that dictate the rate of output. The first of these is the cyclic capacity of the machine or equipment, which establishes the number of units produced per cycle. The second is the cyclic rate or speed of an equipment piece. A truck, for instance, with a capacity of 16 cubic yards, can be viewed as producing 16 yards each time it hauls. The question of capacity is a function of the size of the machine, the state of the material that is to be processed, and the unit to be used in measurement. A hauler such as a scraper pan usually has a rated capacity, "struck," versus its "heaped" capacity. The bowl of the scraper can be filled level (struck), yielding one capacity, or can be filled above the top to a heaped capacity. In both cases, the earth hauled tends to take on air voids and bulks, yielding a different weight per unit volume than it had in the ground when excavated (i.e., its in situ location). The material has a different weight-to-volume ratio when it is placed in its construction location (e.g., a road fill, an airport runway) and is compacted to its final density. This leads to three types of measure: (a) bank cubic yards (cu yd [bank], in situ volume), (b) loose cubic yards (cu yd [loose]), and (c) compacted cubic yards. Payment in the contract is usually based on the placed earth construction, so that the "pay" unit is the final compacted cubic yard. The relationship between these three measures is shown in Figure 14.1.

The relationship between the bulk or loose volume and the bank volume is defined by the percent swell. In Figure 14.1, the percent swell is 30%. Percent swell is given as:

$$\text{Percent swell} = \left[\left(\frac{1}{\text{load factor}} \right) - 1 \right] \times 100$$

where

$$\text{Load factor} = \frac{\text{pounds per cubic yard} - \text{loose}}{\text{pounds per cubic yard} - \text{bank}}$$

Tables such as Table 14.1 give the load factor for various types of materials indicating their propensity for taking on air voids in the loose state. The higher the load factor, the smaller tendency the material has to "bulk up." Therefore, with a high load factor the loose volume and the in situ volume tend to be closer to one another. Each material has

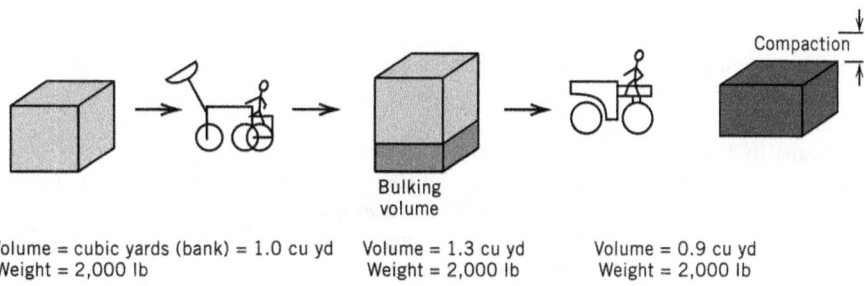

Volume = cubic yards (bank) = 1.0 cu yd   Volume = 1.3 cu yd   Volume = 0.9 cu yd
Weight = 2,000 lb   Weight = 2,000 lb   Weight = 2,000 lb

Figure 14.1 Volume relationships.

Table 14.1 Approximate Material Characteristics[a]

Material	Pounds per Cubic Yard—Bank	Percent of Swell	Load Factor	Pounds per Cubic Yard—Loose
Clay, natural bed	2,960	40	0.72	2,130
Clay and gravel				
Dry	2,960	40	0.72	2,130
Wet	2,620	40	0.72	2,220
Clay, natural bed				
Anthracite	2,700	35	0.74	2,000
Bituminous	2,160	35	0.74	1,600
Earth, loam				
Dry	2,620	25	0.80	2,100
Wet	3,380	25	0.80	2,700
Gravel, $\frac{1}{4}$–2 in.				
Dry	3,180	12	0.89	2,840
Wet	3,790	12	0.89	3,380
Gypsum	4,720	74	0.57	2,700
Iron ore				
Magnetite	5,520	33	0.75	4,680
Pyrite	5,120	33	0.75	4,340
Hematite	4,900	33	0.75	4,150
Limestone	4,400	67	0.60	2,620
Sand				
Dry, loose	2,690	12	0.89	2,400
Wet, packed	3,490	12	0.89	3,120
Sandstone	4,300	54	0.65	2,550
Trap rock	4,420	65	0.61	2,590

[a]The weight and load factor will vary with factors such as grain size, moisture content, and degree of compaction. A test must be made to determine an *exact* material characteristic.

its own characteristic load factor. In the preceding example, the material has a load factor of 0.77.

$$\text{Percent swell} = \left[\left(\frac{1}{0.77}\right) - 1\right] \times 100 = 30\%$$

Therefore, we would expect 10 yards of bank material to expand to 13 yards during transport. The shrinkage factor relates the volume of the compacted material to the volume of the bank material. In the example, the shrinkage factor is 10% because the bank cubic yard is reduced by 10% in volume in the compacted state.

In order to understand the importance of capacity, consider the following situation. A front-end loader has an output of 200 bank cu yd of common earth per hour. It loads a fleet of four trucks (capacity 18 loose cu yd each), which haul the earth to a fill where it is compacted with a shrinkage factor of 10%. Each truck has a total cycle time of 15 min, assuming it does not have to wait in line to be loaded. The earth has a percent swell of 20%. The job requires a volume of 18,000 compacted cu yd. How many hours will be required to excavate and haul the material to the fill? Two types of productive machines are involved: four trucks and a front-end loader. We must see which unit or set of units is most productive. Reference all calculations to the loose cubic yard production per hour. Then the loader productivity (given 20% swell) is:

$$200 \text{ cu yd (bank)}/\text{hr} = 1.2(200) \text{ or } 240 \text{ cu yd (loose)}/\text{hr}$$

The truck fleet production is

$$4 \text{ trucks } \times \frac{60 \text{ min/hr}}{15 \text{ min/cycle}} \times 18 \text{ cu yd (loose) truck}$$

$$= 72 \text{ cu yd (loose)} \times 4 \text{ cycle/hr}$$

$$= 288 \text{ cu yd (loose)/hr for 4 trucks}$$

Because the loader production is lower, it constrains the system to a maximum output of 240 cu yd (loose)/hr. We must now determine how many loose cubic yards are represented by 18,000 cu yd (compacted).

$$18{,}000 \text{ cu yd compacted} = \frac{18{,}000}{0.9} \text{ or } 20{,}000 \text{ cu yd (bank)}$$

$$20{,}000 \text{ cu yd (bank)} = 24{,}000 \text{ cu yd (loose) required}$$

Therefore, the number of hours required is

$$\text{Hours} = \frac{24{,}000 \text{ cu yd (loose)}}{240 \text{ cu yd (loose)/hr}} = 100$$

This problem illustrates the interplay between volumes and the fact that machines that interact with other machine cycles may be constrained or constraining.

14.2   CYCLE TIME AND POWER REQUIREMENTS

The second factor affecting the rate of output of a machine or machine combination is the time required to complete a cycle, which determines the cyclic rate. This is a function of the speed of the machine and, in the case of heavy equipment, is governed by (a) the power required, (b) the power available, and (c) the usable portion of the power available that can be developed to propel the equipment unit.

The power required is related to the rolling resistance (RR) inherent in the machine due to internal friction and friction developed between the wheels or tracks and the traveled surface. The power required is also a function of the grade resistance (GR) inherent in the slope of the traveled way. RR in tracked vehicles is considered to be zero because the track acts as its own roadbed, being laid in place as the unit advances. The friction between track and support idlers is too small to be considered. RR for rubber-wheeled vehicles is a function of the road surface and the total weight on the wheels. Tables such as Table 14.2 are available in equipment handbooks giving the RR in pounds per tons of weight. Figure 14.2 indicates visually the factors influencing

Table 14.2 Typical Rolling Resistance Factors (Caterpillar Tractor Co.)

A hard, smooth, stabilized roadway without penetration under load (concrete or blacktop)	40 lb/ton
A firm, smooth-rolling roadway flexing slightly under load (macadam or gravel-topped road)	65 lb/ton
Snow-packed	50 lb/ton
Loose	90 lb/ton
A rutted dirt roadway, flexing considerably under load; little maintenance, no water (hard clay road, 1 in. or more tire penetration)	100 lb/ton
Rutted dirt roadway, no stabilization, somewhat soft under travel (4–6 in. tire penetration)	150 lb/ton
Soft, muddy, rutted roadway, or in sand	200–400 lb/ton

Internal friction   Tire flexing   Penetration into surface   Weight on wheels

Figure 14.2 Factors influencing rolling resistance.

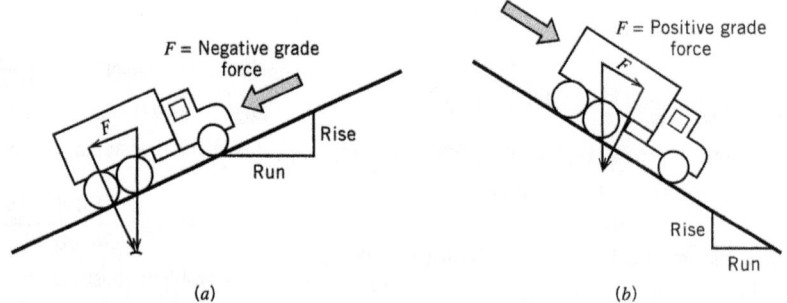

Figure 14.3 Grade resistance: (*a*) negative (resisting) force and (*b*) positive (aiding) force.

RR and therefore contributing to the required power that must be developed to move the machine.

If tables are not available, a rule of thumb can be used. The rule states that the RR is approximately 40 lbs/ton plus 30 lbs/ton for each inch of penetration of the surface under wheeled traffic. If the estimated deflection is 2 inches and the weight on the wheels of a hauler is 70 tons, we can calculate the approximate RR as:

$$RR = [40 + 2(30)] \, \text{lb/ton} \times 70 \, \text{tons} = 7,000 \, \text{lb}$$

The second factor involved in establishing the power required is the grade resistance. In some cases, the haul road across which a hauler must operate will be level and, therefore, the slope of the road will not be a consideration. In most cases, however, slopes (both uphill and downhill) will be encountered and lead to higher or reduced power requirements based on whether gravity is aiding or resisting movement (see Figure 14.3).

The percent grade is calculated by the ratio of rise over run, as depicted in Figure 14.3. If, for instance, a slope rises 6 feet in 100 feet of horizontal distance, the percent grade is 6. Similarly, a slope that increases 1.5 feet in 25 feet also has a percent grade of 6. Percent grade is used to calculate the GR using the following relationship:

$$GR = \text{percent grade} \times 20 \, \text{lb/ton/\% grade} \times \text{weight on wheels (tons)}$$

If the 70-ton piece of equipment referred to previously is ascending a 6% grade, the GR is

$$GR = 6\% \, \text{grade} \times 20 \, \text{lb/ton/\% grade} \times 70 \, \text{tons} = 8,400 \, \text{lb}$$

Assuming the RR calculated above holds for the road surface of the slope and assuming the equipment is wheeled, the total power required to climb the slope will be

$$\text{Power required} = RR + GR = 7,000 \, \text{lb} + 8,400 \, \text{lb} = 15,400 \, \text{lb}$$

If the slope is downward, an aiding force is developed, and the total power required becomes

$$\text{Power required} = RR - GR = 7,000 \, \text{lb} - 8,400 \, \text{lb} = -1,400 \, \text{lb}$$

The sign of the GR becomes negative since it is now aiding and helping to overcome the RR. Because a negative RR has no meaning, the power required on a downward 6% grade is zero. In fact, the 1,400 lb represents a downhill thrust that will accelerate the machine and lead to a braking requirement.

Traveled ways or haul roads normally consist of a combination of uphill, downhill, and level sections. Therefore, the power requirement varies and must be calculated for each section. Knowing the power required for each haul road section, a gear range that will

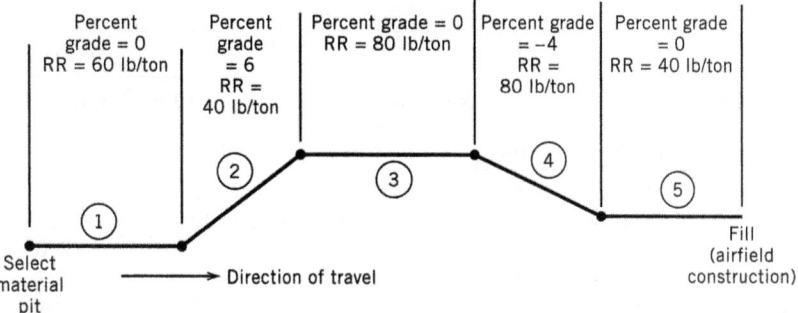

Figure 14.4 Typical haul road profile.

Table 14.3 Calculations for Haul Road Sections[a]

Section	Percent Grade (%)	Grade Resistance (lb)	Rolling Resistance (lb)	Power Required (lb)
1	0	0	4,200	4,200
2	6	8,400	2,800	11,200
3	0	0	5,600	5,600
4	−4	−5,600	5,600	0
5	0	0	2,800	2,800

[a]All calculations assume travel from pit to fill.

provide the required power can be selected. The gear range allows a speed to be developed and, given the speed, we can develop the time required to transit each section and the total time required for a cycle.

Consider the haul road profile shown in Figure 14.4 with RR and percent grade values as shown. The calculation of power required for each section of the road based on a 70-ton machine is shown in Table 14.3. Given the power requirements, the next section indicates how a gear range is selected. As noted previously, this allows determination of the speed across each section and the time required.

14.3 POWER AVAILABLE

The power available is controlled by the engine size of the equipment and the drive train, which allows transfer of power to the driving wheels or power take-off point. The amount of power transferred is a function of the gear being used. Most automobile drivers realize that lower gears transfer more power to overcome hills and rough surfaces. Lower gears sacrifice speed to provide more power. Higher gears deliver less power, but allow higher speeds. Manufacturers publish figures regarding the power available in each gear for individual equipment pieces in equipment handbooks that are updated annually. This information can be presented in a tabular format such as that shown in Table 14.4 or in graphical format such as the nomograph shown in Figure 14.5.

For tracked vehicles, the power available is quoted in drawbar pull. This is the force that can be delivered at the pulling point (i.e., pulling hitch) in a given gear for a given tractor type. Power available for a wheeled vehicle is stated in pounds of rimpull. This is the force that can be developed by the wheel at its point of contact with the road surface. Manufacturers also provide information regarding rated power and maximum power. Rated power is the level of power that is developed in a given gear under normal load and

Table 14.4 Speed and Drawbar Pull (270 hp) (Track-Type Tractor)

Gear	Forward mph	Forward km/h	Reverse mph	Reverse km/h	At Rated rpm lb	At Rated rpm kg	Maximum at Lug lb	Maximum at Lug kg
1	1.6	(2–6)	1.6	(2.6)	52,410	(23,790)	63,860	(28,990)
2	2.1	(3–4)	2.1	(3.4)	39,130	(17,760)	47,930	(21,760)
3	2.9	(4.7)	2.9	(4.7)	26,870	(12,200)	33,210	(15,080)
4	3.7	(6.0)	3.8	(6.1)	19,490	(8,850)	24,360	(11,060)
5	4.9	(7.9)	4.9	(7.9)	13,840	(6,280)	17,580	(7,980)
6	6.7	(10.8)	6.8	(10.9)	8,660	(3,930)	11,360	(5,160)

Drawbar Pull Forward[a]

[a]Usable pull will depend on traction and weight of equipped tractor.

over extended work periods. It is the base or reference level of power that is available for continuous operation. The maximum power is just what it indicates. It is the peak power that can be developed in a gear for a short period of time to meet extraordinary power requirements. For instance, if a bulldozer is used to pull a truck out of a ditch, a quick surge of power would be used to dislodge the truck. This short-term peak power could be developed in a gear using the maximum power available.

Most calculations are carried out using rated power. If, for example, the power required for a particular haul road section is 25,000 pounds based on the procedures described in Section 14.2, the proper gear for the 270-hp track-type tractor is third gear. This is determined by entering Table 14.4 and comparing power required with rated power. Consider the example shown in the shaded box example.

The sum of the rolling resistance and grade resistance that a particular wheel-type tractor and scraper must overcome on a specific job has been estimated to be 10,000 lb. If the "pounds pull-speed" combinations listed below are for this particular machine, what is the maximum reasonable speed of the unit?

Third gear would be selected since the rated rimpull is 12,190 lb. (If the total power required had been in excess of 12,190 lb. we would select second gear because you recall that rated pounds pull should always be used for gear selection. The reserve rimpull of the maximum rating is always available-at reduced speed-to pull the unit out of small holes or bad spots.)

Gear	Speed	Pounds Rimpull Rated	Pounds Rimpull Maximum
1	2.6	38,670	49,100
2	5.0	20,000	25,390
3	8.1	12,190	15,465
4	13.8	7,185	9,115
5	22.6	4,375	5,550

Nomographs are designed to allow quick determination of required gear ranges as well as the maximum speed attainable in each gear. The nomograph shown in Figure 14.5 is typical of those for an off-highway truck. To illustrate the use of this figure, consider the following problem. On a particular road construction job, the operator has to choose between two available routes linking the select material pit with a road site fill. One route is 4.6 miles on a firm, smooth road with a RR = 50 lb/ton. The other route is 2.8 miles (one-way) on a rutted dirt road with RR = 90 lb/ton. The haul road profile in both cases is level so that grade resistance is not a factor. Using the nomograph of Figure 14.5, we are to

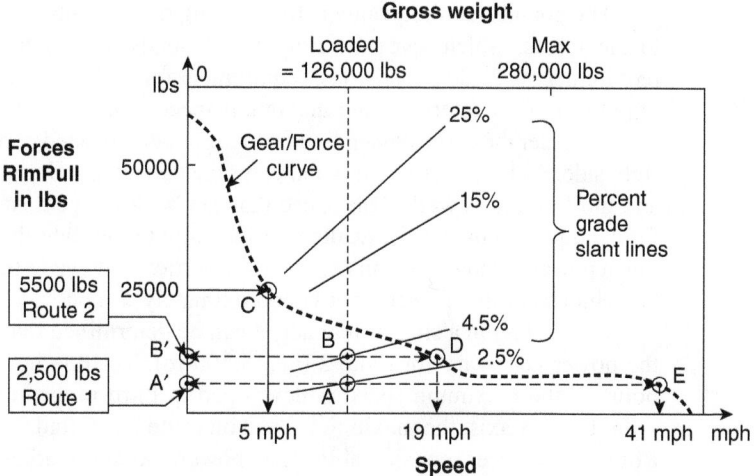

Figure 14.5 Typical Force/Speed nomograph for off Highway Truck.

determine the pounds pull to overcome RR for a loaded off-highway truck. The same chart allows determination of the maximum speed.

In order to use the chart, consider the information in the chart regarding gross weight. The weight in pounds ranges from 0 to 280,000 (140 tons). The weight of the truck with a 70,000-lb load is indicated by a vertical dashed line intersecting the gross weight axis (top of chart) at approximately 126,000 lb (loaded). For this problem, the loaded line is relevant since the truck hauls loads from pit to fill.

Next consider the slant lines sloping from lower left to upper right. These lines indicate the total resistance (i.e., RR + GR) in increments from 0 to 25% grade. In this problem, there is no grade resistance. In dealing with rolling resistance, it is common to convert it to an equivalent percent grade. Then, the total resistance can be stated in percent grade by adding the equivalent percent grade for rolling resistance to the slope percent grade. To convert rolling resistance to equivalent percent grade, the following expression is used:

$$\text{Equivalent percent grade} = \frac{(RR)}{20\,\text{lb/ton/\% grade}}$$

For the rolling resistance values given in the problem, the equivalent percent grades become as follows:

Route	Distance	RR	Equivalent Percent Grade
1	4.6 miles	50 lb/ton	2.5
2	2.8 miles	90 lb/ton	4.5

In order to determine the required pounds pull, the intersection of the slant line representing the equivalent percent grade with the load vertical line is located. This intersection for route 1 is designated point A in Figure 14.5. The corresponding point for route two is labeled B.

The pounds required value is found using these points by reading horizontally across to the y axis, which gives the rimpull in pounds. For route 1, the approximate power requirement is 2,500 pounds. The requirement for route 2 is 5,500 pounds. Points A and B are also used to determine the maximum speed along each route.

Consider the curve descending from the upper left-hand corner of the chart to the lower right side. As labeled, this curve indicates the deliverable rimpull available in first, second, and third ranges as well as the speed that can be developed. At 25,000 pounds of rimpull, for example, on the y axis, reading horizontally to the right the only range delivering this much power is shown by point C. Reading vertically down to the x axis, the speed that can be achieved at this power level is approximately 5 mph.

Proceeding in a similar manner, it can be determined that lower ranges will provide the power necessary for route 2 (i.e., 5,500 lb). Reading horizontally to the right from point B, the maximum speed is developed in third range at point D. Referencing this point to the x axis, the maximum speed on route 2 is found to be approximately 19 mph. Route 1 requires considerably less power. Again reading to the right, this time from point A, the third range provides a maximum speed of approximately 41 mph (see point E).

Now, having established the maximum speeds along each route and knowing the distances involved, it should be simple to determine the travel times required. Knowledge of the speeds, however, is not sufficient to determine the travel times since the requirements to accelerate and decelerate lower the effective speed between pit and fill. Knowing the mass of the truck and the horsepower of the engine, the classic equation, force = mass × acceleration (F = ma), would allow determination of the time required to accelerate to and decelerate from maximum speed. This is not necessary, however, since the equipment handbooks provide time charts that allow direct readout of the travel time for a route and piece of equipment, given the equivalent percent grade and the distance. A travel time chart for the loaded truck is given in Figure 14.6. Inspection of the chart for the loaded truck indicates that the distance to be traveled in feet is shown along the x axis. The equivalent percent grade is again shown as slant lines sloped from lower left to upper right. Converting the mileages given to feet

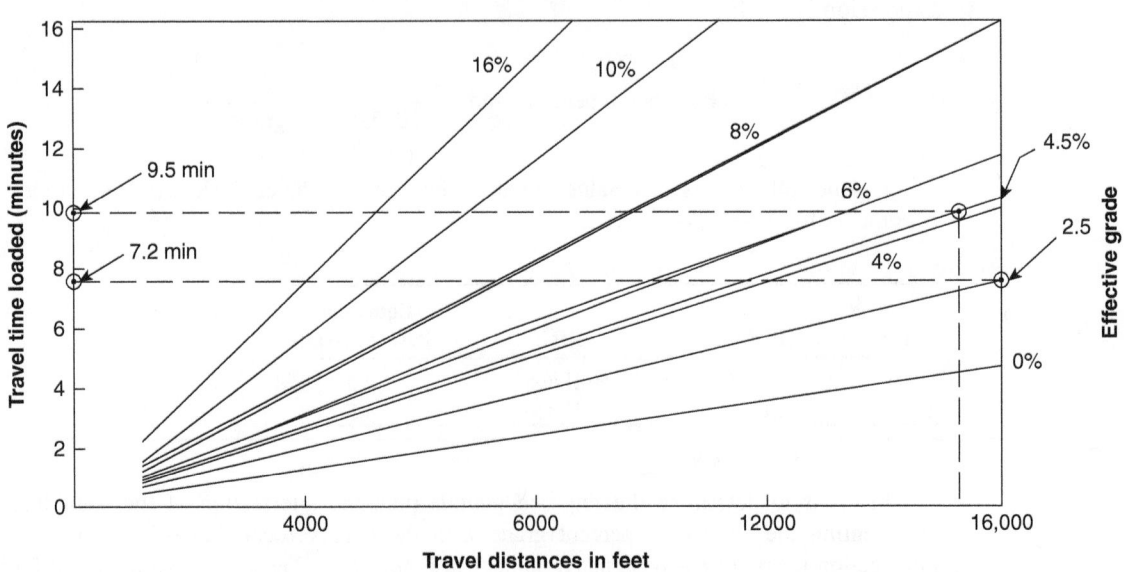

Figure 14.6 Typical Time-Distance Chart.

yields the values 24,344 feet for route 1 and 14,784 feet for route 2. Entering the chart, with an equivalent percent grade of 4.5 for route 2, the travel time can be read (on the y axis), as 9.5 min.

A problem develops in reading the travel time for route 1 since the maximum distance shown on the chart is 16,000. One way of reconciling this problem is to break the 24,344 into two segments: (a) 16,000 ft as shown and (b) the 8,344 ft remaining. The assumption is made that the 8,344 feet is traveled at the maximum speed determined previously to be 41 mph. At this speed, the travel time for this segment is

$$T2 = \frac{8,344 \text{ ft } (60 \text{ mi/hr})}{(41 \text{ mi/hr})(5,280 \text{ ft/mi})} = 2.31 \text{ min}$$

The time for the remaining 16.000 feet is read from the chart as 7.2 min. It is assumed that acceleration and deceleration effects are included in this time. Therefore, the required time for route 1 is also 9.5 min (T1 + T2 = 7.2 + 2.31). Therefore, the decision as to which route to use is based on wear and tear on the machines, driver skill, and other considerations. This problem illustrates the development of time, given information affecting power required and power available. Using the same procedure, the travel time empty returning to the pit from the fill can be determined and total cycle time can be determined.

14.4   USABLE POWER

To this point, it has been assumed that all of the available, power is usable and can be developed. Environmental conditions play a major role in determining whether the power available can be used under operating conditions. The two primary constraints in using the available power are the road surface traction characteristics (for wheeled vehicles) and the altitude at which operations are conducted. Most people have watched the tires of a powerful car spin on a wet or slippery pavement. Although the engine and gears are delivering a certain horsepower, the traction available is not sufficient to develop this power into the ground as a driving force. Combustion engines operating at high altitudes experience a reduction in oxygen available within the engine cylinders. This also leads to reduced power.

Consider first the problem of traction. The factors that influence the usable power that can be developed through the tires of wheeled vehicles are the coefficient of traction of the surface being traveled and the weight of the vehicle on the driving wheels.

The coefficient of traction is a measure of the ability of a particular surface to receive and develop the power being delivered to the driving wheels and has been determined by experiment. The coefficient of traction obviously varies based on the surface being traversed and the delivery mechanism (i.e., wheels, track, etc.). Table 14.5 gives typical values for rubber-tired and tracked vehicles on an assortment of surface materials.

The power that can be developed on a given surface is given by the expression:

Usable pounds pull = (coefficient of traction) × (weight on drivers)

In the consideration of RR and GR, the entire weight of the vehicle or combination was used. In calculating the usable power, only the weight on the driving wheels is used because it is the weight pressing the driving mechanism (e.g., wheels) and surface together. Equipment handbooks specify the distribution of load to all wheels for both empty and

Table 14.5 Coefficients of Traction

Materials	Rubber Tires	Tracks
Concrete	.90	.45
Clay loam, dry	.55	.90
Clay loam, wet	.45	.70
Rutted clay loam	.40	.70
Dry sand	.20	.30
Wet sand	.40	.50
Quarry pit	.65	.55
Gravel road (loose, not hard)	.36	.50
Packed snow	.20	.25
Ice	.12	.12
Firm earth	.55	.90
Loose earth	.45	.60
Coal, stockpiled	.45	.60

loaded vehicles and combinations. The weight to be considered in the calculation of usable power for several types of combinations is shown in Figure 14.7. To illustrate the constraint imposed by usable power, consider the following situation. A 30-yard capacity, two-wheel tractor-scraper is operating in sand and carrying 26-ton loads. The job superintendent is concerned about the high RR of the sand (RR = 400 lb/ton) and the low traction available in sand. The question is: Will the tractors have a problem with 26-ton loads under these conditions? The weight distribution characteristics of the 30-yard tractor-scraper are as follows:

	Empty Weight (lb)	Percentage (%)	Loaded Weight (lb)	Percentage (%)
Drive wheels	50,800	67	76,900	52
Scraper wheels	25,000	33	70,900	48
Total weight	75,800	100	147,800	100

The difference between the total weight empty and loaded is 72,000 pounds, or 36 tons. The loaded weight with 26-ton loads would be 127,800 pounds.

In Determining Weight on Drivers

For track–type tractor	For four–wheel tractor	For two–wheel tractor
Use total tractor weight	Use weight on drivers shown on spec sheet or approximately 40% of vehicle gross weight	Use weight on drivers shown on spec sheet or approximately 60% of vehicle gross weight

Figure 14.7 Determination of driver weights.

Assuming the same weight distribution given above for fully loaded vehicles, the wheel loads would be as follows:

	Percentage (%)	Weight in Pounds
Drive wheels	52	66,456
Scraper wheels	48	61,344
Total	100	127,800

The resisting force (assuming a level haul site) would be

$$\text{Pounds required} = 400\,\text{lb/ton} \times 63.9\,\text{tons} = 25{,}560\,\text{1b}$$

The deliverable or usable power is

$$\text{Usable power} = 0.20 \times 66{,}456\,\text{lb} = 13{,}291.20\,\text{1b}$$

Quite obviously, there will be a problem with traction since the required power is almost twice the power that can be developed. The "underfoot" condition must be improved. A temporary surface (e.g., wood or steel planking) could be installed to improve traction. One simple solution would be to simply wet the sand. This yields an increased usable power:

$$\text{Usable power} = 0.40 \times 66{,}456\,\text{lb} = 26{,}582.41\,\text{lb} > 25{,}560\,\text{1b}$$

The impact of usable power constraints can be shown graphically (see Figure 14.8). Now, if the total resistance of the unit (RR plus GR) is 10,000 pounds, then an operating range for the machine is indicated in Figure 14.8b.

The altitude at which a piece of equipment operates also imposes a constraint on the usable power. As noted previously, the oxygen content decreases as elevation increases, so that a tractor operating in Bogota, Colombia (elevation 8,600 feet), cannot develop the

(a)

(b)

Figure 14.8 Impact of usable power constraints.

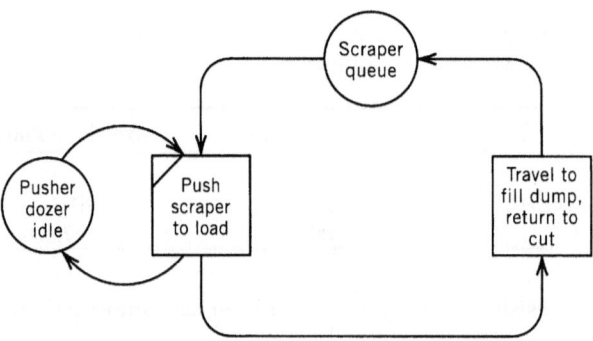

Figure 14.9 Scraper-pusher dual-cycle model.

same power as one operating in Atlanta, Georgia (elevation 1,050 feet). A good rule of thumb to correct this effect is as follows: Decrease pounds pull 3% for each 1,000 feet (above 3,000 feet). Therefore, if a tractor is operating at 5,000 feet above sea level, its power will be decreased by 6%.

14.5   EQUIPMENT BALANCE

In situations in which two types of equipment work together to accomplish a task, it is important that a balance in the productivity of the units be achieved. This is desirable so that one unit is not continually idle waiting for the other unit to "catch up." Consider the problem of balancing productivity within the context of a pusher dozer loading a tractor scraper. A loading simple model of this process is shown in Figure 14.9. The circles represent delay or waiting states, while the squares designate active work activities with associated times that can be estimated. The haul unit is a 30 cubic yard scraper, and it is loaded in the cut area with the aid of a 385-horsepower pusher dozer. The system consists of two interacting cycles.

Assume that in this case the 30-cubic yard tractor scraper is carrying rated capacity and operating on a 3,000-foot level haul where the RR developed by the road surface is 40 lb/ton. Using the standard formula, this converts to

$$\text{Effective grade} = (\text{RR})/(20\,\text{lb/ton/\% grade})$$
$$= (40\,\text{lb/ton})/(20\,\text{lb/ton/\% grade}) = 2\%\ \text{grade}$$

By consulting the charts given in Figure 14.10, the following travel times can be established:

1. Time loaded to fill: 1.4 min

2. Time empty to return: 1.2 min

Assume further that the dump time for the scraper is 0.5 minutes and the push time using a track-type pusher tractor is 1.23 minutes, developed as follows:

Load time = 0.70

Boost time = 0.15

Transfer time = 0.10

Return time = 0.28

Total = 1.23 minutes

Using these deterministic times for the two types of flow units in this system (i.e., the pusher and the scrapers), the scraper and pusher cycle times can be developed, as shown in Figure 14.11.

Figure 14.10 Travel time nomographs (Caterpillar Tractor Co.).

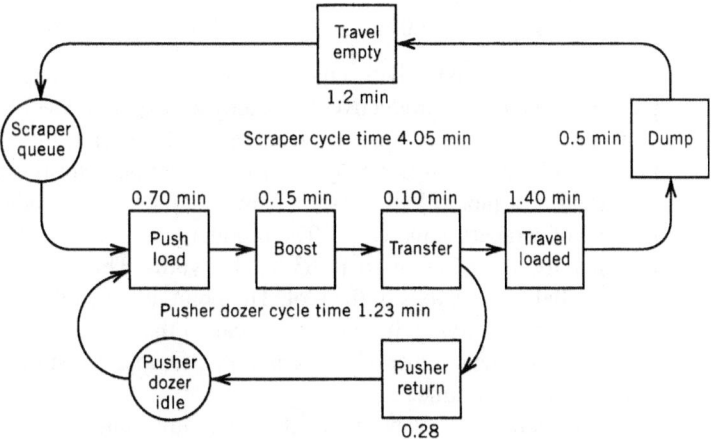

Figure 14.11 Scraper-pusher cycle times.

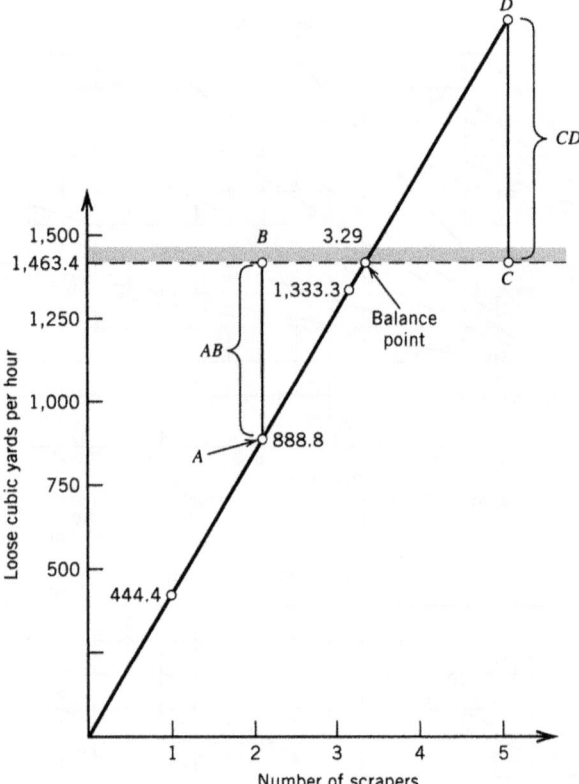

Figure 14.12 Productivity plot.

$$\text{Pusher cycle} = 1.23 \text{ minutes}$$
$$\text{Scraper cycle} = 0.95 + 1.2 + 1.4 + 0.5 = 4.05 \text{ minutes}$$

These figures can be used to develop the maximum hourly production for the pusher unit and for each scraper unit as follows.

Maximum system productivity (assuming a 60-min working hour)

1. Per scraper
Prod (scraper) = $(60 \text{ min/hr } 4.05 \text{ min}) \times 30 \text{ cu yd (loose)} = 444.4 \text{ cu yd (loose)/hr}$

2. Based on single pusher
Prod (pusher) = $(60/1.23) \times 30 \text{ cu yd (loose)} = 1,463.4 \text{ cu yd (loose)/hr}$

Using these productivities based on a 60-minute working hour, it can be seen that the pusher is much more productive than a single scraper and would be idle most of the time if matched to only one scraper. By using a graphical plot, the number of scrapers that are needed to keep the pusher busy at all times can be determined.

The linear plot of Figure 14.12 shows the increasing productivity of the system as the number of scrapers is increased. The productivity of the single pusher constrains the total productivity of the system to 1,463.4 cubic yards. This is shown by the dotted horizontal line parallel to the x axis of the plot. The point at which the horizontal line and the linear plot of scraper productivity intersect is called the balance point. The balance point is the point at which the number of haul units (i.e., scrapers) is sufficient to keep the pusher unit busy 100% of the time.

To the left of the balance point, there is an imbalance in system productivity between the two interacting cycles; this leaves the pusher idle. This idleness results in lost productivity.

The amount of lost productivity is indicated by the difference between the horizontal line and the scraper productivity line. For example, with two scrapers operating in the system, the ordinate AB of Figure 14.12 indicates that 574.6 cubic yard, or a little less than half of the pusher productivity, is lost because of the mismatch between pusher and scraper productivities. As scrapers are added, this mismatch is reduced until, with four scrapers in the system, the pusher is fully used. Now the mismatch results in a slight loss of productivity caused by idleness of the scrapers. This results because, in certain instances, a scraper will have to wait to be loaded until the pusher is free from loading a preceding unit. If five scraper units operate in the system, the ordinate CD indicates that the loss in the productive capacity of the scraper because of delay in being push loaded is

$$\text{Productive loss} = 5(444.4) - 1463.4 = 758.6 \,\text{cu yd}$$

This results because the greater number of scrapers causes delays in the scraper queue of Figures 14.9 and 14.11 for longer periods of time. The imbalance or mismatch between units in dual-cycle systems resulting from deterministic times associated with unit activities is called interference. It is due only to the time imbalance between the interacting cycles. It does not consider idleness or loss of productivity because of random variations in the system activity durations. In most cases, only a deterministic analysis of system productivity is undertaken because it is sufficiently accurate for the purpose of the analyst.

14.6   RANDOM WORK TASK DURATIONS

The influence of mismatches in equipment fleets and crew mixes on system productivity was discussed in the last section in terms of deterministic work task durations and cycle times. In systems in which the randomness of cycle times is considered, system productivity is reduced further. The influence of random durations on the movement of resources causes various units to become bunched together and thus to arrive at and overload work tasks. Resulting delays impact the productivity of cycles by increasing the time that resource units spend in idle states pending release to productive work tasks.

Consider the scraper-pusher problem and assume that the effect of random variation in cycle activity duration is to be included in the analysis.

In simple cases such as the two-cycle system model of Figure 14.9, mathematical techniques based on *queuing theory* can be used to develop solutions for situations in which the random arrival of scrapers to the dozer can be postulated. To make the system amenable to mathematical solution, however, it is necessary to make certain assumptions about the characteristics of the system that are not typical of field construction operations.

Figure 14.13 indicates the influence of random durations on the scraper fleet production. The curved line of Figure 14.13 slightly below the linear plot of production based on deterministic work task times shows the reduction in production caused by the addition of random variation of cycle activity times. This randomness leads to bunching of the haulers on their cycle. With deterministic work task times, the haul units are assumed to be equidistant in time from one another within their cycle.

In deterministic calculations, all three of the haul units shown in Figure 14.14a are assumed to be exactly 1.35 minutes apart. In this system, there are three units, and the hauler cycle time is taken as a deterministic value of 4.05 minutes. In systems that include the effect of random variation of cycle times, "bunching" eventually occurs between the units on the haul cycle. That is, the units do not stay equidistant from one another but are continuously varying the distances between one another. Therefore, as shown in Figure 14.14b, a situation often occurs in which the units on the haul are unequally spaced apart in time from one another. This bunching effect leads to increased idleness and reduced productivity. It is intuitively clear that the three units that are

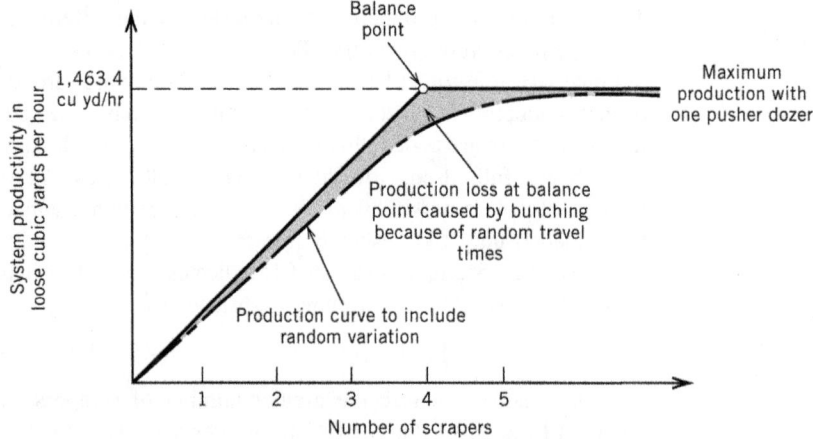

Figure 14.13 Productivity curve to include effect of random cycle times.

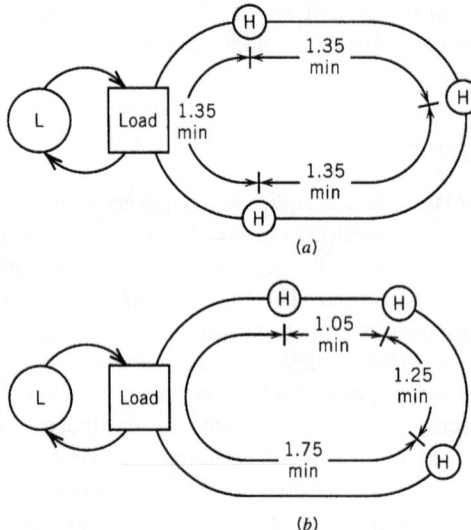

Figure 14.14 Comparison of haul unit cycles.

bunched as shown in Figure 14.14b will be delayed for a longer period at the scraper queue since the first unit will arrive to load only 1.05 minutes instead of 1.35 minutes in advance of the second unit. The bunching causes units to "get into each other's way." The reduction in productivity caused by bunching is shown as the shaded area in Figure 14.13 and occurs in addition to the reduction in productivity caused by mis-matched equipment capacities.

This bunching effect is most detrimental to the production of dual-cycle systems such as the scraper-pusher process at the balance point. Several studies have been conducted to determine the magnitude of the productivity reduction at the balance point because of bunching. Simulation studies conducted by the Caterpillar Tractor Company indicate that the impact of random time variation is the standard deviation of the cycle time distribution divided by the average cycle time. Figure 14.15 illustrates this relationship graphically.

As shown in the figure, the loss in deterministic productivity at the balance point is approximately 10% due to the bunching; this results in a system with a cycle coefficient of variation equal to 0.10. The probability distribution used in this analysis was

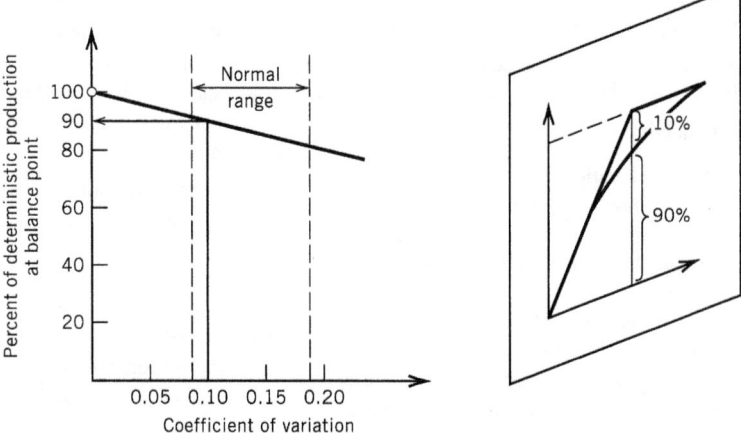

Figure 14.15 Plot of cycle time coefficient of variation.

lognormal. Other distributions would yield slightly different results. The loss in productivity in equipment-heavy operations such as earthmoving is well documented and recognized in the field, mainly because of the capital-intensive nature of the operation and the use of scrapers in both single-unit operations and fleet operations. To some extent, field policies have emerged to counteract this effect by occasionally breaking the queue discipline of the scrapers so that they self-load when bunching effects become severe. The resulting increased load and boost time for the scraper add little to the system productivity, but it does break down the bunching of the scrapers.

REVIEW QUESTIONS AND EXERCISES

14.1 A customer estimates that he is getting 30 cubic yards (loose) of gypsum in his scraper. Determine the percent overload if the load estimate is correct. The maximum load capacity of the scraper is 84,000 pounds.

14.2 Stripping overburden in the Illinois coal belt, the Dusty Coal Company uses 270-horsepower, track-type tractors (with direct-drive transmissions) and drawn scrapers. The overburden is a very soft loam that weighs 2,800 lb/yd (loose). Estimated RR factor for the haul road is 300 lb/ton. If the scraper weighs 35,000 lb (empty) and carries 25 loose cubic yards per trip, what is the RR of the loaded unit? What operating gear and speed do you estimate for the loaded machines on level ground? (See Table 14.4.)

14.3 The ABC Company is planning to start a new operation hauling sand to a ready-mix concrete plant. The equipment superintendent estimates that the company-owned 30-yard wheel tractor scrapers can obtain 26-ton loads. He is concerned about the high RR of the units in the sand (RR factor 250 lb/ton) and the low tractive ability of the tractors on this job. Will traction be a problem? If so, what do you suggest to help?

14.4 Estimate the cycle time and production of a 30-cubic-yard wheel tractor scraper carrying rated capacity, operating on a 4,500-foot level haul. The road flexes under load, has little maintenance, and is rutted. Material is 3000 lb/BCY. The

scrapers are push-loaded by one 385-horsepower, track-type pusher tractor. How many scrapers can be served by this one pusher?

14.5 How many trips would one rubber-tired Herrywampus have to make to backfill a space with a geometrical volume of 5400 cubic yard? The maximum capacity of the machine is 30 cubic yard (heaped), or 40 tons. The material is to be compacted with a shrinkage of 25% (relative to bank measure) and has a swell factor of 20% (relative to bank measure). The material weighs 3,000 lb/cu yd (bank). Assume that the machine carries its maximum load on each trip. Check by both weight and volume limitations.

14.6 You own a fleet of 30-cubic-yard tractor-scrapers and have them hauling between the pit and a road construction job. The haul road is clayey and deflects slightly under the load of the scraper. There is a slight grade (3%) from pit at the fill location. The return road is level. The haul distance to the dump location is 0.5 miles and the return distance is 0.67 miles. Four scrapers are being used.

   a. What is the rimpull required when the scraper is full and on the haul to the fill?

   b. What are the travel times to and from the dump location (see Figure 14.10)?

c. The scrapers are push loaded in the pit. The load time is 0.6 min. What is the cycle time of the pusher dozer?

d. Is the system working at, above, or below the balance point? Explain.

e. What is the production of the system?

14.7 You are excavating a location for the vault shown below. The top of the walls shown are 1 foot below grade. All slopes of the excavation are 2 to 1 to a toe 1 foot outside the base of the walls. The walls sit on a slab 1 foot in depth. Draw a sketch of the volume to be excavated, break it into components, and calculate the volume. The material from the excavation is to be used in a compacted fill. The front-end loader excavating the vault has an output of 200 bank cubic yards of common earth per hour. It loads a fleet of four trucks (capacity 18 loose cubic yards each) that haul the earth to a fill where it is compacted with a shrinkage factor of 10%. Each truck has a total cycle time of 15 minutes, assuming it does not have to wait in line to be loaded. The earth has a swell factor of 20%. How many hours will it take to excavate and haul the material to the fill?

a. What is the production of this system in cubic yards per hour, assuming the trucks carry 35 cubic yard per load?

b. Is the system working at, above, or below its balance point?

c. If there is probability of major delay on the travel elements to and from the fill of 7% and the mean value of delay is 5 minutes, what is the new system production?

14.8 You have four 35-ton, off-highway trucks hauling from a pit to an airfield job. The haul road is maintained by a patrol grader and has a rolling resistance of 80 lb/ton. The road is essentially level and the distance one-way is 2.1 miles. The gross weight of the truck when loaded is 70 tons.

a. What is the power required on the haul to the fill location?

b. What is the maximum speed when hauling to the fill?

c. What are the travel times to and from the fill (i.e., loaded, and empty)? The trucks are being loaded by a shovel with a 5-yard bucket (assume seven load cycles per truck load). The cycle time for the shovel is 0.5 minutes.

d. What is the total truck cycle time?

e. Is the new system above or below the balance point?

14.9 You are given the following information about a dry-batch paving operation. You are going to use one mixer that has a service rate of 30 services per hour. The dry-batch trucks you use for bringing concrete to the paver have an arrival rate of 7.5 arrivals per hour. Each truck carries 6 cubic yard of concrete. You have a total amount of 13,500 cubic yard of concrete to pour. You rent a truck at $15 per hour and the paver at $60 per hour. If the job takes more than 80 hours, you pay a penalty of $140 per hour owing to delays in job completion. On the basis of least cost, determine the number of trucks you should use. Plot the cost versus the number of trucks used.

14.10 The superintendent in a concrete batch plant has measured the following cycle times for a large pouring operation, for which he is using 10 trucks of 9 cubic yards each.

Cycle #	Duration (min)
1	30.5
2	25.4
3	27.3
4	29.1
5	23.5
6	24.3
7	19.8
8	25.8
9	23.7
10	25.1

a. What is the productivity of this system, in cubic yards per hour?

b. What is the approximate loss of productivity due to the bunching effect explained in 14.6? Use the Caterpillar method.

c. Using the results from (b), what would be the theoretical productivity of this system without considering the bunching effect? Hint: If the loss of productivity is 9%, then the theoretical productivity would be [Actual productivity (from a)]/(1–9%)

Chapter 15

Construction Labor

CONTENTS

Biometric Identification Recognition

The access control system plays a major role in site security. These days, only a few engineers and managers are needed to wholly control a 20-story building's physical security objects such as elevators, door locks, lights, security system control, etc. through computerized automatic systems. Controlling and screening personnel entering a facility is almost impossible for a few security people to physically control. Even though visitors, office workers, and residents are given keys or ID tags after being screened, there is still a possibility that unauthorized people could access important property with fake IDs or keys. Recently, biometric technology has arisen as a new, effective, and secure method for identification recognition of personnel.

Biometrics are computerized methods of recognizing people based on physical or behavioral characteristics. The main biometric technologies include face recognition,

Biometric Identification Devices

fingerprint, hand geometry, iris, palm prints, signature, and voice. Biometric technologies can work in two modes—authentication (one-to-one matching) and identification (one-to-many matching). However, only three biometrics are capable of the latter—face, finger, and iris. This technology has great potential for controlling access to construction work sites as well as other facilities.

15.1 LABOR RESOURCE

The manpower component of the four M's of construction is by far the most variable and unpredictable. It is, therefore, the element that demands the largest commitment of time and effort from the management team. Manpower or labor has four major aspects that are of interest to management. To properly understand the management and control of labor as a resource, the manager must be aware of the interplay among the following elements:

1. Labor organization
2. Labor law
3. Labor cost
4. Labor productivity

The cost and productivity components were central to the discussion of equipment management in Chapters 13 and 14. Labor includes the added human factor. This element can only be understood in the context of the prevailing legal and organizational climate that is characteristic of the construction industry.

15.2 SHORT HISTORY OF LABOR ORGANIZATIONS

The history of labor organizations begins in the early 19th century, and their growth parallels the increasing industrialization of modem society. Initially, tradesmen possessing some skill or craft began organizing into groups variously called guilds, brotherhoods, or mechanics societies. Their objectives were to provide members, widows, and children with sickness and death benefits. In addition, these organizations were interested in the development of trade proficiency standards and the definition of skill levels such as apprentice and journeyman. They were often "secret" brotherhoods because such organizations were considered unlawful and illegal conspiracies posing a danger to society.

From the 1840s until the era of the New Deal[1] in the 1930s, the history of labor organizations is the saga of confrontation between management and workers, with the pendulum of power on the management side. With the coming of the New Deal and the need to rejuvenate the economy during the Depression, labor organizations won striking gains that virtually reversed the power relationship between managers and workers. The American Federation of Labor (AFL) was organized by Samuel Gompers in 1886. This was the first successful effort to organize skilled and craft workers such as cabinetmakers, leather tanners, and blacksmiths. Since its inception, the AFL has been identified with skilled craft workers as opposed to industrial "assembly line" type of workers. The

Building and Construction Trades Department of the AFL, which is the umbrella organization representing all construction craft unions, was organized in 1908.

The semiskilled and unskilled factory workers in "sweat shop" plants and mills were largely unorganized at the time Gompers started the AFL. Many organizations were founded and ultimately failed in an attempt to organize the industrial worker. These organizations, with euphonious-sounding names, such as Industrial Workers of the World and the Knights of Labor, had strong political overtones and sought sweeping social reforms for all workers. This was particularly attractive to immigrant workers arriving from the socially repressive and politically stagnant atmosphere in Europe. Such organizations attracted political firebrands and anarchists preaching social change and upheaval at any cost. Confrontation with the police was common, and violent riots often led to maiming and killing. The most famous such riot occurred in the Haymarket in Chicago in 1886.

Gompers was seriously interested in protecting the rights of skilled workers and had little interest in the political and social oratory of the unskilled labor organizations. Therefore, separate labor movements representing skilled craft and semiskilled factory workers developed and did not combine until the 1930s. This led to different national and local organizational structures and bargaining procedures that are still used and strongly influence the labor picture even today.

In the 1930s, industrial (i.e., factory semiskilled) workers began to organize effectively with the support of legislation evolving during the post-Depression period. The AFL, realizing such organizations might threaten its own dominance, recognized these organizations by bringing them into the AFL camp with the special designation of Federal Locals. Although nominally members, the industrial workers were generally treated as second-class citizens by the older and more established craft unions. This led to friction and rivalry that culminated in the formation of the Committee for Industrial Organizations (CIO). This committee was established in 1935 unilaterally by the industrial locals without permission from the governing body of the AFL. The act was labeled treasonous, and the AFL board ordered the committee to disband or be expelled. The AFL suspended the industrial unions in 1936. In response, these unions organized as the Congress of Industrial Organizations (CIO), with John L. Lewis of the United Mine Workers as the first CIO president. Following this rift between the industrial and craft union movements, the need to cooperate and work together was apparent. However, philosophical and personal differences prevented this until 1955, at which time the two organizations combined to form the AFL–CIO. This organization remains the major labor entity in the United States today.

15.3   EARLY LABOR LEGISLATION

The courts and legislative bodies of the land have alternately operated to retard or accelerate the progress of labor organizations. The chronology of major items of legislation and the significant events in the labor movement are shown in Table 15.1. At the outset, the law was generally interpreted to check organization of labor and, therefore, management was successful in controlling the situation. The most classic illustration of this is the application of the Sherman Antitrust Law to enjoin workers from organizing. The Sherman Antitrust Act had originally been enacted in 1890 to suppress the formation of large corporate trusts and cartels, which dominated the market and acted to fix prices and restrain free trade. The oil and steel interests formed separate cartels in

---

[1] The administration of President Franklin D. Roosevelt.

Table 15.1 Chronology of Labor Law and Organization

Labor Law		Labor Movement	
1890	Sherman Antitrust Act	1886	AFL founded by Samuel Gompers; Knights of Labor organized factory workers
1908	Supreme Court supported application to union activity	1905	Industrial Workers of the World
1914	Clayton Act Ineffective—individual basis—as court rules	1908	Building and Construction Trades Department of AFL founded
1931	Davis–Bacon Act	1930s	AFL takes in industrial workers as federal locals
	Applied to federal contracts, wages and fringes paid at prevailing rate	1935	Committee for Industrial Organization-AFL ordered disbanding
1932	Norris–LaGuardia Act (Anti-Injunction Act)	1936	Federal locals (CIO) thrown out of AFL
1935	Wagner Act (National Labor Relations Act)	1938	Congress of Industrial Organizations
1938	Fair Labor Standards Act		
	Minimum wages, maximum hours defined	1940s	Wartime strikes-labor accused of not supporting war effort Criminal activities alleged
1943	Smith–Connolly Act (War-Labor Disputes Act) Reaction to labor in wartime; ineffective	1955	AFL and CIO reconcile differences and recombine as AFL–CIO
1946	Hobbs Act—"Anti-Racketeering law" Protects employer from paying kickbacks to labor		
1947	Taft–Hartley (Labor Management Relations Act)		
1959	Landrum–Griffin Act (Labor Management Reporting and Disclosures)		
1964	Title IV Civil Rights Act		

the late 19th century to manipulate the market. More recently, Microsoft Corporation has been reviewed by the Justice Department to prevent similar dominance in the computer market. To break up such potential market control, the Antitrust Law provides the government with the power to enjoin corporations from combining to control prices and restrict trade.

In 1908, the Supreme Court ruled that the Antitrust Law could be applied to prevent labor from organizing. The argument ran roughly as follows: "If laborers are allowed to organize, they can act as a unit to fix wage prices and restrict free negotiation of wages. This is a restraint of trade and freedom within the labor market." Based on this interpretation, local courts were empowered to issue injunctions to stop labor from organizing. If a factory owner found his workers attempting to organize, he could simply go to the courts and request an injunction forbidding such activity.

In 1914, Congress acted to offset the effect of the Sherman Antitrust Act by passing the Clayton Act. This act authorized employees to organize to negotiate with a particular employer. However, in most cases the employer could demonstrate that the organizing activity was directed by parties outside the employer's shop. This implied that the action was not a local one and, therefore, was subject to action under the Sherman antitrust legislation. Therefore, the injunction remained a powerful management tool in resisting unionization.

15.4 NORRIS-LAGUARDIA ACT

The passing of the Norris–LaGuardia Act heralded the first major movement of the power pendulum away from management and toward labor. This act, sometimes referred to as the Anti-Injunction Act, accomplished what the Clayton Act had failed to do. It specifically stated that the courts could not intercede on the part of management so as to obstruct the formation of labor organizations. It effectively overrode the Supreme Court interpretation that the Sherman Antitrust Act could be applied to labor organizations. It curtailed the power of the courts to issue injunctions and protected the rights of workers to strike and picket peaceably. It also outlawed the use of "yellow-dog" contracts on the part of management. It was a common practice to have an employee sign a contract upon being hired in which he agreed not to join or become active in any union organization. Such yellow-dog contracts were declared illegal by the Norris–LaGuardia Act. This piece of legislation as interpreted by the Supreme Court during the period of the New Deal effectively freed labor from the constraints of the Sherman Antitrust Act.

15.5   DAVIS-BACON ACT

In 1931, a far-reaching piece of legislation was passed that even today has a significant impact on the cost of federally funded projects throughout the United States. The Davis-Bacon act provides that wages and fringe benefits on all federal and federally funded projects shall be paid at the "prevailing" rate in the area. The level of prevailing rates is established by the secretary of labor, and a listing of these rates is published with the contract documents so that all contractors will be aware of the standards. To ensure that these rates are paid, the government requires submittal by all contractors of a certified payroll each month to the federal agency providing the funding. These rates are reviewed to determine whether any violations of the Davis-Bacon pay scale have occurred. This act is so far-reaching in its effect because much of public construction at the state and local level may be funded in part by federal grants. A large municipal mass-transit system or wastewater treatment plant, for instance, may be funded in part by a federal agency. In such cases, the prevailing rates must be paid. Since the Department of Labor generally accepts the most recently negotiated union contract rates as the prevailing ones, this allows union contractors to bid without fear of being underbid by nonunion contractors paying lower wage rates.

15.6   NATIONAL LABOR RELATIONS ACT

The National Labor Relations Act, also referred to as the Wagner Act, is a landmark piece of legislation that established a total framework within which labor management relations are to be conducted. Its central purposes are to protect union-organizing activity and encourage collective bargaining. Employers are required to bargain in good faith with the properly chosen representatives of the employees. Among other things, it establishes the procedures by which labor can organize and elect representatives. Discrimination against an employee for labor-organizing activities or participation in a union is forbidden by this act.

Employer unfair labor practices defining precisely what actions are not acceptable in management dealings with labor are specified. These practices are summarized in Table 15.2. Comparable unfair practices on the part of labor in dealing with management were not defined. It was assumed that labor was the abused party and would act equitably in its dealings with management. This trust had to be specifically spelled out later in the Taft-Hartley Act.

The act also established a "watch dog" organization to ensure its provisions were properly administered. This organization is the National Labor Relations Board (NLRB). The NLRB acts as the clearinghouse for all grievances and issues leading to complaints by labor against management and vice-versa. It is the highest tribunal below the Supreme Court for settling labor disputes and rules on most issues affecting labor–management relationships.

The act also established the concept of a *closed shop*. For years, labor organizations had fought for the right to force all members of a particular work activity (shop) to be

Table 15.2 Employer Unfair Labor Practices

Under the National Labor Relations Act, as amended, an employer commits an unfair
labor practice if he:

1. Interferes with, restrains, or coerces employees in the exercise of rights protected by the act,
   such as their right of self-organization for the purposes of collective bargaining or other mutual
   assistance {Section 8(a)(1)}.

2. Dominates or interferes with any labor organization in either its formation or its administration
   or contributes financial or other support to it {Section 8(a)(2)}. Thus "company" unions
   dominated by the employer are prohibited, and employers may not unlawfully assist any union
   financially or otherwise.

3. Discriminates against an employee in order to encourage or discourage union membership
   {Section 8(a)(3)}. It is illegal for an employer to discharge or demote an employee or to single
   him out in any other discriminatory manner simply because he is or is not a member of a union.
   In this regard, however, it is not unlawful for employers and unions to enter into compulsory
   union-membership agreements permitted by the National Labor Relations Act. This is subject to
   applicable state laws prohibiting compulsory unionism.

4. Discharges or otherwise discriminates against an employee because he has filed charges or given
   testimony under the act {Section 8(a)(4)}. This provision protects the employee from retaliation
   if he seeks help in enforcing his rights under the act.

5. Refuses to bargain in good faith about wages, hours, and other conditions of employment with
   the properly chosen representatives of his employees {Section 8(a)(5)}. Matters concerning rates
   of pay, wages, hours, and other conditions of employment are called mandatory subjects, about
   which the employer and the union must bargain in good faith, although the law does not require
   either party to agree to a proposal or to make concessions.

6. Enters into a hot-cargo agreement with a union {Section 8(e)}. Under a hot-cargo agreement, the
   employer promises not to do business with or not to handle, use, transport, sell, or otherwise deal
   in the products of another person or employer. Only in the garment industry and the construction
   industry (to a limited extent) are such agreements now lawful. This unfair labor practice can be
   committed only by an employer and a labor organization acting together.

Source: From Clough and Sears, *Construction Contracting,* 7th ed., John Wiley & Sons, New York, 2005,
page 369.

members of a union. If the majority voted for union membership, then to work in the shop, a
new employee had to belong to the union. This is in contrast to the *open shop* in which
employees are not organized and do not belong to a union. The Wagner Act endorsed
the concept of the closed shop and made it legal. This concept was later revoked by the
Taft-Hartley Act and replaced by the union shop. The closed shop was attacked as illegal on
the basis that it infringed upon a person's "right to work" and freedom of choice regarding
union membership. The union shop will be discussed later in the section on the Taft-
Hartley Act.

15.7 FAIR LABOR STANDARDS ACT

The Fair Labor Standards Act is commonly referred to as the minimum wage law. It was
originally passed in 1938 and establishes the minimum wages and maximum hours for all
workers. The minimum wage level is periodically changed to be consistent with changing
wage rates. The law defines the 40-hour work week and time over this amount as overtime.
It is, generally, an outgrowth of the child labor abuses that occurred in the 19th century. It
also forbids discrimination by establishing the concept of "equal pay for equal work."
Recent arguments against increasing the minimum wage have hinged on the ideas that
certain menial and domestic tasks that could provide unskilled workers with employment
have become so expensive that it no longer is reasonable to perform them. The clearing of

refuse and cutting of grass along roadways were done in former times by hand labor at low wages. Increasing minimum wages make this too expensive.

15.8   UNION GROWTH

Under the provisions of the union legislation of the 1930s, the labor unions began to flourish. As is often the case during periods of transition, in which inflexible barriers previously existed, a vacuum in favor of labor developed. The hard line of management was broken, and labor rushed in to organize and exploit the new situation. Along with the benefits accruing to the worker from these events, the inevitable abuses of the unstructured and unrestricted growth soon became apparent. In 1938, the unbridled actions of the unions and their leaders started to swing public opinion against them. Some unions flaunted their newfound power by introducing *restrictive labor practices* and *wartime strikes*, which shut down plants producing critical military supplies. Criminal activities within the unions were widespread and virtually unchecked. In 1943, Congress responded to this changing public perception of unions by passing the War Labor Disputes Act (Smith Connolly Act). This reflected public displeasure with the high-handed tactics and unpatriotic stance of the labor unions. It was designed to limit strikes in critical wartime industries and expedite settlement of disputes. It was largely ineffective but did reflect increasing public support of legislation that would control the prerogatives of labor unions. By 1947, thirty-seven states had enacted some form of labor control bill.

The inroads made by criminal elements active in union activities were recognized by the Hobbs (Anti-Racketeering) Act of 1946. This legislation was enacted to protect employers from threats, force, or violence by union officials extorting payments for "services rendered." Payments requested included commissions for various types of aid and assistance, gifts for controlling labor trouble, and equipment rentals forced on the employers at exorbitant costs. These laws and the continuing difficulties developing from abuse of power on the part of unions set the stage for the enactment of the Taft-Hartley Act of 1947.

15.9   LABOR MANAGEMENT RELATIONS ACT

The Labor Management Relations (Taft-Hartley) Act together with the Wagner Act form the two cornerstones of U.S. labor relations legislation. The Taft-Hartley Act amended the Wagner Act and reversed the swing of the power pendulum once more, still leaving labor in a strong position, but pushing the pendulum more toward center. It is the first post-Depression law to place effective constraints on the activities of labor. It restructured the makeup and operation of the NLRB, attempting to give management a stronger voice and to balance representation of labor and management. Section 7 of the bill defines the rights of workers to participate in or refrain from union activities. Section 8 provides the counterpoint to the Employer Unfair Practices Section of the Wagner Act. It defines *Union Unfair Labor Practices*, which specify tactics on the part of the labor that are illegal (see Table 15.3). The law also established the Federal Mediation and Conciliation Service, which acts as a third party in trying to expedite a meeting of the minds between unions and management involved in a dispute. This service has been visible in meeting with representatives of players unions and sports team owners to work out the terms of player contracts.

Under the Taft-Hartley legislation, the president is empowered to enjoin workers on strike (or preparing to strike) to work for an 80-day cooling-off period during which time negotiators attempt to reach agreement on contractual or other disputes. This strike moratorium may be invoked in industries in which a strike endangers the health of the national economy. The president has used his powers under this provision of the law on numerous occasions since 1947.

Table 15.3 Union Unfair Labor Practices

Under the National Labor Relations Act, as amended, it is an unfair labor practice for a labor organization or its agents:

1. **a.** To restrain or coerce employees in the exercise of their rights guaranteed in Section 15.8 of the Taft–Hartley Act {Section 8(b)(1)(A)}. In essence Section 15.8 gives an employee the right to join a union or to assist in the promotion of a labor organization or to refrain from such activities. This section further provides that it is not intended to impair the right of a union to prescribe its own rules concerning membership.
   b. To restrain or coerce an employer in his selection of a representative for collective bargaining purposes {Section 8(b)(1)(B)}.
2. To cause an employer to discriminate against an employee in regard to wages, hours, or other conditions of employment for the purpose of encouraging or discouraging membership in a labor organization {Section 8(b)(2)}. This section includes employer discrimination against an employee whose membership in the union has been denied or terminated for cause other than failure to pay customary dues or initiation fees. Contracts or informal arrangements with a union under which an employer gives preferential treatment to union members are violations of this section. It is not unlawful, however, for an employer and a union to enter an agreement whereby the employer agrees to hire new employees exclusively through a union hiring hall so long as there is no discrimination against nonunion members. Union security agreements that require employees to become members of the union after they are hired are also permitted by this section.
3. To refuse to bargain in good faith with an employer about wages, hours, and other conditions of employment if the union is the representative of his employees {Section 8(b)(3)}. This section imposes on labor organizations the same duty to bargain in good faith that is imposed on employers.
4. To engage in, or to induce or encourage others to engage in, strike or boycott activities, or to threaten or coerce any person, if in either case an object thereof is:
   a. To force or require any employer or self-employed person to join any labor or employer organization, or to enter into a hot-cargo agreement that is prohibited by Section 8(e) {Section 8(b)(4)(A)}.
   b. To force or require any person to cease using or dealing in the products of any other producer or to cease doing business with any other person {Section 8(b)(4)(B)}. This is a prohibition against secondary boycotts, a subject discussed further in Section 14.18 of this text. This section of the National Labor Relations Act further provides that, when not otherwise unlawful, a primary strike or primary picketing is a permissible union activity.
   c. To force or require any employer to recognize or bargain with a particular labor organization as the representative of his employees that has not been certified as the representative of such employees {Section 8(b)(4)(C)}.
   d. To force or require any employer to assign certain work to the employees of a particular labor organization or craft rather than to employees in another labor organization or craft, unless the employer is failing to conform with an order or certification of the NLRB {Section 8(b)(4)(D)}. This provision is directed against jurisdictional disputes, a topic discussed in Section 14.12 of this text.
5. To require of employees covered by a valid union shop membership fees that the NLRB finds to be excessive or discriminatory {Section 8(b)(5)}.
6. To cause or attempt to cause an employer to pay or agree to pay for services that are not performed or not to be performed {Section 8(b)(6)}. This section forbids practices commonly known as featherbedding.
7. To picket or threaten to picket any employer to force him to recognize or bargain with a union:
   a. When the employees of the employer are already lawfully represented by another union {Section 8(b)(7)(A)}.
   b. When a valid election has been held within the past 12 months {Section 8(b)(7)(B)}.
   c. When no petition for a NLRB election has been filed within a reasonable period of time, not to exceed 30 days from the commencement of such picketing {Section 8(b)(7)(C)}.

Source: From Clough and Sears, *Construction Contracting,* 6th ed., John Wiley & Sons, New York, 1994.

All present employees who are members of the Union on the effective date of this agreement shall be required to remain members in good standing of the Union as a condition of their employment.

All present employees who are not members of the Union shall, from and after the 7th day following the date of execution of this agreement, be required to become and remain members in good standing of the Union as a condition of their employment.

All employees who are hired thereafter shall be required to become and remain members in good standing of the Union as a condition of their employment from and after the 7th day of their employment or the effective date of this Agreement, whichever is later, as long as Union membership is offered on the same terms as other members.

Any employee who fails to become a member of the Union or fails to maintain his membership therein in accordance with provisions of the paragraphs of this Section, shall forfeit his rights of employment and the employer shall within two (2) working days of being notified by the Union in writing as to the failure of an employee to join the Union or maintain his membership therein, discharge such employee. For this purpose, the requirements of membership and maintaining membership shall be consistent with State and Federal Laws. The Employer shall not be deemed in default unless he fails to act within the required period after receipt of registered written notice.

(Excerpted from Agreement Between Central Illinois Builders and The United Brotherhood of Carpenters and Joiners of America Local Union No. 44, Champaign-Urbana, Illinois.)

Figure 15.1 Contract typical member clause.

Section 14(b) is significant in that it redefines the legality of closed-shop operations and defines the *union shop*. A totally closed shop is one in which the worker must be a union member before he or she is considered for employment. As already stated this is declared illegal by the Taft-Hartley Act. The union shop *is* legal. A union shop is one in which a nonmember can be hired. The worker is given a grace period (usually 30 days in manufacturing shops and a shorter period in the construction industry), during which time he or she must become a union member. If the new employee does not become a member, the union can request that the candidate employee be released. Under the closed-shop concept, it was much easier for the unions to block a worker from gaining employment. This could be used to discriminate against a potential employee. The union shop gives the worker a chance to join the union (see Figure 15.1). If the worker requests membership and the union refuses after 30 days,[2] management can ask the union to show cause why the employee has not been admitted to membership.

The law also recognizes the concept of "agency" shop. In such facilities a worker can refuse to join the union. The employee, therefore, has no vote in union affairs. The worker must, however, pay union dues because he or she theoretically benefits from the actions of the union and the union acts as his or her "agent." If the union, for instance, negotiates a favorable pay increase, all employees benefit and all are required to financially support the labor representation (i.e., the union negotiators).

Because of the way in which the law regarding closed shop under the Wagner Act was implemented, many workers felt that their constitutional right to work was being abrogated. That is, unless they were already union members, they were not free to work in certain firms. They had no choice. They were forced to either join the union or go elsewhere. The Taft-Hartley Act allows the individual states to enact right-to-work

---

[2] The period shown in Figure 15.1 is 7 days. This is typical in the construction industry and recognizes the more transient nature of construction work.

laws that essentially forbid the establishment of totally union shops. States in the South and the Southwest where unions are relatively weak have implemented this feature at the state level. Clough and Sears explain this as follows.

> *Section 14(b) of the Taft-Hartley Act provides that the individual states have the right to forbid negotiated labor agreements that require union membership as a condition of employment. In other words, any state or territory of the United States may, if it chooses, pass a law making a union-shop labor agreement illegal. This is called the "right-to-work" section of the act, and such state laws are termed right-to-work statutes. At the present writing, 21 states have such laws in force.[3] It is interesting to note that most of these state right-to-work laws go beyond the mere issues of compulsory unionism inherent in the union shop. Most of them outlaw the agency shop, under which nonunion workers must pay as a condition of continued employment the same initiation fees, dues, and assessments as union employees, but are not required to join the union. Some of the laws explicitly forbid unions to strike over the issue of employment of nonunion workers.[4]*

The fact that a right-to-work provision has been implemented can be detected by reading the language of the labor agreements within a given state. In states in which no right-to-work law is in effect, a clause is included indicating that a worker must join the union within a specified period. Such a clause taken from an Illinois labor contract is shown in Figure 15.1. This clause would be illegal in Georgia.

15.10   OTHER LABOR LEGISLATION

The Labor Management Reporting and Disclosure (Landrum–Griffin) Act was passed in 1959 to correct some of the deficiencies of previous legislation. Among its major objectives were (a) the protection of the individual union member, (b) improved control and oversight of union elections, and (c) an increased government role in auditing the records of unions. Misappropriation of union funds by unscrupulous officials and apparent election fraud were the central impetus in enacting this law. Under this law, all unions must periodically file reports with the Department of Labor regarding their organization finances and other activities. The act provides that employers cannot make payments directly to union officials. They can, however, pay dues and fringe benefits to qualified funds of the union for things such as health and welfare, vacation, apprenticeship programs, and the like. Records regarding these funds are subject to review by government auditors.

Title IV of the Civil Rights Act (enacted in 1964) establishes the concept of equal employment opportunity. This legislation was expanded by the Civil Right Act of 1991. It forbids discrimination on the basis of race, color, religion, sex, or national origin. It is administered by the Equal Employment Opportunity Commission (EEOC) and applies to discrimination in hiring, discharge, conditions of employment, and classification. Its application in the construction industry has led to considerable controversy. Individual workers can file an *unfair labor practice* charge against a union because of alleged discrimination. Unions found guilty face *cease-and-desist* orders as well as possible rescission of their mandate to act as the authorized employee representative.

Executive Order 11246 issued by President Lyndon B. Johnson in 1965 further amplified the government position on equal opportunity. It establishes affirmative action requirements on all federal government or federally funded construction work. It is administered by the Office of Federal Contract Compliance (OFCC). This office is

---

[3] Alabama, Arizona, Arkansas, Florida, Georgia, Idaho, Iowa, Kansas, Louisiana, Mississippi, Nebraska, Nevada, North Carolina, North Dakota, South Carolina, South Dakota, Tennessee, Texas, Utah, Virginia, and Wyoming now have right-to-work legislation in effect.

[4] Clough, R. and Sears, G. (1994) *Construction Contracting* (6th ed.). New York: John Wiley and Sons.

instrumental in establishing the level of minority participation in government work. It has spawned a number of plans for including minority contractors in federally funded projects. Executive Order 11375 (1968) extends Order 11246 to include sex discrimination. Contractors working on federally funded work are required to submit affirmative action reports to the OFCC. If the plan is found to be deficient, the OFCC can suspend or terminate the contract for noncompliance.

15.11   VERTICAL VERSUS HORIZONTAL LABOR ORGANIZATION STRUCTURE

The traditional craft unions are normally referred to as horizontally structured unions. This is because of the strong power base that is located in the union local. Contract negotiations are conducted at the local level and all major decisions are concentrated at the local level. Construction unions are craft unions with a strong local organization. The local normally is run on a day-to-day basis by the *business agent*. Representatives at the individual job sites are called *job stewards*. The local elects officers and a board of directors on a periodic basis. The local president and business agent may be the same individuals. The bylaws of the local define the organizational structure and particulars of union structure. At the time of contract negotiations, representatives from the local meet with representatives of the local union contractors to begin discussions. The Associated General Contractors (AGC) in the local area often acts as the contractor's bargaining unit. This horizontal structure leads to a proliferation of contracts and a complex bargaining calendar for the contractors' association. If a contractors' group generally deals with 12 craft unions in the local area and renegotiates contracts on an annual or biennial basis, it is obvious that the process of meeting and bargaining can become complicated. Contracts are signed for each union operating in a given area. The national headquarters organizations for construction craft unions normally coordinate areas of national interest to the union, such as congressional lobbying, communication of information regarding recently negotiated contracts, national conventions, printing of newsletters and magazines, seminars, workshops, and other general activities. The real power in most issues, however, is concentrated at the local level. The horizontal organization then is similar to a confederation, with strength at the bottom and coordination at the top.

Vertically structured unions tend to concentrate more of the power at the national level. Significantly, labor contracts are negotiated at the national level. This means a contract is signed at the national level covering work throughout the country. This is considerably more efficient than the hundreds of locally negotiated contracts that are typical of horizontally structured unions. The industrial unions of the CIO have traditional organization in a vertical structure, whereas the construction unions of the AFL maintain the strong local horizontal structure. The construction elements within industrial unions usually follow the example of the parent union. The construction workers of the United Mine Workers (UMW) are an example of this. They sign a single contract with the mine owners covering all of the crafts from operating engineers to electricians. A list of scales covering all specialties (i.e., craft disciplines) is contained in the national contract. Since the members of the union are mine construction workers first and carpenters, operators, or electricians second, the jealousy regarding so-called craft lines and jurisdiction is less pronounced. It is not uncommon to see an equipment operator in a vertically structured union get down from a tractor and do some small carpentry. This would be impossible in a horizontally organized craft union situation because the carpenters would immediately start a jurisdictional dispute.

15.12   JURISDICTIONAL DISPUTES

In addition to the fragmentation of contracts by craft and local area, one of the major difficulties inherent in the horizontal craft structured union is the problem of craft jurisdiction. Job jurisdiction disputes arise when more than one union claims jurisdiction over a given item of work. This is true primarily because many unions regard a certain type of work as a proprietary right and jealously guard against any encroachment of their traditional sphere by other unions. As technology advances and new products are introduced, the question of which craft most appropriately should perform the work involved inevitably arises. A classical example in building construction is provided by the introduction of metal window and door frames. Traditionally, the installation of windows and doors had been considered a carpentry activity. However, the introduction of metal frames led to disputes between the carpenters and the metal workers as to which union had jurisdiction in the installation of these items. Such disputes can become heated and lead to a walkout by one craft or the other. This may shut down the job. The contractor is sometimes simply an innocent bystander in such instances. If these disputes are not settled quickly, the repercussions for client and contractor can be serious, as indicated by the following excerpt from *Engineering News Record*:

> The nozzle-dispute on the $1-billion Albany, N.Y., mall project has caused hundreds of stoppages on that job, which employs over 2,000 persons. The argument revolves around whether the teamster driving a fuel truck or the operating engineer running a machine shall hold the nozzle during the fueling operation. Both unions claim the job. Because holding the nozzle involves a certain amount of work, the question is why either union should want it, since regardless of which man does the job, the other still gets paid. The answer undoubtedly is that the union that gets jurisdiction will eventually be able to claim the need for a helper. This particular dispute has been reported as plaguing contractors in many states, including West Virginia, Oklahoma, Missouri, California and Washington.[5]

Although this is a rather extreme example, it is indicative of the jealousies that can arise between crafts.

Concern on the part of unions for jurisdiction is understandable because rulings that erode their area of work ultimately can lead to the craft slowly dwindling into a state of reduced work responsibilities and, eventually, into extinction. Therefore, the craft unions jealously protect their craft integrity. The following clause from a contract indicates how comprehensive the definition of craft responsibility can become.

> *Scope of Work*
> *This Agreement shall cover all employees employed by the Employer engaged in work coming under all classifications listed under the trade autonomy of the United Brotherhood of Carpenters and Joiners of America.*
> *The trade autonomy of the United Brotherhood of Carpenters and Joiners of America consists of the milling, fashioning, joining, assembling, erection, fastening or dismantling of all material of wood, plastic, metal, fiber, cork and composition, and all other substitute materials and the handling, cleaning, erecting, installing and dismantling of machinery, equipment and all materials by members of the United Brotherhood.*
> *Our claim of jurisdiction, therefore, extends over the following divisions and subdivisions of the trade: Carpenters and Joiners; Millwrights; Pile Drivers; Bridge, Dock and Wharf Carpenters; Divers; Underpinners; Timbermen and Core Drillers; Shipwrights, Boat Builders, Ship Carpenters; Joiners and Caulkers; Cabinet Makers, Bench Hands, Stair Builders, Millmen; Wood and Resilient Floor Layers, and Finishers; Carpenter Layers; Shinglers; Siders; Insulators; Acoustic and Dry Wall Applicators; Shorers and House Movers; Loggers, Lumber and Sawmill Workers; Furniture Workers, Reed and Rattan Workers; Shingle Weavers; Casket and Coffin Makers; Box Makers, Railroad Carpenters and Car Builders, regardless of material used; and all those engaged in the operation of woodworking or other machinery required in the fashioning, milling or manufacturing of*

---

[5] "Low Productivity: The Real Sin of High Wages," *Engineering News Record*, February 24, 1972.

products used in the trade, or engaged as helpers to any of the above divisions or subdivisions' burning, welding, rigging and the use of any instrument or tool for layout work incidental to the trade. When the term "carpenter and joiner" is used, it shall mean all the subdivisions of the trade. The above occupational scope shall be subject to all agreements between International Representatives.[6]

Jurisdictional disputes present fewer problems in vertically structured unions, since craft integrity is not a matter that determines the strength of the union. All major automobiles are assembled by members of the United Automobile Workers (UAW). The UAW is a typical vertically structured union. Technological changes do not mean the work could be shifted to another union. Therefore, UAW workers can be installing windows today and can be moved to installation of electrical wiring next month. Craft integrity does not have to be jealously protected.

European construction workers are organized into vertically structured unions. National agreements in countries such as Germany cover all workers and are signed periodically defining wage scales and general labor management procedures. Each worker has a primary specialty and is paid at the rate established in the national agreement. Because craft jurisdiction is not a major issue, it is not unusual to see a worker who is operating a backhoe get down and work as part of a crew installing shoring. Similar mobility back and forth across craft lines is common in the UMW, since it is also vertically structured.

15.13   UNION STRUCTURE

The largest labor organization in the United States is the AFL–CIO. The building and construction trade unions are craft unions and as such are affiliates of the Building and Construction Trades Department of the AFL–CIO. The structure of affiliates from local to national level is shown schematically in Figure 15.2. A list of the construction unions that

Figure 15.2 Structure typical of an affiliate of the Building and Construction Trades Department, AFL–CIO.

[6] Excerpted from Agreement Between The United Brotherhood of Carpenters and Joiners of American Local No. 44 Champaign-Urbana, Illinois, and the Central Illinois Builders Chapter of Associated General Contractors of America.

Table 15.4 AFL–CIO Construction Unions

1. International Association of Bridge, Structural, Ornamental, and Reinforcing Iron Workers
2. International Association of Heat and Frost Insulators and Asbestos Workers
3. International Brotherhood of Boilermakers, Iron Ship Builders, Blacksmiths, Forgers, and Helpers
4. International Brotherhood of Electrical Workers
5. International Brotherhood of Painters, and Allied Trades
6. International Union of Bricklayers and Allied Craft Workers
7. International Union of Elevator Constructors
8. International Union of Operating Engineers
9. International Brotherhood of Teamsters
10 Laborers International Union of North America
11. Operative Plasterers and Cement Masons' International Association
12. Sheet Metal Workers' International Association
13. United Association of Journeymen and Apprentices of the Plumbing and Pipe Fitting Industry of the United States and Canada
14. United Brotherhood of Carpenters and Joiners of America
15. United Union of Roofers, Water proofers and Allied Workers

are within the AFL–CIO is given in Table 15.4. Most construction-related unions are presently affiliated with the AFL–CIO.

There are two ways a national union may join the AFL–CIO. The first is for an already established union to apply for a charter. The other is for the federation to create a new union from a related group of locals that are not members of any national union but are directly associated with the AFL–CIO.

The top governing body of the AFL–CIO is the biennial convention. Between conventions, the executive council runs the affairs of the federation. The members are the president, secretary-treasurer, and several vice-presidents elected by the majority at the convention (usually from among the presidents of the national unions). The president has the authority to rule on any matters concerning the constitution or a convention decision between meetings of the council.

The AFL–CIO maintains trade departments at the level directly below the executive council. The mission of these major sections is to further unionization in the appropriate industry or trade. They also aid in the settlement of jurisdictional disputes between the members in their department. Disputes with a union in another department are appealed to the executive council. Departments also represent their members before Congress and other government agencies. The Building and Construction Trades Department is responsible for all construction craft unions.

15.14 NATIONAL UNIONS

National unions are defined as those unions having collective bargaining agreements with different employers in more than one state and federal employee unions with exclusive bargaining rights. Because of their assumed role of collective bargaining in many areas, the national unions have become increasingly powerful. In construction unions, however, the locals still play the most important role in collective bargaining and, therefore, power still resides at the local level.

Each union has exclusive jurisdiction to function as the workers' representative in its trade or branch of industry. The jurisdiction of most unions is at least partially set forth in their charter and constitution. As the unions' outlook and purposes have changed or as their members' jobs have altered, many unions have changed their jurisdiction as well.

The daily conduct of union business is in the hands of the national president, whose influence is a big factor in deciding what issues the union executive board will discuss and vote on. What the president decides will have an effect on the general public as well as on the union. The president's more important powers are to decide on constitutional matters, issue or revoke local charters, hire or fire union employees, and sanction strikes. Most actions involving the powers of the president can be appealed to the board or to the convention.

The organizer or representative of the union provides contact between the locals and the national headquarters and attempts to gain new members for the union and to set up new locals. He or she is the union advisor to all of the locals within his/her area and must explain national policies to them. Moreover, the organizer informs the national level of local problems.

15.15   STATE FEDERATIONS AND CITY CENTRALS

State federations are concerned mostly with lobbying for needed legislation and public relations on the state level. They are composed of locals whose national union is a member of the AFL–CIO. Conventions are held annually in which programs of interest to all of the state's workers are concerned.

City centrals are concerned more with economics, serving as a clearinghouse for locals and aiding in dealings with employers. They have become increasingly involved in general community affairs and activities that may indirectly benefit their members.

Joint boards and trade councils are composed of locals involved in similar trades or industries. Their principal duty is to ensure that workers present a unified front in collective bargaining and obtain uniform working conditions in their area. A joint board or council is usually required for unions with more than three locals in the same region. The joint board is made up of all locals of the same national union, whereas the trades council is composed of locals of different national unions in related trades in the same industry.

The prototype for local trades councils is the Building and Construction Trades Council, which has its higher-level counterpart in the Building and Construction Trades Department of the AFL–CIO. Its problems are not limited to labor–management relations; it is often involved in settling ticklish jurisdictional disputes. The Building and Construction Trades Council provides craft unions with an important advantage characteristic to the industrial unions: the ability to present a united front in dealings with management. Some councils negotiate city-wide agreements with employers or see that the agreements of their member locals all expire on the same date. They have a great deal of influence with the locals but may not make them act against national union policy.

15.16   UNION LOCALS

The locals are the smallest division of the national union. They provide a mechanism through which the national union can communicate with its members at the local level. Locals provide for contact with other workers in the same trade and area means by which better working conditions are obtained, grievances are settled, and educational and political programs are implemented. They may be organized on an occupational or craft basis or on a plant or multiplant basis. In the building industry, it is common to have locals for each craft in large cities. The local officials who preside over the committees and the general meeting are the president, vice-president, treasurer, and various secretaries. They are usually unpaid or paid only a small amount and continue to work at their trade. They perform their union duties in their spare time. In small locals, a financial secretary will take care of the local books and records; but in large locals a trained bookkeeper is employed for this purpose.

The most important local official is the business agent, a full-time employee of the local. He exercises a great deal of leadership over the local and its affairs through the advice he provides to the membership and elected officials. He is usually trained and experienced in labor relations and possesses a large amount of knowledge of conditions on which other members are poorly informed.

The business agent's duties cover the entire range of the local's activities. He or she helps settle grievances with employers, negotiates agreements, points out violations of trade agreements, and operates the union hiring hall. The business agent is also an organizer, trying to get unorganized workers into the union. Only locals with a large membership can afford a full-time agent, and over one-half of the locals employing agents are in the building trades where there is a greater need due to the transient nature of the work. For the locals who do not have enough money to employ their own business agent, an agent is usually maintained by the city central or state federation.

The shop steward is not a union official but is the representative who comes in closest contact with the members. He or she must see that union conditions are maintained on the job and handle grievances against the employer. The steward is a worker on the job site elected by his or her peers.

15.17   UNION HIRING HALLS

One of the salient features of construction labor is its transient nature. Construction workers are constantly moving from job site to job site and company to company. It is not uncommon for a construction worker to be employed by five or six different contractors in the same year. The union hiring hall provides a referral service that links available labor with contractor's requests. Following each job, a worker registers with the union hall and is referred to a new job site as positions become available. The procedures governing operation of the union hiring hall constitute an important part of the agreement between the union and the contractor. Articles of the labor contract specify precisely how the hiring hall is to operate. Although there are small variations from craft to craft and region to region, similar procedures are commonly used for referring workers through the union hall.

15.18   SECONDARY BOYCOTTS

The legality of boycotts to influence labor disputes has been an issue of primary importance throughout the history of labor–management relations. A boycott is an action by one party to exert some economic or social pressure on a second party with the intent of influencing the second party regarding some issue. A *secondary* boycott is one in which party A who has a dispute with party B attempts to bring pressure on B by boycotting party C who deals with B and who can bring strong indirect pressure on B to agree to some issue. This is shown schematically in Figure 15.3. If the electrical workers in a plant fabricating small

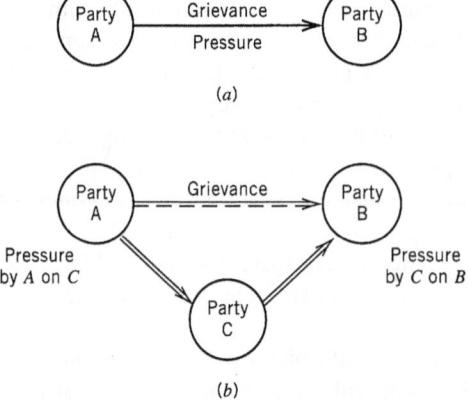

Figure 15.3 Types of boycotts: (*a*) primary boycott and (*b*) secondary boycott.

appliances go to the factory and form a picket line to get an agreement, there is a primary boycott in progress. If, however, the workers send some of their members into the town and put pickets up at stores selling appliances from the plant, a secondary boycott is established. The store owners are a third party (C) being pressured to influence the factory to settle with the workers. The Taft-Hartley Act declared the use of a secondary boycott to be illegal.

In the construction industry, such secondary boycotts occur on sites with both union and nonunion workers when a union attempts to force a nonunion subcontractor to sign a union contract. In such cases, the union will put up a picket line at the entrance to the work site, in effect, to picket or boycott the nonunion subcontractor. Tradition among labor unions, however, demands that no union worker can cross another union's picket line. Therefore, the actual effect of the union picket line will be to prevent all union workers from entering the site. This may cause the shutdown of the entire site pending resolution of the nonunion subcontractor's presence on the site. In this situation, the general or prime contractor is a third party being pressured by the union to influence the nonunion subcontractor. This is called *common situs* picketing. In 1951, the U.S. Supreme Court ruled this practice to be a secondary boycott and, therefore, illegal under the Taft-Hartley Act. The high court made this ruling in the case of the Denver Building and Construction Trades Council.

Following this decision, the doctrine of "separate gates" was developed to deal with secondary boycott problems. Under this policy, the prime contractor establishes a separate or alternate gate for the nonunion subcontractor with whom the union has a dispute. The union is then directed to place its picket line at this gate rather than the main project gate. If it fails to comply, it can be enjoined from boycotting. Other union personnel entering the site can enter at the main gate without crossing the picket line of another union.

Certain interpretations of the secondary boycott have essentially provided exceptions in the construction industry. Unions normally have attempted to refrain from handling goods or products from nonunion shops. Such materials are called "hot cargo," and unions have bargained for hot-cargo contract clauses that, in effect, prevent a contractor from handling such materials from nonunion fabricators. This is a secondary boycott in the sense that the contractor becomes an innocent third party in the dispute between the union and the fabricator or product supplier. The Landrum–Griffin Act provides that such hot-cargo or subcontractor clauses that ban use of these materials or contact with these open shop units are illegal. An exemption is made, however, for the construction industry. As noted by Clough and Sears:

> Subcontractor agreements typically require the general contractor to award work only to those subcontractors who are signatory to a specific union labor contract or who are under agreement with the appropriate union.[7]

The Supreme Court also ruled in 1967 that prefabrication clauses that ban the use of certain prefabricated materials are exempted from the secondary boycott legislation if such prefabricated items threaten the craft integrity and eliminate work that would normally be done on site. Union carpenters, for instance, might refuse to install prefabricated door units because the doors and the frames are preassembled in a factory off site. This eliminates assembly work that could be done on site and endangers the union scope of work. Use of such prefabricated items could lead to the decay of the craft's jurisdiction and integrity.

---

[7] Clough and Sears, *op. cit.,* p. 376.

Therefore, the use of such clauses in labor agreements is not considered to be an unlawful practice in these instances.

15.19 OPEN-SHOP AND DOUBLE-BREASTED OPERATIONS

Restrictive work rules and high wages have made it difficult for union contractors to be competitive in some market areas. In an open-shop firm, there is no union agreement and workers are paid and advanced on a merit basis. The largest group of open-, or merit, shop contractors is represented by the Associated Builders and Contractors (ABC). Traditionally, open-shop contractors have bid successfully in the housing and small building market in which the required skill level is not high. Union contractors have dominated the more sophisticated building and heavy construction markets based on their ability to attract skilled labor with higher wages and benefits.

Large open-shop contractors have been willing to meet or exceed the union wage rates to avoid the costly work delays associated with jurisdictional disputes and restrictive work rules. In some cases, the unions have responded by signing project agreements that relax certain work rules for the duration of a given job.

To be able to bid in both open-shop and union formats, some firms have organized as *double-breasted* contractors. Large firms will have one subsidiary that operates with no union contracts. A separately managed company will be signed to all union contracts. In this way, the parent firm can bid both in union shop markets[8] and in markets in which the lower-priced open shop encourages more cost-competitive bidding.

15.20 LABOR AGREEMENTS

Just as the contractor enters into a contract with the client, with vendors supplying materials (i.e., purchase orders), and with subcontractors working under his direction, if union labor is used, the contractor also enters into contracts, or labor agreements, with each of the craft unions with whom he or she deals. These contracts usually cover a 1- or 2-year period and include clauses governing the reconciliation of disputes, work rules, wage scales, and fringe benefits. The wages are normally defined in step increases throughout the period of the contract. These step increases are usually contained in the addendum to the labor contract.

The opening sections of the agreement typically provide methods for reconciling disputes that can arise between the contractors and the union during the life of the contract. To handle disputes, articles in the contract set up a joint conference committee to reconcile disputes and provide for arbitration procedures for disputes that cannot be settled by the committee. Typical contracts also include provisions governing:

1. Maintenance of membership
2. Fringe benefits
3. Work rules
4. Apprentice program operation
5. Wages (addendum)
6. Hours
7. Worker control and union representation
8. Operation of the union hiring hall

---

[8] The owner may specify that union labor is to be used, or Davis-Bacon rulings may dictate that union rates will prevail.

9. Union area

10. Subcontractor clauses (see Section 15.18)

11. Special provisions

Fringe benefits are economic concessions gained by unions covering vacation pay, health and welfare, differentials in pay due to shift, contributions by the contractor to apprenticeship programs, and so-called industrial advancement funds. These are paid by the contractor in addition to the base wage and garnish the salary of the worker. The building and construction trades councils for each union area normally disseminate summaries of contract wage and fringe benefit provisions that assist in the preparation of payroll. Such a summary is shown in Figure 15.4.

Work rules are an important item of negotiation and have a significant effect on the productivity of workers and the cost of installed construction. A typical work rule might require that all electrical materials on site will be handled by union electricians. Another might require that all trucks moving electrical materials on site be driven by union electricians. Such provisions can lead to expensive tradesmen doing work that could be done by less-expensive crafts or laborers. Therefore, work rules become major topics of discussion during the period of contract negotiation.

15.21   LABOR COSTS

The large number of contributions and burdens associated with the wage of a worker makes the determination of a worker's cost to the contractor a complex calculation. The contractor must know how much cost to put in the bid to cover the salary and associated contributions for all of the workers. Assuming that the number of carpenters, ironworkers, operating engineers, and other craft workers required is known and the hours for each can be estimated, the average hourly cost of each craft can be multiplied by the required craft hours to arrive at the total labor cost. The hourly average cost of a worker to the contractor consists of the following components:

1. Direct wages

2. Fringe benefits

3. Social security contributions (FICA)

4. Unemployment insurance

5. Workmen's compensation (WC) insurance

6. Public liability and property damage insurance

7. Subsistence pay

8. Shift pay differentials

The direct wages and fringe benefits can be determined by referring to a summary of wage rates such as the one shown in Figure 15.4.

All workers must pay social security and Medicare on a portion of their salary. For every dollar the worker pays, the employer must pay a matching dollar. The worker pays a fixed percent on every dollar earned up to a cutoff level. After the annual income has exceeded the cutoff level, the worker (and the worker's employer) need pay no more. The FICA and Medicare contribution in 2010 was required on the first $106,800 of annual income at the rate of 7.65%. Therefore, a person making $106,800 or more in annual income would contribute $8170.20 and the person's employer or employers, the contractor, would contribute a like amount.

Unemployment insurance contributions are required of all employers. Each state sets a percentage rate that must be paid by the employer. The premiums are escrowed on a

CRAFT AND BUSINESS REPRESENTATIVE	WAGE RATE PER HOUR	FOREMAN	OVERTIME RATE	W—WELFARE P—PENSION A—APPRENTICE V—VACATION	TRAVEL PAY SUBSISTENCE	AUTOMATIC WAGE INCREASE	AUTOMATIC FRINGE INCREASES	EXPIRATION DATE
Asbestos Workers Local No. 18 Robert J. Scott, BR 946 North Highland Indianapolis, Indiana 46202	$21.57		Double	W—$1.95 P—$1.95 A—$0.38 V—$3.85 Deduct	$45 per day			5-31-08
Boilermakers Local No. 60 George Williams, BR 400 North Jefferson Peoria, Illinois 61603	$22.30	$1.50—F $3.00—GF	Double	W—$5.21 P—$1.95 A—$0.10	$40 per day	$3.00 9-1-07		8-31-08
Carpenters Local No. 44 Gene Stirewalt, BR 212 W. Hill St. Champaign, Illinois 61820	$19.70	12%	Double	W—$0.52 P—$0.90 A—$0.20 IAF—$0.10		$1.00 10-15-08	$0.35 4-15-08	4-15-09
Cement Finishers Local No. 143 Francis E. Ducey, BR 212 ½ South First St. Champaign, Illinois 61820	$16.89	$1.50 15% GF	Double	W—$0.55		$0.90 1-24-09		7-24-08
Electricians Local No. 601 Jack Hensler, BR 212 South First St. Champaign, Illinois 61820	$22.25	10% 20%—GF	Double	W—$0.70 A—0.2%		$1.00—11-1-08 $0.50—5-1-08 $0.50—11-1-08		4-30-09

Figure 15.4 Labor organizations and wage rates. (Figures provided are for demonstration only and do not purport to be accurate.)

monthly or quarterly basis and sent periodically to the state unemployment agency. The amount to be paid is based on certified payrolls submitted by the employer at the time of paying this contribution. The fund established by these contributions is used to pay benefits to workers who are temporarily out of work through no fault of their own.

The states also require employers to maintain WC insurance for all workers in their employ. This insurance reimburses the worker for injuries incurred in the course of employment. Labor agreements also specifically state this requirement. This recognizes the employer's responsibility to provide a safe working environment and the employer's obligation to provide support to disabled workers. Without this insurance, workers injured in the course of their work activity could become financially dependent on the state. The rates paid for WC area function of the risk associated with the work activity. The contribution for a pressman in a printing plant is different from that of a worker erecting steel on a high-rise building. A typical listing of construction specialties and the corresponding rates is given in Table 15.5. Similar summaries of WC rates are printed in the *Quarterly Cost Roundup* issues of *Engineering News Record*. The rates are quoted in dollars of premium per $100 of payroll. The rate for an ironworker, for example, is $29.18 (or 29.2%) per hundred dollars of payroll paid to ironworkers and structural steel erectors. The premium paid for public liability and property damage (PL and PD) insurance is also tied to the craft risk level and is given in Table 15.5.

When a construction project is underway, accidents occurring as a result of the work can injure persons in the area or cause damage to property in the vicinity. If a bag of cement falls from an upper story of a project and injures persons on the sidewalk below, these persons will normally seek a settlement to cover their injuries. The public liability (PL) arising out of this situation is the responsibility of the owner of the project. Owners, however, normally pass the requirement to insure against such liability to the contractor in the form of a clause in the General Conditions of the construction contract. The General Conditions direct the contractor to have sufficient insurance to cover such public liability claims. Similarly, if the bag of cement falls and breaks the windshield of a car parked near the construction site, the owner of the car will seek to be reimbursed for the damage. This is a property damage situation that the owner of the construction project becomes liable to pay. PD insurance carried by the contractor (for the owner) covers this kind of liability. Insurance carriers normally quote rates for PL and PD insurance on the same basis as for WC insurance. Therefore, to provide PL and PD insurance, the contractor must pay $3.00 for PL and $1.88 for each $100 of steel erector salary paid on the job. These rates vary over time and geographical area and can be reduced by maintaining a safe record of operation. The total amount of premium is based on a certified payroll submitted to the insurance carrier.

Subsistence is paid to workers who must work outside of the normal area of the local. As a result, they incur additional cost because of their remoteness from home and the need to commute long distances or perhaps live away from home. If an elevator constructor in Chicago must work in Indianapolis for 2 weeks, he will be outside of the normal area of his local and will receive subsistence pay to defray his additional expenses.

Shift differentials are paid to workers in recognition that it may be less convenient to work during one part of the day than during another. Typical provisions in a sheet metal worker's contract are given in Figure 15.5. In this example, the differential results in an add-on to the basic wage rate. Shift differential can also be specified by indicating that a worker will be paid for more hours than he works. A typical provision from a California ironworkers contract provides the following standards for shift work: (a) If two shifts are in effect, each shift works 7.5 hours for 8 hours of pay and (b) if three shifts are in effect, each shift works 7 hours for 8 hours of pay. This means that if a three-shift project is being worked the ironworker will receive overtime for all time worked over 7 hours. In addition, he or she will be paid 8 hours pay for 7 hours work. Calculation of shift pay will be demonstrated in the following section.

Table 15.5 Building Craft Wage and Insurance Rates[a]

Locals	Wages	Pension	Health and Welfare	Vacation	Apprentice Training	Misc	Workmen's Compensation[b]	Public Liability[c]	Property Damage[c]
Asbestos Workers	$20.30	$1.20	$1.10		$0.20		$12.18	$2.00	$1.10
Boilermakers	$20.50	$1.50	$2.10		$0.04		$12.92	$0.74	$0.72
Bricklayers	$18.70	$1.00	$1.10	$1.30		$0.16 prom.	$7.10	$0.76	$0.54
Carpenters	$18.90	$.90	$1.00		$0.04		$11.34	$0.80	$0.52
Cement Masons	$17.80	$1.10	$.80			$0.40 bldg.	$5.02	$0.82	$0.58
Electricians	$20.90	1.1%	0.9%	0.8%	0.05%		$4.38	$0.34	$0.42
Operating Engineers	$18.70	$1.50	$1.00		$0.14	$0.20 admin.	$11.22	$1.86	$2.00
Iron Workers	$19.20	$1.14	$1.30	$1.00	$0.14		$29.18	$3.00	$1.88
Laborers	$12.50	$0.66	$0.40			$0.10 educ.	$7.50	$0.38	$0.40
Painters	$18.90	$1.30	$1.30		$500/yr		$7.18	$0.26	$0.88
Plasterers	$18.34	$1.10	$.80			$0.40 bldg. $0.20 prom.	$6.96	$0.78	$0.54
Plumbers	$21.50	$1.00	$1.30		$0.22	$0.12 prom. $0.04 natl.	$5.60	$0.58	$1.18
Sheet Metal	$20.40	$1.40	$1.00		$0.08	$0.18 ind.	$ 7.14	$0.41	$0.40

Unemployment 5.0%
Social Security 7.65%

[a]These rates are only indicative of wage and insurance rates. They are not representative of current data.

[b]Rates are applied per $100 of pay.

[c]Public liability. Maximum coverage under these rates—$5000/person, $10,000 per accident

    For higher coverage—$10,000/$20,000: 1.26 × basic rate

                $25,000/50,000: 1.47 × basic rate

                $50,000/100,000: 1.59 × basic rate

                $300,000/300,000: 1.78 × basic rate

Property damage: Maximum coverage under these rates—$5000/person, $25,000 per accident

    For higher coverage—$25,000/100,000: 1.23 × basic rate

                $50,000/100,000: 1.30 × basic rate

A shift differential premium of twenty (20) cents per hour will be paid for all time worked on the afternoon or second shift, and a shift differential of thirty (30) cents per hour will be paid for all time worked on the night or third shift as follows:

(1) *First Shift.* The day, or first, shift will include all Employees who commence work between 6 A.M. and 2 P.M. and who quit work at or before 6 P.M. of the same calendar day. No shift differential shall be paid for time worked on the day, or first, shift.

(2) *Second Shift.* The afternoon, or second, shift shall include all Employees who commence work at or after 2 P.M. and who quit work at or before 12 midnight of the same calendar day. A shift differential premium of twenty (20) cents per hour shall be paid for all time worked on the afternoon, or second, shift.

(3) *Third Shift.* The night, or third, shift shall include all Employees who commence work at or after 10 P.M. and who quit work at or before 8 A.M. of the next following calendar day. A shift differential premium of thirty (30) cents per hour shall be paid for all time worked on the night, or third, shift.

(4) *Cross Shift.* Where an Employee starts work during one shift, as above defined, and quits work during another shift, as above defined, said Employee shall not be paid any shift differential premium for time worked, if any, between the hours of 7 A.M. and 3 P.M.; but shall be paid a shift differential of twenty (20) cents per hour for all time worked, if any, between the hours of 3 P.M. and 11 P.M. and a shift differential premium of thirty (30) cents per hour for all time worked, if any, between the hours of 11 P.M. and 7 A.M.

Figure 15.5 Shift work provision.

15.22 AVERAGE HOURLY COST CALCULATION

A typical summary[9] of data regarding trade contracts in given areas is presented in Table 15.5. A worksheet showing the calculation of an ironworker's hourly cost to a contractor is shown in Figure 15.6. It is assumed that the ironworker is working in a subsistence area on the second shift of a three-shift job during June and will be paid 8 hours for 7 hours of work (i.e., shift differential).

The ironworker works 10-hour shifts each day for 6 days, or 60 hours for the week. It is important to differentiate between those hours that are straight-time hours and those that are premium hours. Insurance premiums and fringe benefit contributions are based on straight-time hours. Social security and unemployment insurance contributions are calculated using the total income figure. The Hours Worked column breaks the weekday and Saturday hours into straight-time and premium-time components. Since the worker receives a shift differential, the first 7 hours are considered straight time and the other 3 hours are paid at overtime rate. The straight-time hours corresponding to the hours worked are shown in the second column. Eight hours are paid for the first 7 hours worked. The overtime is double time. The single-time portion, or first half of the double time, is credited to straight time. The second half of the double time is credited to the premium-time column. Based on the column totals the worker works 60 hours and will be paid 66 straight-time hours and 26 premium hours.

By consulting Table 15.5, it can be determined that the base wage rate for ironworkers is $19.20 per hour. This yields a straight-time wage of $1,267.20 (66 hours) and premium pay of $499.20 (26 hours). Total gross pay is $1,766.40.

---

[9] Although representative, data in this table is not current. Such information is dynamic and changes continuously.

Compute the average hourly cost to a contractor of an ironworker involved in structural steel erection in a subsistence area. The ironworker works on the second shift of a three-shift job and works six 10-hour days per week. The workers work 7 hours and are paid 8 hours under the shift pay agreement. Additional PL and PD insurance for $50,000/$100,000 coverage is desired. Use 6.2% FICA and 5.0% for unemployment insurance.

	Hours Worked	Straight Time-Hours (ST)	Premium Time (PT)
Monday–Friday	$5 \times 7 = 35$	$5 \times 8 = 40$	
	$5 \times 3 = 15$	$5 \times 3 = 15$	$1 \times 5 \times 3 = 15$
Saturday	$1 \times 7 = 7$	$1 \times 8 = 8$	$1 \times 1 \times 8 = 8$
	$1 \times 3 = 3$	$1 \times 3 = 3$	$1 \times 1 \times 3 = 3$
	60	66	26

Base Rate = $19.20
ST 66 hours @ $19.20 = $1267.20
PT 26 hours @ $19.20 = $ 499.20

Gross Pay $1766.40

Fringes:	Health and Welfare	$1.30 \times 66 = \$85.80$
	Pension	$1.14 \times 66 = \$75.24$
	Vacation	$1.00 \times 66 = \$66.00$ (deferred wage)
	Apprenticeship training	$0.14 \times 66 = \$9.24$
		$3.58 \times 66 = \$236.28$

WC = $29.18

$$\text{WC, PL, and PD} = \$36.39 \times \frac{1267.20}{100} = \$461.13$$

PL $1.59 \times 3.00 = \$4.77$
PD $1.30 \times 1.88 = \$2.44$
    Total $= \$36.39$ per $100.00 of Payroll

FICA = $0.0765 \times (\$1766.40 + \$66) = \$140.18$
Unemployment = $0.05 \times (\$1766.40 + \$66) = \$91.62$
Subsistence = 6 days $\times$ $20.00/day = $120.00

Total Cost = Base + Fringes + WC, PL, PD + UNEMPL + FICA + SUBS = $2815.61
$$\text{Average Hourly Cost (to contractor)} = \frac{\$2815.61}{60} = \$46.93$$

Figure 15.6 Sample wage calculation.

Fringes are based on straight-time hours, and the rates are given in the contract wage summary. The fringes paid by the contractor to union funds amount to $3.58 per hour. The vacation portion of the fringe is considered to be a deferred income item and, therefore, is subject to FICA. It is also used in the calculation of unemployment insurance contribution.

The amounts to be paid to the insurance carrier for WC, PL, and PD can be taken from Table 15.5. The contract calls for increased PL and PD rates. The bodily injury (PL) portion and the property damage coverage are to be increased to cover $50,000 per person/$100,000 per occurrence. This introduces a multiplier of 1.59 for the PL rate and 1.30 for the PD rate (see notes to Table 15.5). The total rate per $100 of payroll for WC, PL, and PD is $36.39. This is applied to the straight-time pay of $1,267.20 and gives a premium to be escrowed of $461.13.

Both FICA and unemployment insurance are based on the total gross pay plus the deferred vacation fringe. Subsistence is $20.00 per day not including travel pay and time to travel to the site (not included in this calculation). By summing all of these cost components, the contractor's total cost becomes:

Gross pay	$1766.40
Fringes	$ 236.28
WC, PL, PD	$ 461.13
FICA	$ 140.18
Unemployment	$ 91.62
Subsistence	$ 120.00
Total	$2815.61

Hourly rate = $2815.61/60 = $46.93 or approx. $47.00

This is considerably different from the base wage rate of $19.20 per hour. A contractor relying on the wage figure only to come up with an estimated price will grossly underbid the project and "lose his shirt."

It is particularly important to verify that the WC, PL, and PD rates being used for a worker are the correct ones. Particularly hazardous situations result in rates as high as $44 per $10 of payroll (e.g., tunneling). However, if a worker is simply installing miscellaneous metals he should not be carried as a structural steel erector. The difference in the rates between the two specialties can be significant. It should also be noted that the rates given in Table 15.5 are for a particular geographical area and are the so-called manual rates. The manual rate is the one used for a firm for which no safety or experience records are available. These rates can be substantially reduced for firms that evidence over years of operation that they have an extremely safe record. This provides a powerful incentive for contractors to be safe. If the WC, PL, and PD rates can be reduced by 30%, the contractor gains a significant edge in bidding against the competition.

The calculation of the hourly average wage indicates the complexity of payroll preparation. A contractor may deal with anywhere from 5 to 14 different crafts, and each craft union has its own wage rate and fringe benefit structure. Union contracts normally require that the payroll must be prepared on a weekly basis, further complicating the situation. In addition, all federal, state, and insurance agencies to which contributions or premiums must be paid require certified payrolls for verification purposes. Because of this, most contractors with a work force of any size use the computer for payroll preparation.

REVIEW QUESTIONS AND EXERCISES

15.1 What is meant by the following terms?

  a. Yellow-dog contract

  b. Agency shop

  c. Subcontractor clause

15.2 What is a secondary boycott? Name two types of secondary boycotts. Does the legislation forbidding secondary boycotts apply to construction unions? Explain.

15.3 What is a jurisdictional dispute? Why does this kind of dispute present no problem in District 50 locals?

15.4 What are the basic differences between the AFL as a labor union and the CIO type of union?

15.5 What will be the impact on double-breasted operations and the right-to-work provision of the Taft-Hartley legislation if labor is able to revoke existing practices regarding common sites picketing?

15.6 Answer the following questions true (T) or false (F):

  a. _____ Some state laws authorize use of closed shops.

  b. _____ A union can legally strike a job site in order to enforce the provision of a subcontractor clause.

  c. _____ The Teamsters union is the largest member of the AFL–CIO.

  d. _____ Open-shop operations have caused construction labor unions to rethink their position vis a vis union contractors.

e. _____ The right-to-work clause of the Taft-Hartley law allows the individual states to determine whether union shops are legal.

f. _____ The unit-price contract is an incentive-type negotiated contract.

g. _____ The local AFL craft unions have very little authority and are directed mainly by the national headquarters of AFL–CIO.

h. _____ The Sherman Antitrust Law was originally designed to prevent the formation of large corporations or cartels that could dominate the market.

i. _____ The business agent is the representative of the union charged with enforcing the work rules of the labor agreement.

j. _____ A submittal must be verified for accuracy in accordance with contract plans and specifications.

k. _____ The Sherman Antitrust Act, enacted in 1890, was used to suppress the formation of large trusts and cartels, which dominated the market and acted to fix prices and restrain free trade.

l. _____ Yellow-dog contracts were used by employers to encourage their employees to join and become active in union organizations.

m. _____ Under the Taft-Hartley legislation, the president of the United States is empowered to enjoin workers on strike (or preparing to strike) to work for a 90-day cooling-off period during which time negotiators attempt to reach agreement on contractual or other disputes.

n. _____ The National Labor Relations Act was enacted to protect union-organizing activity and encourage collective bargaining.

o. _____ In an open-shop working environment, workers are paid based on which union hall they belong to.

p. _____ The calculation of fringe benefits is based on gross pay, whereas FICA is based on straight-time hours.

q. _____ If a general contractor does not feel like paying worker's compensation fees, then the contractor does not have to. Each state has appropriated funds that will cover this option.

15.7 Compute the average hourly cost of a carpenter to a contractor. Assume the work is in a subsistence area and the daily subsistence rate is $19.50. The carpenter works the second shift on a two-shift project where a project labor contract establishes a "work 7 pay 8 hour" pay basis for straight time. He works 6 days, 10 hours a day. In addition to time and a half for overtime Monday through Friday, the contract calls for double time for all work on weekends. Use 6.2% FICA and 5.0% for unemployment insurance. Assume all data relating to the WC, PL, PD, fringes, and wage are as given in Table 15.5.

15.8 Identify the local labor unions that operate in your region. List the relevant business agents and the locations of the hiring halls.

15.9 List the labor unions that you consider would be involved in a project similar to the gas station project of Appendix I.

15.10 Visit a local contractor and a local hiring hall and determine the procedure to be followed in the hiring of labor.

15.11 Discuss the effects of the increasing popularity of nonunion contractors.

15.12 Ask a union tradesperson the monthly amount of his or her current dues for union membership. Does he or she consider that the benefits offset the membership fees?

Chapter 16

Estimating Process

CONTENTS

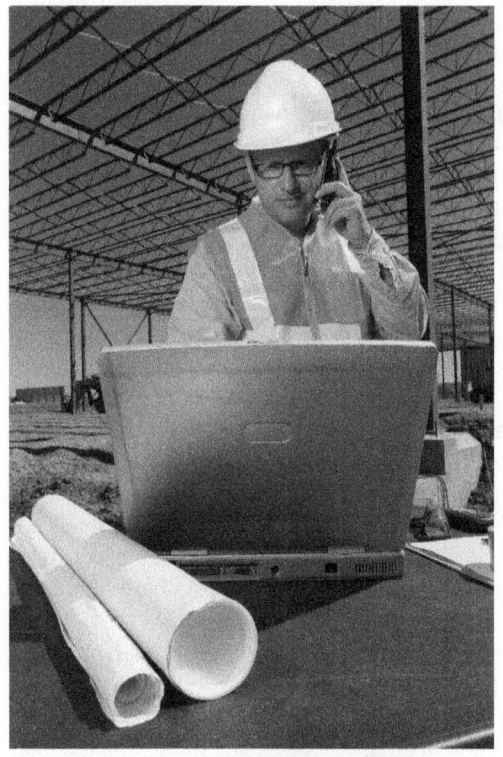

Engineer on the Phone.
(© David Jones/iStockphoto).

Estimating Using Handheld Devices

Estimators need a way to increase efficiency and lessen the chance for error when collecting estimate information in the field. With handheld devices, users can have access to the same precision estimating data they use in the office. Cover page information, item numbers and descriptions, assembly numbers and descriptions, VMS codes, variables, and variable help can all be stored on a handheld device. This ensures that estimators have the information they need to collect all project details necessary to deliver an accurate, complete bid.

Designed for handheld devices, estimating programs on personal digital assistants (PDAs) equip estimators with all the tools needed to perform detailed takeoffs remotely. Then, when convenient, data can be transferred to desktop software to instantly generate a detailed estimate or change order. With this type of mobile data acquisition, there's a much better chance of collecting all the necessary dimensional information to create a bid the first time. Specialized programs allow the user to take information from the desktop into the field to use as a checklist for takeoff.

PDA estimating lets users access existing estimate information or create brand new estimates directly in the field. All the project information can be logged in one place,

279

keeping estimators better organized and ready to make additions and changes as needed.

16.1 ESTIMATING CONSTRUCTION COSTS

The key to a good job and successful cost control is the development of a good estimate as the basis for bid submittal. The estimate represents the cost "flight plan" that will be followed by the constructor and that will aid him or her in achieving profit. If the flight plan is unrealistic or contains basic errors, the contractor will lose money on the job. If the estimate is well-thought out and correctly reflects the costs that will be encountered in the field, the chances of a profitable job are greatly increased.

Estimating is the process of looking into the future and trying to predict project costs and resource requirements. Studies indicate that one of the major reasons for the failure of construction contracting firms is incorrect and unrealistic estimating and bidding practices. If 20 estimators or contractors were furnished the same set of plans and specifications and told to prepare an estimate of cost and resources, it would be safe to assume there would not be more than two estimates prepared on the same basis or from the same units. Therefore, a consistent procedure or set of steps for preparing an estimate is needed to minimize errors and achieve reliable results.

16.2 TYPES OF ESTIMATES

Estimating methods vary in accordance with the level of design detail that is available to the estimator. Prior to the commencement of design, when only conceptual information is available, a comprehensive unit such as a square foot of floor space or a cubic foot of usable space is used to characterize the facility being constructed. The representative unit is multiplied by a price per unit to obtain a gross estimate ($\pm$ 10% accuracy) of the facility cost. A table of square foot and cubic foot building costs as given in *Building Construction Cost Data* published by the R. S. Means Company is shown in Figure 16.1. Such information is available in standard references and can be used for preliminary cost projections based on minimal design data. This conceptual estimate is useful in the schematic or budgetary phase, when design details are not available. The figures developed are of limited use for project control, and their use should be discontinued as soon as design data are available. These estimates are based on documents such as that given in Figure 2.2.

As the level of design detail increases, the designer typically maintains estimates of cost to keep the client informed of the general level of costs to be expected. The production of the plans and specifications usually proceeds in two steps. As noted in Chapter 2, the first step is called *preliminary design* and offers the owner a pause in which to review construction before detail design commences. A common time for this review to take place is at 40% completion of the total design. The preliminary design extends the concept documentation. At this point in the design process, a preliminary estimate is prepared by the architect or architect/engineer to reflect expected costs based on more definitive data.

Once the preliminary design has been approved by the owner, final or detail design is accomplished. The detail design phase culminates in the plans and specifications that are given to the constructor for bidding purposes. In addition to these detailed design documents, the architect/engineer produces a final engineer's estimate indicating the total job cost minus markup. This estimate should achieve approximately $\pm$3% accuracy since total design documentation is now available. The owner's estimate is used (a) to ensure that the design produced is within the owner's financial resources to construct (i.e., that the

50 17 | Square Foot Costs

		50 17 00 | S.F. Costs	UNIT	UNIT COSTS 1/4	MEDIAN	3/4	% OF TOTAL 1/4	MEDIAN	3/4	
01	0010	APARTMENTS Low Rise (1 to 3 story)	S.F.	62.50	78.50	105				01
	0020	Total project cost	C.F.	5.60	7.40	9.15				
	0100	Site work	S.F.	5.35	7.30	12.80	6.05%	10.55%	14.05%	
	0500	Masonry		1.23	2.86	4.94	1.54%	3.67%	6.35%	
	1500	Finishes		6.65	9.10	11.25	9.05%	10.75%	12.85%	
	1800	Equipment		2.04	3.09	4.60	2.73%	4.03%	5.95%	
	2720	Plumbing		4.86	6.25	7.95	6.65%	8.95%	10.05%	
	2770	Heating, ventilating, air conditioning		3.10	3.81	5.60	4.20%	5.60%	7.60%	
	2900	Electrical		3.61	4.79	6.45	5.20%	6.65%	8.40%	
	3100	Total: Mechanical & Electrical	↓	12.50	15.95	19.90	15.90%	18.05%	23%	
	9000	Per apartment unit, total cost	Apt.	58,000	89,000	131,000				
	9500	Total: Mechanical & Electrical	"	11,000	17,300	22,600				
02	0010	APARTMENTS Mid Rise (4 to 7 story)	S.F.	83	100	124				02
	0020	Total project costs	C.F.	6.45	8.90	12.20				
	0100	Site work	S.F.	3.31	6.55	11.80	5.25%	6.70%	9.15%	
	0500	Masonry		5.50	7.60	10.80	5.10%	7.25%	10.50%	
	1500	Finishes		10.40	14.50	17.10	10.55%	13.45%	17.70%	
	1800	Equipment		2.51	3.88	4.94	2.54%	3.48%	4.31%	
	2500	Conveying equipment		1.87	2.27	2.75	1.94%	2.27%	2.69%	
	2720	Plumbing		4.85	7.80	8.60	5.70%	7.20%	8.95%	
	2900	Electrical		5.45	7.80	9	6.65%	7.20%	8.95%	
	3100	Total: Mechanical & Electrical	↓	17.45	22	26.50	18.50%	21%	23%	
	9000	Per apartment unit, total cost	Apt.	93,500	110,500	183,000				
	9500	Total: Mechanical & Electrical	"	17,700	20,400	25,800				
03	0010	APARTMENTS High Rise (8 to 24 story)	S.F.	94	114	138				03
	0020	Total project costs	C.F.	9.10	11.15	13.55				
	0100	Site work	S.F.	3.40	5.50	7.70	2.58%	4.84%	6.15%	
	0500	Masonry		5.45	9.90	12.30	4.74%	9.65%	11.05%	
	1500	Finishes		10.40	13	15.35	9.75%	11.80%	13.70%	
	1800	Equipment		3.02	3.71	4.91	2.78%	3.49%	4.35%	
	2500	Conveying equipment		2.13	3.24	4.40	2.23%	2.78%	3.37%	
	2720	Plumbing		6.90	8.15	11.40	6.80%	7.20%	10.45%	
	2900	Electrical		6.45	8.15	11	6.45%	7.65%	8.80%	
	3100	Total: Mechanical & Electrical	↓	19.30	24.50	29.50	17.95%	22.50%	24.50%	
	9000	Per apartment unit, total cost	Apt.	97,500	107,500	149,000				
	9500	Total: Mechanical & Electrical	"	21,100	24,100	25,500				
04	0010	AUDITORIUMS	S.F.	96.50	132	189				04
	0020	Total project costs	C.F.	6.10	8.50	12.20				
	2720	Plumbing	S.F.	6.15	8.55	10.80	5.85%	7.20%	8.70%	
	2900	Electrical		7.55	10.90	14.65	6.80%	8.95%	11.30%	
	3100	Total: Mechanical & Electrical	↓	21.50	38	52	24.50%	30.50%	31.50%	
05	0010	AUTOMOTIVE SALES	S.F.	72	98	121				05
	0020	Total project costs	C.F.	4.75	5.70	7.40				
	2720	Plumbing	S.F.	3.29	5.70	6.25	2.89%	6.05%	6.50%	
	2770	Heating, ventilating, air conditioning		5.10	7.75	8.40	4.61%	10%	10.35%	
	2900	Electrical		5.80	8.95	12.15	7.30%	9.80%	12.15%	
	3100	Total: Mechanical & Electrical	↓	18.20	26	31	20.50%	21.50%	26%	
06	0010	BANKS	S.F.	141	176	223				06
	0020	Total project costs	C.F.	10.10	13.75	18.15				
	0100	Site work	S.F.	16.15	24.50	35	7.75%	12.95%	17%	
	0500	Masonry		7.35	14.10	25	3.36%	7.65%	10.35%	
	1500	Finishes		12.50	17.10	22	5.80%	8.35%	11.05%	
	1800	Equipment		5.55	11.70	25	1.34%	5.95%	10.65%	
	2720	Plumbing		4.43	6.35	9.25	2.82%	3.90%	4.93%	
	2770	Heating, ventilating, air conditioning		8.45	11.25	15	4.86%	7.15%	8.50%	
	2900	Electrical		13.35	17.85	23.50	8.20%	10.20%	12.20%	
	3100	Total: Mechanical & Electrical	↓	31.50	42.50	51	16.55%	19.45%	23%	
	3500	See also division 11020 & 11030 (MF2004 11 16 00 & 11 17 00)								

Figure 16.1 Costs based on a representative unit. From *Building Construction Cost Data 2008.* Copyright Reed Construction Cost Data, R. S. Means (All rights reserved.)

architect/engineer has not designed a gold-plated project) and (b) to establish a reference point in evaluating the bids submitted by the competing contractors.

On the basis of the final drawings and specifications the contractor prepares his or her estimate of the job's cost to include a markup for profit. This is the bid estimate. Both the engineer's and bid estimates require a greater level of effort and a considerable number of estimator hours to prepare. A rough rule of thumb states that the preparation of a bid estimate by the contractor will cost one-fourth of 1% of the total bid price. From the contractor's point of view this cost must be recovered as overhead on jobs that are won. Therefore, a prorate based on the number of successful bids versus total bids must be included in each quotation to cover bid costs on unsuccessful bids.

In building construction, these four levels of estimates are the ones most commonly encountered. To recapitulate, the four types of estimates are:

1. Conceptual estimate.
2. Preliminary estimate.
3. Engineer's estimate.
4. Bid estimate.

These four levels of precision reflect the fact that as the project proceeds from concept through preliminary design to final design and the bidding phase, the level of detail increases, allowing the development of a more accurate estimate. Estimating continues during the construction phase to establish whether the actual costs agree with the bid estimate. This type of "estimating" is what allows the contractor to project profit or loss on a job after it is in progress.

The definitive estimate can be prepared when all components comprising the project scope definition have been quantitatively determined and priced by using actual anticipated material and labor costs. This estimate is normally prepared when the project scope is defined in terms of firm plot plans, mechanical and process flow diagrams, equipment and material specifications, and engineering and design layouts. The pricing is based on formal vendor quotations for all major items and current predictable market costs for all commodity accounts.

AACE International has developed a Cost Estimate Classification System (Recommended Practice No. 18R-97, 2005 revision). A summary of this system is shown in Figure 16.2. It includes five "classes" which reflect the engineering/procurement/construction (EPC) projects typical of this association. As Figure 16.2 shows, the amount of variability inherent in each class of estimate is quite high at the concept screening level and decreases to the 3 to 10% range as bid-level documents become available.

16.3   DETAILED ESTIMATE PREPARATION

The preparation of a detailed bid-level estimate requires that the estimator break the project into cost centers or cost subelements. That is, the project is broken down into subcomponents that will generate costs. It is these costs that the estimator must develop on the basis of the characteristic resources required. The word *resource* is used here in the broad sense and applies to the man hours, materials, subcontracts, equipment hours, and dollars needed to accomplish the work or meet the requirements associated with the cost center. Typically in construction, the cost center relates to some physical subcomponent of the project, such as foundation piles, excavation, steel erection, interior dry wall installation, and the like. Certain nonphysical components of the work generate costs, however, and these cost centers must also be considered.

Many of the items listed as "indirects" are typical of costs that are not directly connected with physical components or end items in the facility to be constructed. Such items do, however, generate cost that must be recovered. These costs include insurance and bonding premiums, fees for licenses and permits required by contract, expense for special items relating to safety and minority participation programs, and home office overheads projected as allocated to the job.

	Primary Characteristic	Secondary Characteristic			
ESTIMATE CLASS	**LEVEL OF PROJECT DEFINITION** Expressed as % of complete definition	**END USAGE** Typical purpose of estimate	**METHODOLOGY** Typical estimating method	**EXPECTED ACCURACY RANGE** Typical variation in low and high ranges [a]	**PREPARATION EFFORT** Typical degree of effort relative to least cost index of 1 [b]
Class 5	0% to 2%	Concept Screening	Capacity Factored, Parametric Models, Judgment, or Analogy	L: -20% to -50% H: +30% to +100%	1
Class 4	1% to 15%	Study or Feasibility	Equipment Factored or Parametric Models	L: -15% to -30% H: +20% to +50%	2 to 4
Class 3	10% to 40%	Budget, Authorization, or Control	Semi-Detailed Unit Costs with Assembly Level Line Items	L: -10% to -20% H: +10% to +30%	3 to 10
Class 2	30% to 70%	Control or Bid/ Tender	Detailed Unit Cost with Forced Detailed Take-Off	L: -5% to -15% H: +5% to +20%	4 to 20
Class 1	50% to 100%	Check Estimate or Bid/Tender	Detailed Unit Cost with Detailed Take-Off	L: -3% to -10% H: +3% to +15%	5 to 100

Notes: [a] The state of process technology and availability of applicable reference cost data affect the range markedly. The +/- value represents typical percentage variation of actual costs from the cost estimate after application of contingency (typically at a 50% level of confidence) for given scope.

[b] If the range index value of "1" represents 0.005% of project costs, then an index value of 100 represents 0.5%. Estimate preparation effort is highly dependent upon the size of the project and the quality of estimating data and tools.

Figure 16.2 Cost estimate classification system (Reprinted with permission of AACE International, 1265 Suncrest Towne Centre Dr., Morgantown, WV 25605-1876 USA. Phone 800-858-Cost/304-296-8444. Fax: 304-291-5728. Internet: http://www.aacei.org, Email: info@aacei.org, Copyright © 2005 by AACE International; All rights reserved).

These items are sometimes referred to as general conditions, or general requirements, although they may or may not be specifically referred to in the contract documents. Accounts relating to these items fall into the categories for conditions of contract and general requirements of the contract. As estimators prepare bids, they have a general framework for cost recovery in mind. In addition, they have a knowledge of the technologies involved in building the project, which allow them to divide projects into individual pieces of work (i.e., physical subcomponents, systems, etc.). These work packages consume resources, generating costs that must be recovered from the client. Typically, a chart of cost accounts specific to the company acts as a guide or checklist as the estimator reviews the plans and specifications to highlight what cost centers are present in the contract being estimated.

Although the process of estimating is part art, part science, the estimator generally follows certain steps in developing the estimate:

1. Break the project into cost centers.
2. Estimate the quantities required for cost centers that represent physical end items (e.g., cubic yards of earth, lineal feet of pipe, etc.). For physical systems this procedure is commonly called *quantity takeoff*. For those cost centers that relate to nonphysical items, determine an appropriate parameter for cost calculation (e.g., the level of builder's risk insurance required by the contract or the amounts of the required bonds).
3. Price out the quantities determined in step 2 using historical data, vendor quotations, supplier catalogs, and other pricing information. This pricing may be based on a price per unit (unit cost) basis or a lump-sum (one job) basis. Price development for

physical work items may require an analysis of the production rates to be achieved based on resource analysis. If this analysis is used, the estimator must:

a. Assume work team composition to include number of workers (skilled and unskilled) and equipment required.

b. On the basis of team composition, estimate an hourly production rate based on the technology being used.

c. Make an estimate of the efficiency to be achieved on this job, considering site conditions and other factors.

d. Calculate the effective unit price.

4. Calculate the total price for each cost center by multiplying the required quantity by the unit price. This multiplication is commonly called an *extension*, and this process is called *running the extensions*.

The estimator usually summarizes the values for each cost center on a summary sheet, such as that shown in Figure 16.3.

16.4   DEFINITION OF COST CENTERS

The subdivisions into which the project is divided for detailed cost estimation purposes are variously referred to as:

1. Estimating accounts.

2. Line items.

3. Cost accounts.

4. Work packages.

The estimating account is typically defined so as to provide target values for the cost accounts that will be used to collect as-built costs while the job is in progress. Therefore, the end item that is the focus of cost development in the estimating account is linked to a parallel cost account for actual cost information collecting during construction. The cost account expenditures developed from field data are compared with the estimated cost as reflected by the estimating account to determine whether costs are exceeding, under-running, or coming in on the original estimate values. Therefore, the use of the term *cost account* is not strictly correct during the preparation of bid because this account is not active until the job is in progress and actual cost data are available.

The term *work package* is commonly used to indicate a subdivision of the project that is used both for cost control and scheduling (i.e., time control). When both cost and time control systems are combined into an integrated project management system, work packages are controlled to determine cost versus estimate and time versus schedule.

The subdividing of the project into work packages results in the definition of a work breakdown structure (WBS).

A work package is a well-defined scope of work that usually terminates in a deliverable product. Each package may vary in size, but must be a measurable and controllable unit of work to be performed. It also must be identifiable in a numerical accounting system in order to permit capture of both budgeted and actual performance information. A work package is a cost center.[1]

---

[1] James N. Neil. (1982) *Construction Cost Estimating for Project Control*, Englewood Cliffs, NJ: Prentice-Hall, p. 73.

Fudd Associates Contractors, Inc.
ESTIMATE SUMMARY
Estimate No. 6692 By: DWH

					Date: 1 August 2xxx	
Owner: NASA					**Project: Admin Building**	
Code Description	**LH**	**Labor**	**Material**	**Sub**	**Owner**	**Total**
01 Site improvements						
02 Demolition						
03 Earthwork						
04 Concrete						
05 Structural steel	8,265	93,840	75,665			169,505
06 Piling						
07 Brick & masonry						
08 Buildings						
09 Major equipment	11,240	130,295	8,970			139,265
10 Piping	14,765	172,590	287,085	7,500	172,705	639,880
11 Instrumentation				165,000		165,000
12 Electrical				632,710		632,710
13 Painting				70,170		70,170
14 Insulation				21,150		21,150
15 Fireproofing			2,650	5,550		8,200
16 Chemical cleaning						
17 Testing						
18 Const. equipment					178,330	178,330
19 Misc. directs	5,040	53,040	10,250		10,000	73,290
20 Field extra work						
Sub Total Direct Cost	39,310	449,765	384,620	902,080	361,035	2,097,500
21 Con. tools/sup.			36,805			36,805
22 Field payroll/burden					82,900	82,900
23 Start-up asst.						
24 Ins. & taxes					26,340	26,340
25 Field sprvsn.	2,400	36,000			10,190	46,190
26 Home off. exp.					12,270	12,270
27 Field emp. ben.					51,975	51,975
Sub Total Indirect Cost	2,400	36,000	36,805		183,675	256,480
Adjustment Sheets						
Total Field Cost	41,710	485,765	421,425	902,080	544,710	2,353,980
28 Escalation						
29 Overhead & profit		41,710	25,285	45,105	50,950	163,050
30 Contingency						90,380
31 Total Project Cost						2,607,410

Figure 16.3 Typical estimate summary.

   The breakdown of a project into estimating accounts or work packages (depending on the sophistication of the system) is aided by a comprehensive chart of cost accounts or listing of typical work packages that can be used as a checklist. This checklist or template can be matched to the project being estimated to determine what types of work are present. That is, accounts in the general chart are compared to the project being estimated to determine which ones apply.

16.5 QUANTITY TAKEOFF

The development of the quantities of work to be placed in appropriate units (e.g., square feet cubic yards, etc.) is referred to as the quantity takeoff or quantity surveying.[2] The procedures employed by the estimator to calculate these quantities should incorporate steps to minimize errors. Five of the most common errors experienced during quantity takeoff are:

1. Arithmetic: Errors in addition, subtraction, and multiplication.
2. Transposition: Mistakes in copying or transferring figures, dimensions, or quantities.
3. Errors of omission: Overlooking items called for or required to accomplish the work.
4. Poor reference: Scaling drawings rather than using the dimensions indicated.
5. Unrealistic waste or loss factors.

The first step in the quantity takeoff procedure is to identify the materials required by each estimating account or work package. Once the types of materials are identified, relevant dimensions are recorded on a spreadsheet so that quantity calculations in the required unit of measure can be made. Calculation of quantities by estimating account or work package has several advantages, not the least of which is the fact that it allows the estimating process to be performed by several estimators, each with a well-defined area of responsibility. No matter how competent an estimator may be in his own field, it is not reasonable to expect him to have an intimate knowledge of all phases of construction. This method enables one estimator to check another estimator's work. It also facilitates computations required to develop the financial "picture" of the job and processing of progress payment request. When changes occur, only those activities affected must be recalculated. Other procedures require a completely new takeoff.

Before the calculations for the quantity takeoff are performed, detailed working drawings are sometimes required to clarify the contract drawings and specifications or the chosen construction method (e.g., forming techniques). Such a drawing for a small wall is given in Figure 16.4. During construction these details are of tremendous value to the

Prepare: Activity list
Activity material list (estimate)–include work sheets
Material recap sheets

Figure 16.4 Small wall construction (CMU = concrete masonry unit).

[2]This term is commonly used in the United Kingdom and the Commonwealth countries.

person in the field who is trying to perform the work within the cost guidelines provided. From these drawings and details, a checklist should be developed to indicate all of the materials required for each work package. After this checklist has been made, it should be checked against a standard checklist to identify errors of omission.

The actual calculations should be performed on a standard spread sheets to allow for independent check and self-checks. As the calculations for those items shown on the plans progress, each item taken off should be highlighted by a color marker so that those items remaining to be considered are obvious at a glance. Arithmetic should be performed on a calculator or computer that produces hard copy output. This output should be used by the estimator to identify errors. All supporting documentation should be attached to the estimate to aid in checking by other sources or at a later date. The quantities calculated should be exact. Waste and loss factors will be applied later. A materials takeoff sheet for the small wall (Figure 16.4) is given in Figure 16.5.

A summary, or "recap sheet," should be made. This recap sheet should consist of a listing, by material type, of all the materials required for the entire work item or package. The listing should include total quantities as well as subquantities identified by activity

Project _____

Activity code	Activity description	Material description	Quantity	Unit	Cost code
1	Layout	Stakes 2 x 4 x 24 8 ea.	10.3	BF	0100
3	Place rebar	#5 st. 2 PCS 16 – 2	32.3	LF	0320
		Tie wire	1	Roll	0320
4	Cost and cure	footing			
		Concrete	1.23	CY	0330
		Curing compound	.25	Gal	0337
5	Erect CMU wall				
		CMU 8 x 8 x 16 stretcher	143	Ea	0412
		CMU 8 x 8 x 16 corner	14	Ea	0412
		CMU 8 x 8 x 16 corner	16	Ea	0412
		Scaffolding 4' x 4' x 6'	2	Sec.	0100
		Mortar	.27	CY	0412
7	Form bond beam				
		2 x 4 (4 – 15' – 0")	43.5	BF	0310
		2 x 2	12.7	BF	0310
		1 x 2	2.0	BF	0310
		3/4" ext ply	60.3	SF	0310
		Snapties 8"	24	Ea	0310
		Nails 8d	1.5	Lb	0310
		Nails 6d	.4	Lb	0310
		Form oil	.07	Gal	0310
8	Place bond beam rebar				
		#6 rebar (str.)	28.67	LF	0320
9	Cost and cure	Bond beam			
		Concrete	.35	CY	0330
		Curing compound	.05	Gal	0337
10	Strip forms and rub bond beam				
		Grout	1	CF	0339.2

Figure 16.5 Activity material list.

Project _____ Wall _____

Description	Activity code	Sub-quantity	Waste	Total quantity	Unit		Cost code
2 x 4 Lumber	Total	53.8	10%	60.0	BF		
	1	10.3					0100
	7	43.5					0310
2 x 4 Lumber	7	12.7	10%	14.0	BF		0310
1 x 2 Lumber	7	2.0	10%	2.25	BF		0310
3/4" Exterior plywood	7	60.3	10%	66	SF		0310
Curing compound	Total	.30		1	Gal		0337
	4	.25					
	9	.05					
Snap ties 8"	7	24.	5%	25	Ea		0310
Nails 8d	7	1.5		3	LB		0310
Nails 6d	7	.4		1	LB		0310
Form oil	7	.07		.25	Gal		0310

Figure 16.6 Construction support materials recap sheet.

code. The listing should also include appropriate waste and loss factor calculations. An example of a recap sheet is given in Figure 16.6. This example is simple and is included only to demonstrate the nature of quantity development. In practice, most companies will use computerized databases and spreadsheet programs to prepare final estimates. The basic principles of estimating must be well understood, however, to avoid omissions that could prove disastrous at the time of bid submittal.

16.6 METHODS OF DETAILED COST DETERMINATION

After quantities have been determined for accounts that are relevant to the project at hand, the method by which costs will be assigned can be selected. The two methods of cost determination most frequently used are (a) unit pricing and (b) resource enumeration. If the work as defined by a given estimating account is fairly standard, the cost can be calculated by simply taking dollar per unit cost from company records and applying this cost with a qualitative correction factor to the quantity of work to be performed. For instance, if the project calls for 100 lineal feet of pipe and historical data in the company indicate that the pipe can be placed for $65 a lineal foot to include labor and materials, the direct cost calculation for the work would yield a value of $6,500. This value can then be adapted for special site conditions.

03 11 13	Concrete Forming	CREW	DAILY OUTPUT	UNIT	BARE COSTS			TOTAL INCL. O&P
					MAT.	INST.	TOTAL	
13.25	FORMS IN PLACE, COLUMNS							
6500	24" X 24" plywood columns, 1 use	C-1	190	S.F.C.A	2.64	6.10	8.74	12.35

General work description

Square Foot Contact Area

Crew C-1

Crew No.	Bare Costs		Incl. Subs O&P	
Crew C-1	Hr.	Daily	Hr.	Daily
3 Carpenters	38.10	914.40	59.30	1423.20
1 Laborer	30.25	242.00	47.05	376.40
32 L.H. Daily Total		$1156.40		$ 1799.60

Line item determination:

Major UCI subdivision = 03.11

Major classification within subdivision =13.25

Line item number = 6500

Complete line number: 03 11 13.25 6500

(These figures are printed above)
BARE COSTS are developed as follows:
MAT. is 2.64/SFCA

INST. $\dfrac{\text{Crew C-1 Daily Bare Cost}}{\text{Daily Output}} = \dfrac{1156.40}{190} = \$6.10/\text{SFCA}$ (rounded)

TOTAL = MAT. + INST. = $2.64+ $6.10= $8.74per S.F.C.A.

(Only the total of these figures is printed above)
TOTAL COSTS INCL. O&P are developed as follows·
MAT. is BARE MAT. + 10% = 2.64 + 0.26 = $2.90/SFCA

INST. $\dfrac{\text{Crew C-1 Daily Bare Cost}}{\text{Daily Output}} = \dfrac{\$ 1799.60}{190} = 9.45/\text{SFCA}$ (rounded)

TOTAL = MAT. + INST. = $2.90+ $9.45= $12.35/SFCA

Figure 16.7 Line item cost development using R. S. Means data. From *Means Building Construction Cost Data 1981.* Copyright Reed Construction Data, Kingston, MA 781-585-7880. (All rights reserved.)

Unit pricing values are available in many standard estimating references. The standard references normally give a nationally averaged price per unit. A multiplier is used to adjust the national price to a particular area. These references are updated on an annual basis to keep them current, and usually are available in printed and online versions. Examples of these services are:

1. R. S. Means Company's *Building Construction Cost Data*

2. R. Walker's *Building Estimator's Reference Book*

3. *Richardson Construction Estimating Standards*

These references contain listings of cost line items similar to the cost account line items a contractor would maintain.

The development of direct costs to include overhead and profit for a particular line item using the R. S. Means system is shown in Figure 16.7.

The line items, specified in the R. S. Means *Construction Cost Data* are defined by using the Construction Specifications Institute (CSI) MasterFormat numerical designators. The system assumes a given crew composition and production rate for each line item. In the case illustrated a standard crew designated C-1 can construct 190 square foot contact area

(SFCA) of plywood column form per shift (daily output). This underlines the fact that unit pricing data must make some assumption regarding the resource group (i.e., crew, equipment fleet, etc.) and the production rate being used. That is, although unit pricing data are presented in dollars-per-unit format, the cost of the resource group and the rate of production achieved must be considered.

The dollars-per-unit value is calculated as follows:

Cost of resources per unit time/Production rate of resources = unit/hr = \$/unit

The unit cost is the ratio of resource costs to production rate. The crew composition and assumed cost for the crew are shown in the middle of Figure 16.7.

In the R. S. Means system, two costs are specified for each line item. The bare cost is the direct cost for labor and materials. The total cost includes the cost of burdens, taxes, and subcontractor overhead and profit (inclusive O & P). In Figure 16.7, the bare cost of the C-1 crew is calculated as \$1156.40 per shift. Therefore; the bare unit installation cost is

(\$1156.40/shift)/(190 SFCA/shift) = \$6.09, rounded to \$6.10/SFCA

Combining this installation cost with the material cost per unit of \$2.64 yields a bare unit cost for materials and installation of \$8.74/SFCA.

The O&P charges associated with labor (as considered in the Means system) are:

1. Fringe benefits (included in bare costs).
2. Workmen's compensation.
3. Average fixed overhead.
4. Subcontractor overhead.
5. Subcontractor profit

To adjust the bare costs of installation to include subcontractor's O & P, the appropriate craft values for the members of the craft are located and applied. For the carpenters the total correction is 55.6%, or \$21.20 (rounded) per hour. Therefore, the carpenter rate to include O & P is \$59.30 per hour. Similarly, the laborer rate is adjusted to \$47.05 per hour to include O & P. A markup of 10% is applied to the materials, yielding a cost of \$2.90/SFCA. The new installation rate to include O & P is

\$1799.60/190 = \$12.35/SFCA

16.7 PROBLEMS WITH UNIT-COST METHOD

The data that the contractor has available from company records are presented as dollars per unit, and in most cases no records of the crew composition, cost, and production rates is maintained. In fact, the dollars-per-unit value is typically an average of the values obtained on recent jobs. Because on each job the crew composition, costs, and production rates achieved are probably unique to the individual job, the figure represents an aggregate cost per unit. The actual number of man hours used and the productivity achieved are masked by the dollar per-unit figure. Unless the resource (i.e., man hour, etc.) information is kept separately, it has been lost. Therefore, the unit price available from averaging values on previous jobs has to be treated with some caution.

Since every job is unique, some of the estimator's intuition must be applied to ensure that the value is adapted to the conditions of the job being estimated. If the conditions of jobs vary little, however, the application of the unit pricing approach is both practical and efficient.

Clearly, the numerator (cost of resources per unit time) of the unit-cost ratio will vary significantly over time as the costs of labor and machines vary. The costs of all components of the construction process have risen sharply over the past 20 years. This is shown dramatically in the *Engineering News Record* construction and building cost indexes shown in Chapter 2. To factor out the inflationary escalation inherent in resource costs, some contractors maintain the ratio of man hours or resource hours (RHs) per hour to production. This establishes a company database tied to resource hours required rather than dollars per unit. Therefore, the contractor can retrieve a man hour or RH per unit value for each line item. The value is calculated as:

$$\text{Resource hours per hour/Units per hour} = \text{RH/unit}$$

The cost per unit can then be calculated by multiplying the RHs per unit value by the average hourly cost per resource. If it takes 25 RHs per unit and the average cost of a RH is $20.00, the unit cost will be $500.00 per unit. This method recognizes that the number of RHs required per unit is much more stable over the years than the cost per unit. Therefore, data on RHs per unit collected over several years will not be affected by inflationary trends and escalation in the cost of goods and services.

Use of the unit-pricing approach assumes that historical data have been maintained for commonly encountered cost accounts. Data are collected and linked to a reference unit such as a cubic yard or square foot. The costs of materials and installation are aggregated and then presented as a cost per unit. Companies typically accumulate such data either manually or on the computer as a by-product of the job cost system. On a typical job 80 to 90% of the work to be accomplished can be estimated by calculating the number of reference units and multiplying this number by the unit price. Typically the estimator will intuitively adjust this price to reflect special characteristics of the job, such as access restrictions, difficult management environment, and the like. One historical approach to the quantification of these site and job unique factors was proposed by Louis Dallavia (*Estimating General Construction Costs*, 1957). Although not currently used, it elegantly frames the factors that are considered by an estimator in adjusting general unit prices to a given project. The system defines a percent efficiency factor based on a production range index for each of eight job characteristics. The method of calculating the percent efficiency factor and the table production range indices are shown in Figure 16.8.

16.8   RESOURCE ENUMERATION

Although the unit-pricing approach is sufficiently accurate to estimate the common accounts encountered on a given project, almost every project has unique or special features for which unit-pricing data may not be available. Unusual architectural items that are unique to the structure and require special forming or erection procedures are typical of such work. In such cases, the price must be developed by breaking the special work item into its subfeatures and assigning a typical resource group to each subfeature. The productivity to be achieved by the resource group must be estimated by using either historical data or engineering intuition. The breakdown of the cost center into its subelements would occur much in the same fashion in which the wall of Section 16.5 was subdivided for quantity development purposes. The steps involved in applying the resource enumeration approach are shown in Figure 16.9.

An example of resource enumeration applied to a concrete-placing operation is shown in Figure 16.10. In this example a concrete placement crew consisting of a carpenter foreman, two cement masons, a pumping engineer (for operation of the concrete pump), and seven laborers for placing, screeding, and vibrating the concrete has been selected. A concrete pump (i.e., an equipment resource) has also been included in the crew. Its hourly cost has been determined using methods described previously. The total hourly rate for the crew is found to be $370.00. The average assumed rate of production for the crew is 12 cubic yards per hour. This results in an average labor cost per cubic yard of concrete of

Production Range Index			
	Production Efficiency (%)		
	25 35 45 55 65 75 85 95 100		
Production Elements	Low	Average	High
1. General Economy	Prosperous	Normal	Hard times
Local business trend	Stimulated	Normal	Depressed
Construction volume	High	Normal	Low
Unemployment	Low	Normal	High
2. Amount of Work	Limited	Average	Extensive
Design areas	Unfavorable	Average	Favorable
Manual operations	Limited	Average	Extensive
Mechanized operations	Limited	Average	Extensive
3. Labor	Poor	Average	Good
Training	Poor	Average	Good
Pay	Low	Average	Good
Supply	Scarce	Average	Surplus
4. Supervision	Poor	Average	Good
Training	Poor	Average	Good
Pay	Low	Average	Good
Supply	Scarce	Average	Surplus
5. Job Conditions	Poor	Average	Good
Management	Poor	Average	Good
Site and materials	Unfavorable	Average	Favorable
Workmanship required	First rate	Regular	Passable
Length of operations	Short	Average	Long
6. Weather	Bad	Fair	Good
Precipitation	Much	Some	Occasional
Cold	Bitter	Moderate	Occasional
Heat	Oppressive	Moderate	Occasional
7. Equipment	Poor	Normal	Good
Applicability	Poor	Normal	Good
Condition	Poor	Fair	Good
Maintenance, repairs	Slow	Average	Quick
8. Delay	Numerous	Some	Minimum
Job flexibility	Poor	Average	Good
Delivery	Slow	Normal	Prompt
Expediting	Poor	Average	Good

Example: After studying a project on which he is bidding, a contractor makes the following evaluations of the production elements involved:

Production Element	*% Efficiency*
1. Present economy	75
2. Amount of work	90
3. Labor	70
4. Supervision	80
5. Job conditions	95
6. Weather	85
7. Methods and equipment	55
8. Delays	75
Total	625

As the total of the eight elements is 625, the average value will be 625/8, or 78%.

Figure 16.8 Dallavia method.

Start

Figure 16.9 Resource enumeration method of estimating.

$30.83. The line items requiring concrete are listed with the quantities developed from the plans and specifications. Consider the first item that pertains to foundation concrete. The basic quantity is adjusted for material waste. The cost per unit is adjusted to $34.25 based on an efficiency factor for placement of foundation concrete estimated as 90%.

The resource enumeration approach has the advantage over unit pricing in that it allows the estimator to stylize the resource set or crew to be used to the work in question. The rates of pay applied to the resource group reflect the most recent pay and charge rates, and therefore incorporate inflationary or deflationary trends into the calculated price. The basic equation for unit pricing is

$$\text{Resource cost per unit time/Production rate} = \$/\text{hr/unit/hr} = \$/\text{unit}$$

In the unit-pricing approach the resource costs and the production rates are the aggregate values of resources and rates accumulated on a number of jobs over the period of historical data collection. With the resource enumeration approach, the estimator specifies a particular crew or resource group at a particular charge rate and a particular production level for the specific work element being estimated. This should yield a much more precise cost-per-unit definition. The disadvantage with such a detailed level of cost definition is the fact that it is time consuming. Therefore, resource enumeration would be used only on

Concrete Placing Crew

Quantity	Member	Rate	Total/Hour
1	Carpenter foreman	$40.00	$ 40.00
2	Cement masons	$36.00	$ 72.00
1	Pumping engineer	$38.00	$ 38.00
7	Laborers	$28.00	$196.00
1	Concrete pump	$24.00	$ 24.00
		Crew hourly rate	$370.00

Production rate of crew under normal circumstances (efficiency factor 1) = 12 cu yd/hr.
Average labor cost/cubic yard = $370/12 = $30.83.

Area	Quantity	Percent Waste	Efficiency Factor	Labor Cost/ Cubic Yard	Activity Cost
1. Foundation	53.2	15	0.9	$34.25	$ 1,822
2. Wall to elevation 244.67	52.9	12	0.8	38.54	2,039
3. Slab 10 in.	1.3	30	0.3	102.77	134
4. Beams elevated 244.67	10.5	15	0.7	44.04	462
5. Beams elevated 245.17	9.1	15	0.7	40.44	401
6. Slab elevation 244.67	8.7	10	0.7	40.44	383
7. Interior wall to 244.67	5.5	15	0.4	77.07	424
8. Slab elevation 254.17	6.3	10	0.75	41.11	259
9. Walls 244.67 −254.17	57.2	10	0.8	38.54	2,205
10. Walls 254.17 −267	42.0	10	0.8	38.54	1,619
11. Floors elevated 267	8.9	10	0.9	34.25	305
12. Manhole walls	27.3	10	0.85	36.27	990
13. Roof	14.0	15	0.7	44.04	617
14. Headwall	8.5	10	0.8	38.59	328
Total direct labor cost for concrete				$11,988 say $12,000	

Figure 16.10 Labor resource enumeration.

(a) items for which no unit cost data are available (b) "big-ticket" items, which constitute a large percentage of the overall cost of the job and for which such a precise cost analysis may lead to cost savings that may provide the winning margin at bid time, or (c) extremely complex work items on complicated and unique projects for which the use of the unit-pricing approach is deemed inadequate.

16.9   WORK PACKAGE OR ASSEMBLY-BASED ESTIMATING

In this approach to estimate development, a work package or assembly that is commonly encountered in construction is viewed as an estimating group, and appropriate dimensional and cost-related parameters are defined for the package. The wall of Figure 16.4 could be

Figure 16.11 Construction systems concept—concrete footing.

considered an assembly. In this case, the height, width, and depth of the footer, block portion, and cap beam would be specified each time the assembly is encountered. Pricing information for the defined wall would be retrieved from a pricing catalog. Since the reference subelement in this approach is the work package, an extensive listing of assemblies or packages into which the work can be subdivided is maintained.

A concrete footer assembly is shown in Figure 16.11. The relevant data required for takeoff are the dimensional values shown as items A through K. Data regarding the methodology of placement and the relevant specifications are indicated by items (1) to (9) in the figure. Such work-package-based systems can be considered a structured extension of the resource enumeration approach and can be calculated manually. In general, most of these system-based (i.e., assembly-based) systems are computerized and are based on presenting the estimator with individual assemblies. The estimator is interrogated by the computer and provides the dimensional and methodology information in a question-and-answer format. This procedure is shown schematically in Figure 16.12. The estimator goes through the construction systems sequentially, selecting those that are relevant and providing the required data. These data are integrated with information from a pricing catalog. The pricing catalog allows for price, resource, and productivity adjustment. The

Figure 16.12 Work package concept.

WORK PACKAGE COLLECTION SHEET

System/Structure Identifier: `0 2 • 1 3 3` Crew Level Work Package Identifier: `0 3 • 1 3 1` Description: *CONCRETE PLACEMENT, FLOAT FINISH GROUND SLAB, BUILDING 2*

Permanent Materials (PM)

Resource Code	Description	Unit	QTY	Unit Cost Low	Unit Cost Target	Unit Cost High	Extension
1 3 2 5	CONCRETE, 2500 PSI	CY	135		30 90		4171 50
	+5% WASTE INCLUDED						
							4171 50

Other Materials and Supplies (M&S)

							4171 50

Installed Equipment (IE)

Crew Labor (L)

	NR.	Cost/Hour Low	Cost/Hour Target	Cost/Hour High	
0 2 9 1 FOREMAN	1		10 90		87 20
0 2 9 3 LABORERS	4		10 40		332 80
0 2 9 2 FINISHER	4		12 85		411 20
					831 20

Equipment Not Charged As Job Indirects

	NR.	AVAILABILITY Periods	$/Period	USE Hours	$/Hour	
7 3 1 1 VIBRATOR, GED	2	(☐ x	☐) + (8)x(	3.25)=		52 00
7 3 1 9 HAND TOOLS	1	(☐ x	☐) + (8)x(	1.00)=		8 00
7 3 1 2 FINISHER, GED	1	(☐ x	☐) + (8)x(	4.50)=		36 00
		(☐ x	☐) + (☐)x(	☐)=		
						96 00

Productivity

Base Unit for Productivity: *CY CONCRETE*

Total Quantity Base Unit: *128 CY*

	Low	Target	High
Duration (Crew–Hours)		8	

Escalation Rates (%)

Materials ☐

Labor ☐

Equipment ☐

Notes

$$DURATION = \frac{128\ CY}{22\ CY/HR.} = 6\ HRS$$

ALLOW 8 HRS WITH STARTUP AND CLEANUP

Cost Summary:
PM = 4171.50
M&S = _____
IE = _____
L = 831.20
CE = 96.00
TOTAL = 5098.70

Figure 16.13 Work package collection sheet—concrete slab (Neil, James M., *Construction Cost Estimating for Project Control*, 1st edition, ©1982. Reprinted by permission of Pearson Education, Inc. Upper Saddle River, NJ).

WORK PACKAGE COLLECTION SHEET

System/Structure Identifier: 0 1 . 3 0 0

Crew Level Work Package Identifier: 0 2 . 1 1 1

Description: EXCAVATION, AREA 3, AND DISPOSAL W/O COMPACTION

Permanent Materials (PM)

Resource Code	Description	Unit	QTY	Unit Cost Low	Unit Cost Target	Unit Cost High	Extension

Other Materials and Supplies (M&S)

Installed Equipment (IE)

Crew Labor (L)				NR.	Cost/Hour Low	Cost/Hour Target	Cost/Hour High	
0 8 1 1	FOREMAN			1		13.90		444.80
0 8 1 2	EQUIP OPER, MEDIUM			10		13.40		4288.00
0 8 1 3	SPOTTER			2		10.40		665.60
								5398.40

Equipment Not Charged As Job Indirects				NR.	AVAILABILITY Periods	AVAILABILITY $/Period	USE Hours	USE $/Hour	
7 3 2 2	SCRAPER, ELEV, 22 CY			6	(1 x	1850–)	+ (32)x(	24.00)=	15,708.00
7 2 0 7	DOZER, D7			3	(1 x	1600–)	+ (32)x(	11.25)=	5,808.00
7 4 1 2	GRADER, 12'			1	(1 x	1000–)	+ (32)x(	8.10)=	1,259.20
					(x	)	+ ( )x(	)=	
									22,775.20

Productivity

Base Unit for Productivity: BANK CY

Total Quantity Base Unit: 24,000 BCY

	Low	Target	High
Duration (Crew-Hours)		32	

Escalation Rates (%)

Materials ☐☐
Labor ☐☐
Equipment ☐☐

Notes

L.F. = 0.8
CYCLE=6.3 MIN; ASSUME 45 MIN HOUR

$$BCY/HR= \left(\frac{45\ MIN}{6.3\ MIN}\right)(22CY)(0.8LF)(6\ SCRAPERS)$$

= 754 BCY/HR

$$DURATION = \frac{24,000\ BCY}{754\ BCY} = 32\ HOURS$$

ADD 1 DAY FOR BAD WEATHER
∴ PROJECT WILL LAST ONE WORK WEEK

Cost Summary:
PM = _____
M&S = _____
IE = _____
L = 5,398.40
CE = 22,775.20
TOTAL = 28,173.60

Figure 16.14 Work package collection sheet—excavation (Neil, James M., *Construction Cost Estimating for Project Control*, 1st edition, ©1982. Reprinted by permission of Pearson Education, Inc. Upper Saddle River, NJ).

manual or computer program integration of these data produces the estimate reports required for bid preparation.

If a manual approach is used to estimate each work package, a work package takeoff sheet is helpful in organizing the collection of data. Such a work package or assembly collection sheet can be organized as shown in Figure 16.13. This form illustrates the development of an estimate for slab on grade in a building project. Material, labor, and equipment resources required for the package are shown on the left side of the sheet, together with target prices for each resource. It is interesting to note that the equipment resources are normally charged on a period basis because partial-day allocation of equipment is not a common practice. The right side of the sheet considers the productivity rate to be used, special notes or characteristics relating to the package, and a total cost summary for the package. This sheet is quite versatile and can be used for earthwork, masonry, and virtually any assembly encountered in the construction of a project. A similar sheet for earthwork is shown in Figure 16.14.

16.10 SUMMARY

The estimate is the basis for the contractor's bid, and as such, has a significant effect on whether or not a given project is profitable. In building construction the four levels of estimate preparation are (a) conceptual, (b) preliminary, (c) engineer's, and (d) bid.

The first three of these estimates are typically prepared by the architect/engineer and reflect the increasing refinement of the design. Large and complex projects include a magnitude and definitive estimate in addition to those noted previously. The bid estimate is

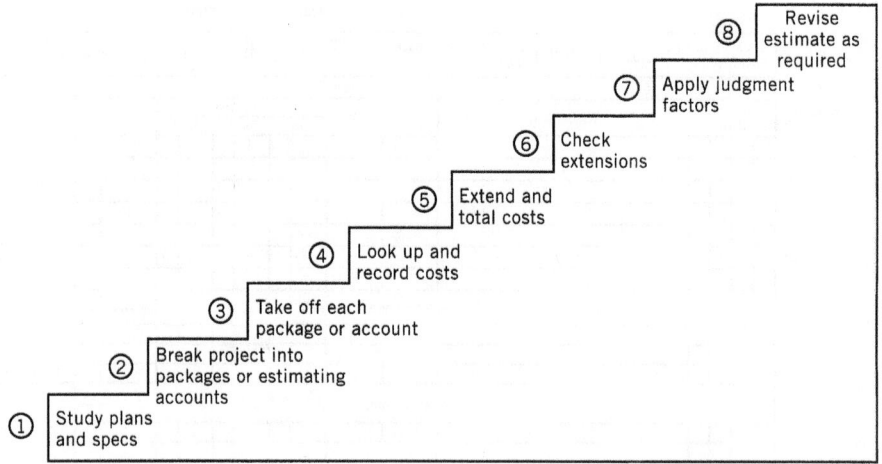

Figure 16.15 Steps in the estimating process.

a detailed estimate prepared by the contractor. The steps involved in preparing a detailed estimate are shown graphically in Figure 16.15. The project to be estimated is subdivided for cost analysis purposes into estimating accounts or work packages. Quantities for each package or account are developed. These quantities are priced, and the extensions are calculated and checked for errors.

At this stage, professional judgment and engineering intuition are used to adjust the bid to reflect special or unique factors peculiar to the particular job. Profit margin is also applied at this point, and the bid is revised as required and finalized. Steps 3 through 6 are quantitative in nature and involve the application of formulas and arithmetical concepts. Steps 1 and 2 require professional expertise. Steps 7 and 8 require a good deal of experience and engineering judgment.

In the development of the estimate three methods are commonly used. These are:

1. Unit-pricing or catalog lookup method.

2. Resource enumeration.

3. Work package/assembly method.

The work package method can be thought of as an extension of the resource enumeration method. Different methods may be used on different parts of the job. On portions of the job that are cost sensitive and constitute a large portion of the overall project, methods 2 and 3 may be appropriate. On parts of the work that are standard and straightforward, the unit-pricing approach is normally acceptable. Selection of methods is a tradeoff between the need for accuracy and the cost of obtaining that accuracy. The keys to a successful estimate are (a) the ability to assess the required level of accuracy and (b) the ability to achieve the required level of accuracy at minimal cost.

REVIEW QUESTIONS AND EXERCISES

16.1 Explain the difference between unit-cost estimating methods and resource enumeration methods. When would you use unit cost? When would you use resource enumeration?

16.2 What is meant by contractor O & P? Give three components that are considered in the O & P.

16.3 What is the difference between labor cost and labor productivity? Use a sketch to illustrate.

16.4 The partition wall shown below is to be constructed of 8 × 16 × 6 block. Estimate the cost of the wall to include labor, materials, and contractor O & P using the R. S. Means building cost data or other appropriate estimating reference. The job is located in Cincinnati, Ohio.

16.5 Given the bridge abutment shown here, determine the number of man hours required to form the structure using data from R. S. Means or other appropriate reference.

16.6 Find the labor cost, including overhead and profit for the formwork for 48 columns of 24″ × 24″ × 9′. Use the data shown in Figure 16.7.

16.7 Revise the cost for the formwork for the columns of the previous problem if the carpenters earn $42.00/hr.

Chapter 17

Cost Control

CONTENTS

Leadership in Energy and Environmental Design Certification

The Leadership in Energy and Environmental Design (LEED) certification program provides tangible recognition to project owners for their interest and investment in sustainability. It is hosted by the Green Building Council (USGBC) and examines five aspects of sustainability: (1) sustainable site development, (2) water use efficiency, (3) energy and atmospheric impact, (4) choice of materials and resources, and (5) indoor environmental quality. The achievements in each of these aspects are evaluated by a panel of experts chosen by the USGBC, and quantified in a point scale. Depending on the total points achieved, the project is either certified or denied certification.

There are several possible levels of certification: Certified, Silver, Gold, and Platinum. Each category is more stringent than the preceding one (e.g., Gold needs to fulfill more requisites than Silver). A LEED certified project receives a formal certificate of recognition and a plaque, is registered in an online directory of registered and certified projects, and included in the U.S. Department of Energy High Performance Buildings Database.

Professional consultants are frequently used by the owner to help attain LEED certification. These consultants are accredited by the Green Building Certification Institute, at the levels of associate, accredited professional (AP), and fellow. The prerequisites, examination, and continuing professional education are similar to those of any licensed profession.

In addition to the intrinsic importance of creating sustainable construction, LEED certification can be an excellent marketing tool for both owner occupants and developers.

Leed Data Processes   Leed Data Collection

This side of LEED is important because it helps the owner and design teams compare in financial terms the initial costs necessary to build a high performance building with a frequently less expensive but less sustainable traditional design.

17.1   COST CONTROL AS A MANAGEMENT TOOL

The early detection of actual or potential cost overruns in field construction activities is vital to management. It provides the opportunity to initiate remedial action and increases the chance of eliminating such overruns or minimizing their impact. Because cost overruns increase project costs and diminish profits, it is easy to see why both project management and upper-level management must become sensitive to the costs of all project activities.

An important byproduct of an effective cost reporting system is the information that it can generate for management on the general cost performance of field construction activities. This information can be brought to bear on problems of great interest to project management. The determination of current project status, effectiveness of work progress, and preparation of progress payment requests require data generated by both project planning and cost control reporting systems. Project cost control data are important not only to project management in decision-making processes but also to the company's estimating and planning departments because these data provide feedback information essential for effective estimates and bids on new projects. Thus, a project control system should both serve current project management efforts and provide the field performance database for estimating future projects.

17.2   PROJECT COST CONTROL SYSTEMS

The design, implementation, and maintenance of a project cost control system can be considered a multistep process. The five steps, shown schematically in Figure 17.1, form the basis for establishing and maintaining a cost control system. The following questions regarding each step in the implementation of the cost control system must be addressed.

1. Chart of Cost Accounts: What will be the basis adopted for developing estimated project expenditures, and how will this basis be related to the firm's general accounts and accounting functions? What will be the level of detail adopted in defining the project cost accounts, and how will they interface with other financial accounts?

Step 1 Chart of cost accounts
Step 2 Project cost plan
Step 3 Cost data collection
Step 4 Project cost reporting
Step 5 Cost engineering

Figure 17.1 Steps in cost control.

2. Project Cost Plan: How will the cost accounts be used to allow comparisons between the project estimate and cost plan with actual costs as recorded in the field? How will the project budget estimate be related to the construction plan and schedule in the formation of a project cost control framework?

3. Cost Data Collection: How will cost data be collected and integrated into the cost reporting system?

4. Project Cost Reporting: What project cost reports are relevant and required by project management in its cost management of the project?

5. Cost Engineering: What cost engineering procedures should project management implement in its efforts to minimize costs?

These are basic questions that management must address in setting up the cost control system. The structure of cost accounts will be discussed in this chapter.

17.3 COST ACCOUNTS

The first step in establishing a cost control system for a construction job is the definition of project-level cost centers. The primary function of the cost account section of a chart of accounts is to divide the total project into significant control units, each consisting of a given type of work that can be measured in the field (see Figure 17.2). Once job cost accounts are established, each account is then assigned an identifying code known as a cost code. Once segregated by associated cost centers, all the elements of expense (i.e., direct labor, indirect labor, materials, supplies, equipment costs, etc.) constituting work units can be properly recorded by cost code.

MASTER LIST OF PROJECT COST ACCOUNTS
Subaccounts of General Ledger Account 80.000
PROJECT EXPENSE

		Project Work Accounts 100–699			Project Overhead Accounts 700–999
100		Clearing and grubbing	700		Project administration
101		Demolition		.01	Project manager
102		Underpinning		.02	Office engineer
103		Earth excavation	701		Construction supervision
104		Rock excavation		.01	Superintendent
105		Backfill		.02	Carpenter foreman
115		Wood structural piles		.03	Concrete foreman
116		Steel structural piles	702		Project office
117		Concrete structural piles		.01	Move in and move out
121		Steel sheet piling		.02	Furniture
240		Concrete, poured		.03	Supplies
	.01	Footings	703		Timekeeping and
	.05	Grade beams		.01	security
	.07	Slab on grade		.02	Timekeeper
	.08	Beams		.03	Watchmen
	.10	Slab on forms	705		Guards
	.11	Columns		.01	Utilities and services
	.12	Walls		.02	Water
	.16	Stairs		.03	Gas
	.20	Expansion joint		.04	Electricity
	.40	Screeds			Telephone
	.50	Float finish	710		Storage facilities
	.51	Trowel finish	711		Temporary fences
	.60	Rubbing	712		Temporary bulkheads
	.90	Curing	715		Storage area rental
245		Precast concrete	717		Job sign
260		Concrete forms	720		Drinking water
	.01	Footings	721		Sanitary facilities
	.05	Grade beams	722		First-aid facilities
	.07	Slab on grade	725		Temporary lighting
	.08	Beams	726		Temporary stairs
	.10	Slab	730		Load tests
	.11	Columns	740		Small tools
	.12	Walls	750		Permits and fees
270		Reinforcing steel	755		Concrete tests
	.01	Footings	756		Compaction tests
	.12	Walls	760		Photographs
280		Structural steel	761		Surveys
350		Masonry	765		Cutting and patching
	.01	8-in. block	770		Winter operation
	.02	12-in. block	780		Drayage
	.06	Common brick	785		Parking
	.20	Face brick	790		Protection of adjoining property
	.60	Glazed tile	795		Drawings
400		Carpentry	796		Engineering
440		Millwork	800		Worker transportation
500		Miscellaneous metals	805		Worker housing
	.01	Metal door frames	810		Worker feeding
	.20	Window sash	880		General clean-up
	.50	Toilet partitions	950		Equipment
560		Finish hardware		.01	Move in
620		Paving		.02	Set up
680		Allowances		.03	Dismantling
685		Fencing		.04	Move out

Figure 17.2 List of typical project expense (cost) accounts.

The design, structure, and development of a cost coding system and its associated set of expense accounts have a significant impact on the cost management of a company or project. The job cost accounting system is essentially an accounting information system. Therefore, management is free to establish its own chart of accounts in any way that helps it in reaching specific financial and cost control objectives, whether these objectives are related to general company performance, to the control of a specific project, or to specific contract requirements.

17.4 COST CODING SYSTEMS

A variety of cost coding systems exist in practice, and standard charts of accounts are published by organizations such as the American Road Builders Association, Associated General Contractors, and the Construction Specifications Institute. In many industries, cost codes have a company-wide accounting focus emphasizing expense generation based on a departmental breakdown of the firm. In some construction firms, cost systems have a structured sequence corresponding to the order of appearance of the various trades or types of construction processes typical of the company's construction activity. In most construction companies, detailed project cost accounts such as those shown in Figure 17.2 are used. This method recognizes the fact that construction work is project oriented and that to achieve the cost management goal of maximizing profit, projects must be accounted for individually. One project may be a winner while another is losing money. Such situations may be masked in the accounting system unless job cost accounts are maintained on a project-by-project basis. Therefore, both billings (revenue) and cost (work in progress) accounts are typically maintained for each project. The actual account descriptions or designations vary in accordance with the type of construction and the technologies and placement processes peculiar to that construction. Building contractors, for instance, are interested in accounts that describe the cost aspects of forming and casting structural concrete as used in building frames. Heavy construction contractors, on the other hand, are interested in earthwork-related accounts such as grading, ditching, clearing and grubbing, and machine excavation. Standard cost accounts published by the American Road Builders Association emphasize these accounts, whereas the MasterFormat classification, published by the Construction Specifications Institute, emphasizes building-oriented accounts. A breakdown of the major classifications within the MasterFormat account system is shown in Table 17.1. A portion of the second level of detail for classifications 0 to 3 is shown in Figure 17.3.

17.5 PROJECT COST CODE STRUCTURE

The MasterFormat code as used by the R. S. Means *Building Construction Cost Data* identifies four levels of detail, an example of which is shown in Figure 17.4. At the highest level the major work classification as given in Table 17.1 is defined. In Figure 17.4, the first two digits 03 apply to all concrete items. Level 2 identifies a major subdivision within the work of level 1. In this case, 03 11 is assigned to concrete forming. There are many types of concrete forms, and they are defined at the third level. In the case of Figure 17.4, the item is structural cast-in-place concrete forming. This level adds two additional digits to the identifier, resulting in 03 11 13. The fourth level adds a two-digit decimal to level 3 and gives even more specificity to the item. In this case, 03 11 13.25 refers to forms in place, columns. MasterFormat does not provide further detail in its system, and it is left to each user to add more digits to the identifier. R. S. Means adds four digits in a fifth level. In Figure 17.4, the complete description of the item is 03 11 13.25 5000, for structural cast-in place concrete forms made of plywood and used once for job-built 8" × 8" columns. This

0 Conditions of the Contract		0270.	Site Improvements
0000-0099.	unassigned	0271.	Fences
		0272.	Playing fields
1 General Requirements		0273.	Fountains
0.100.	Alternates of Project	0274.	Irrigation systems
	Scope	0275.	Yard improvements
0.101-0109.	unassigned	0276-0279.	unassigned
0110.	Schedules and Reports	0280.	Lawns and Planting
0111-0119.	unassigned	0281.	Soil Preparation
0120.	Samples and Shop	0282.	Lawns
	Drawings	0283.	Ground covers and other
0121-0129.	unassigned		plants
0130.	Temporary Facilities	0284.	Trees and shrubs
0131-0139.	unassigned	0285-0289.	unassigned
0140.	Cleaning Up	0290.	Railroad Work
0141-0149.	unassigned	0291-0294.	unassigned
0150.	Project closeout	0295.	Marine Work
0151-0159.	unassigned	0296.	Boat Facilities
0160.	Allowances	0297.	Protective Marine Struc-
0161-0169.	unassigned		tures
		0298.	Dredging
2 Site Work		0299.	unassigned
0200.	Alternates		
0210-0209.	unassigned	*3 Concrete*	
0120.	Clearing of Site	0300.	Alternates
0211.	Declination	0301-0309.	unassigned
0212.	Structures moving	0310.	Concrete Formwork
0213.	Clearing and grubbing	0311-0319.	unassigned
0214-0219.	unassigned	0320.	Concrete Reinforcement
0220.	Earthwork	0321-0329.	unassigned
0221.	Site grading	0330.	Cast-in-Place Concrete
0222.	Excavating and backfilling	0331.	Heavyweight aggregate
0223.	Dewatering		concrete
0224.	Subdrainage	0332.	Lightweight aggregate
0225.	Soil poisoning		concrete
0226.	Soil compaction control	0333.	Post-tensioned concrete
0227.	Soil stabilization	0334.	Nailable concrete
0228-0229.	unassigned	0335.	Specially finished con-
0230.	Piling		crete
0231-0234.	unassigned	0336.	Specially placed concrete
0235.	Caissons	0337-0339.	unassigned
0236-0239.	unassigned	0340.	Precast Concrete
0240.	Shoring and bracing	0341.	Precast concrete panel
0241.	Sheeting	0342.	Precast structural con-
0242.	Underpinning		crete
0243-0249.	unassigned	0343.	Precast prestressed con-
0250.	Site drainage		crete
0251-0254.	unassigned	0344-0349.	unassigned
0255.	Site utilities	0350.	Clementitious Decks
0256-0259.	unassigned	0351.	Poured gypsum deck
0260.	Roads and Walks	0352.	Insulating concrete roof
0261.	Paving		decks
0262.	Curbs and gutters	0353.	Cementitious unit
0263.	Walks		decking
0264.	Road and parking	0354-0399.	unassigned
Appurtenances			
0265-0269.	unassigned		

Figure 17.3 Detailed codes for classification within Uniform Construction Index.

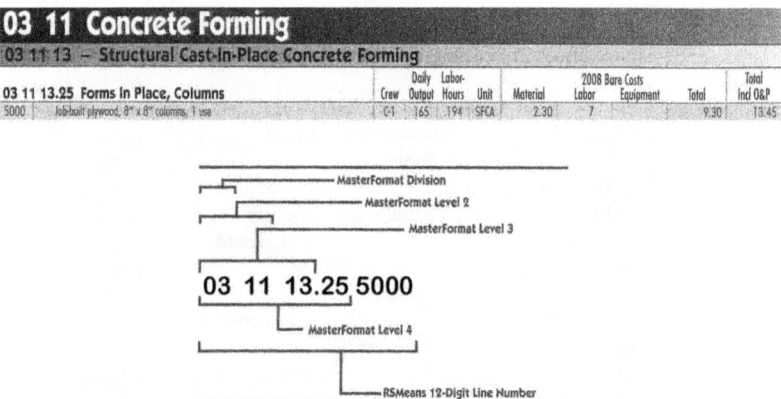

Figure 17.4 UCI cost (line item) structure in the master format code. (R. S. Means *Building Construction Cost Data*)

precise description requires of 12 digits. At this level, the cost engineer and construction manager have a great deal of flexibility in reflecting unique aspects of the placement technology that lead to cost fluctuations and thus must be considered in defining cost centers.

Large and complex projects in industrial and energy-related construction may require cost codes that reflect additional information, such as the project designation, the year in which the project was started, and the type of project. Long and complex codes in excess of 10 digits can result. An example of such a code is shown in Figure 17.5. This code, consisting of 13 digits, specifically defines the following items:

1. Year in which project was started (2014).
2. Project control number (15).
3. Project type (5 for power station).
4. Area code (16 for boiler house).
5. Functional division (2, indicating foundation area).
6. General work classification (0210, indicating site clearing).
7. Distribution code (6, indicating construction equipment).

The distribution code establishes what type of resource is being costed to the work process (i.e., clearing), the physical subelement (i.e., foundations) in what area of which project. Typical distribution codes might be as follows:

1. Labor.
2. Permanent materials.
3. Temporary materials.
4. Installed equipment.
5. Expendables.
6. Construction equipment.
7. Supply.
8. Subcontract.
9. Indirect.

A high concentration of information can be achieved by proper design of the cost code. Such codes are also ideally suited for data retrieval, sorting, and assembly of reports on the

Table 17.1 Major Divisions in CSI's MasterFormat Uniform Construction Index 2004

Procurement ands Contracting subgroup
  00 – Procurement and Contracting Requirements
General Requirements subgroup
  01 – General Requirements
Facilities Construction subgroup
  02 – Existing Conditions
  03 – Concrete
  04 – Masonry
  05 – Metals
  06 – Wood, Plastics, and Composites
  07 – Thermal and Moisture Protection
  08 – Openings
  09 – Finishes
  10 – Specialties
  11 – Equipment
  12 – Furnishings
  13 – Special Construction
  14 – Conveying Equipment
  15 – Reserved for future use in Mechanical
  16 – Reserved for future use in Electrical
Facilities Services subgroup
  21 – Fire Suppression
  22 – Plumbing
  23 – Heating Veritilating and Air Conditioning
  25 – Integrated Automation
  26 – Electrical
  27 – Communications
  28 – Electronic Safety and Security
Site and Infrastructure subgroup
  31 – Earthwork
  32 – Exterior Improvements
  33 – Utilities
  34 – Transportation
  35 – Waterway and Marine Construction
Process Equipment subgroup
  40 – Process Integration
  41 – Material Processing and Handling Equipment
  42 – Process Heating, Cooling, and Drying Equipment
  43 – Process Gas and Liquid Handling, Purification and Storage Equipment
  44 – Pollution Control Equipment
  45 – Industry-Specific Manufacturing Equipment
  48 – Electrical Power Generation

Figure 17.5 Classification of accounts: typical data structure for a computerized cost code.

basis of selected parameters (e.g., all construction equipment costs for concrete forming on project 10 started in a given year). The desire to cram too much information into cost codes, however, can make them so large and unwieldy that they are confusing to upper-level management.

17.6 COST ACCOUNTS FOR INTEGRATED PROJECT MANAGEMENT

In large and complex projects, it is advantageous to break the project into common building blocks for control both of cost and time. The concept of a common unit within the project that integrates both scheduling and cost control has led to the development of the work breakdown approach. The basic common denominator in this scheme is the work package, which is a subelement of the project on which both the cost and time data are collected for project status reporting. The collection of time and cost data based on work packages has led to the term *integrated project management*. That is, the status reporting function has been integrated at the level of the work package. The set of work packages in a project constitutes its work breakdown structure (WBS).

The WBS and work packages for control of a project can be defined by developing a matrix similar to the one shown in Figure 17.6. The columns of this matrix are defined by breaking the down project into physical subcomponents. Thus, we have a hierarchy of levels that begins with the project as a whole and, at the lowest level, subdivides the project into physical end items such as foundations and areas. As shown in Figure 17.6, the project is subdivided into systems. The individual systems are further divided into disciplines (e.g., civil, mechanical, electrical). The lowest level of the hierarchy indicates physical end items (foundation 1, etc.). Work packages at this lowest level of the hierarchy are called control accounts.

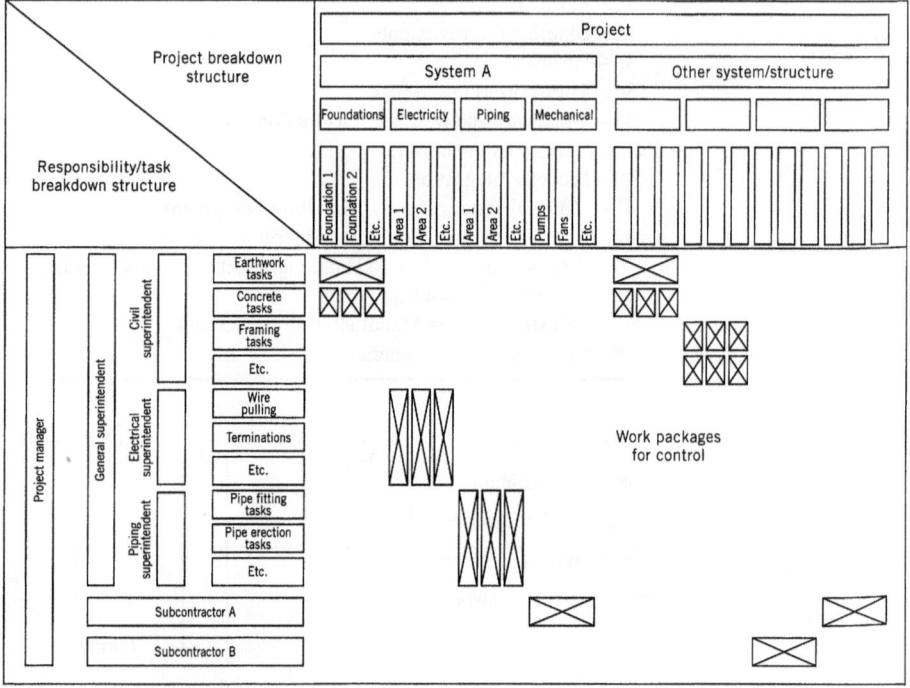

Figure 17.6 Project control matrix.

Figure 17.7 Three-dimensional visualization of work-package-oriented cost accounts.

The rows of the matrix are defined by technology and responsibility. At the lowest level of this hierarchy, the responsibilities are shown in terms of tasks, such as concrete, framing, and earthwork. These tasks imply various craft specialties and technologies. Typical work packages then are defined as concrete tasks on foundation 1 and earthwork on foundations 1 and 2.

This approach can be expanded to a three-dimensional matrix by considering the resources to be used on each work package (see Figure 17.7). Using this three-dimensional breakdown, we can develop definition in terms of physical subelement, task, and responsibility, as well as resource commitment. A cost code structure to reflect this matrix structure is given in Figure 17.8. This 15-digit code defines units for collecting information in terms of work package and resource type. Resource usage in terms of monetary units, quantities, man hours, and equipment hours for a foundation in the boiler building would be collected under work package code 121002. If this work relates to placement and vibration of concrete by using a direct chute, the code is expanded to include the alphanumeric code

Figure 17.8 Basic cost code structure.

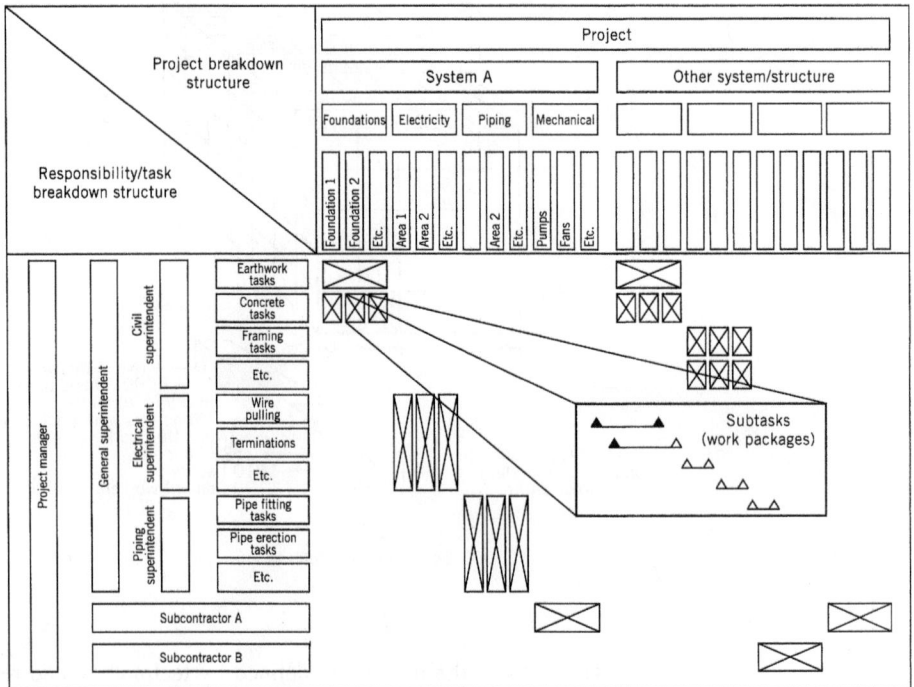

Figure 17.9 Project control matrix with scheduling of subtasks.

DF441. The resource code for the concrete is 2121. Therefore, the complete code for concrete in the boiler building foundations placed by using a chute would be 121002-DF441-2121.

This code allows collection of cost data at a fine level. Scheduling of this work is also referenced to the work package code as shown in Figure 17.9. The schedule activities are shown in this figure as subtasks related to the work package.

17.7   EARNED VALUE METHOD

One widely accepted way of calculating progress on complex projects using a work or account based breakdown system is the *earned value* approach. This system of determining project progress addresses both schedule status (i.e., on schedule, behind schedule or ahead of schedule.) and cost status (i.e., on budget, over budget or under budget). This method of tracking cost and schedule was originally implemented by the Department of Defense in the late 1970s to help better control complex projects. The system was called the *Cost and Schedule Control Systems Criteria* or C/SCSC. This method of monitoring contracts proved to be so effective that other government agencies (e.g., Department of Energy, Veterans Administration) adopted C/SCSC as a means of maintaining oversight on complex projects such as nuclear and conventional power plants. Private owners such as power companies implemented similar systems since reporting to various government authorities encouraged or required the use of C/SCSC and earned value concepts. Ultimately, owners of complex industrial projects began to use the system as well.

The idea of earned value is based upon a rigorous development of percentage complete of the budgeted costs associated with individual work packages or line items. Each work package has an initial budget or estimate which is defined as the budgeted cost at completion or BCAC. As work proceeds on an individual work package or account, assessment of the percentage complete is made at various study dates. The initial schedule

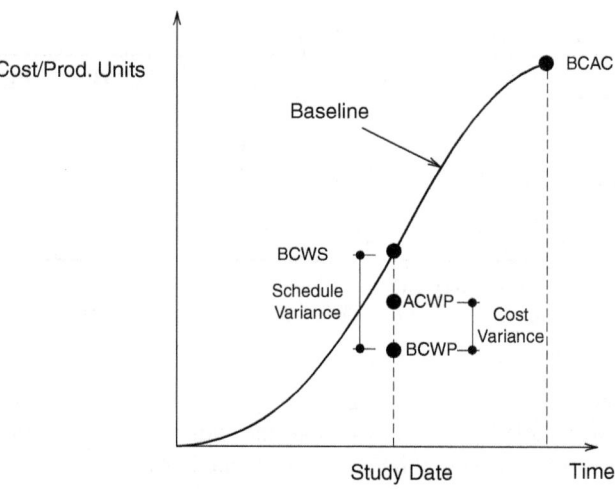

Figure 17.10 Control Values for Earned Value Analysis.

establishes an expected level of work completion as of the study date. The level of expected production is often shown as an S-Curve plotting the cost or units of production (e.g., units produced, work hours expended, and so on) against time. This cost/production curve is referred to as the baseline. At any given time (study date), the units of cost/production indicated by the baseline are called the budgeted cost of work scheduled (BCWS).

The tracking system requires that field reports provide information about the actual cost of work performed (ACWP) and the actual quantity of work performed (AQWP). The earned value is the budgeted cost of work performed (BCWP). The relative values for a given work package or account at a given point in time (see Figure 17.10) provide information about the status in terms of cost and schedule variance. The six parameters which form the foundation of the earned value concept are:

BCWS: Budgeted Cost of Work Scheduled = Value of the baseline at a given time

ACWP: Actual Cost of Work Performed. Measured in the field

BCWP: Budgeted Cost of Work Performed = [% Complete] × BCAC

BCAC: Budgeted Cost At Completion = Contracted total cost for the work package

AQWP: Actual Quantity of Work Performed. Measured in the field

BQAC: Budgeted Quantity At Completion. Value of the quantity baseline as projected at a given point.

To put these terms into context, consider the small project shown in Figure 17.11. The project consists of two control accounts—"A" and "B." A consists of two subaccounts, A.1 and A.2. The information for these work packages depends on its cutoff date (the

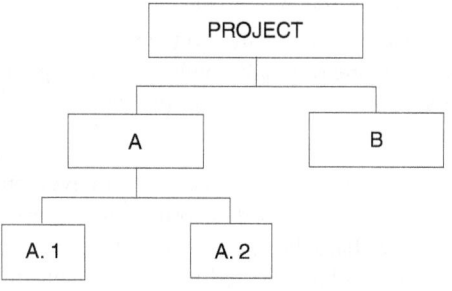

Figure 17.11 A Simple Project Hierarchy.

Table 17.2 Study Date Data for Simple Project

	BCAC	ACWP	BQAC	AQWP	PC (%)	BCWP	ECAC
A							
A.1	100	40	105	35	33.3	33.3	120
A.2	50	35	77	60	78.0	39.0	45
B	65	50	125	100	80.0	52.0	62.5
TOTAL	215	125	—	—	57.8	124.3	227.5

Project PC (PPC) = Total BCWP ÷ Total BCAC = 124.3 ÷ 215 = 57.8%
$ECAC_i$ = Estimated Cost at Completion for Work Package i = $ACWP_i \div PC_i$

"now" date – obviously, each date may have different values), and for the example it is given in Table 17.2. In this example, the budget is expressed in worker hours so the baseline for control is in worker hours. The estimated total number of worker hours for this scope of work is 215 (the sum of the estimated worker hours for A.1, A.2, and B). The BCWP or earned value for a given work package is given as:

$$BCWP_i = PC_i \times BCAC_i$$

where i is the work package or account label, and PC is the percentage complete as of the study date.

The percent complete (PC) for each package is based on the ratio of the actual quantities (i.e., its AQWP) divided by its Budgeted Quantity at Completion (i.e., BQAC) based on the latest quantity assessment. If we know the original quantity estimate is 100 units but updated information indicates that a total of 120 units will be required to complete the work, completion of 50 units would not indicate 50 percentage complete. The correct PC would be 50/120 (i.e., AQWP/BQAC).

Based on the information in Table 17.2, the PC for each work package in the small project would be:

$$PC\ (A.1)\ 35/105 = 0.333$$
$$PC(A.2)\ 60/77 = 0.780$$
$$PC(B) = 100/125 = 0.800$$

Then

$$BCWP\ (Project) = .333\ (100) + .78\ (50) + .8\ (65)\ 33.3 + 39 + 52 = 124.3$$

Therefore, the project percentage complete (PPC) for the small project is:

$$PPC\ \{124.3/215\} \times 100 = 57.8\ \text{percent}$$

This simple example illustrates several points:

1. The PC for a given package is based on the ratios of the AQWP/BQAC.
2. The PPC is calculated by relating the total BCWP (i.e., earned value) to the total BCAC for the project scope of work.
3. The total work earned is compared to the work required. The values of units to be earned are based on the originally budgeted units in an account/work package and the percentage earned is based on the latest projected quantity of units at completion.

Worker hours are used here to demonstrate the development of the PPC. However, other cost or control units may be used according to the needs of management.

It is important to know that schedule and cost objectives are being achieved. Schedule and cost performance can be characterized by cost and schedule variances as well as cost performance and schedule performance indices. These values in C/SCSC are defined as follows:

$$CV, \text{ Cost Variance} = BCWP - ACWP$$
$$SV, \text{ Schedule Variance} = BCWP - BCWS$$
$$CPI, \text{ Cost Performance Index} = BCWP/ACWP$$
$$SPI, \text{ Schedule Performance Index} = BCWP/BCWS$$

Figures 17.12 a, b, and c plot the values of BCWP, ACWP, and BCWS for the small project data given in Table 17.2. At any given study date, management will want to know what are the cost and schedule variance for each work packages. The variances can be calculated as follows:

$$CV (A.1) = BCWP (A.1) - ACWP (A.1) = 33.3 - 40 = -6.7$$
$$CV (A.2) = BCWP (A.2) - ACWP (A.2) = 39 - 35 = +4$$
$$CV (B) = BCWP (B) - ACWP (B) = 52 - 50 = +2$$

Since the CV values for A.2 and B are positive, those accounts are within budget (i.e., the budgeted cost earned is greater than the actual cost). In other words, less is being paid in the field than was originally budgeted. The negative variance for A.1 indicates it is overrunning budget. That is, actual cost is greater than the cost budgeted.

This is confirmed by the values of the CPI for each package.

$$CPI (A.1) = 33/40 < 1.0 \text{ A value less that 1.0 indicates cost overrun of budget}$$
$$CPI (A.2) = 39/35 > 1.0$$
$$CPI (B) = 52/50 > 1.0 \text{ Values greater than 1.0 indicate actual cost less than budgeted cost}$$

The schedule variances for each package are as follows:

$$SV (A.1) = BCWP (A.1) - BCWS (A.1) = 33.3 - 50 = -16.7$$
$$SV (A.2) = BCWP (A.2) - BOWS (A.2) = 39 - 32 = +7$$
$$SV (B) = BCWP (B) - BCWS (B) = 52 - 45 = +7$$

The positive values for A.2 and B indicate that these items are ahead of schedule. The negative value for A.1 indicates a scheduling problem. The calculation of the SPI values will confirm this assessment. Overall, it can be stated that A.2 and B are ahead of schedule and below cost while A.1 is behind schedule and over cost.

Six scenarios for permutations of AQWP, BCWP, and BCWS are possible as established by Singh (1991). The various combinations are shown in Figure 17.13 and Table 17.3. The reader is encouraged to verify the information in this table.

The earned value approach requires a comprehensive knowledge of work packaging, budgeting, and scheduling. It is a data intensive procedure and requires the acquisition of current data on the ACWP and AQWP for each work package or account. It is a powerful tool, however, when management is confronted with complex projects consisting of hundreds of control accounts. In large projects consisting of thousands of activities and control accounts, it is a necessity. Without it, projects can quickly spiral out of control. A more detailed presentation of this topic is beyond the scope of this chapter. The interested reader should refer to current government publications that describe the earned value management system (EVMS) and the inherent procedures associated with its implementation.

17.8   LABOR COST DATA COLLECTION

The purpose of the payroll system is to (a) determine the amount of and disburse wages to the labor force, (b) provide for payroll deductions, (c) maintain records for tax and other

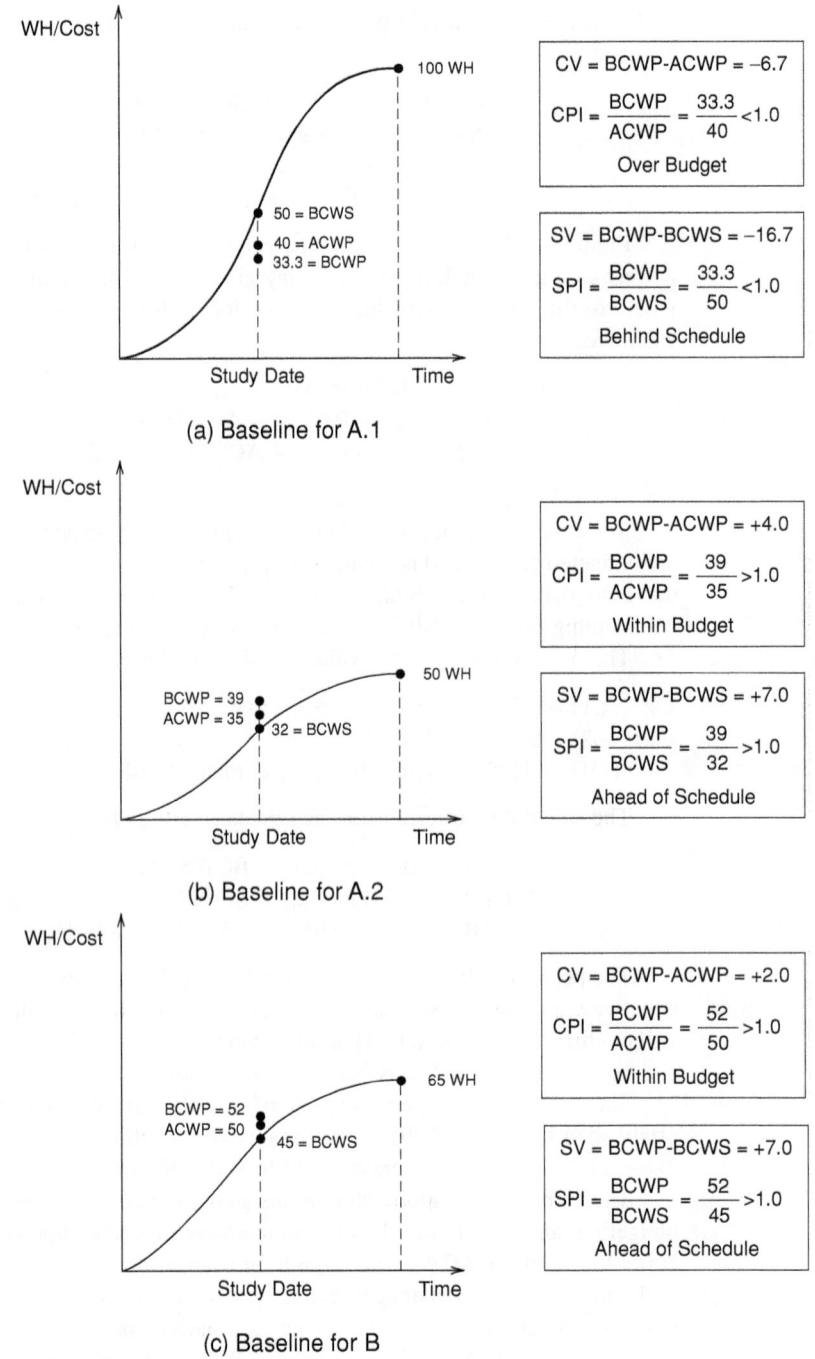

Figure 17.12 States of Control Account for Single Project.

purposes, and (d) to provide information regarding labor expenses. The source document used to collect data for payroll is a daily or weekly time card for each hourly employee similar to that shown in Figure 17.14. This card is usually prepared by foremen, checked by the superintendent or field office engineer, and transmitted via the project manager to the head office payroll section for processing. The makeup of the cards is such that the foreman or timekeeper has positions next to the name of each employee for the allocation of the

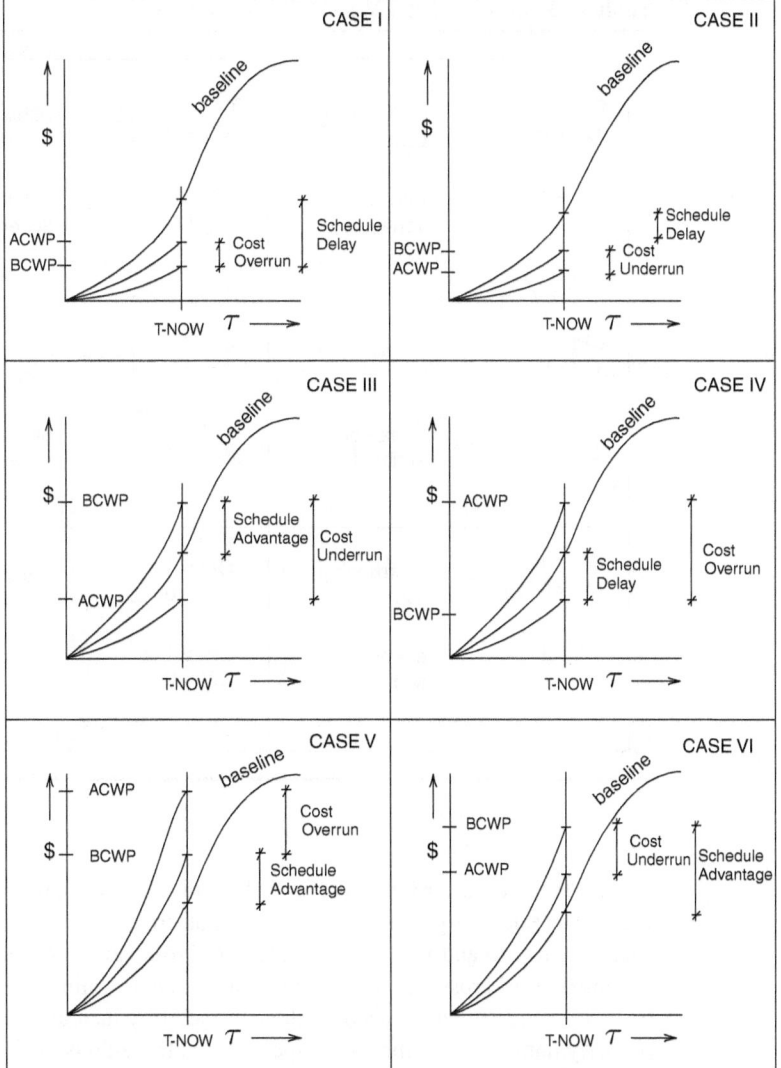

Figure 17.13 Scenarios for Permutations Between ACWP, BCWP, and BCWS (Singh, 1991).

time worked on appropriate cost subaccounts. The foreman in the distribution made in Figure 17.14 has charged 4 hours of A. Apple's time to an earth excavation account and 4 hours to rock excavation. Apple is a code 15 craft, indicating that he is an operating engineer (equipment operator). As noted, this distribution of time allows the generation of management information aligning work effort with cost center. If no allocation is made, these management data are lost.

The flow of data from the field through preparation and generation of checks to cost accounts and earnings accumulation records is shown in Figure 17.15.

This data structure establishes the flow of raw data or information from the field to management. Raw data enter the system as field entries and are processed to service both payroll and cost accounting functions. Temporary files are generated to calculate and produce checks and check register information. Simultaneously, information is derived from the field entries to update project cost accounts. These quantity data are not required by the financial accounting system and can be thought of as management data only.

Table 17.3 Values of CPI, CV, and SPI, SV for the Six Scenarios (Singh, 1991)

CPI < 1 CV < 0 → overrunning cost SPI < 1 SV < 0 → behind schedule **CASE I**	CPI > 1 CV > 0 → within budget SPI < 1 SV < 0 → behind schedule **CASE II**
CPI > 1 CV > 0 → within budget SPI > 1 SV > 0 → ahead of schedule **CASE III**	CPI < 1 CV < 0 → overrunning cost SPI < 1 SV < 0 → behind schedule **CASE IV**
CPI < 1 CV < 0 → overrunning cost SPI > 1 SV > 0 → ahead of schedule **CASE V**	CPI > 1 CV > 0 → within budget SPI > 1 SV > 0 → ahead of schedule **CASE VI**

From the time card, the worker's ID (badge number), pay rate, and hours in each cost account are fed to processing routines that cross check them against the worker data (permanent) file and use them to calculate gross earnings, deductions, and net earnings. Summations of gross earnings, deductions, and net earnings are carried to service the legal reporting requirements placed on the contractor by insurance carriers (public liability and property damage, workmen's compensation), the unions, and government agencies (e.g., Social Security and Unemployment Compensation).

17.9   CHARGES FOR INDIRECT AND OVERHEAD EXPENSE

Contractor-incurred expenses associated with the construction of a given facility relate to:

1. Direct cost consumed in the realization of a physical subelement of the project (e.g., labor and material costs involved in pouring a slab).

2. Production support costs incurred by the project-related support resources or required by the contractor (e.g., superintendent's salary, site office costs, various project related insurances) costs associated with the operation and management of the company as a viable business entity (e.g., home office overhead, such as the costs associated with preparation of payroll in the home office, preparation of the estimate, marketing, salaries of company officers).

The production support costs are typically referred to as project indirect costs. The home office charges are normally referred to as home office overhead. All of these costs must be recovered before income to the firm is generated. The home office overhead, or general and administrative (G & A) expense, can be treated as a period cost and charged

					80.103	80.104	80,260.01	80,260.07				

Dewey, Cheatum, and Howe
Company

Report No. __16__   **DAILY LABOR DISTRIBUTION REPORT**   Date __12 September XX__

__Dan Duck__   Location __Peachtree Corners Shopping Mall__   Job No. __101__
Foreman's signature

Description of work performed and cost account number

Employee or badge number	Name	Code	Craft or union	Rate	Hours	Hours	Hours	Hours	Hours	Hours	Total hours ST	Total hours PT
65	Adam Apple	ST	15	16.50	4	4					8	0
		PT										
14	Ella Del Fabbro	ST	10	12.50			8				8	0
		PT										
22	Charles Hoarse	ST	10	12.50			6	2			8	0
		PT										
		ST										
		PT										
		ST										
		PT										
		ST										
		PT										
		ST										
		PT										
		ST										
		PT										
		ST										
		PT										
Approved by				Totals	4	4	14	2			24	0

Figure 17.14 Foreman's daily labor distribution report.

separately from the project (direct costing). On the other hand, they may be prorated to the job and charged to the job cost overhead accounts and the work-in-progress expense ledger accounts (absorption costing).

17.10 PROJECT INDIRECT COSTS

Job-related indirect costs such as those listed in the labor cost report of Figure 17.16 (e.g., haul trash) are typically incurred as part of the on-site related cost associated with realizing the project. As such, they are charged to appropriate accounts within the job cost system. The level and amount of these costs should be projected during the estimating phase and included in the bid as individual estimate line items. Although it is recommended that job indirect be precisely defined during estimate development, many contractors prefer to

Figure 17.15 Payroll data structure.

handle these charges by adding a flat rate amount to cover them. Under this approach, the contractor calculates the direct costs (as defined previously) and multiplies these charges by a percentage factor to cover both project indirects and home office fixed overhead. To illustrate, assume that the direct costs for a given project are determined to be $200,000. If the contractor applies a fixed factor of 20% to cover field indirects and home office overhead, the required flat charge would be $40,000. If he adds 10% for profit, his total bid amount would be $264,000.

The estimate summary shown in Figure 16.3 establishes line items for indirect charges and calculates them on an -item-by-item basis (rather than applying a flat rate). Typical items of job-related indirect cost that should be estimated for recovery in the bid are those listed in Figure 17.2 as project overhead accounts (700-999). This is the recommended procedure because it is felt that sufficient information is available to the contractor at the time of bid to allow relatively precise definition of these job-related indirect costs. The R. S. Means method of developing overhead and profit (illustrated in Figure 16.7) represents a percentage rate approach that incorporates a charge into the estimate to cover overhead on a

CENTURY CENTER BLDG #5

ATLANTA, GA

LABOR COST REPORT
HALCON CONSTRUCTORS, INC.
ATLANTA DIVISION

WEEK 57   **PAGE 1**

WEEK ENDING 10/11   **PROJECT NUMBER 13-5265**

Cost Code	Description	Units	% Comp	Quantity Estimated	Quantity Actual	Unit Estimated	Unit Actual	Price Estimated	Price Actual	Cost Actual	Projected To Date Over/Under	Projected To Complete Over/Under
111	**This Week**		2		1		248.000			248	48	
	Haul Trash	Wk	68	50	34	200.000	92.765	10,000		3,154	3,646-	1,716-
112	**This Week**		1		1		543.000			543	326	
	Daily Clean	Wk	68	69	47	217.391	311.596	15,000		14,645	4,428	2,072
115										1,747	1,747	No Budg.
130	**This Week**		1		1		13.000			13	204-	
	Safety	Wk	81	69	56	217.391	206.304	15,000		11,553	621-	144-
131	Protect Trees	Ls						500		31		
132	Shoring	Ls	100	18	18	1,388.889	1,345.389	25,000		24,217	783-	Comp.
307	**This Week**		1		5		19.600			98	4	
	Hand Exc	Cy	97	725	705	18.793	19.569	13,625		13,796	547	15
310	Dewater	Ls	100					2,000		2,060	60	Comp.
312	Bkfl Hand	Cy	83	6,000	5,000	1.500	1.291	9,000		6,453	1,047-	209-
316	Fine Gr	Sf	99	15,000	14,830	.167	.135	2,500		1,999	501-	Comp.

Figure 17.16 Labor cost report (some typical line items).

line item-by-item basis. This is essentially a variation of the flat rate application described previously.

17.11 FIXED OVERHEAD

Whereas the project indirect charges are unique to the job and should be estimated on a job-by-job basis, home office overhead is a more or less fixed expense that maintains a constant level not directly tied to individual projects. In this case, the application of a percentage rate to prorate or allocate home office expense to each project is accepted practice, since it is not reasonable to try to estimate the precise allocation of home office to a given project. Rather, a percentage prorate or allocation factor is used to incorporate support of home office charges into the bid.

The calculation of this home office overhead allocation factor is based on:

1. The G & A (home office) expenses incurred in the past year.
2. The estimated sales (contract) volume for the coming year.
3. The estimated gross margin (i.e., markup) for the coming year.

This procedure is illustrated in the following example (Adrian, 1998).

Step 1: Estimate of Annual Overhead (G & A Expense)

Last year's G & A	$270,000
10% inflation	27,000
Firm growth	23,000
Estimated G & A	$320,000

Step 2: Estimate of $ of Cost Basis for Allocation

Estimated volume	$4,000,000
Gross margin	20% = $800,000
Labor and material	$3,200,000

Step 3: Calculate Overhead Percentage

Overhead costs estimated (G & A)/Labor and material estimate
$= 320,000/3,200,000 = 10\%$

Step 4: Cost to Apply to a Specific Project

Estimated labor and material costs	$500,000
Overhead to apply (@ 10 percent)	50,000
	$550,000

In the example, the anticipated volume for the coming year is $4,000,000. The G & A expense for home office operation in the previous year was $270,000. This value is adjusted for inflation effects and expected expansion of home office operations. The assumption is that the overhead allocation factor will be applied to the direct labor and materials costs. These direct costs are calculated by factoring out the 20% gross margin. Gross margin, in this case, refers to the amount of overhead and profit anticipated.

Direct costs amount to $3,200,000. The $320,000 in G & A costs to be recovered indicate a 10% prorate to be applied against the $3,200,000 of direct costs. This means that an overhead amount of $50,000 would be added to a contract bid based on $500,000 of direct cost to provide for G & A cost recovery. The profit would be added to the $550,000 base recovery amount.

17.12   CONSIDERATIONS IN ESTABLISHING FIXED OVERHEAD

In considering costs from a business point of view, it is common to categorize them either as variable costs or fixed costs. Variable costs are costs directly associated with the production process. In construction they are the direct costs for labor, machines, and materials as well as the field indirect costs (i.e., production support costs). These costs are considered variable since they vary as a function of the volume of work underway. Fixed costs are incurred at a more or less constant rate independent of the volume of work in progress. To be in business, a certain minimum of staff in the home office, space for home office operations, telephones, supplies, and the like must be maintained, and costs for these items are incurred. These central administrative costs are generally constant over a given range of sales/construction volume. If volume expands drastically, home office support may have to be expanded also. For purposes of analysis, however, these costs are considered fixed or constant over the year. Fixed costs are essentially the general and administrative costs referred to previously.

As described in Section, 17.11, the level of G & A (fixed) costs can be estimated by referring to the actual costs incurred during the previous year's operation. The method of projecting fixed overhead as a percentage of the estimated total direct costs projected for the coming year is widely used. Since the fixed overhead incurred in the previous year is typically available as a percentage of the previous year's total sales volume, a simple conversion must be made to reflect it as a percentage of the total direct cost. The formula for this conversion is

$$P_C = P_S/(100 - P_S)$$

where

P_C = percentage applied to the project's total direct cost for the coming year

P_S = percentage of total volume in the reference year incurred as fixed or G & A expense

If, for instance, $800,000 is incurred as home office G & A expense in a reference year in which the total volume billed was $4,000,000, the P value would be 20% ($800,000/ $4,000,000 × 100). The calculated percentage to be added to direct costs estimates for the coming year to cover G & A fixed overhead would be

$$P_C = 20/(100 - 20) = 25\%$$

If the direct cost estimate (e.g., labor, materials, equipment, and field indirects) for a job is $1 million, $250,000 would be added to cover fixed overhead. Profit would be added to the total of field direct and indirects plus fixed overhead. The field (variable) costs plus the fixed overhead (G & A) charge plus profit yield the bid price. In this example, if profit is included at 10%, the total bid would be $1,375.000. It is obvious that coverage of the field overhead is dependent on generating enough billings to offset both fixed and variable costs.

Certain companies prefer to include a charge for fixed overhead that is more responsive to the source of overhead support. The assumption here is that home office support for management of certain resources is greater or smaller, and this effect should be included in charging for overhead. For instance, the cost of preparing payroll and support for labor in the field maybe considerably higher than the support needed in administering materials procurement and subcontracts. Therefore, a 25% rate for fixed overhead is applied to labor and equipment direct cost, while a 15% rate on materials and subcontract costs is used. If differing fixed overhead rates are used on various subcomponents of the field (variable) costs in the bid, the fixed overhead charge will reflect the mix of resources used. This is shown in Table 17.4 in which a fixed rate of 20% on the total direct costs for

Table 17.4 Comparison of Fixed Overhead Rate Structures

			20% on Total Direct	25% on Labor and Equipment; 15% on Material and Subcontracts
Job 101	Labor and equipment	$ 800,000	$160,000	$200,000
	Materials and subcontracts	$1,200,000	240,000	180,000
			$400,000	$380,000
Job 102	Labor and equipment	200,000	$ 40,000	$ 50,000
	Materials and subcontracts	2,000,000	400,000	300,000
			$440,000	$350,000
Job 103	Labor and equipment	700,000	$140,000	$175,000
	Materials and subcontracts	700,000	140,000	105,000
			$280,000	$280,000

three jobs is compared to the use of a 25% rate on labor and equipment and a 15% rate on materials and subcontracts.

It can be seen that the fixed overhead amounts using the 25/15% approach are smaller on jobs 101 and 102 than the flat 20% rate. This reflects the fact that the amount of labor and equipment direct cost on these projects is smaller than the materials and subcontract costs. The assumption is that support requirements on labor and equipment will also be proportionately smaller. On job 102, for instance, it appears that most of the job is subcontracted with only $200,000 of labor and equipment in house. Therefore, the support costs for labor and equipment will be minimal, and the bulk of the support cost will relate to management of materials procurement and subcontract administration. This leads to a significant difference in fixed overhead charge when the 20% flat rate is used, as opposed to the 25/15% modified rates (i.e., $440,000 vs. $350,000).

On job 103, the fixed overhead charge is the same with either of the rate structures because the amount of labor and equipment cost is the same as the amount of the materials and subcontract cost.

It should be obvious that in tight bidding situations use of the stylized rate system, which attempts to better link overhead costs to the types of support required, might give the bidder an edge in reducing his bid. Of course, in the example given (i.e., the 25/15% rate vs. 20%) the 20% flat rate would yield a lower overall charge for fixed overhead on labor- and equipment-intensive jobs. The main point is that the charge for fixed overhead should be reflective of the support required. Because the multiple rate structure tends to reflect this better, some firms now arrive at fixed overhead charges by using this approach rather than the flat rate applied to total direct cost.

REVIEW QUESTIONS AND EXERCISES

17.1   As a construction project manager, what general categories of information would you want to have on a cost control report to properly evaluate what you think is a developing overrun on an operation, "place foundation concrete," that is now under way and has at least 5 weeks to go before it is completed?

17.2   What are the major functions of a project coding system?

17.3   List advantages and disadvantages of the Uniform Construction Index (UCI) coding system.

17.4   Assume you are the cost engineer on a new $12 million commercial building project. Starting with your company's standard cost code, explain how you would develop a project cost code for this job. Be sure the differences in purpose and content between these two types of cost codes are clear in your explanation. Specify any additional information that may be needed to draw up the project cost code.

17.5 Develop a cost code system that gives information regarding:

a. When project started

b. Project number

c. Physical area on project where cost accrued

d. Division in UCI

e. Subdivision

f. Resource classification (labor, equipment)

17.6 The following planned figures for a trenching job are available:

Quantity	Resources (hours)	Cost
Excavation—second hauling	Machines 1000	$100,000
	Labor 5000	$100,000
100,000 cubic yard	Trucks 2000	$ 62,500

At a particular time during the construction, the site manager realizes that the actual excavation will be in the range of 110,000 cubic yards. Based on the new quantity, he figures that he will have 30,000 cubic yards left.

From the main office, the following job information is available:

Resources		Cost
Machines	895 hours	85000
Labor	6011 work hours	79000
Trucks	1684 hours	50140

What would concern you as manager of this job?

17.7 Categorize the following costs as (a) direct, (b) project indirect, or (c) fixed overhead:

Labor

Materials

Main office rental

Tools and minor equipment

Field office

Performance bond

Sales tax

Main office utilities

Salaries of managers, clerical personnel, and estimators

17.8 The following data are available on Del Fabbro International, Inc. The fixed (home office) overhead for the past year was $365,200. Total volume was $5,400,000. It is assumed that G & A costs will account for $1,080,000 of this volume. Del Fabbro uses a profit markup of 10%. The estimating department has indicated that the direct and field indirects for a renovation job will be $800,000. What bid price should be submitted to ensure proper coverage of fixed overhead? Assume a 5% inflation factor and a 12% growth factor in the calculation.

17.9 Calculate the cost and scheduling variances for each of the work packages shown. What is the percentage complete for the entire package?

	WORK HOURS			QUANTITIES		
	EST	ACT	FORECAST	EST	ACT	FORECAST
A	15000	8940	15500	1000	600	1100
B	2000	1246	1960	200	93	195
C	500	356	510	665	540	680

17.10 Draw a hierarchical diagram of the work packages given, using the WBS code values. Calculate the BCWP and percent complete for all codes and work packages to include A.00 and B.00. Finally, compute the total percentage complete of the project.

		WORK HOURS			QUANTITIES		
CODE	DESCRIPTION	EST	ACT	FORECAST	EST	ACT	FORECAST
A.00	E/W Duct	440					
A.10	Partitions	230	150	225	25	14	25
A.20	Hangers	210	130	220	3	2.2	3.8
B.00	N/W Duct	645					
B.10	Partitions	370	75	390	50	12	48
B.20	Hangers	275	85	260	16	4.5	16

17.11   Given the following diagrams of progress on individual work packages of a project answer the following questions:

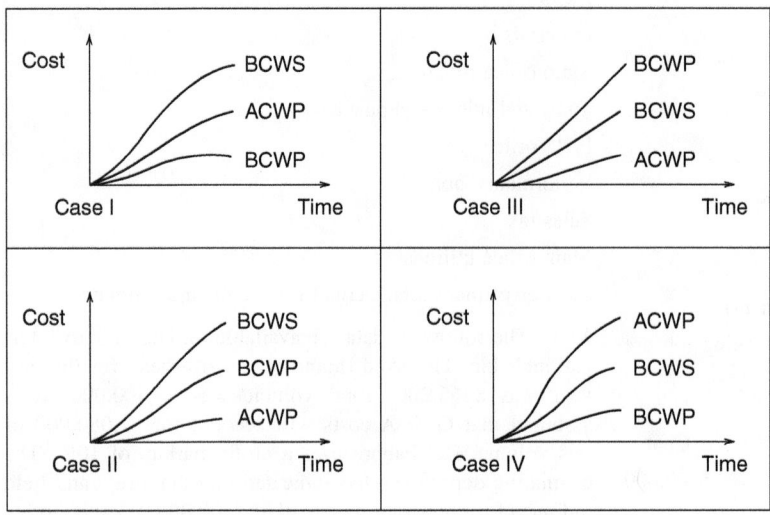

a. For Case 1, is the project ahead or behind schedule?

b. For Case 2, is the project over or under budget?

c. For Case 3, is the CPI greater than 1?

d. For Case 1, is the SPI greater than 1? Explain by calculation.

e. For Case 4, is the project on schedule and budget? Explain.

17.12   A project with an SPI < 1 is, according to the discussions in this chapter, behind schedule. How would the critical path in its critical path method schedule look? Is there any possibility that the SPI leads to an erroneous conclusion about the project's schedule?

Chapter 18

Materials Management

CONTENTS

(a) Dry Dock #4 Pearl Harbor, Hawaii

Fiber Reinforced Polymer Rebar

Reinforced concrete is a common building material for the construction of facilities and structures. As a complement to concrete's very limited tensile strength, steel rebar has been an effective and cost-efficient reinforcement. However, insufficient concrete cover, poor design or workmanship, and the presence of large amounts of aggressive agents in the concrete as well as environmental factors can all lead to cracking of the concrete and corrosion of the steel rebar. For instance, in the United States, almost 40% of bridges are structurally deficient or functionally obsolete largely due to cracking and corrosion.

Composite materials made of fibers embedded in a polymeric resin, also known as fiber-reinforced polymers (FRPs), have become an alternative to steel reinforcement for concrete structures. Aramid fiber-reinforced polymer (AFRP), carbon fiber-reinforced polymer (CFRP), and glass fiber-reinforced polymer (GFRP) rods are commercially available products for use in the construction industry. They have been proposed for use in lieu of steel reinforcement or steel prestressing tendons in non-prestressed or prestressed concrete structures. The problems of steel corrosion are avoided with the use of FRPs because FRP materials are nonmetallic and noncorrosive. In addition, FRP materials exhibit several properties including high tensile strength, which make them suitable for use as structural reinforcement. Fiberglass rebar may be a suitable alternative to steel reinforcing in architectural concrete, in concrete exposed to de-icing or marine salts, and in concrete used near electromagnetic equipment.

(b) Caissons and port facilities

18.1   MATERIAL MANAGEMENT PROCESS

In the traditional contractual relationship, the owner contracts with a general contractor or construction manager to build a facility and with an architect to perform the design. The general contractor, through this contract with the owner, is obligated to perform the work in accordance with the architect's instructions, specifications, and drawings. Thus, the architect is the owner's agent during the design and construction of a project. The lines of communication between the three parties are established as shown in Figure 18.1.

The materials that comprise facilities in building construction are subject to review by the architect or design professional. The contractor usually delegates responsibility for some of the categories of work involved in the project to subcontractors and suppliers. This delegation is accomplished through subcontracts and purchase orders. As a result of this delegation, a distinct life cycle evolves for the materials that makeup the project. The four main phases of this cycle are depicted in Figure 18.2.

18.2   THE ORDER

When the contract for construction is awarded, the contractor immediately begins awarding subcontracts and purchase orders for the various parts of the work. How much of the work is subcontracted depends on the individual contractor. Some contractors subcontract virtually all of the work in an effort to reduce the risk of cost overruns and to have every cost item assured through stipulated-sum subcontract quotations. Others perform almost all the work with their own field forces.

The subcontract agreement defines the specialized portion of the work to be performed and binds the contractor and subcontractor to certain obligations. The subcontractor,

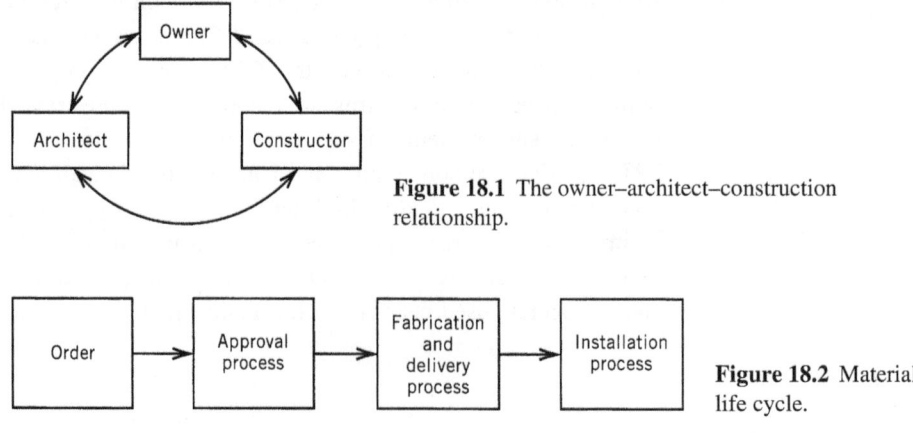

Figure 18.1 The owner–architect–construction relationship.

Figure 18.2 Material life cycle.

through the agreement, must provide all materials and perform all work described in the agreement. Most contractors either adopt a standard agreement, such as that provided by the Associated General Contractors of America (AGC), or implement their own agreement.

In most cases, a well-defined and legally-tested subcontract is used for subcontracting work. All provisions of the agreement between the owner and contractor are made part of the subcontract agreement by reference. The most important referenced document in the subcontract agreement is the General Conditions. Procedures for the submittal of shop drawings and samples of certain materials are established in the General Conditions. The General Conditions provide that

> *Where a Shop Drawing or Sample is required by the Contract Documents . . . any related Work performed prior to ENGINEER'S review and approval of the pertinent submittal will be at the sole expense and responsibility of the CONTRACTOR.*

The purchase order is a purchase contract between the contractor and the supplier. This document describes the materials to be supplied, their quantities, and the amount of the purchase order.

Purchase orders vary in complexity and can be as simple as a mail order house order form, or as complex as the construction contract itself. When complex and specially fabricated items are to be included in the construction, detailed specifications and drawings become part of the purchase order. Some typical purchase order forms are shown in Figures 18.3 and 18.4. Figure 18.3 shows a form for field-purchased items procured from locally available sources. These items are usually purchased on a cash-and-carry basis. The purchase order in this case is used primarily to document the purchase for record-keeping and cost accounting purposes (rather than as a contractual document). A more formal purchase order used in a contractual sense is shown in Figure 18.4. It is used in the purchase of more complex items from sources that are remote to the site.

Special Purchase Order

HCB HENRY C BECK COMPANY

VENDOR:

MAIL INVOICE TO:
HENRY C. BECK COMPANY
1210 S. Old Dixie Highway
Jupiter, Florida 33468

DATE:

CHG. TO JOB #21330 ____

SHIP TO: 1210 S. Old Dixie Highway / Jupiter, Florida 33468

QUANTITY	ARTICLE	U.P.	AMOUNT	COST CODE

STATE AND LOCAL SALES TAXES MUST BE SET OUT SEPARATELY ON INVOICE

Invoice in Triplicate
To Above Address
No Later Than 25th of Month —— Vendor's Acceptance (when required)   SUPT. OR PROJECT MGR.
Show S.P.O. Number On Invoice   WHITE (ORIGINAL) - VENDOR'S COPY   PINK   - SUPERINTENDENT'S COPY
                                  CANARY   –JOB OFFICE COPY   GOLDENROD -PROJECT MANAGER'S COPY
                                  (MAIL TO DALLAS WITH INVOICE)

Figure 18.3 Field purchase order (Courtesy of Henry C. Beck Co.).

Letter or transmittal form accompanying this order when mailed to Vendor should show the number of shop drawings and/or samples to be furnished and the address to which they must be sent; also the address to which Vendor is to mail correspondence relating to this order.

PURCHASE ORDER

HCB HENRY C BECK COMPANY   No.

VENDOR
ADDRESS

_____19_____

JOB:
Job Mailing Address:

Please ship the following to HENRY C. BECK COMPANY, at

SHIP VIA:

It is agreed that shipment will be made on or before   or right is reserved to cancel order.

IMPORTANT NOTE: It is IMPERATIVE in the interest of prompt payment that all invoices be rendered in the original with two (2) copies. Mail together with two (2) copies of bills of lading and/or other papers to JOB at address above.

ITEM NO.	QUANTITY	DESCRIPTION	UNIT	AMOUNT

SALES or USE TAX (is) (is not) included in amounts shown above.   HENRY C. BECK COMPANY
F. O. B.
TERMS: By_____

See above IMPORTANT NOTE for invoicing instructions. They MUST be complied with.

Accepted:_____

Show above order number on invoices,
and on the outside of each package
          containing Shipment.   By_____

Figure 18.4 Formal purchase order (Courtesy of Henry C. Beck Co.).

Regardless of the complexity of the transaction, certain basic elements are present in any purchase order. Five items can be identified as follows:

1. Quantity or number of items required.

2. Item description. This may be a standard description and stock number from a catalog or a complex set of drawings and specifications.

3. Unit price.

4. Special instructions.

5. Signatures of agents empowered to enter into a contractual agreement.

For simple purchase orders, the buyer normally prepares the order. If the vendor is dissatisfied with some element of the order, he may prepare his own purchase order document as a counterproposal.

The special instructions normally establish any special conditions surrounding the sale. In particular, they provide for shipping and invoicing procedures. An invoice is a billing document that states the billed price of shipped goods. When included with the shipped goods, it also constitutes an inventory of the contents of the shipment. One item of importance in the order is the basis of the price quotation and responsibility for shipment. Price quotations normally establish a *free on board* (FOB) location at which point the vendor will make the goods available to the purchaser. FOB means that the vendor will be responsible for presenting the goods at some mutually agreed-upon point such as the vendor's sales location, factory, or the purchaser's yard or job site. This is important because if the FOB location is other than the vendor's location, the vendor is indicating that the price includes shipment. The vendor may quote the price as cost, insurance, and freight (CIF). This indicates that the quoted price includes item cost plus the shipment cost to include freight and insurance expenses to the FOB location.

In the event the vendor ships the goods, it is of interest to establish at what point in time title of ownership passes from the vendor to the purchaser. This is established by the *bill of lading*. The bill of lading is a contractual agreement between a common carrier and a shipper to move a specified item or group of goods from point A to point B at a contracted price. If ownership passes to the purchaser at the vendor's location, the contract for shipment is made out between the purchaser and the common carrier. In cases in which the vendor has quoted a CIF price, he acts as the agent of the purchaser in retaining a carrier and establishing the agreement on behalf of the purchaser. The bill of lading is written to pass title of ownership at the time of pickup of the goods by the common carrier at the vendor's location. In such cases, if the common carrier has an accident and damages the goods during transfer, the purchaser must seek satisfaction for the damage because he or she is the owner at that point.

If goods are to be paid cash on delivery (COD), the title of ownership passes at the time of payment. In such cases, the bill of lading is between vendor and common carrier. If damage should occur during shipment, recovery of loss falls to the vendor as owner.

The sequence of events in CIF and COD transactions is shown in Figure 18.5. This figure also indicates the relationship between order, bill of lading, and invoice. A typical bill of lading memorandum and invoice are shown in Figures 18.6 and 18.7.

Figure 18.5 Procurement documents and title transfer sequence.

This Shipping Order

Carbon, and retained by the Agent.

Shipper's No. _____

Carrier's No. _____

_____
(Name of Carrier)

RECEIVED, subject to the classifications and tariffs in effect on the date of the issue of this Bill of Lading.

at **FT. COLLINS, CO** Date _____ 20xx From **SENIOR STEEL PRODUCTS, INC.**

The property described below, in apparent good order, except as noted (contents and condition of contents of packages unknown), marked consigned, and destined as indicated below, which said carrier (the word carrier being understood throughout this contract as meaning any person or corporation in possession of the property under the contract) agrees to carry to its usual place of delivery of said destination, if on its own route, otherwise to deliver to another carrier on route to said destination. It is mutually agreed, as to each carrier of all or any of said property over all or any portion of said route to destination, and as to each party at any time interested in all or any of said property, that every service to be performed hereunder shall be subject to all the terms and conditions of the Uniform Domestic Straight Bill of Lading set forth (1) in Official Southern, Western and Illinois Freight Classification in effect on the date thereof, if this is a rail or rail-water shipment, or (2) in the applicable motor carrier classification or tariff if this is a motor carrier shipment.

Shipper hereby certifies that he is familiar with all the terms and conditions of the said bill of lading, including those on the back thereof, set forth in the classification or tariff which governs the transportation of this shipment, and the said terms and conditions are hereby agreed to by the shipper and accepted for himself and his assigns.

Consigned to _____

(Mail or street address of consignee—For purposes of notification only.)

Delivery Address* _____

Destination _____ State _____ Zip _____ County _____

(*To be filled in only when shipper desires and governing tariffs provide for delivery thereof.)

Route _____

Delivering Carrier _____ Car or Vehicle Initials _____ No. _____

330

No. Packages	Kind of Package, Description of Articles, Special Marks, and Exceptions	*WEIGHT (Subject to Correction)	Class or Rate	Check Column
	REINFORCING STEEL ACCESSORIES		50	

*If the shipment moves between two ports by a carrier by water, the law requires that the bill of lading shall state whether it is carrier's or shipper's weight.

NOTE: Where the rate is dependent on value, shippers are required to state specifically in writing the agreed or declared value of the property. Agreed or declared value of the property is hereby specifically stated by the shipper to be not exceeding

per

+ The fibre boxes used for this shipment conform to the specifications set forth in the box maker's certificate thereon, and all other requirements of the Consolidated Freight Classification.

Subject to Section 7 of Conditions of applicable bill of lading, if this shipment is to be delivered to the consignee without recourse on the consignor the consignor shall sign the following statement:

The carrier shall not make delivery of this shipment without payment of freight and all other lawful charges.

_____
(Signature of Consignor)

If charges are to be prepaid, write or stamp here: "To Be Prepaid."

TO BE PREPAID

Received $_____
to apply in prepayment of the charges on the property described hereon.

Agent or Cashier

Per_____
(The signature here acknowledges only the amount prepaid.)

Charges Advanced:

$_____

+ Shipper's imprint in lieu of stamp; not a part of Bill of Lading approved by the Interstate Commerce Commission.

Agent must detach and retain this Shipping

SENIOR STEEL PRODUCTS, INC. _____ Shipper, Per _____ **Order and must sign the Original Bill of Lading.**

Permanent post-office address of shipper ___2000 JEFFERSON RD., FT. COLLINS CO. 80525___

Figure 18.6 Typical bill of lading.

BASB Steel and Supply Company   INVOICE

4111 Broadway
Ft Collins, CO 80566
www.basbsteel.com

DATE: December 13, 2008
INVOICE # 801

Bill To:	Ship To:
Belian Brothers Contracting Co., Inc. 218 6th St Windsor, CO 80585 USA (555) 970-1234	BBCC Bridge #1 CO406 Station 344 Scott Glick

P.O. #	Sales Rep. Name	Ship Date	Ship Via	Terms	Due Date
1453	Guggemos	12/9/2008	WEXC	1/10 Net 30	1/31/2009

Product ID	Description	Quantity	Unit Price	Line Total
QW6780	Brace for Temp const Model T-#	10	156.88	1,568.80

SUBTOTAL		1,568.80
PST	6.50%	101.97
GST	3.20%	50.20
SHIPPING & HANDLING		-
TOTAL		1,720.97
PAID		-
TOTAL DUE		1,720.97

THANK YOU FOR YOUR BUSINESS!

Figure 18.7 Typical invoice.

The invoice normally states the payment procedures and establishes trade discounts that are available to the purchaser if payment is made in a timely fashion. Trade discounts are incentives offered by the vendor for early payment. If the purchaser pays within a specified period, he or she must pay the stated price minus a discount. Failure to pay within the discount period means that the full price is due and payable. Terminology relating to trade discounts is as follows:

1. ROG/AOG: The discount period begins upon receipt of goods (ROG) or arrival of goods (AOG).
2. 2/10 NET 30 ROG: This expression appearing on the invoice means 2% can be deducted from the invoiced amount if the contractor pays within 10 days of AOG/ROG. Full payment is due within 30 days of AOG/ROG.
3. 2/10 PROX NET 30: A 2% cash discount is available if invoice is paid not later than the 10th of the month following ROG. Payment is due in full by the end of the following month.

4. 2/10 E.O.M.: The discount (2%) is available to the 11th of the month following ROG. Payment in full is due thereafter.

Trade discounts received are treated as earned income in financial statements.

The special conditions of the purchase order may include a "hold harmless" clause. Such clauses protect one of the parties to the purchase order from liability arising out of damages resulting from the conditions of the purchase order. A transit concrete mix company, for instance, may have the contractor submit orders on their forms holding the vendor harmless for damages arising out of delivery of the concrete to the site. Thus, if the transit mix truck should back across a gas main on the site, rupturing it during normal delivery, liability for repair costs will accrue to the contractor because the concrete vendor is "held harmless." The converse could, of course, occur if the contractor uses his own purchase order form, which holds him harmless in such an event. These situations are not covered by normal liability insurance since such "contractually accruing" liability is considered to be outside the realm of normal liability. If the language of the order is prepared by the contractor, the hold harmless clause will operate to protect him. If the vendor's language is used, the special conditions will hold him harmless in these damages situations.

For the contractor's protection, reference is made in complex purchase orders (requiring special fabrication) to the contractor specifications and other documents that define the materials to be supplied. Specifications detail the required *shop drawings*, *product data*, and *samples* that must be submitted for approval prior to fabrication and delivery. The provisions of the purchase order and the subcontract agreement require the subcontractor and supplier to obtain approval for their materials.

18.3 APPROVAL PROCESS

The contract drawings prepared by the architect are generally not specific enough to facilitate accurate fabrication of the materials involved. Therefore, to produce the necessary materials for a project, subcontractors and suppliers must provide details that further amplify the contract drawings. These details can be classified into three groups: (a) shop drawings, (b) product data, and (c) samples.

Shop drawings are defined in the AGC *General Conditions* as

> *All drawings, diagrams, illustrations, schedules, and other data or information which are specifically prepared or assembled by or for CONTRACTOR and submitted by CONTRACTOR to illustrate some portion of the Work.*

The detailing, production, and supplying of shop drawings are the sole responsibility of the contractor or the contracted agent. However, the design professional is responsible for verification that the supplied shop drawings correctly interpret the contract documents. Dimensions, quantities, and coordination with other trades are the responsibility of the contractor. Approved shop drawings become the critical working drawings of a project and are considered a part of the contract documents. Typically, shop drawings are submitted for materials such as reinforcing steel, formwork, precast concrete, structural steel, millwork, casework, metal doors, and curtain walls.

Product data may be submitted to illustrate the performance characteristics of the material items described by the shop drawings or may be submitted as verification that a standard product meets the contract specifications. Product data are illustrations, standard schedules, performance charts, instructions, brochures, diagrams, and other information furnished by the contractor to illustrate a material, product, or system for some portion of the work. Mill test reports, concrete mix designs, masonry fire rating tests, curtain wall wind test reports, and mechanical equipment performance tests are examples of product data.

Product data are particularly important when a subcontractor or supplier is submitting data on a product that is a variance from the contract specifications. The architect carefully analyzes the submitted data prior to rendering an approval of the substitution. Also, the product data are used extensively to coordinate the materials used by the mechanical and electrical subcontractors. The contractor must communicate the product data between these major subcontractors to ensure proper performance of their portion of the work.

Samples usually involve the finishes of a project and are physical examples of materials to be supplied. The architect may require samples of plastic laminate finishes for doors and counters, flooring, wall coverings, paint, stucco, precast concrete, ceilings, and other items. These are used by the architect in developing the overall building finish scheme.

The approval process involving shop drawings, product data, and samples has several substages that are critical to the material life cycle. These are: (a) submission by the subcontractor or supplier, (b) review of the submittal by the contractor, (c) review by the architect or design professional, and (d) return of submittal to the subcontractor or supplier.

At the time of awarding subcontracts and purchase orders, the contractor usually establishes the quantity, size, and other requirements for all submittals. In most cases, several blue line prints (usually six) are required when shop drawings are submitted for approval. The product data quantities required may range from three to six copies. The copies of a submittal may vary depending on the number of other subcontractors or vendors that must receive approved copies to coordinate their work. In all cases, careful planning of the quantity of submittals will expedite the other substages by eliminating the handling of unnecessary copies of submittals.

Timing of submittals is of utmost importance in the effective processing of material submittals. Subcontracts and purchase orders often contain language such as "all submittals must be made immediately" or "fifteen (15) days after execution of this agreement, all submittals must be made." In most cases, contractors do not preplan in detail the required submittal data from a subcontractor or supplier. The result is a landslide of submittals, most of which are not necessary, in the early stages of the project. Thus, field office personnel waste time sorting and determining the most critical submittals. A well-planned approach to scheduling submittals will ensure timely processing and better control of required submittals.

Once a submittal is received by the contractor, the process of checking for conformance with the intent of the contract documents is performed. A submittal, whether it is a shop drawing, product data, or sample, is governed by the contract drawings and specifications. The contractor's field or main office personnel in charge of submittals may make notations and comments to the designer or his engineers to clarify portions of the submittal or to correct the submittal. The submittal represents specific details of the project and is of primary importance in coordination, as well as depicting exactly what a supplier or subcontractor is providing. The contractor is required by the general conditions to clearly note to the architect or design professional any variation from the contract documents.

The amount of time involved in the contractor's review of submittals may vary from 1 to 5 days, depending on the nature of the submittal and its correctness. Reinforcing steel and structural steel shop drawings typically require the greatest amount of time. Also, schedules such as doors, hardware, and door frames consume a great deal of time because of the minute details that must be checked. However, the time expended in submittal processing by the contractor can most easily be controlled at this substage. It must be remembered that time spent in reviewing, checking, and coordinating submittals is one of the most effective methods of ensuring a highly coordinated and smooth running project.

Once the contractor has completed the review of a submittal, the document is transmitted to the architect for approval. The contractor may indicate on the transmittal the date when approval is needed. Here again, the amount of time required for the architect to review a submittal depends on its complexity and whether or not other engineers (i.e., mechanical, electrical, or structural) must participate in the review. As a general rule, 2 to 3 weeks is a good estimate for the time required by the architect to complete the review and return the submittal.

The period when a submittal is in the hands of the architect is probably the most critical substage of the approval process for materials. During this critical substage, the contractor's submittal can be "lost in the shuffle" if the architect's activities are not monitored daily. The most common method of monitoring submittals is through the use of a submittal log, which indicates the date, description, and quantity of each submittal. From this log the contractor can develop a listing of critical submittals to monitor on a daily basis. Once the submittal leaves the contractor's control in the field office, its return must be followed constantly or valuable time will be wasted.

The final substage of the approval process for a material item is the return of the submittal to the supplier or subcontractor. The submittal may be in one of the following four states when returned by the architect:

1. Approved.

2. Approved with noted corrections; no return submittal needed.

3. Approved with noted corrections; however, a final submittal is required.

4. Not approved; resubmit.

The first through third designations would release the vendor or subcontractor to commence fabrication and delivery. The fourth stage would require that the approval process be repeated. In some cases the disapproval by the architect is due to a subcontractor or supplier not communicating clearly through the submittal of the information needed. A meeting between all parties may then be arranged to seek a reasonable solution.

When the approval process is completed, the material has been accepted as part of the project. Its details have been carefully reviewed for conformance with the contract documents. Also, through this process, the item has been coordinated with all trades involved in its installation and verified for inclusion into the project. The material is now ready for fabrication and delivery.

18.4 FABRICATION AND DELIVERY PROCESS

As a submittal is returned to the subcontractor or supplier, the needed delivery date to meet the construction schedule is communicated on the transmittal, verbally, or through other correspondence. In any event, delivery requirements are established and agreed on. The supplier or subcontractor may be required to return to the contractor corrected file and field-use drawings, product data, or samples. These are used to distribute to the contractor's field personnel (i.e., superintendent or foreman) and the other subcontractors and suppliers that must use these final submittals.

Of the four phases of a material's life cycle the fabrication and delivery process is the most critical. Generally, the largest amount of time is lost or gained in this phase. The duration of the fabrication and delivery process depends directly on the nature of the material and the amount of physical transformation involved. For these reasons, the contractor must employ every available method of monitoring materials throughout the fabrication and delivery process.

Contractors generally devote the largest amount of time and effort to controlling and monitoring the fabrication and delivery phase. The term *expediting* is most commonly used to describe monitoring methods in this phase of a material item's life. Methods used to ensure timely fabrication and delivery may range from using checklists developed from the job schedule to actually including this phase as a separate activity on a job schedule. Unfortunately, the fabrication and delivery usually only become activities on the job schedule when the delivery becomes a problem. Extremely critical items requiring extended fabrication times often warrant visits by the contractor to the fabrication facility to ensure the material is actually in fabrication, and proceeding on schedule.

At the completion of fabrication, the delivery of the material is made and the final phase of the life cycle is begun. Materials delivered are checked for compliance with the approved submittal as regards quality, quantity, dimensions, and other requirements. Discrepancies are reported to the subcontractor or supplier. These discrepancies, whether they are shortages or fabrication errors, are subjected to the same monitoring and controlling processes as the entire order. Occasionally they become extremely critical to the project and must be given a great deal of attention until delivery is made.

18.5 INSTALLATION PROCESS

The installation process involves the physical incorporation into the project of a material item. Depending on how effectively materials are scheduled and expedited, materials arriving at the job site may be installed immediately, partially installed and partially stored, or completely stored for later installation. When storage occurs, the installation process becomes directly dependent on the effective storage of materials.

One of the most important aspects of the effective storage of materials is the physical protection of material items. Careful attention must be given to protection from weather hazards such as prevention of water damage or even freezing. Another important aspect is protection against vandalism and theft. Finish hardware, for instance, is generally installed over a considerable time period. A secure hardware room is usually set aside where it is sorted, shelved, and organized to accommodate the finish hardware installation process.

Location of materials stored outside the physical building on the project site or within the building must be carefully planned and organized to facilitate effective installation. In high-rise building construction material storage, each floor can be disastrous if careful planning is not used. For instance, materials stored concurrently on a floor may include plumbing and electrical rough-in materials, ductwork, window wall framing, glazing materials, drywall studs, and other items. The magnitude of the amount of materials involved warrants meticulous layout of materials. Equally important is the storage of materials to facilitate hoisting with a minimal amount of second handling. Reinforcing steel, for instance, may be organized in a "lay-down" area and then directly hoisted as needed. Adequate lay-down areas must be provided within reach of vertical hoisting equipment.

18.6 MATERIAL TYPES

Building construction materials can be logically grouped into three major categories: (a) bulk materials that require little or no fabrication, (b) manufacturer's standard items that require some fabrication, and (c) items that are fabricated or customized for a particular project. Grouping materials into categories can be of value in determining which materials warrant major contractor control efforts. Obviously, material items that require fabrication have longer life cycles because of submittal requirements and fabrication. These materials require a great deal of control by the contractor.

The bulk material category includes those materials that require little vendor modification and can be delivered from vendor storage locations to the job site with little fabrication delay. Table 18.1 lists examples of typical bulk materials in building construction projects. These materials usually require only a 1- to 5-day delivery time, fallowing execution of purchase order or subcontract and approved submittals. Submittal requirements generally include only product and performance data.

Table 18.1 Typical Bulk Materials

Paving materials
Fill materials—crushed stone, soil, sand, etc.
Damproofing membrane
Lumber and related supplies
Form materials—plywood, post shores, etc.
Ready-mix concrete
Wire mesh
Stock reinforcing steel and accessories
Masonry
Stock miscellaneous metals
Soil and waste piping
Water piping
Electrical conduit
Electrical rough-in materials—outlet boxes, switch boxes, etc.
Caulking and sealants

Table 18.2 Typical Standard Material Items

General Materials
    Fencing materials
    Formwork systems—metals and fiberglass pans, column forms, etc.
    Brick paving
    Brick or ceramic veneers
    Standard structural steel members
    Metal decking
    Waterproofing products
    Insulation products
    Built-up roof materials
    Caulking and sealants
    Standard casework and millwork
    Special doors
    Metal-framed windows
    Finish hardware and weather-stripping
    Ceramic and quarry tile
    Flooring materials
    Acoustical ceilings
    Paints and wall coverings
    Lath and plaster products
    Miscellaneous specialties
    Equipment—food service, bank, medical, incinerators, etc.
    Building furnishings
    Special construction items—radiation protection, vaults, swimming pools, integrated ceilings
    Elevators, escalators, dumbwaiters, etc.

Mechanical and Plumbing Equipment and Materials
    Fire protection equipment
    Water supply equipment
    Valves
    Drains
    Clean-outs
    Plumbing fixtures
    Gas-piping accessories
    Pumps
    Boilers
    Cooling towers
    Control systems
    Air-handling equipment
    Refrigeration units (chillers)

Manufacturer's standard material items include materials that are usually stocked in limited quantities and are manufactured for the project after the order is executed and submittals are approved. Table 18.2 illustrates typical materials that are included in this category. Submittal requirements include detailed shop drawings, product and performance data, and samples. Finish materials such as paints, wall coverings, floor coverings, and plastic laminates require a fully developed finish design for the project. Development of the finish design can have serious consequences on ordering and delivery of finish materials. Manufacturing and delivery times generally range from 3 to 12 weeks for these materials. These extended manufacturing and delivery times place considerable importance on planning and controlling these materials.

The fabricated category of construction materials must conform to a particular project's unique requirements. The fabricated item, however, is composed of or results from modification of standard components. Table 18.3 illustrates materials that fall into this category. Submittals required include highly detailed shop drawings, product data, and samples. Fabrication and delivery times range from 2 weeks for items such as reinforcing steel and precast concrete to 10–12 weeks for curtain wall systems, doors and frames, and similar items.

Table 18.3 Typical Fabricated Materials Items

Electrical Equipment and Materials
  Busduct
  Special conduit
  Switchboards and panels
  Transformers
  Wire
  Trim devices
  Lighting fixtures
  Underfloor duct
  Communications devices
  Motors and starters
  Motor control centers
  Electric heaters
  Fire alarm equipment
  Lightning protection equipment
  Concrete reinforcement
  Structural steel
  Precast panels and decks
  Stone veneers
  Miscellaneous and special formed metals
  Ornamental metals
  Millwork
  Custom casework and cabinetwork
  Sheet metal work
  Sheet metal veneers
  Hollow metal doors and frames
  Wood and plastic laminate doors
  Glass and glazing
  Storefront
  Window walls and curtain walls

REVIEW QUESTIONS AND EXERCISES

18.1 Name four important items of information that should be on a typical purchase order.

18.2 What are four good sources of price information about construction materials?

18.3 What is meant by the following expressions?

    a. CIF

    b. 2/10 E.O.M.

    c. 2/10 net 30

    d. ROG

    e. Bill of lading

18.4 Visit a local architect's office and ascertain how product data are obtained and used.

18.5 Visit a local building contractor and determine how he or she handles control of submittals from subcontractors to architect/engineer. What system does he or she use to ensure the job will not be held up due to procurement and approval delays?

18.6 Visit a construction site and determine what procedures are used for verifying receipt arrival and ensuring proper storage of materials at the site.

18.7 Determine what procedures are used for removing waste materials from a local construction (building) site. Is there any scrap value in these materials? Explain.

18.8 Determine the local prices for some bulk materials such as concrete, sand, cement, steel mesh, bricks, and lumber and compare them to the periodically published prices in *Engineering News Record.*

18.9 Select a particular material item (e.g., concrete) and follow its material handling process from the local source through final installation in the building. What special equipment is needed (if any)?

18.10 What types of special materials handling equipment can be identified on local building sites? Do they take advantage of certain properties of the material being handled (e.g., the fluidity of concrete)?

18.11 A 9-cubic-yard concrete truck gets stuck in traffic and arrives just after the maximum time allowed for fresh concrete. The project superintendent rejects the concrete, but the driver insists that it is visible still in good shape. However, because the superintendent refuses to accept the concrete, it is dumped nearby the project. Who should pay for this concrete?

Chapter 19

Safety

CONTENTS

Pipe manipulator mounted on the excavator

Safety in Trenches

Even though heavy construction equipment, such as a crane or backhoe excavator, is used to perform the task of pipe laying in the trench, workers are required to be inside the trench to guide the excavation, pipe laying, and final alignment. Workplace safety has become a major concern in the construction industry over the past few decades, and trench cave-ins have caused serious and often fatal injuries to workers in the United States. It has become of crucial importance to implement the use of new technologies to prevent accidents in trench excavation and pipe installation.

Diverse approaches such as shoring, shielding, and sloping have been applied to protect workers from cave-ins in trenching and pipe laying operations. However, even when support systems are used, the danger of cave-ins still exists due to the nature of the soil and unexpected circumstances. One method which has been developed involves advanced new technology: the prototype robotic excavation and pipe installation system called Pipeman.

The basic Pipeman concept consists of a three-dimensional (3-D) spatial positioning system (SPS), which is interfaced with an excavator to provide the location of the excavator and a beam laser. A pipe manipulator prototype is attached to the bucket of the

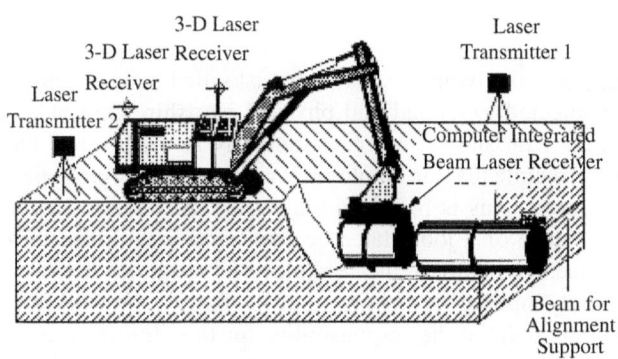

Layout of the robotic trenching and pipe installation system (Huang & Bernold, 1993).

Overview of pipe manipulator

excavator, which is capable of handling pipes of various sizes. A beam laser is also used to help the operator align pipes.

Integration of SPS with a computer-aided design (CAD) system updates the excavator position in real time and provides an as-built drawing of pipe laying.

The main components of the concept are the man-machine interface, actuation system, laser beam, and feedback system. The man-machine interface is used to keep the operator in a safe area and allow him to guide the Pipeman intelligently while the excavator works in a hazardous environment.

19.1   NEED FOR SAFE PRACTICE

A disabling injury or fatal accident on the job site has negative impact on operations at many levels. Accidents cost money and affect worker morale. Because of the type of work involved in construction, many dangers exist both for the workers and for the public. For this reason, the subject of safety offers one area of noncontroversial mutual interest between management and the workforce. The necessity of safe operations and of protecting and conserving lives by preventing accidents is understood by all.

Although the fatality rate in construction has been reduced within recent years, the improvement in safety record achieved by the construction industry still lags seriously behind that achieved in other hazardous industries. The annual number of fatalities in the construction industry in 2003 exceeded the number of combat deaths during the first 18 months of armed conflict in Iraq (period 2003–2004). Construction is a dangerous business.

It is the contractor's responsibility to see that everything possible is done to provide a safe working environment for the work force and the public in general. The factors that motivate safe practices at the job site are generally identified as follows:

1. Humanitarian concerns.

2. Economic costs and benefits.

3. Legal and regulatory considerations.

Society has taken the position that because of the high health and accident potential intrinsic to the construction industry, the contractor must accept the liabilities associated with this hazardous environment and make an appropriate commitment to safe practice and accident prevention.

19.2 HUMANITARIAN CONCERNS

It is normally accepted that day-to-day living has intrinsic risks that may result in members of the society being subjected to mental and physical hardship. One of the functions of society is to minimize pain and suffering. Particularly at the level of the work site, society has defined the principle that the employer is responsible for providing a safe environment for the workforce. This is based on humanitarian concern. If, for instance, a worker loses a leg because of a job-related accident and is confined to a wheelchair, the worker is, in a sense, a casualty of the workplace. Through his desire to be a participating member of society and support members of his family, the worker is injured. Society has traditionally shouldered the responsibility for this limitation on a worker's abilities. Over the past 130 years, the principle of employer liability for death and injury resulting from accidents or health hazards occurring at the workplace has been firmly established in common law. The courts have further charged the employer with the following five responsibilities:[1]

1. To provide a reasonably safe workplace.
2. To provide reasonably safe appliances, tools, and equipment.
3. To use reasonable care in selecting employees.
4. To enforce reasonable safety rules.
5. To provide reasonable instructions regarding the dangers of employment.

Mandatory requirements for the employer to make formal provision for injuries and deaths on the job resulted in the enactment of Workmen's Compensation laws in all 50 states during the first half of the 20th century.

In 1884, Germany enacted the first workmen's compensation act, followed by Austria in 1887 and England in 1897. The U.S. federal government passed the first U.S. compensation act in 1908 covering government employees. Following several legal battles, the Supreme Court, in 1917, declared that states could enact and enforce compulsory Workmen's Compensation laws under their power to provide for the public health, safety, and welfare.

19.3 ECONOMIC COSTS AND BENEFITS

Safety costs can be broken into three categories as follows:

1. Direct cost of previous accidents
   a. Insurance premiums and ratings
   b. Mandatory accident prevention methods
   c. Records, safety personnel
2. Direct cost of each accident occurrence
   a. Delay to project
   b. Uninsured damages
3. Indirect cost
   a. Investigation
   b. Loss of skilled workers
   c. Loss of equipment
   d. Lost production

---

[1] Lee E. Knack, in *Handbook of Construction Management and Organization*, Bonny and Frein (eds.), Van Nostrand Reinhold, New York, 1973, Chapter 25.

Direct costs from previous accidents come primarily in the form of insurance premiums, which have a significant effect on a contractors operating expense. Workmen's compensation and liability insurance premiums can be calculated using either manual or merit rating systems.

Manual rating is based on the past losses of the industry as a whole. The premium rate for compensation is normally set by the individual state Compensation Rating Bureaus. Many states are guided by or actually have their rates set by the National Council on Compensation Insurance (NCCI). The premium rates are based on factors such as classification of operations, rates of pay, the frequency and severity of accidents in a particular classification, increases in the cost of cases, and the attitudes of various industrial compensation commissions. The rates as set and approved by each state insurance commissioner are known as the manual (standard) rates. These manual rates are published periodically in the *Engineering News Record* (ENR) Quarterly Cost Roundup issues. A listing of some of the rates as reported in the R. S. Means *Building Construction Cost Data* is given in Figure 19.1.

The merit rating system bases premiums on a particular company's safety record. High-risk (high-accident-rated) companies are therefore penalized with higher premiums than those paid by companies with low accident rates. In this way, a good safety program can result in substantial financial savings to a company. Higher returns on jobs can be realized, and the ability to bid lower and win more jobs is greatly enhanced.

Once the premiums reach a value of $1,000, the contractor is eligible for a merit system rating. That is, the cost of the premium will be individually calculated with the safety record of the company being the critical consideration. Under the merit system, there are two basic methods used to incorporate the safety record into the final cost of the premium. These are referred to as the experience rating and retrospective rating methods.

Most insurance carriers use the experience rating method, which is based on the company's record for the past 3 years not including the most recent preceding year. In this system, an experience modification rate is multiplied by the manual rate to establish the premium for a given firm. Data on losses, the actual project being insured, and other variables are considered in deriving the experience modification rate (EMR). If the company has an EMR of 75%, it will pay only 75% of the manual premium. Good experience ratings (EMRs) can lead to significant savings. Clough and Sears (Wiley, 1994) illustrate this with the following example:

> Assume that a building contractor does an annual volume of $10 million worth of work. Considering a typical amount of subcontracting and the cost of materials, this general contractor's annual payroll will be of the order of magnitude of $2.5 million. If his present workmen's compensation rate averages about 8%, his annual premium cost will be about $200,000. Now assume that an effective accident prevention program results in an experience modification rate (EMR) of 0.7. This will result in a reduction of the annual premium cost to about $140,000 for this contractor. Annual savings on the order of $60,000 are thereby realized on the cost of this one insurance coverage alone.

Retrospective rating is somewhat like self-insurance. It is basically the same as experience rating except for one point. It uses the loss record of the contractor for the previous year or other defined retrospective period to compute the premium. This can raise or lower the premium cost based on performance during the retrospective period. The starting point or basis for this method is again the manual premium. A percentage (usually 20%) of the standard premium resulting from applying the experience modification factor to the manual rate is used to obtain the basic premium. The retrospective rate is then calculated as

$$\text{Retrospective rate} = (\text{Tax multiplier}) \times \{\text{Basic premium} + [(\text{Incurred loss}) \times (\text{Loss conversion factor})]\}$$

State	CA	CO	DC	FL	GA	IN	KY	NY	WY	AVG.
Carpentry — general	30.29	12.76	10.82	18.25	24.91	7.45	22.00	12.39	8.93	18.31
Concrete Work — NOC	11.42	13.19	9.43	18.38	14.49	5.04	15.75	17.19	8.93	15.22
Concrete Work — flat (flr., sdwk.)	11.42	7.46	12.00	8.91	10.83	3.48	8.00	11.26	8.93	9.42
Electrical Wiring — inside	7.84	5.18	6.35	7.48	9.87	2.74	6.00	6.69	8.93	6.66
Excavation — earth NOC	13.27	10.35	12.14	9.33	17.01	4.62	10.40	8.37	8.93	10.46
Excavation — rock	13.27	10.35	12.14	9.33	17.01	4.62	10.40	8.37	8.93	10.46
Glaziers	17,86	9.79	17.96	14.41	15.94	5.59	20.44	10.18	8.93	14.15
Insulation Work	12.98	13.91	9.67	11.19	22.44	6.77	20,69	9.91	8.93	15.12
Masonry	17.53	14.30	15.03	13.96	19.68	5.15	9.50	15.70	16.00	14.80
i Painting & Decorating	16.24	10.29	8.37	13.51	18.93	4.77	12.50	10.41	8.93	12.86
Pile Driving	15.03	15.08	14.12	38.24	26.79	8.47	25.37	15.68	8.93	21.48
Plastering	27.23	9.00	12.98	25.08	18.55	3.83	15.33	9.04	8.93	14.39
Plumbing	12.47	8.45	11.77	8,13	10.89	2.85	7.00	7.33	8.93	8.11
Roofing	51.93	23.95	21.47	28.05	44.56	11.36	45.00	33.39	8.93	32.94
Sheet Metal Work (HVAC)	18.17	11.46	8.93	12.68	20.76	4.45	19.58	12.79	8.93	11.88
Steel Erection — structure	17.31	35.52	43.78	35.41	52.69	15.75	46.98	21,77	8.93	40.12
Steel Erection — NOC	20.57	15.85	18.27	23.49	41.28	7.98	23.88	15.20	8.93	23.82
Tile Work — (interior ceramic)	8.69	8.60	21.67	8.34	10.07	3.09	11.89	8.72	8.93	9.47
Waterproofing	16.24	5.74	4.05	6.22	8.65	2.27	6.75	6.92	8.93	7.07
Wrecking	20.57	15.85	43.78	35.41	52.69	15.75	46.98	10.56	8.93	36.28

Figure 19.1 Compensation insurance base rates for construction workers (selected states and crafts). *Source*: R. S. Means *Building Construction Cost Data*. Copyright Reed Contstruction Data.

The incurred loss is the amount paid out to settle claims over the retrospective period. The loss conversion factor is a percentage loading used to weight the incurred losses to cover general claims investigation and adjustment expenses. The tax multiplier covers premium taxes that must be paid to the state. If the data for a given company are:

Manual premium $25,000
EMR   0.75
(25% credit)

then

Standard premium = 0.75($25,000) = $18,750
Basic premium@20% of standard = 0.20($18,750) = $3,750
Loss conversion factor = 1.135 (derived from experience)
Tax multiplier = 1.03 (based on state tax)
Incurred losses = $10,000

Then

Retrospective premium = $1.03[\$3,750 + (1.135 \times \$10,000)] = \$15,553$

This is a nice savings over the standard premium of $18,750 and provides the contractor with a clear incentive to minimize the incurred losses. By so doing, the contractor can expect a large premium rebate at the end of the year.

19.4   UNINSURED ACCIDENT COSTS

In addition to the cost of insurance premiums, additional direct costs for things such as the salary of the safety engineer and his or her staff as well as costs associated with the implementation of a good safety program can be identified. The precise amount of the costs associated with other safety cost categories is more difficult to assess, and these costs can be thought of as additional uninsured costs resulting from accidents. Typical uninsured costs associated with an accident are shown in Table 19.1.

Although varying slightly from source to source, hidden losses of this variety have been estimated to be as much as nine times the amount spent on comprehensive insurance. In addition to the costs noted in Table 19.1, another cost is that of paying an injured employee to show up for work even if he or she cannot perform at his or her best. This is a common, if questionable, practice for relatively minor injuries. It is done to avoid recording a lost time accident, which might impact their employer's insurance premium. Although it is tempting for workers with small injuries to return quickly to work, it is important that they refrain from returning too soon to avoid their being reinjured or injured more seriously.

The following situation illustrates the additional losses resulting from hidden costs. At a large industrial construction site, the survey party chief was on the way to the office to get a set of plans. The survey crew was to lay out four machine foundations that morning. The wooden walkway beneath the party chief collapsed. Due to the confusion resulting from the accident, work activity on the entire site was impacted. The party chief was in the hospital for 5 weeks with a shattered pelvis. Another party chief who was unfamiliar with the site was assigned to the surveying crew. As a result, the four machine foundations were constructed 2 feet farther west than called for in the plans. After this was discovered, six laborers worked for 20 hours removing the reinforced concrete. The survey crew of four spent another 5 hours laying out the foundations.

Four carpenters worked 20 more hours preparing new forms. Five more hours were required for the ironworkers to place the steel reinforcement. The total indirect cost was approximately $5,000. Although this activity was not on the critical path, if it had been,

Table 19.1 Uninsured Costs

Injuries	Associated Costs
1. First-aid expenses 2. Transportation costs 3. Cost of investigations 4. Cost of processing reports	1. Difference between actual losses and amount recovered 2. Rental of equipment to replace damaged equipment 3. Surplus workers for replacement of injured workmen 4. Wages or other benefits paid to disabled workers 5. Overhead costs while production is stopped 6. Loss of bonus or payment of forfeiture of delays
Wage Losses	**Off the Job Accidents**
1. Idle time of workers whose work is interrupted 2. Man-hours spent in cleaning up accident area 3. Time spent repairing damaged equipment 4. Time lost by workers receiving first aid	1. Cost of medical services 2. Time spent on injured workers' welfare 3. Loss of skill and experience 4. Training replacement worker 5. Decreased production of replacement 6. Benefits paid to injured worker or dependents
Production Losses	**Intangibles**
1. Product spoiled by accident 2. Loss of skill and experience 3. Lowered production of worker replacement 4. Idle machine time	1. Lowered employee morale 2. Increased labor conflict 3. Unfavorable public relations

Source: From Lee E. Knack, in *Handbook of Construction Management and Organization*, Bonny and Frein (eds.), Van Nostrand Reinhold, New York, 1973, Chapter 25.

liquidated damages might have been charged to the contractor. Still, the accident resulted in costs amounting to 1 week's pay for the employees affected and the cost of material that had to be replaced.

19.5   FEDERAL LEGISLATION AND REGULATION

The federal government implemented a formal program of mandatory safety practices in 1969 with the passage of the Construction Safety Act as an amendment to the Contract Work Hours Standard Act. This legislation requires contractors working on federally funded projects to meet certain requirements to protect the worker against health and accident hazards. In addition, certain reporting and training provisions were established. This program of required procedures has been referred to as a physical approach to achieving safety. That is, regulations are prescribed that are designed to minimize the possibility of an unsafe condition arising. A typical physical measure of this type is the requirement to install guard rails around all open floors of a multistory building during construction. Guard rails are needed anytime there is change in elevation of 6 feet and the worker is not protected by a personnel fall arrest system, warning line, or warning attendant (used to watch workers and warn them if they're too close to falling).

Furthermore, physical measures are implemented to minimize injury in the event of an accident.

An example of this is the requirement to wear a safety belt when working with high steel, and the installation of safety nets to protect a man who slips and falls. This physical approach is in contrast to the behavioral approach that is designed to make all levels of the workforce from top management to the laborer think in a safe way and thus avoid unsafe situations.

Shortly after the passage of the Construction Safety Act, a more comprehensive approach to mandatory safe practices was adopted in the form of the Williams-Steiger Occupational Safety and Health Act (OSHA) passed by Congress in 1970. This act established mandatory safety and health procedures to be followed by all firms operating in interstate commerce.

Under this act, all employers are required to provide "employment and a place of employment which are free from recognized hazards that are causing or are likely to cause death or serious physical harm to his employees." The provisions of the Construction Safety Act were included in the act by reference. The provisions of OSHA fall within the jurisdiction of the secretary of labor. In 1971, the law was implemented by publishing the Code of Federal Regulation (CFR) 1926 that specifically refers to the construction industry and CFR 1910 that pertains to General Standards. Many existing standards issued by various standards organizations, including the American National Standards Institute (ANSI), were included in the basic law by reference. The schematic development of the legislation is shown in Figure 19.2. OSHA has a service for providing update information on standards that is designed to aid in keeping the six volumes of regulations current. The regulations are divided as follows:

Volume I: *General Industry Standards*

Volume II: *Maritime Standards Volume III: Construction Standards*

Volume IV: *Other Regulations and Procedures*

Volume V: *Field Operation Manual*

Volume VI: *Industrial Hygiene Manual*

Under the OSHA legislation, the assistant secretary of labor for occupational safety and health administers and enforces OSHA through the labor department's Occupational Safety and Health Administration and its regional offices around the country. The Occupational Safety and Health Review Commission is designed by OSHA as the review body to which citations for alleged violations and proposed penalties can be

Figure 19.2 Development of OSHA legislation.

appealed. Research on safety topics is under the control of the National Institute for Occupational Safety and Health (NIOSH), which is part of the Department of Health and Human Services.

The OSHA legislation allows individual states to establish programs that operate in place of the federal program. Many contractors prefer a state-operated program since this affords a closer contact with the inspector and the regulatory agency. Appeals and requests for clarification are handled at a state level. This expedites the process of responding to citations and rectification of disputes if differences of opinion arise. The state agency must establish that it is able to administer the law as effectively as the federal government. Presently, 26 states operate approved state plans.

19.6   OCCUPATIONAL SAFETY AND HEALTH ADMINISTRATION REQUIREMENTS

Employers must "make, keep and preserve, and make available to representatives of the Secretaries of Labor, and Health and Human Services" records of recordable occupational injuries and illnesses (Williams-Steiger Act, 1970). Any fatal or serious accidents must be reported to OSHA within 8 hours. Certain records of job-related fatalities, injuries, and illnesses must be maintained by firms having eight or more employees. The two key forms that must be available for review when a compliance officer makes an inspection are:

1. OSHA 300: This is a log that summarizes each reportable case as a single line entry and must be posted for employee inspection (see Figure 19.3). The instructions that accompany the OSHA record-keeping forms do include the following instructions. "You must post the Summary (Form 300A) only–not the Log (Form 300)–by February 1 of the year following the year covered by the form and keep it posted until April 30 of that year." OSHA allows the use of industry-generated forms similar to the OSHA forms as long as they have at least the same information as the OSHA forms.

2. First report of injury: Example in Figure 19.4.

These records must be preserved for 5 years.

The employer is also required to post at the work site records of citations and notices of employees' rights. There have been strong drives to change OSHA so that employers with scattered work sites, as in construction, can be allowed to maintain the required records at a central location (e.g., home office).

19.7   HOW THE LAW IS APPLIED

The OSHA regional and area offices employ inspectors whose duties include visits to active projects to determine if the builders are conforming to the regulations. As noted, there are "State Plan States" that have their own laws, rules, and regulations. The OSHA legislation allows this as long as the program is "at least as effective" as the federal OSHA provisions. Indiana, for instance, operates a state system that simply adopted all of the federal OSHA rules, regulations, changes, and so forth. Michigan, on the other hand, writes and publishes its own rules, some of which go above and beyond what federal OSHA rules require.

An inspection can be initiated at random by OSHA or state safety inspectors or by an employee (or his union) who submits a written statement to the labor department that he believes there is a violation that threatens physical harm or imminent danger. All inspections must be on an unannounced basis during the working day. Rulings by the Supreme Court, however, require that an inspection warrant be obtained from the proper authority if access to the site is denied. It is not unusual that the contractor requires a

V. How to Fill Out the Log

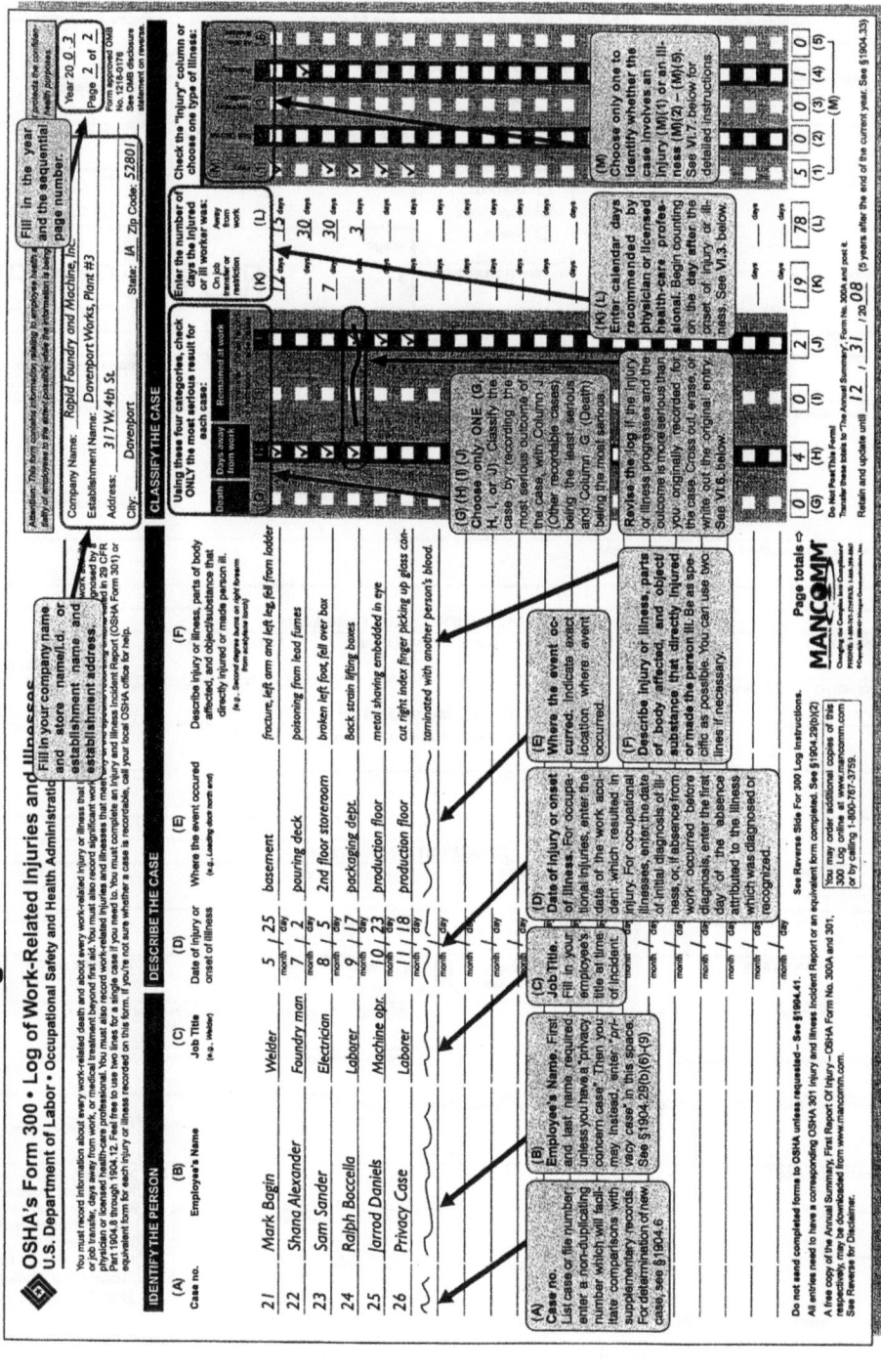

Figure 19.3 OSHA Form 300.

349

INDIANA WORKER'S COMPENSATION
FIRST REPORT OF EMPLOYEE INJURY, ILLNESS
State Form 34401 (R9 / 3-01)

Please return completed form electronically by an approved EDI process.

NOTE: Your Social Security number is being requested by this state agency in order to pursue its statutory responsibilities. Disclosure is voluntary and you will not be penalized for refusal.

FOR WORKER'S COMPENSATION BOARD USE ONLY		
Jurisdiction	Jurisdiction claim number	Process date

PLEASE TYPE or PRINT IN INK

EMPLOYEE INFORMATION

Social Security number	Date of birth	Sex ☐ Male ☐ Female ☐ Unknown	Occupation / Job title		NCCI class code
Name (last, first, middle)		Marital status ☐ Unmarried ☐ Married ☐ Separated ☐ Unknown	Date hired	State of hire	Employee status
Address (number and street, city, state, ZIP code)			Hrs / Day Days / Wk Avg Wg / Wk		☐ Paid Day of Injury ☐ Salary Continued
			Wage Per		
Telephone number (include area code)		Number of dependents	$	☐ Hour ☐ Day ☐ Week ☐ Month ☐ Year ☐ Other	

EMPLOYER INFORMATION

Name of employer	Employer ID#	SIC code	Insured report number
Address of employer (number and street, city, state, ZIP code)	Location number	Employer's location address (if different)	
	Telephone number		
	Carrier / Administrator claim number	Report purpose code	
Actual location of accident / exposure (if not on employer's premises)			

CARRIER / CLAIMS ADMINISTRATOR INFORMATION

Name of claims administrator	Carrier federal ID number	Check if appropriate ☐ Self Insurance
Address of claims administrator (number and street, city, state, ZIP code)	☐ Insurance Carrier ☐ Third Party Admin.	Policy / Self-Insured number
Telephone number		Policy period From To
Name of agent	Code number	

OCCURRENCE / TREATMENT INFORMATION

Date of Inj./ Exp.	Time of occurrence ☐ AM ☐ PM	Date employer notified	Type of injury / exposure	Type code
Last work date	Time workday began	Date disability began	Part of body	Part code
RTW date	Date of death	Injury / Exposure occurred on employer's premises? ☐ Yes ☐ No	Name of contact	Telephone number
Department or location where accident / exposure occurred		All equipment, materials, or chemicals involved in accident		
Specific activity engaged in during accident / exposure		Work process employee engaged in during accident / exposure		
How injury / exposure occurred. Describe the sequence of events and include any relevant objects or substances.				Cause of injury code

Name of physician / health care provider				**INITIAL TREATMENT** ☐ No Medical Treatment ☐ Minor: By Employer ☐ Minor: Clinic / Hospital ☐ Emergency Care ☐ Hospitalized > 24 Hours ☐ Future Major Medical / Lost Time Anticipated
Name of witness	Telephone number	Date administrator notified		
Date prepared	Name of preparer	Title	Telephone number	

An employer's failure to report an occupational injury or illness may result in a $50 fine (IC 22-3-4-13).

Figure 19.4 Typical first report of injury/illness form.

warrant prior to permitting entry to the work site. This essentially changes the "surprise" nature of the inspection and allows the work site supervisor to prepare for the inspection.

The inspection is divided into four parts:

1. An opening conference with the employer.

2. Selection of a representative of the employees and of the employer to accompany the inspector on a tour of the workplace.

3. The walk-around inspection. The inspector is allowed to talk with any employees.

4. The closing conference during which the inspector discusses the conditions and practices he observed that might be safety or health violations. Only "alleged violations" are discussed. Any Safety Orders will follow via U.S. mail 2 weeks to 3 months after the actual inspection. However, correction of any items mentioned by the safety officer in the closing conference for which there is no basis for appeal should be corrected immediately. Fines may be proposed with the citations. The employer is allowed 15 days to appeal a penalty.

The appeals of citations, penalties, or abatement periods are made through a procedure to the Occupational Safety and Health Review Commission or appropriate state board. The commission, after a hearing and review, can affirm, modify, or vacate the citation, proposed penalty, and abatement period.

Table 19.2 is a list of OSHA standards representing job site physical hazards that are commonly cited as violations. In addition to those listed, the 1988 provision requiring firms to inventory and label hazardous materials is the most highly cited violation. This Hazard Communication (HazCom) Standard also requires that employees be trained in the safe use of such materials. Material Safety Data Sheets (MSDS) must be maintained for each hazardous substance. The program addresses the need to make employees more aware of the chemical hazards in the workplace.

If, during the course of an inspection, a violation is noted, a written citation is given to the employer and the area in which the violation occurs will be posted. A reasonable length of time shall be granted the employer for correction of the violation.

These violations, and failure to abate in the given time, incur monetary violations up to $70,000. Serious violations incur a mandatory fine of $7,000. Failure to abate within the given time period can result in a fine of $7,000 a day for the period the violation persists.

Table 19.2 OSHA Standards Most Commonly Cited for Violations

Section	Subject	Section	Subject
1926.500	Guardrails, Handrails, Covers	1926.100	Head Protection
.451	Scaffolding	.552	Materials, Hoists, Personnel Hoists, Elevators
.450	Ladders	.50	Medical Services, First Aid
.350	Gas Welding and Cutting	.501	Stairways
.401	Grounding and Bending	.300	General Requirements, Hand and
.550	Cranes and Derricks	.651	Excavation
.25	Housekeeping		Power Tools
.152	Flammable and Combustible Liquids	.51	Sanitation
.400	General Electrical	.28	Personal Protective Equipment
.402	Electrical Equipment Installation and Maintenance	.102	Eye and Face Protection
.150	Fire Protection	.302	Power-operated Hand Tools
.652	Trenching	.351	Arc Welding and Cutting
.601	Motor Vehicles	.105	Safety Nets

19.8   SAFETY RECORD KEEPING

Documentation under the Williams-Steiger Act is required as follows: "Every employer who is covered under this act must keep occupational injury and illness records for his employees in the establishment in which his employees usually report to work."

The OSHA laws require employers to keep both a log of recordable occupational injuries and illnesses and a supplementary record of each injury or illness. These records must be kept up to date and should be available to government representatives.

These records are also used to compile the annual accident report (OSHA 300), which must be posted in a prominent place in the establishment available to the employees.[2] Also the poster titled "Safety and Health Protection on the Job" shall be posted in a similar manner.

The only employers excluded from this portion of the act are those who are already reporting this material under the Federal Coal Mine Health and Safety Act or the Federal Metal and Nonmetallic Mine Safety Act.

Recordable occupational illnesses and injuries are those that result from a work accident or from exposure to the work environment and lead to fatalities, lost workdays, transfer to another job (temporary or permanent), termination/limitation of employment, or treatment beyond simple first aid measures. Also, those cases involving loss of consciousness or restriction of work or motion are recordable.

Reporting at the job site level breaks into six reporting levels as follows:

1. First aid log.

2. First report of injury log.

3. Supervisor's accident investigation report.

4. Project accident report.

5. OSHA required Injury Report (OSHA 300).

6. Fatality or major accident report.

The first aid log is kept on the job and lists every treatment given. The first report of injury log is required by the workmen's compensation laws in most states. It is prepared to record every personal injury that requires off-site medical treatment regardless of whether the employee lost time from work or not. The supervisor's accident investigation report is prepared by the foreman for each recordable accident and places special emphasis on identifying methods by which the accidents could be prevented in the future. A typical project accident report form is shown in Figure 19.5. It is a monthly summary of disabling injuries and lost time sent to the home office. The form shown is a report of information on each disabling injury required by OSHA and kept at the job site. Finally, as noted previously, any fatality or accident that hospitalizes three or more employees must be reported to the OSHA area director within 8 hours.

19.9   SAFETY PROGRAM

A good job site safety program should be founded on:

1. Safety training of all new personnel arriving on the site.

2. Continuous inspection for possible safety hazards.

3. Regular briefings to increase the safety awareness of personnel at all levels.

4. Written programs and documentation specifying all safety activities.

---

[2] The OSHA 300 (or equivalent) must be kept up to date at all times.

```
Job Name Peachtree Shopping Mall Job No. 10-100 Location Atlanta, Georgia Month
April 20xx

This report should be completed and mailed to the Safety Branch of the Industrial Re-
lations Department in the Atlanta office by the fifth day of the month.
                                        Project Superintendent _____

This figure may be taken from payroll
  records. In the case of fractions use
  the nearest whole number. Do not in-
  clude subcontractors or others.             1. Average number of employees _____

Figure actual hours worked whether
  straight time or overtime. Includes         2. Total hours worked by all
  only those on our payroll.                     employees _____

Record only those injuries that cause
  death, permanent disability (loss of a
  finger, etc.), or loss of time beyond
  the day on which the accident oc-
  curred. No matter what time of day
  the injury may occur, if the employee       3. Number of:
  returns to his regular job at the start        Temporary disabling injuries _____
  of his next regular shift, the injury is       Permanent disabling injuries _____
  not counted. If he does not return at          Deaths _____
  that time, it must be counted as a disa-       Total disabling injuries for this
  bling injury.                                  month _____

For temporary injuries, count the actual
  calendar days lost, excluding the day
  of injury. If the injured employee has      4. Number of days lost as a result of:
  not returned by the end of the month,          Temporary disabling injuries _____
  make an estimate of projected num-             Permanent disabling injuries _____
  ber of lost days. For deaths and per-          Death _____
  manent injuries, use the number of             Total days lost attributable to this
  days specified in the standard table.          month _____
```

Figure 19.5 Project accident report.

If workers or supervisors flagrantly neglect safety rules and regulations, warnings should be considered.

It is good practice to personally brief each employee arriving on site regarding job procedures. A briefing sheet such as the one shown in Figure 19.6 is an effective aid for conducting this type of briefing. This focuses the worker's attention at the outset on the importance of safety and indicates management's interest in this phase of the job. Safety rules and regulations such as those shown in Figure 19.7 should be available in "handout" form and be conspicuously posted around the job site.

General safety meetings conducted by the safety engineer should be held at least once a month with supervisors at the foreman and job steward level. A typical report for such a meeting is shown in Figure 19.8. The objective of these meetings is primarily to heighten the safety awareness of supervisors directly in charge of workers. These foreman-level personnel in turn should hold at least one "tool box" safety meeting each week to transfer this awareness to the work force and discuss safety conditions with their crew. The report format includes a record of those in attendance, the first aid report, and a description of the

August 2, 20xx

PEACHTREE SHOPPING MALL
Atlanta, Georgia

Welcome to the job! ABC Construction Company is interested in you, and during your employment with us, we will exert every effort to make this job pleasant, with a good working atmosphere. On the other hand, your skills, ability, and performance are most important and essential to the successful completion of the project. To set up and complete a good job, certain rules and regulations must be established. For our mutual benefit, these rules and regulations are as follows:

WORKING RULES AND REGULATIONS

Employment

The Project Manager, or his duly authorized representative, will do all the hiring on the job.

Identification

Employees shall wear a company badge at all times, in full view, above the waist, on an outer garment. Badge numbers will be used in gate clearance, payroll, and timekeeping identification.

Hours of Work

The regular workday will begin as per individual instructions, with a lunch period of one-half hour at a designated time. The workweek shall be five days, Monday through Friday.

All employees will be at their work locations, ready to start work at work time. All employees are expected to remain at work until the authorized quitting time, at which time they may put up their tools and leave their place of work. Loitering in the change rooms and/or other places during working hours, or late starting of work and early quitting of work will be subject to proper disciplinary measures.

Checking In and Out

Employees are to check in and out at starting and quitting time. Infractions of this rule will be treated with appropriate disciplinary measures. Employees authorized to leave the project during regular working hours must check out with the timekeeper.

Issuing, Care, and Use of Tools

Certain company tools will be issued to journeymen and apprentices, or the foreman on a check or receipt system. Tools (while issued) must be properly used and maintained. A toolroom clearance will be required on termination. Loss of or damage to tools will be noted on the employee's record.

A Day's Work

Each employee on the job is expected to perform a full day's work. Your willingness, cooperation, and right attitude will go a long way in accomplishing this objective.

Conduct on the Job

Good conduct on the job is essential to the overall welfare of all employees and the daily progress of the job. Therefore, conduct including, but not limited to, the following violations will be subject to appropriate disciplinary action or discharge.

Figure 19.6 Job briefing sheet.

> Theft of company's or employees' property
> Recurring tardiness
> Leaving company's premises without proper authorization
> Possession and/or use of intoxicants and/or narcotics on company's premises
> Willful damage to company's materials, tools, and/or equipment
> Engaging in horseplay (including shouting to passers-by)
> Insubordination
> Gambling
> Fighting on company premises
> Sleeping on the job
> Failure to observe established safety rules and regulations

Housekeeping

Good housekeeping is essential to the safe and efficient construction of the job and is the responsibility of each employee. Work areas, stairways, walkways, and change rooms shall be kept clean at all times.

Safety Rules

Established safety rules and regulations will be observed and followed by all employees in the best interest of accident-free operations.

All unsafe working conditions should be reported to your immediate foreman, who in turn reports it to the company safety engineer.

All employees will be required to wear proper clothing above and below the waist. Hard hats must be worn by all employees and visitors while on the construction site.

Pay Period

Wednesday thru Tuesday is the pay period, with pay day on Friday of each week.

Use of First Aid Facilities

First aid facilities are available at the job site and direct contracts have been established with local doctors, hospitals, and emergency crews for accidents of a serious nature. All injuries, regardless of severity, must be reported to the employee's supervisor, field safety supervisor, and/or first aid immediately upon occurrence. Insurance regulations make this requirement mandatory.

Sanitary Facilities

Adequate sanitary facilities are provided on the job site and are to be used by all employees. We request your cooperation in maintaining these facilities in a clean and orderly condition.

Raincoats and Boots

Raincoats and boots are supplied to employees where the conditions of the job being performed require them.

Remaining in Work Areas

Each employee must remain on the job site and at his work location at all times during regular working hours, unless authorized to leave by his supervisor.

Absenteeism

Unauthorized absenteeisms will result in termination of employment. An employee who must be absent or late should call 999-9000 and report to timekeeper.

Your cooperation in observing the rules and regulations for the job will show proper consideration for other employees and will be appreciated by the company.

If you agree to and will abide by the above, please sign and return to our field supervisor, Charles Hoarse.

cc: Employee File

Figure 19.6 (*Continued*)

ABC Contractors and Engineers

760 Spring Street, N.W., Atlanta, Georgia 30308
(404) 999-9000

July 30, 20xx

Re: OCCUPATIONAL SAFETY & HEALTH ACT 1970 (Construction) (OSHA)

Employers, owner, contractors, subcontractors, superintendents, or foremen in charge shall not direct or permit an employee to work under conditions that are not in compliance with the above code.

Where one contractor is selected to execute the work of the project, he shall assure compliance with the requirements of this code from his employees as well as all subcontractors.

Every employee shall observe all provisions of the above codes that directly concern or affect his conduct. He shall use the safety devices provided for his personal protection and he shall not tamper with or render ineffective any safety device or safeguard.

1. *Overhead Hazards*   All employees shall be provided with HARD HATS and shall use HARD HATS.
2. *Falling Hazards*   Every hole or opening in floors, roofs, platforms, etc., into or through which a person may fall shall be guarded by a barrier sufficient to PREVENT FALLS.
3. *Slipping Hazards*   Scaffolds, platforms, or other elevated working surfaces covered with ice, snow, grease, or other substances causing slippery footing shall be removed, turned, sanded, etc., to ensure safe footing.
4. *Tripping*   Areas where employees must work shall be kept *reasonably free* from accumulations of dirt, debris, scattered tools, materials, and sharp projections.
5. *Projecting Nails*   Projecting nails in boards, planks, and timbers shall be *removed,* hammered, or *bent over* in a safe way.
6. *Riding of Hoisting Equipment*   No employee shall ride on or in the load bucket, sling, platform, ball, or hook.
7. *Lumber & Nail Fastenings*   Lumber used for temporary structures must be sound. Nails shall be driven full length and shall be of the proper size, length, and number. The proper use of double-headed nails is not prohibited.
8. *Guard Rail or Safety Rail*   Should be 2 × 4 at a height of 35″–37″ plus a midrail of 1 × 4. The hand rail shall be smooth and free from splinters and protruding nails. Other material or construction may be used provided the assembly *assures* equivalent *safety*.
9. *Toe Boards*   Shall extend 4″ above platform level and shall be installed *where needed* for the safety of those working below.
10. *Protection Eye Equipment*   Eye protection shall be provided by employers and *shall* be used for cutting, chipping, drilling, cleaning, buffing, grinding, polishing, shaping, or surfacing masonry, concrete, brick, metal, or similar substances. Also for the use and handling of corrosive substances.
11. *Protective Apparrel*   *Waterproof boots* where required shall have safety insoles unless they are the overshoe type.
    Waterproof clothing shall be supplied to the employee required to work in the rain.

Figure 19.7 Job safety rules and regulations.

12. *Safety Belts & Lines*	Shall be *arranged* so that a free fall of no more than 6″ will be allowed.
13. *Stairways*	Temporary stairways shall not be less than 3 feet in width and shall have treads of no less than 2 inch × 10 inch plank. Must have hand rails. (See #8.)
14. *Smoking*	Prohibited in areas used for gasoline dispensing and fueling operations or other *high hazard fire areas.*
15. *Flammable*	*Flammable liquid shall be kept in safety cans* or approved use and storage containers. Suitable grounding to prevent the buildup of static charges shall be provided on all flammable liquid transfer systems.
16. *Sanitation*	*Toilet facilities* shall be provided and made available in sufficient number to accommodate all employees.
17. *Drinking Water*	A supply of *clean* and *cool* potable water shall be provided in readily accessible locations on all projects.
18. *Salt Tablets*	Shall be made available at *drinking stations* when required.
19. *Excavations*	Material and other superimposed loads shall be placed at least 3 feet back from the edge of any excavation and shall be piled or retained so as to prevent them from falling into the excavation. Sides and slopes of excavation shall be stripped of loose rocks or other material. Slopes shall be at an angle of 45 degrees or less (*1 on 1 slope*).
20. *Structural Steel Erection*	When erection connections are made, 20% of the bolts in each connection must be drawn up wrench tight. At least 2 *bolts* must be used at each end of the member. *No loads* shall be placed on a framework until the permanent bolting is complete. Only employees of the structural steel erector engaged in work directly involved in the steel erection shall be permitted to work under any single-story structural steel framework that is not in true alignment and *permanently bolted.*
21. *Use of Ladders*	Ladders shall be provided to give access to floors, stagings, or platforms. Ladders shall be maintained in a safe condition at all times. Ladders shall be securely *fastened top* and *bottom* as well as braced where required. Ladders leading to floors, roofs, stagings, or platforms shall extend at least 3 feet above the level of such floors, stagings, or platforms.
22. *Scaffolds*	All scaffolding shall be constructed so as to *support* 4 times the anticipated working load, and shall be braced to prevent lateral movement. Planks shall overhang their end supports not less than 6″ or more than 12″. 2″ planking may span up to and including 10′. The minimum *width* of any planked platform shall be 18 inches. *Guard rails* and *toe rails* shall be provided on the open sides and ends of all scaffold platforms more than 8′ high (see #8).
23. *Rigging, Ropes, and Chains*	All rope, chains, sheaves, and blocks shall be of sufficient strength, condition, and size to safely raise, lower, or sustain the imposed load *in any position.* *Wire rope* shall be used with power-driven hoisting machinery. No rope shall be used when visual inspection of the rope shows marked signs of corrosion, misuse, or *damage.* All load hooks shall have *safety clips.* Loads that tend to swing or turn during hoisting shall be controlled by a *tag line* whenever practicable.

Figure 19.7 (*Continued*)

24. *Welding and Cutting*	Oxygen from a cylinder or torch shall never be used for *ventilation.*
	Shields or goggles must be worn where applicable. *Cradles* shall be used for lifting or lowering cylinders.
25. *Cranes & Derricks*	All cranes and derricks shall be equipped with a properly operating boom angle *indicator* located within the normal view of the operator.
	Every derrick and crane shall be operated by a designated person.
	A copy of the *signals in use* shall be posted in a conspicuous place on or near each derrick or crane. Cranes and derricks shall have a *fire extinguisher* attached.
26. *Trucks*	Trucks shall not be backed or dumped in places where men are working nor backed into a hazardous location unless guided by a person so stationed on the side where he can see the truck driver and the space in back of the vehicle.

The above items do not encompass all the construction safety regulations as they pertain to OSHA but are intended as a guide to the ever present hazards and primary causes of accidents in our industry.

Figure 19.7 (*Continued*)

ABC Construction Company
Job 10-100
Peachtree Shopping Mall
Atlanta, Georgia
Sept. 1, 20xx

GENERAL SAFETY MEETING #7

Safety Slogan for the Week:

"Be Alert, Don't Get Hurt."

C. Hoarse—Safety Supervisor
A. Apple—Carpenter Foreman
D. Duck—Surveyor
M. Maus—Laborer
D. Halpin—Field Engineer
R. Woodhead—Tool Room

Subcontractors Present:

Live Wire Electric
Henry Purcell
James Wallace

The First Aid Report for August 15 to August 31 Was Given. There Were:

First Aid	7
Doctor's Cases	0
Lost Time Injuries	0

Figure 19.8 Safety meeting minutes.

SHORTCUTS

All of us, supposedly, at one time or another, have been exposed to possible injury by short cutting when a few extra steps would have meant the safe way. We did so as kids when we jumped the fence instead of using the gate and we do so as men when we cross streets by jaywalking instead of using the intersection. Accident statistics plainly indicate the fact that people disregard the fact that minor safety violations may have very serious results.

In construction work, short cutting can be deadly. All of us know of cases in which this kind of thoughtless act resulted in a serious injury. For instance, an ironworker tried to cross an opening by swinging on reinforcing rods, his hands slipped, and he fell about 20 feet to a concrete floor. If he had bothered to take a few moments to walk around the building, he would still be tying rods.

The safe way is not always the shortest way and choosing the safe way is your *Personal Responsibility*. When you are told to go to work in a particular area, you are expected to take the safe route, not an unsafe short route. We cannot be your guardian angel; that is one thing you will have to do for yourself.

If you are told to go to work in some place that has no safe access, report this fact to your foreman so that necessary means of access can be provided.

Ladders and scaffolds are provided for high work; use them. Even though a high job may take only a few moments, DO NOT CLIMB ON FALSE WORK, or on some improvised platform.

Your first responsibility is to yourself. Remember that ladders, steps, and walkways have been built to save you trouble and to save your neck, too. Use them always.

Gambling a few minutes and a little energy against a possible lifetime of pain and misery is a poor bet.

GENERAL DISCUSSION

Flagmen must control all the back-up operations on this job.

Traffic—Be on the alert for moving vehicles, our area is slippery. *Don't* walk beside moving equipment.

Injuries—Report all injuries to your foreman immediately.

Charlie Hoarse
_____
C. Hoarse, Safety Supervisor

Figure 19.8 (*Continued*)

safety topics discussed. In addition to the general safety meetings, each job should have a designated safety committee that meets regularly. The members of the safety committee should include key supervisory personnel and craftsmen with an alertness to potential danger and a genuine desire to prevent accidents and injuries. One of the purposes of the safety committee should be to make suggestions as to how to improve overall job safety. Therefore, the members appointed should be sensitive to safety and innovative in devising safe methods.

REVIEW QUESTIONS AND EXERCISES

19.1   What factors should motivate a contractor to have a safe operation and a good safety program?

19.2   What factors influence the rate assigned to a contractor for Workmen's Compensation Insurance?

19.3 What are two major economic benefits of a good construction program?

19.4 Explain organizing for safety.

19.5 What actions could you as the contractor take to instill a sense of safety among your workers?

19.6 Observe several construction sites and ascertain details of their safety program. If possible attend a tool box safety meeting. Then prepare a list of both good and bad examples of safety practice.

19.7 Using OSHA regulations as a guide, determine what are the accepted safety standards for:

(a) Guard rails

(b) Exposed reinforcing steel

(c) Protection of openings

(d) Man hoists

19.8 Many construction workers resist the use of safety helmets, goggles, and protective mittens and clothing despite the fact that they are designed to protect them. Give several reasons why this practice persists.

19.9 Investigate and report the main safety issues emphasized in a country of your choice (not the United States). You can search the Internet for this exercise.

Appendices

---

[1] Reprinted by permission of the Engineers Joint Contract Documents Committee (EJCDC). Copyright 2002. National Society of Professional Engineers, 1420 King Street, Alexandria, VA 22314; American Consulting Engineers Council, 1015 15th Street N.W., Washington, DC 20005; American Society of Civil Engineers, 1801 Alexander Bell Drive, Reston, VA 20191-4400. To order AGC contract documents, phone 1-800-AGC-1767 or fax your request to 703-837-5405, or visit the AGC website at www.agc.org.
[2] Builders Association of Chicago, Inc.

Appendix A

Typical Considerations Affecting the Decision to Bid

TYPICAL CONSIDERATIONS: THE DECISION TO BID (OR NOT)[1]

A. Goals and Present Capabilities of Your Company (Plans for Growth, Type of Work, Market Conditions)

1. It is quite reasonable to actually want to stay where you are if you are satisfied with a situation of making a good living and staying active in work.
   - If so, is this job the kind you like doing? Does it have a good profit potential?
2. If you wish to grow larger, how fast do you wish to grow? Do you have the people and capital to do so?
   - Will the project to be bid help you in your growth?
   - Or will you have to bid it low just to keep your present men and equipment working, thus tying them up and postponing growth? (If you prefer type 1 goals, this latter strategy may be fine.)
3. *Type of work.* Which type of work do you presently have the capability and experience to do? What types of work do you want to do in the future? Can you handle this particular project now? Will it give you good experience for the type of work you want to do in the future?
4. Consider the present and future competitive market conditions in this type of work.
   - Is it possible to earn a fair and reasonable profit? Or is the competition heavy?
   - Think of the job as an investment of your time, your talent, and your money. It should earn a good return—in money, in satisfaction, and pride; or provide some other return.

B. Location of the Work

1. Is the project located in an area in which you normally like to operate?
2. If not, would too large a portion of your time be consumed traveling to and from this job?
3. Do you have an associate or assistant who you believe can do a good job of supervising the job if you cannot often visit the site yourself?
4. Do you plan to expand your area of operations anyway, and if so, is this job in an area in which you want to expand?

C. Time and Place for Bid

1. When is the bid due (day and hour)? Will you have time to prepare an accurate and careful estimate? (For example, if you need 2 weeks to prepare a good bid and only 4 days remain, don't bid the job.)

---

[1] Based on material prepared by Prof. Boyd C. Paulson, Jr., Stanford University.

2. Where is the bid to be submitted? How will you get it there? Do you have to allow 2 or 3 days for the mail?

3. Are there special rules for late delivery? For faxing last-minute changes?

D. How to Obtain Plans and Specifications

1. If you are a prime contractor, you must find out who will provide the plans and specifications.

   • Is there a fee? How much?

   • Is there a deposit? How much? Is it refundable?

   • Is a plans room open and available? Where? What hours?

2. If you are a subcontractor, you want to know which prime contractors have plans and specifications.

   • Will they give you a copy of those that apply to your work?

   • Do they have a plans room for subcontractors? Where? What hours?

   • Can you get your plans and specifications directly from the owner? Fees? Deposits? How much? Refunds?

E. Legal and Other Official Requirements

1. *Licensing*. Some states, counties, cities, and towns require that a contractor have a license to work in their area.

   • If required, it is a legal necessity.

   • In some cases, unlicensed contractors can be fined without it.

   • Unlicensed contractors may not have recourse to the courts, even if wronged.

   • Especially note this when working on local government-funded projects.

2. *Prequalification* may be required. If so, documents such as a financial statement, a statement of work in progress and experience, as well as a past litigation and performance history will be required.

3. *Bonding*

   • Does project require (a) bid bond? (b) performance bond? (c) payment bond?

   • What is your bonding limit? Can you qualify for bonds on this project?

F. Scope of Work

1. What is the approximate size of the project (or subcontract):
   (a) In dollars—is it within your financial and bonding limits?

   (b) In major units of work (e.g., earth-moving equipment, cubic yards of concrete, pounds of steel, etc.) is it within the capacity of your available manpower and equipment resources?

2. What are the major types of work on the project or subcontract?
   (a) Are they the kind your company prefers to do?

   (b) Are they the kind your company is qualified to do?

3. How much time is available to complete the work?
   (a) When does it start; when does it finish?

   (b) How much other work do you plan to have going at that time? Can you handle this job as well?

G.   Comparison of Resources

Compare the resources available to you to those that will be needed (order of magnitude only) on the job to be bid.

1. *Men:* Do you have a supervisor or foreman for the job? Can you get the laborers and craftsmen that who be needed?

2. *Equipment:* What major items of equipment (truck, crane, loader, etc.) will be needed? Do you own it already? Will it be available? Can you purchase new equipment? Can you rent or lease the equipment you will need?

3. *Money:* Will loans or credit be needed? How much? Can you get the financing needed?

H.   Summary

All of these items should be considered in making the decision to bid or not bid on a particular job.

- This is an *executive decision*.
- It is a decision *you* as the contractor must make.

Appendix B

Performance and Payment Bonds

CONTRACT PERFORMANCE BOND*

Bond No. 31-0120-42879-96-2

KNOW ALL MEN: That we   Ryan Construction Corp.
P. O. Box 16, Zionsville, IN 46077-0493

(here insert the name and address or legal title of the Contractor) hereinafter called the Principal, and

United States Fidelity and Guaranty Company
135 N. Pennsylvania Street, Indianapolis, IN 46204

hereinafter called the Surety or Sureties, are held and firmly bound unto The Trustees of Indiana University, hereinafter called the Owner, in the sum of:

Eight Hundred Twenty Two Thousand and 00/100 Dollars ($822,000.00)

for payment whereof the Principal and the Surety or Sureties bind themselves, their heirs, executors, administrator, successors and assigns, jointly and severally, firmly, by these presents.

WHEREAS, the Principal has, by means of a written Agreement, dated: June 10, XXXX, entered into a contract with the Owner for

Lilly Clinic Expansion at Adult Outpatient Center
Indiana University Medical Center, Indianapolis, IN
Bid Package No. 3
IUPUI#961-5262-3

a copy of which Agreement is by reference made a part hereof:

NOW THEREFORE, the condition of this Obligation is such that, if the Principal shall faithfully perform the Contract on his part and shall fully indemnify and save harmless the Owner from all cost and damage which he may suffer by reason of failure to do so and shall fully reimburse and repay the Owner all outlay and expense which the Owner may incur in making good any such default, then this Obligation shall be null and void, otherwise it shall remain in full force and effect.

CONTRACT PERFORMANCE BOND
(Page 1 of 2 Pages)

Disclaimer:
*This document is representative. Language in actual bonding documents should be verified by a legal professional.

The said surety for value received hereby stipulates and agrees that no change, extension of time, alteration or addition to the terms of the contract, or to the work to be performed thereunder or the specifications accompanying them, shall in any way affect its obligations on this bond, and it does hereby waive notice of any such change, extension of time, alteration, or addition to the terms of the contract, or to the work or to the specifications.

PROVIDED, however that no suit, action or proceeding by reason of any default whatever shall be brought on this Bond after two years from the date of final payment.

AND PROVIDED, that any alterations which may be made in the terms of the Contract, or in the work to be done under it, or the giving by the Owner of any extension of time for the performance of the Contract, or any other forbearance on the part of either the Owner or the Principal to the other shall not in any way release the Principal and the Surety or Sureties, or either or any of them, their heirs, executors, administrators, successors or assigns from their liability hereunder, notice to the Surety or Sureties of any such alterations, extension or forbearance being hereby waived.

Signed and Sealed this 10th day of June, XXXX

In presence of:

Michael Ryan   )
Michael Ryan   ) as to
Corporate Secretary   )

Ryan Construction Corp. (SEAL)

By: _Daniel Ryan_

Daniel Ryan, President

_____   )
) as to
_____   )

United States Fidelity and Guaranty
Company   (SEAL)

By: _U. S. Grant_

U. S. Grant, Attorney-In-Fact

_____   )
) as to
_____   )

_____(SEAL)

_____

CONTRACT PERFORMANCE BOND
(Page 2 of 2 Pages)

LABOR AND MATERIAL PAYMENT BOND*

Bond No. 31-0120-42879-96-3

KNOW ALL MEN BY THESE PRESENTS, THAT _____
Ryan Construction Corp., P. O. Box 16, Zionsville, IN 46077-0493

as Principal, hereinafter called Principal, and _____
United States Fidelity and Guaranty Company, 135 N. Pennsylvania St., Indianapolis, IN 46204

as Surety, hereinafter called Surety, are held and firmly bound unto <u>The Trustees of Indiana University</u>

as Obligee, for the use and benefit of claimants as hereinbelow defined, in the amount of <u>Eight Hundred Twenty Two Thousand and 00/100 Dollars ($822,000)</u>, for the payment whereof Principal and Surety bind themselves, their heirs, executors, administrators, successors and assigns, jointly and severally, firmly by these presents.

WHEREAS, Contractor has by written agreement dated <u>June 10, XXXX</u>, entered into a contract with the Obligee, for

> Lilly Clinic Expansion at Adult Outpatient Center
> Indiana University Medical Center, Indianapolis, IN
> Bid Package No. 3
> IUPUI#961-5262-3

which contract is by reference made a part hereof, and is hereinafter referred to as the contract.

NOW, THEREFORE, THE CONDITION OF THIS OBLIGATION is such that, if principal shall promptly make payment to all claimants as hereinafter defined, for all labor and material used or reasonably required for use in the performance of the Contract, then this obligation shall be void; otherwise, it shall remain in full force and effect, subject, however, to the following conditions:

1. A claimant is defined as one having a direct contract with the Principal or with a Subcontractor of the Principal for labor, material, or both, used or reasonably required for use in the performance of the Contract, labor and material being construed to include that part of water, gas, power, light, heat, oil, gasoline, telephone service or rental of equipment directly applicable to the Contract.

2. The above named Principal and Surety hereby jointly and severally agree with that every claimant as herein defined, who has not been paid in full before the expiration of a period of ninety (90) days after the date on which the last of such claimant's work or labor was done or performed, or materials were furnished by such claimant, may sue on this bond for the use of such claimant, prosecute the suit to final judgment for such sum or sums as may be justly due claimant, and have execution thereon. Obligee shall not be liable for the payment of any costs or expenses of any such suit.

Disclaimer:
*This Document is representative. Language in actual bonding documents should be verified by a legal professional.

3. No suit action shall be commenced hereunder by any claimant:

    a) Unless claimant, other than one having a direct contract with the Principal, shall have given written notice to any two of the following: the Principal, Obligee or the Surety above named, within ninety (90) days after such claimant did or performed the last of the work or labor, or furnished the last of the materials for which said claim is made, stating with substantial accuracy the amount claimed and the name of the party to whom the materials were furnished, or for whom the work or labor was done or performed. Such notice shall be served by mailing the same by registered mail or certified mail, postage prepaid, in an envelope addressed to the Principal, Obligee or Surety, at any place where an office is regularly maintained for the transaction of business, or served in any manner in which the legal process may be served in the state in which the aforesaid project is located, save that such service need not be made by a public officer.

    b) After the expiration of one (1) year following the date on which Principal ceased Work on said Contract, it being understood, however, that if any limitation embodied in this bond is prohibited by any law controlling the construction hereof such limitation shall be deemed to be amended so as to equal to the minimum period of limitation permitted by such law.

    c) Other than in a state court of competent jurisdiction in and for the county or other political subdivision of the state in which the Project, or any part thereof, is situated, or in the United States District Court for the District in which the Project, or any part thereof, is situated, and not elsewhere.

4. The amount of this bond shall be reduced by and to the extent of any payment or payments made in good faith hereunder, inclusive of the payment by Surety of Mechanics' Liens which may be filed of record against said improvement, whether or not claim for the amount of such lien be presented under and against this bond.

Signed and sealed this <u>10th</u> day of <u>June</u>, **XXXX**

Michael Ryan	(RYAN CONSTRUCTION CORPORATION
(Witness)	((Principal) (Seal)
	(
	(
	(BY: _Daniel Ryan_
	Title: Daniel Ryan, President
	(UNITED STATES FIDELITY & GUARANTY
	CO.
	((Surety) (Seal)
	(
_____	(
(Witness)	(BY: _U.S.Grant_
	Title: U.S. Grant,
	Attorney-In-Fact

Appendix C

Standard Form of Agreement Between Owner and Contractor on the Basis of a Stipulated Price

This document has important legal consequences; consultation with an attorney is encouraged with respect to its use or modification.

SUGGESTED FORM OF AGREEMENT BETWEEN OWNER AND CONTRACTOR FOR CONSTRUCTION CONTRACT (STIPULATED PRICE)

Prepared by

ENGINEERS JOINT CONTRACT DOCUMENTS COMMITTEE

and

Issued and Published Jointly By

PROFESSIONAL ENGINEERS IN PRIVATE PRACTICE
a practice division of the
NATIONAL SOCIETY OF PROFESSIONAL ENGINEERS

AMERICAN COUNCIL OF ENGINEERING COMPANIES

AMERICAN SOCIETY OF CIVIL ENGINEERS

This document has been approved and endorsed by

The Associated General Contractors of America

Knowledge for Creating
and Sustaining
the Built Environment

Construction Specifications Institute

EJCDC C-520 Suggested Form of Agreement Between Owner and Contractor for Construction Contract (Stipulated Price)
Copyright©2002 National Society of Professional Engineers for EJCDC. All rights reserved.

This Suggested Form of Agreement has been prepared for use with the Standard General Conditions of the Construction Contract (C-700, 2002 Edition). Their provisions are interrelated, and a change in one may necessitate a change in the other. The language contained in the Suggested Instructions to Bidders (C-200, 2002 Edition) is also carefully interrelated with the language of this Agreement. Their usage is discussed in the Commentary on EJCDC Construction Documents. See also Guide to the Preparation of Supplementary (C-800, 2002 Edition).

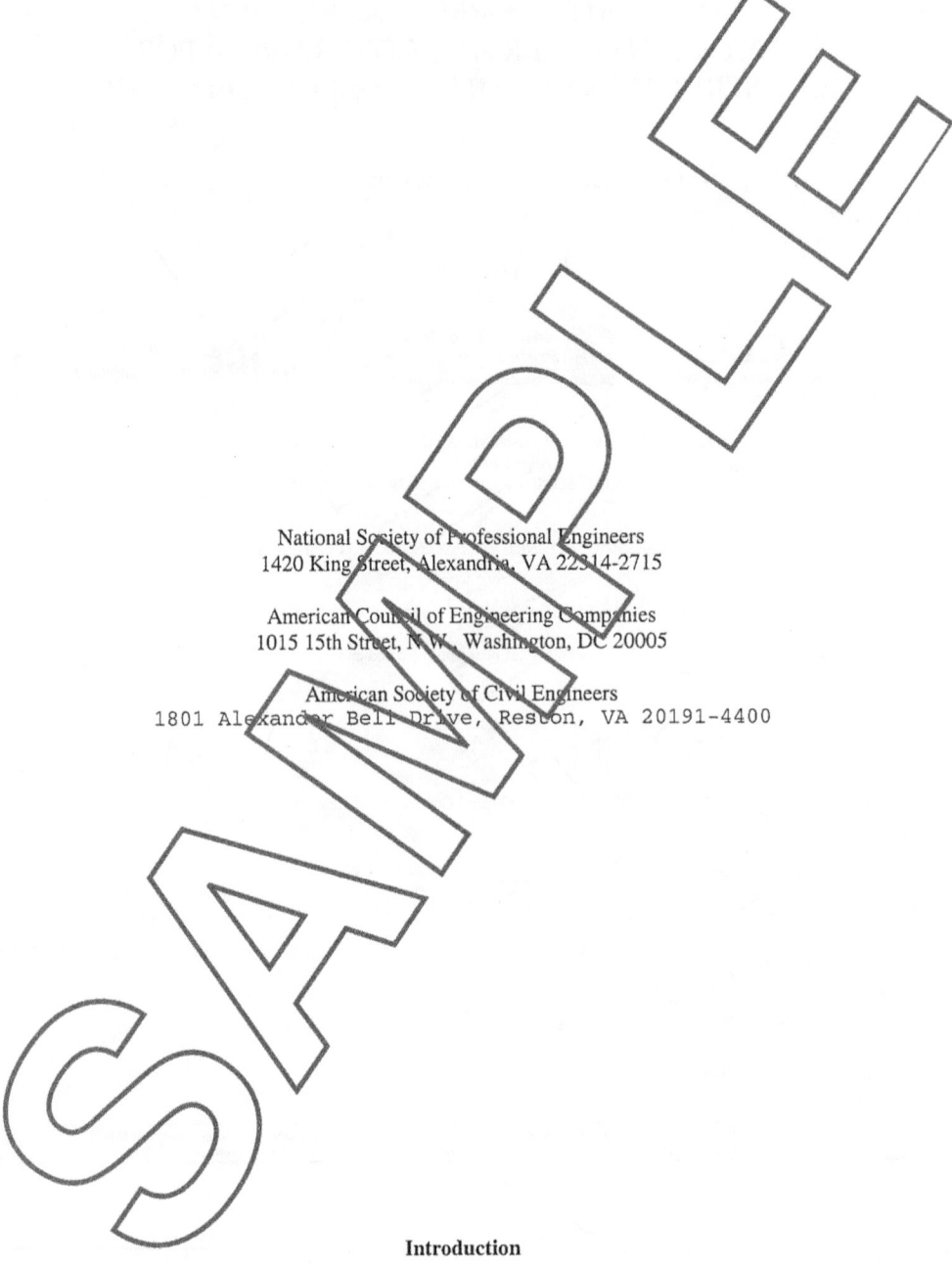

National Society of Professional Engineers
1420 King Street, Alexandria, VA 22314-2715

American Council of Engineering Companies
1015 15th Street, N.W., Washington, DC 20005

American Society of Civil Engineers
1801 Alexander Bell Drive, Reston, VA 20191-4400

Introduction

This Suggested Form of Agreement between Owner and Contractor for Construction Contract (Stipulated Price) ("Agreement") has been prepared for use with the Guide to the Preparation of Instructions to Bidders ("Instructions")(C-200, 2002 Edition) and

EJCDC C-520 Suggested Form of Agreement Between Owner and Contractor for Construction Contract (Stipulated Price)
Copyright©2002 National Society of Professional Engineers for EJCDC. All rights reserved.
00520-1

with the Standard General Conditions of the Construction Contract ("General Conditions")(C-700, 2002 Edition). Their provisions are interrelated, and a change in one may necessitate a change in the others. For guidance in the preparation of Supplementary Conditions and coordination with Instructions to Bidders, see Guide to the Preparation of Supplementary Conditions ("Supplementary Conditions")(C-800, 2002 Edition). See also Suggested Bid Form ("Bid Form") (C-410, 2002 Edition). The EJCDC has not prepared a suggested form of Advertisement or Invitation to Bid because such documents will vary widely to conform to statutory requirements.

This form and the other Bidding Documents prepared and issued by the EJCDC assume acceptance of the Project Manual concept of the Construction Specifications Institute which provides for an organizational format for location of all bound documentary information for a construction project, namely: Bidding Requirements (which term refers to the Advertisement or Invitation to Bid, the Instructions, and any Bid Form that may be suggested or prescribed, all of which provide information and guidance for all Bidders) and the Contract Documents (defined in Article 1 of the General Conditions) which include the Agreement, bonds and certificates, the General Conditions, the Supplementary Conditions, the Drawings, and the Specifications. The Bidding Requirements are not considered part of the Contract Documents because much of their substance pertains to the relationships prior to the award of the Contract and has little effect or impact thereafter and because many contracts are awarded without going through the bidding process. In some cases, however, the actual Bid may be attached as an exhibit to the Agreement to avoid extensive retyping. (The terms "Bidding Documents" and "Bidding Requirements" are defined in Article 1 of the General Conditions.) The Project Manual concept is explained in the Manual of Practice issued by the Construction Specifications Institute.

Suggested language is presented herein with "Notes to User" to assist in preparing the Agreement. Much of the language should be usable on most projects, but modifications and additional provisions will often be necessary. The suggested language has been coordinated with the other standard forms produced by the EJCDC. When modifying the suggested language or writing additional provisions, the user must check the other documents thoroughly for conflicts and coordination of language usage and make appropriate revisions in all affected documents.

Refer to the discussions in EJCDC's Recommended Competitive Bidding Procedures for Construction Projects ("Bidding Procedures") (No. 1910-9-D, 1987 Edition) (to be reissued in 2002) on the particular paragraphs of which frequent reference is made below.

For brevity, paragraphs of the Instructions to Bidders are referenced with the prefix "I," those of the Bid Form are referenced with the prefix "BF," and those of this Agreement are referenced with the prefix "A."

NOTES:

1. EJCDC publications may be ordered from:

NSPE headquarters
1420 King Street
Alexandria VA 22314-2715
703-684-2800
www.nspe.org

ASCE headquarters
1801 Alexander Bell Drive
Reston, VA 20191-4400
800-548-2723
www.asce.org

ACEC headquarters
1015 15th Street NW
Washington DC 20005
202-347-7474
www.acec.org

EJCDC
SUGGESTED FORM OF AGREEMENT
BETWEEN OWNER AND CONTRACTOR FOR

EJCDC C-520 Suggested Form of Agreement Between Owner and Contractor for Construction Contract (Stipulated Price)
Copyright ©2002 National Society of Professional Engineers for EJCDC. All rights reserved.
00520-2

CONSTRUCTION CONTRACT(STIPULATED PRICE)

THIS AGREEMENT is by and between _____

(Owner) and _____

(Contractor).

Owner and Contractor, in consideration of the mutual covenants set forth herein, agree as follows:

ARTICLE 1 - WORK

1.01   Contractor shall complete all Work as specified or indicated in the Contract Documents. The Work is generally described as follows:

ARTICLE 2 - THE PROJECT

2.01   The Project for which the Work under the Contract Documents may be the whole or only a part is generally described as follows:

ARTICLE 3 - ENGINEER

3.01   The Project has been designed by

(Engineer), who is to act as Owner's representative, assume all duties and responsibilities, and have the rights and authority assigned to Engineer in the Contract Documents in connection with the completion of the Work in accordance with the Contract Documents.

EJCDC C-520 Suggested Form of Agreement Between Owner and Contractor for Construction Contract (Stipulated Price)
Copyright©2002 National Society of Professional Engineers for EJCDC. All rights reserved.
00520-3

ARTICLE 4 - CONTRACT TIMES

4.01   Time of the Essence

A.   All time limits for Milestones, if any, Substantial Completion, and completion and readiness for final payment as stated in the Contract Documents are of the essence of the Contract.

4.02   Dates for Substantial Completion and Final Payment

A.   The Work will be substantially completed on or before _____, _____, and completed and ready for final payment in accordance with Paragraph 14.07 of the General Conditions on or before _____, _____.

[or]

4.02   Days to Achieve Substantial Completion and Final Payment

A.   The Work will be substantially completed within _____ days after the date when the Contract Times commence to run as provided in Paragraph 2.03 of the General Conditions, and completed and ready for final payment in accordance with Paragraph 14.07 of the General Conditions within _____ days after the date when the Contract Times commence to run.

4.03   Liquidated Damages

A.   Contractor and Owner recognize that time is of the essence of this Agreement and that Owner will suffer financial loss if the Work is not completed within the times specified in Paragraph 4.02 above, plus any extensions thereof allowed in accordance with Article 12 of the General Conditions. The parties also recognize the delays, expense, and difficulties involved in proving in a legal or arbitration proceeding the actual loss suffered by Owner if the Work is not completed on time. Accordingly, instead of requiring any such proof, Owner and Contractor agree that as liquidated damages for delay (but not as a penalty), Contractor shall pay Owner $_____ for each day that expires after the time specified in Paragraph 4.02 for Substantial Completion until the Work is substantially complete. After Substantial Completion, if Contractor shall neglect, refuse, or fail to complete the remaining Work within the Contract Time or any proper extension thereof granted by Owner, Contractor shall pay Owner $_____ for each day that expires after the time specified in Paragraph 4.02 for completion and readiness for final payment until the Work is completed and ready for final payment.

NOTES TO USER

1.   Where failure to reach a Milestone on time is of such consequence that the assessment of liquidated damages for failure to reach one or more Milestones on time is to be provided, appropriate amending or supplementing language should be inserted here.

ARTICLE 5 - CONTRACT PRICE

5.01   Owner shall pay Contractor for completion of the Work in accordance with the Contract Documents an amount in current funds equal to the sum of the amounts determined pursuant to Paragraphs 5.01.A, 5.01.B, and 5.01.C below:

A.   For all Work other than Unit Price Work, a Lump Sum of:

_____   ($_____)
(words)   (numerals)

All specific cash allowances are included in the above price and have been computed in accordance with paragraph 11.02 of the General Conditions.

B.   For all Unit Price Work, an amount equal to the sum of the established unit price for each separately identified item of Unit Price Work times the estimated quantity of that item as indicated in this paragraph 5.01.B:

As provided in Paragraph 11.03 of the General Conditions, estimated quantities are not guaranteed, and determinations of actual quantities and classifications are to be made by Engineer as provided in Paragraph 9.07 of the General Conditions. Unit prices have been computed as provided in Paragraph 11.03 of the General Conditions.

UNIT PRICE WORK

Item No.	Description	Unit	Estimated Quantity	Unit Price	Estimated

TOTAL OF ALL ESTIMATED PRICES _____ $_____
(words)   (numerals)

C.   For all Work, at the prices stated in Contractor's Bid, attached hereto as an exhibit.

NOTES TO USER

1.   *If adjustment prices for variations from stipulated Base Bid quantities have been agreed to, insert appropriate provisions. See BF-4.*

2.   *Depending upon the particular Bid Form used, use A-5.01.A alone, A-5.01.A and A-5.01.B together, A-5.01.B alone, or A-5.01.C alone, deleting those not used and renumbering accordingly. If A-5.01.C is used, Contractor's Bid is attached as an exhibit and listed in A-9.*

ARTICLE 6 - PAYMENT PROCEDURES

6.01   Submittal and Processing of Payments

A.   Contractor shall submit Applications for Payment in accordance with Article 14 of the General Conditions. Applications for Payment will be processed by Engineer as provided in the General Conditions.

6.02   Progress Payments; Retainage

A.   Owner shall make progress payments on account of the Contract Price on the basis of Contractor's Applications for Payment on or about the _____ day of each month during performance of the Work as provided in Paragraphs 6.02.A.1 and 6.02.A.2 below. All such payments will be measured by the schedule of values established as provided in Paragraph 2.07.A of the General Conditions (and in the case of Unit Price Work based on the number of units completed) or, in the event there is no schedule of values, as provided in the General Requirements:

1.   Prior to Substantial Completion, progress payments will be made in an amount equal to the percentage indicated below but, in each case, less the aggregate of payments previously made and less such amounts as Engineer may determine or Owner may withhold, including but not limited to liquidated damages, in accordance with Paragraph 14.02 of the General Conditions:

a.   _____ percent of Work completed (with the balance being retainage). If the Work has been 50 percent completed as determined by Engineer, and if the character and progress of the Work have been satisfactory to Owner and Engineer, Owner, on recommendation of Engineer, may determine that as long as the character and progress of the Work remain satisfactory to them, there will be no additional retainage; and

b.   _____ percent of cost of materials and equipment not incorporated in the Work (with the balance being retainage).

2.   Upon Substantial Completion, Owner shall pay an amount sufficient to increase total payments to Contractor to _____ percent of the Work completed, less such amounts as Engineer shall determine in accordance with Paragraph 14.02.B.5 of the General Conditions and less _____ percent of Engineer's estimate of the value of Work to be completed or corrected as shown on the tentative list of items to be completed or corrected attached to the certificate of Substantial Completion.

6.03   Final Payment

A.   Upon final completion and acceptance of the Work in accordance with Paragraph 14.07 of the General Conditions, Owner shall pay the remainder of the Contract Price as recommended by Engineer as provided in said Paragraph 14.07.

ARTICLE 7 - INTEREST

7.01   All moneys not paid when due as provided in Article 14 of the General Conditions shall bear interest at the rate of _____ percent per annum.

ARTICLE 8 – CONTRACTOR'S REPRESENTATIONS

8.01   In order to induce Owner to enter into this Agreement Contractor makes the following representations:

A.   Contractor has examined and carefully studied the Contract Documents and the other related data identified in the Bidding Documents.

B.   Contractor has visited the Site and become familiar with and is satisfied as to the general, local, and Site conditions that may affect cost, progress, and performance of the Work.

C.   Contractor is familiar with and is satisfied as to all federal, state, and local Laws and Regulations that may affect cost, progress, and performance of the Work.

D.   Contractor has carefully studied all: (1) reports of explorations and tests of subsurface conditions at or contiguous to the Site and all drawings of physical conditions in or relating to existing surface or subsurface structures at or contiguous to the Site (except Underground Facilities) which have been identified in the Supplementary Conditions as provided in Paragraph 4.02 of the General Conditions and (2) reports and drawings of a Hazardous Environmental Condition, if any, at the Site which has been identified in the Supplementary Conditions as provided in Paragraph 4.06 of the General Conditions.

NOTES TO USER
1.   *If the reports and/or drawings referred to in A-8.01.D do not exist, either modify A-8.01.D or delete A-8.01.D and renumber accordingly.*

EJCDC C-520 Suggested Form of Agreement Between Owner and Contractor for Construction Contract (Stipulated Price)
Copyright©2002 National Society of Professional Engineers for EJCDC. All rights reserved.
00520-6

E.   Contractor has obtained and carefully studied (or assumes responsibility for doing so) all additional or supplementary examinations, investigations, explorations, tests, studies, and data concerning conditions (surface, subsurface, and Underground Facilities) at or contiguous to the Site which may affect cost, progress, or performance of the Work or which relate to any aspect of the means, methods, techniques, sequences, and procedures of construction to be employed by Contractor, including any specific means, methods, techniques, sequences, and procedures of construction expressly required by the Bidding Documents, and safety precautions and programs incident thereto.

NOTES TO USER

1.   *If the reports and/or drawings referred to in A-8.01.D do not exist, delete the phrase "additional or supplementary" in the first sentence of A-8.01.E.*

F.   Contractor does not consider that any further examinations, investigations, explorations, tests, studies, or data are necessary for the performance of the Work at the Contract Price, within the Contract Times, and in accordance with the other terms and conditions of the Contract Documents.

G.   Contractor is aware of the general nature of work to be performed by Owner and others at the Site that relates to the Work as indicated in the Contract Documents.

H.   Contractor has correlated the information known to Contractor, information and observations obtained from visits to the Site, reports and drawings identified in the Contract Documents, and all additional examinations, investigations, explorations, tests, studies, and data with the Contract Documents.

I.   Contractor has given Engineer written notice of all conflicts, errors, ambiguities, or discrepancies that Contractor has discovered in the Contract Documents, and the written resolution thereof by Engineer is acceptable to Contractor.

J.   The Contract Documents are generally sufficient to indicate and convey understanding of all terms and conditions for performance and furnishing of the Work.

ARTICLE 9 - CONTRACT DOCUMENTS

9.01   Contents

A.   The Contract Documents consist of the following:

1.   This Agreement (pages 1 to _____, inclusive).

2.   Performance bond (pages _____ to _____, inclusive).

3.   Payment bond (pages _____ to _____, inclusive).

4.   Other bonds (pages _____ to _____, inclusive).

a.   _____ (pages _____ to _____, inclusive).

b.   _____ (pages _____ to _____, inclusive).

c.   _____ (pages _____ to _____, inclusive).

5.   General Conditions (pages _____ to _____, inclusive).

6.   Supplementary Conditions (pages _____ to _____, inclusive).

7.   Specifications as listed in the table of contents of the Project Manual.

8.   Drawings consisting of _____ sheets with each sheet bearing the following general title: _____ [or] the Drawings listed on attached sheet index.

9.   Addenda (numbers _____ to _____, inclusive).

10.   Exhibits to this Agreement (enumerated as follows):

   a.   Contractor's Bid (pages _____ to _____, inclusive).

   b.   Documentation submitted by Contractor prior to Notice of Award (pages _____ to _____, inclusive).

   c.   _____.

11.   The following which may be delivered or issued on or after the Effective Date of the Agreement and are not attached hereto:

   a.   Notice to Proceed (pages _____ to _____, inclusive).

   b.   Work Change Directives.

   c.   Change Order(s).

B.   The documents listed in Paragraph 9.01.A are attached to this Agreement (except as expressly noted otherwise above).

C.   There are no Contract Documents other than those listed above in this Article 9.

D.   The Contract Documents may only be amended, modified, or supplemented as provided in Paragraph 3.04 of the General Conditions.

ARTICLE 10 - MISCELLANEOUS

10.01 Terms

A.   Terms used in this Agreement will have the meanings stated in the General Conditions and the Supplementary Conditions.

10.02 Assignment of Contract

A.   No assignment by a party hereto of any rights under or interests in the Contract will be binding on another party hereto without the written consent of the party sought to be bound; and specifically but without limitation, moneys that may become due and moneys that are due may not be assigned without such consent (except to the extent that the effect of this restriction may be limited by law), and unless specifically stated to the contrary in any written consent to an assignment, no assignment will release or discharge the assignor from any duty or responsibility under the Contract Documents.

10.03 Successors and Assigns

A.   Owner and Contractor each binds itself, its partners, successors, assigns, and legal representatives to the other party hereto, its partners, successors, assigns, and legal representatives in respect to all covenants, agreements, and obligations contained in the Contract Documents.

10.04 Severability

A.   Any provision or part of the Contract Documents held to be void or unenforceable under any Law or Regulation shall be deemed stricken, and all remaining provisions shall continue to be valid and binding upon Owner and Contractor, who agree that the Contract Documents shall be reformed to replace such stricken provision or part thereof with a valid and enforceable provision that comes as close as possible to expressing the intention of the stricken provision.

10.05 Other Provisions

SAMPLE

NOTES TO USER

1. If Owner intends to assign a procurement contract (for goods and services) to the Contractor, see Notes to User at Article 23 of Suggested Instructions to Bidders for Procurement Contracts (EJCDC No. P-200, 2000 Edition) for provisions to be inserted in this Article.

2. Insert other provisions here if applicable.

IN WITNESS WHEREOF, Owner and Contractor have signed this Agreement in duplicate. One counterpart each has been delivered to Owner and Contractor. All portions of the Contract Documents have been signed or identified by Owner and Contractor or on their behalf.

NOTES TO USER

1. See I-21 and correlate procedures for format and signing between the two documents.

This Agreement will be effective on _____, _____ (which is the Effective Date of the Agreement).

OWNER: _____   CONTRACTOR: _____

By: _____   By: _____

Title: _____   Title: _____

[CORPORATE SEAL]   [CORPORATE SEAL]

Attest: _____   Attest: _____

Title: _____   Title: _____

Address for giving notices: Address for giving notices:

_____   _____

_____   _____

License No.: _____
(Where applicable)

(If Owner is a corporation, attach evidence of authority to sign. If Owner is a public body, attach evidence of authority to sign and resolution or other documents authorizing execution of Owner-Contractor Agreement.)

Agent for service or process: _____

(If Contractor is a corporation or a partnership, attach evidence of authority to sign.)

EJCDC C-520 Suggested Form of Agreement Between Owner and Contractor for Construction Contract (Stipulated Price)
Copyright©2002 National Society of Professional Engineers for EJCDC. All rights reserved.
00520-ii

Engineers Joint Documents Committee
Design and Construction Related Documents
Instructions and License Agreement

Instructions

Before you use any EJCDC document:

1. Read the License Agreement. You agree to it and are bound by its terms when you use the EJCDC document.

2. Make sure that you have the correct version for your word processing software.

How to Use:

1. While EJCDC has expended considerable effort to make the software translations exact, it can be that a few document controls (e.g., bold, underline) did not carry over.

2. Similarly, your software may change the font specification if the font is not available in your system. It will choose a font that is close in appearance. In this event, the pagination may not match the control set.

3. If you modify the document, you must follow the instructions in the License Agreement about notification.

4. Also note the instruction in the License Agreement about the EJCDC copyright.

License Agreement

You should carefully read the following terms and conditions before using this document. Commencement of use of this document indicates your acceptance of these terms and conditions. If you do not agree to them, you should promptly return the materials to the vendor, and your money will be refunded.

The Engineers Joint Contract Documents Committee ("EJCDC") provides **EJCDC Design and Construction Related Documents** and licenses their use worldwide. You assume sole responsibility for the selection of specific documents or portions thereof to achieve your intended results, and for the installation, use, and results obtained from **EJCDC Design and Construction Related Documents**.

You acknowledge that you understand that the text of the contract documents of **EJCDC Design and Construction Related Documents** has important legal consequences and that consultation with an attorney is recommended with respect to use or modification of the text. You further acknowledge that EJCDC documents are protected by the copyright laws of the United States.

License:

You have a limited nonexclusive license to:

1. Use **EJCDC Design and Construction Related Documents** on any number of machines owned, leased or rented by your company or organization.

2. Use **EJCDC Design and Construction Related Documents** in printed form for bona fide contract documents.

3. Copy **EJCDC Design and Construction Related Documents** into any machine readable or printed form for backup or modification purposes in support of your use of **EJCDC Design and Construction Related Documents**.

You agree that you will:

1. Reproduce and include EJCDC's copyright notice on any printed or machine-readable copy, modification, or portion merged into another document or program. All proprietary rights in **EJCDC Design and Construction Related Documents** are and shall remain the property of EJCDC.

2. Not represent that any of the contract documents you generate from **EJCDC Design and Construction Related Documents** are EJCDC documents unless (i) the document text is used without alteration or (ii) all additions and changes to, and deletions from, the text are clearly shown.

You may not use, copy, modify, or transfer EJCDC Design and Construction Related Documents, or any copy, modification or merged portion, in whole or in part, except as expressly provided for in this license. Reproduction of EJCDC Design and Construction Related Documents in printed or machine-readable format for resale or educational purposes is expressly prohibited.

If you transfer possession of any copy, modification or merged portion of EJCDC Design and Construction Related Documents to another party, your license is automatically terminated.

Term:

The license is effective until terminated. You may terminate it at any time by destroying **EJCDC Design and Construction Related Documents** altogether with all copies, modifications and merged portions in any form. It will also terminate upon conditions set forth elsewhere in this Agreement or if you fail to comply with any term or condition of this Agreement. You agree upon such termination to destroy **EJCDC Design and Construction Related Documents** along with all copies, modifications and merged portions in any form.

Limited Warranty:

EJCDC warrants the CDs and diskettes on which **EJCDC Design and Construction Related Documents** is furnished to be free from defects in materials and workmanship under normal use for a period of ninety (90) days from the date of

SAMPLE

delivery to you as evidenced by a copy of your receipt.

There is no other warranty of any kind, either expressed or implied, including, but not limited to the implied warranties of merchantability and fitness for a particular purpose. Some states do not allow the exclusion of implied warranties, so the above exclusion may not apply to you. This warranty gives you specific legal rights and you may also have other rights which vary from state to state.

EJCDC does not warrant that the functions contained in **EJCDC Design and Construction Related Documents** will meet your requirements or that the operation of **EJCDC Design and Construction Related Documents** will be uninterrupted or error free.

Limitations of Remedies:
EJCDC's entire liability and your exclusive remedy shall be:
1. the replacement of any document not meeting EJCDC's "Limited Warranty" which is returned to EJCDC's selling agent with a copy of your receipt, or

2. if EJCDC's selling agent is unable to deliver a replacement CD or diskette which is free of defects in materials and workmanship, you may terminate this Agreement by returning EJCDC Document and your money will be refunded.

In no event will EJCDC be liable to you for any damages, including any lost profits, lost savings or other incidental or consequential damages arising out of the use or inability to use **EJCDC Design and Construction Related Documents** even if EJCDC has been advised of the possibility of such damages, or for any claim by any other party.

Some states do not allow the limitation or exclusion of liability for incidental or consequential damages, so the above limitation or exclusion may not apply to you.

General:
You may not sublicense, assign, or transfer this license except as expressly provided in this Agreement. Any attempt otherwise to sublicense, assign, or transfer any of the rights, duties, or obligations hereunder is void.

This Agreement shall be governed by the laws of the State of Virginia. Should you have any questions concerning this Agreement, you may contact EJCDC by writing to:

Arthur Schwartz, Esq.
General Counsel
National Society of Professional Engineers
1420 King Street
Alexandria, VA 22314

Phone: (703) 684-2845
Fax: (703) 836-4875
e-mail: aschwartz@nspe.org

You acknowledge that you have read this agreement, understand it and agree to be bound by its terms and conditions. You further agree that it is the complete and exclusive statement of the agreement between us which supersedes any proposal or prior agreement, oral or written, and any other communications between us relating to the subject matter of this agreement.

Appendix D

Standard Form of Agreement Between Owner and Contractor on the Basis of Cost-Plus

This document has important legal consequences; consultation with an attorney is encouraged with respect to its use or modification.

SUGGESTED FORM OF AGREEMENT BETWEEN OWNER AND CONTRACTOR FOR CONSTRUCTION CONTRACT (COST-PLUS)

Prepared by

ENGINEERS JOINT CONTRACT DOCUMENTS COMMITTEE

and

Issued and Published Jointly By

PROFESSIONAL ENGINEERS IN PRIVATE PRACTICE
A practice division of the
NATIONAL SOCIETY OF PROFESSIONAL ENGINEERS

AMERICAN COUNCIL OF ENGINEERING COMPANIES

AMERICAN SOCIETY OF CIVIL ENGINEERS

This document has been approved and endorsed by

The Associated General Contractors of America

Knowledge for Creating
and Sustaining
the Built Environment

Construction Specifications Institute

EJCDC C-525 Suggested Form of Agreement Between Owner and Contractor for Construction Contract (Cost-Plus)
Copyright©2002 National Society of Professional Engineers for EJCDC. All rights reserved.

This Suggested Form of Agreement has been prepared for use with the Standard General Conditions of the Construction Contract (C-700, 2002 Edition). Their provisions are interrelated, and a change in one may necessitate a change in the other. The language contained in the Suggested Instructions to Bidders (C-200, 2002 Edition) is also carefully interrelated with the language of this Agreement. Their usage is discussed in the Commentary on EJCDC Construction Documents. See also Guide to the Preparation of Supplementary (C-800, 2002 Edition).

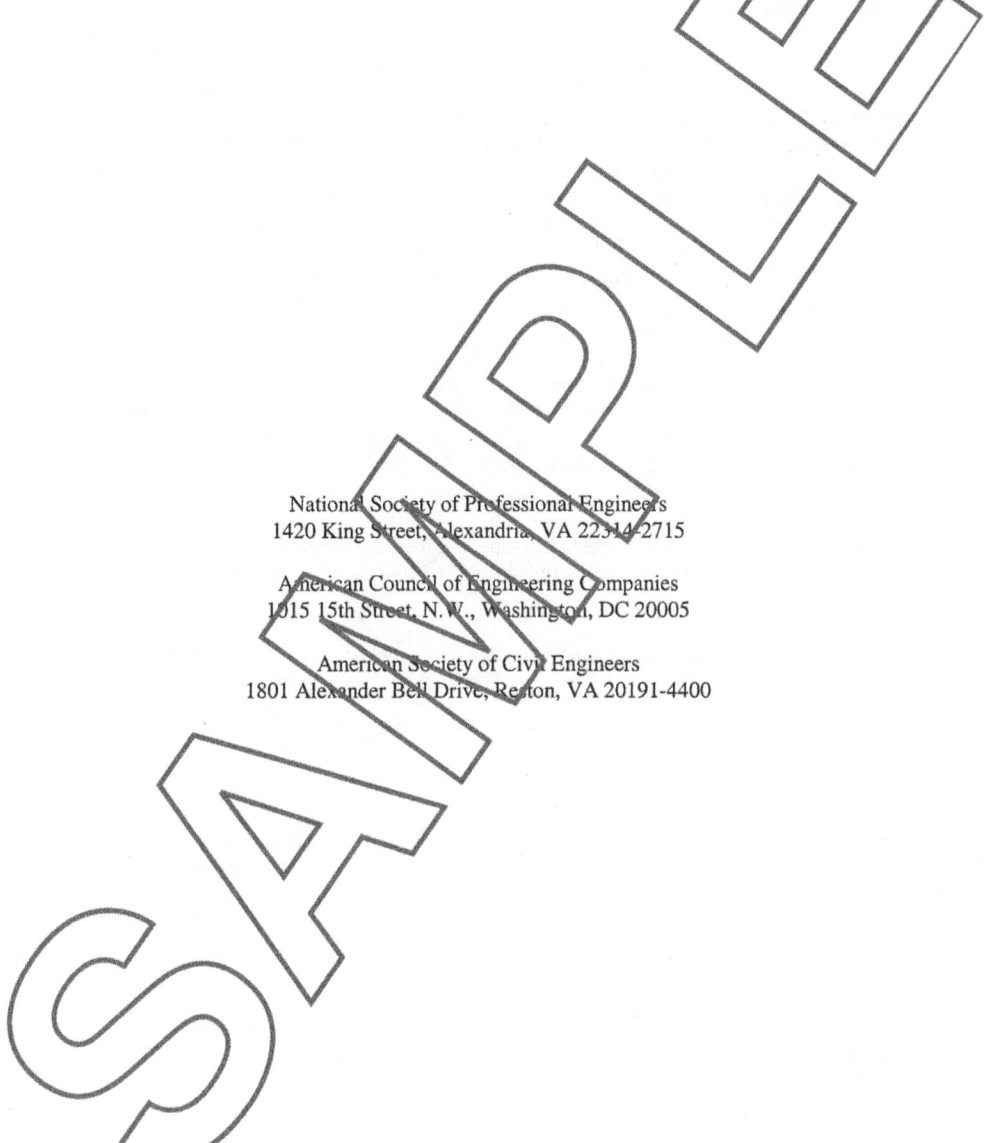

National Society of Professional Engineers
1420 King Street, Alexandria, VA 22314-2715

American Council of Engineering Companies
1015 15th Street, N.W., Washington, DC 20005

American Society of Civil Engineers
1801 Alexander Bell Drive, Reston, VA 20191-4400

EJCDC C-525 Suggested Form of Agreement Between Owner and Contractor for Construction Contract (Cost-Plus)
Copyright©2002 National Society of Professional Engineers for EJCDC. All rights reserved.

Introduction

This Suggested Form of Agreement between Owner and Contractor (Cost-Plus) ("Agreement") has been prepared for use with the Guide to the Preparation of Instructions to Bidders ("Instructions")(C-200, 2002 Edition) and with the Standard General Conditions of the Construction Contract ("General Conditions")(C-700, 2002 Edition). Their provisions are interrelated, and a change in one may necessitate a change in the others. For guidance in the preparation of Supplementary Conditions and coordination with Instructions to Bidders, see Guide to the Preparation of Supplementary Conditions ("Supplementary Conditions")(C-800, 2002 Edition). See also Guide to Preparation of the Bid Form ("Bid Form")(C-400, 2002 Edition). The EJCDC has not prepared a suggested form of Advertisement or Invitation to Bid because such documents will vary widely in response to statutory requirements.

This form and the other Bidding Documents prepared and issued by the EJCDC assume acceptance of the Project Manual concept of the Construction Specifications Institute which provides for an organizational format for location of all bound documentary information for a construction project, namely: Bidding Requirements (which term refers to the Advertisement or Invitation to Bid, the Instructions, and any Bid Form that may be suggested or prescribed, all of which provide information and guidance for all Bidders) and the Contract Documents (defined in Article 1 of the General Conditions), which include the Agreement, bonds and certificates, the General Conditions, the Supplementary Conditions, the Drawings, and the Specifications. The Bidding Requirements are not considered part of the Contract Documents because much of their substance pertains to the relationships prior to the award of the Contract and has little effect or impact thereafter and because many contracts are awarded without going through the bidding process. In some cases, however, the actual Bid may be attached as an exhibit to the Agreement to avoid extensive retyping. (The terms "Bidding Documents" and "Bidding Requirements" are defined in Article 1 of the General Conditions.) The Project Manual concept is explained in the Manual of Practice issued by the Construction Specifications Institute.

Suggested language is presented herein with "Notes to User" to assist in preparing the Agreement. Much of the language should be usable on most projects, but modifications and additional provisions will often be necessary. The suggested language has been coordinated with the other standard forms produced by the EJCDC. When modifying the suggested language or writing additional provisions, the user must check the other documents thoroughly for conflicts and coordination of language usage and make appropriate revisions in all affected documents.

Refer to the discussions in EJCDC's Recommended Competitive Bidding Procedures for Construction Projects ("Bidding Procedures") (No. 1910-9-D, 1987 Edition) (to be reissued in 2002) on the particular paragraphs of which frequent reference is made below.

For brevity, paragraphs of the Instructions to Bidders are referred to with the prefix "I," and those of this Agreement with the prefix "A."

NOTES:

1.   EJCDC publications may be ordered from:

NSPE headquarters
1420 King Street
Alexandria VA 22314-2715
703-684-2800
www.nspe.org

ASCE headquarters
1801 Alexander Bell Drive
Reston, VA 20191-4400
800-548-2723
www.asce.org

ACEC headquarters
1015 15th Street NW
Washington DC 20005
202-347-7474
www.acec.org

EJCDC
SUGGESTED FORM OF AGREEMENT
BETWEEN OWNER AND CONTRACTOR FOR CONSTRUCTION
CONTRACT (COST-PLUS)

THIS AGREEMENT is by and between _____ (Owner)

and _____ (Contractor).

Owner and Contractor, in consideration of the mutual covenants set forth herein, agree as follows:

ARTICLE 1 - WORK

1.01   Contractor shall complete all Work as specified or indicated in the Contract Documents. The Work is generally described as follows:

ARTICLE 2 - THE PROJECT

2.01   The Project for which the Work under the Contract Documents may be the whole or only a part is generally described as follows:

ARTICLE 3 - ENGINEER

3.01   The Project has been designed by

(Engineer), who is to act as Owner's representative, assume all duties and responsibilities, and have the rights and authority assigned to Engineer in the Contract Documents in connection with the completion of the Work in accordance with the Contract Documents.

EJCDC C-525 Suggested Form of Agreement Between Owner and Contractor for Construction Contract (Cost-Plus)
Copyright© 2002 National Society of Professional Engineers for EJCDC. All rights reserved.
00525-2

ARTICLE 4 - CONTRACT TIMES

4.01   Time of the Essence

   A.   All time limits for Milestones, if any, Substantial Completion, and completion and readiness for final payment as stated in the Contract Documents are of the essence of the Contract.

4.02   Dates for Substantial Completion and Final Payment

   A.   The Work will be substantially completed on or before _____, _____, and completed and ready for final payment in accordance with Paragraph 14.07 of the General Conditions on or before _____,_____.

   [or]

4.02   Days to Achieve Substantial Completion and Final Payment

   A.   The Work will be substantially completed within _____ days after the date when the Contract Times commence to run as provided in Paragraph 2.03 of the General Conditions, and completed and ready for final payment in accordance with Paragraph 14.07 of the General Conditions within _____ days after the date when the Contract Times commence to run.

4.03   Liquidated Damages

   A.   Contractor and Owner recognize that time is of the essence of this Agreement and that Owner will suffer financial loss if the Work is not completed within the times specified in Paragraph 4.02 above, plus any extensions thereof allowed in accordance with Article 12 of the General Conditions. The parties also recognize the delays, expense, and difficulties involved in proving in a legal or arbitration proceeding the actual loss suffered by Owner if the Work is not completed on time. Accordingly, instead of requiring any such proof, Owner and Contractor agree that as liquidated damages for delay (but not as a penalty), Contractor shall pay Owner $_____ for each day that expires after the time specified in Paragraph 4.02 for Substantial Completion until the Work is substantially complete. After Substantial Completion, if Contractor shall neglect, refuse, or fail to complete the remaining Work within the Contract Time or any proper extension thereof granted by Owner, Contractor shall pay Owner $_____ for each day that expires after the time specified in Paragraph 4.02 for completion and readiness for final payment until the Work is completed and ready for final payment.

NOTES TO USER

   1.   Where failure to reach a Milestone on time is of such consequence that the assessment of liquidated damages for failure to reach one or more Milestones on time is to be provided, appropriate amending or supplementing language should be inserted here.

ARTICLE 5 - CONTRACT PRICE

5.01   Owner shall pay Contractor for completion of the Work in accordance with the Contract Documents an amount in current funds equal to the sum of the amounts determined pursuant to Paragraphs 5.01.A, 5.01.B, and 5.01.C below:

   A.   For all Work other than Unit Price Work, the Cost of the Work plus a Contractor's fee for overhead and profit, both of which shall be determined as provided in Articles 6 and 7 below, subject to additions and deletions as provided in the Contract Documents and subject to the limitations set forth in Article 8 below.

   B.   For all Unit Price Work, an amount equal to the sum of the established unit price for each separately identified item of Unit Price Work times the estimated quantity of that item as indicated in this Paragraph 5.01.B:

UNIT PRICE WORK

Item No.	Description	Unit	Estimated Quantity	Unit Price	Estimated

TOTAL OF ALL ESTIMATED PRICES _____ $_____

(words)   (numerals)

        As provided in Paragraph 11.03 of the General Conditions, estimated quantities are not guaranteed, and determinations of actual quantities and classifications are to be made by Engineer as provided in Paragraph 9.07 of the General Conditions. Unit prices have been computed as provided in Paragraph 11.03 of the General Conditions.

   C.   For all Work, at the prices stated in Contractor's Bid, attached hereto as an exhibit.

NOTES TO USER
   1.   Depending upon the particular project bid form used, use A-5.01.A alone, A-5.01.A and A-5.01.B together, A-5.01.B alone, or A-5.01.C alone, deleting those not used and renumbering accordingly. If A-5.01.C is used, Contractor's Bid is attached as an exhibit and listed in A-14.

ARTICLE 6 - COST OF THE WORK

6.01   Cost of the Work shall be determined as provided in Paragraph 11.01 of the General Conditions.

EJCDC C-525 Suggested Form of Agreement Between Owner and Contractor for Construction Contract (Cost-Plus)
Copyright ©2002 National Society of Professional Engineers for EJCDC. All rights reserved.
00525-4

ARTICLE 7 – CONTRACTOR'S FEE

7.01   Contractor's fee shall be determined as follows:

A.   A fee based on the following percentages of the various portions of the Cost of the Work:

   1.   Payroll costs (see Paragraph 11.01.A.1 of General Conditions) _____.

   2.   Material and equipment costs (see Paragraph 11.01.A.2 of General Conditions) _____.

   3.   Amounts paid to subcontractors (see Paragraph 11.01.A.3 of General Conditions) _____.

   4.   Amounts paid to special consultants (see Paragraph 11.01.A.4 of General Conditions) _____.

   5.   Supplemental costs (see Paragraph 11.01.A.5 of General Conditions) _____.

   6.   No fee will be payable on the basis of costs itemized in Paragraph 11.01.B of the General Conditions.

   7.   The provisions in Paragraph 11.01.C of the General Conditions will apply only to changes in the Work.

B.   Contractor guarantees that the maximum amount payable by Owner in accordance with Paragraph 7.01.A as a percentage fee will not exceed $_____, subject to increases or decreases for changes in the Work as provided in Paragraph 9.01.B below.

   [or]

C.   A fixed fee of $_____, which shall be subject to increases or decreases for changes in the Work as provided in Paragraph 9.01.A below.

NOTES TO USER

Depending on the fee agreement selected or negotiated, select 7.01.A (percentage fee), or 7.01.A and 7.01.B (percentage fee up to guaranteed maximum), or 7.01.C (fixed fee).

ARTICLE 8 - GUARANTEED MAXIMUM PRICE

8.01   Contractor guarantees that the maximum amount payable (Guaranteed Maximum Price) by Owner for the sum of the Cost of the Work plus Contractor's fee under Article 7 will not exceed $_____, subject to increases or decreases for changes in the Work. The Guaranteed Maximum Price will not apply to Unit Price Work.

ARTICLE 9 - CHANGES IN THE CONTRACT PRICE

9.01   The amount of any increases or decreases in Contractor's fee, in any Guaranteed Maximum Price, or in any guaranteed maximum fee which results from a Change Order shall be set forth in the applicable Change Order subject to the following:

A.   If Contractor's fee is a fixed fee, any increase or decrease in the Contractor's fee resulting from net additions or decreases in the Cost of the Work shall be determined in accordance with Paragraph 12.01.C of the General Conditions.

   [or]

A.   If Contractor's fee is a percentage fee not subject to any guaranteed maximum limitation, Contractor's fee will adjust automatically as the Cost of the Work changes.

NOTES TO USER

Select one of the two provisions above for Paragraph 9.01.A.

B.   Wherever there is a Guaranteed Maximum Price or Fee:

   1.   In the case of net additions in the Work, the amounts of any increase in either guaranteed maximum shall be determined in accordance with Paragraphs 11.01 through 11.02, inclusive, of the General Conditions.

   2.   In the case of net deletions in the Work, the amount of any such decrease shall be determined in accordance with Paragraph 11.02.C of the General Conditions, and any Guaranteed Maximum (Price or Fee) shall be reduced by mutual agreement.

ARTICLE 10 - PAYMENT PROCEDURES

10.01   Submittal and Processing of Payments

A.   Contractor shall submit Applications for Payment in accordance with Article 14 of the General Conditions. Applications for Payment will indicate the amount of Contractor's fee then payable. Applications for Payment will be processed by Engineer as provided in the General Conditions.

10.02   Progress Payments; Retainage

A.   Owner shall make progress payments on account of the Contract Price on the basis of Contractor's Applications for Payment as recommended by Engineer on or about the _____ day of each month during construction as provided in Paragraphs 10.02.A.1 and 10.02.A.2 below. All such payments will be measured by the schedule of values established as provided in Paragraph 2.07.A of the General Conditions (and in the case of Unit Price Work based on the number of units completed) or, in the event there is no schedule of values, as provided in the General Requirements:

   1.   For Cost of Work: Progress payments on account of the Cost of the Work will be made:

    a.   Prior to Substantial Completion, progress payments will be made in an amount equal to the percentage indicated below but, in each case, less the aggregate of payments previously made and less such amounts as Engineer may determine or Owner may withhold, including but not limited to liquidated damages, in accordance with Paragraph 14.02 of the General Conditions:

    (1)   _____ percent Cost of Work completed (with the balance being retainage). If the Work has been 50 percent completed as determined by Engineer, and if the character and progress of the Work have been satisfactory to Owner and Engineer, Owner, on recommendation of Engineer, may determine that as long as the character and progress of the Work remain satisfactory to them, there will be no retainage; and

    (2)   _____ percent of cost of materials and equipment not incorporated in the Work (with the balance being retainage).

    b.   Upon Substantial Completion, Owner shall pay an amount sufficient to increase total payments to Contractor to _____ percent of the Work completed, less such amounts as Engineer shall determine in accordance with Paragraph 14.02.B.5 of the General Conditions and less _____ percent of Engineer's estimate of the value of Work to be completed or corrected as shown on the tentative list of items to be completed or corrected attached to the certificate of Substantial Completion.

2. For Contractor's fee: Progress payments on account of the Contractor's fee will be made:

a. If Contractor's fee is a fixed fee, payments prior to Substantial Completion will be in an amount equal to _____ percent of such fee earned to the date of the approved Application for Payment (less in each case payments previously made on account of such fee) based on the progress of the Work measured by the schedule of values established as provided in Paragraph 2.07.B of the General Conditions (and in the case of Unit Price Work on the number of units completed), and upon Substantial Completion in an amount sufficient to increase total payments to Contractor on account of his fee to _____ percent of Contractor's fee. In the event there is no schedule of values the progress of the Work will be measured as provided in the General Requirements.

b. If Contractor's fee is a percentage fee, payments prior to Substantial Completion will be in an amount equal to _____ percent of such fee (less in each case payments previously made on account of such fee) based on the Cost of the Work completed, and upon Substantial Completion in an amount sufficient to increase total payments to Contractor on account of that fee to _____ percent of Contractor's fee.

10.03 Final Payment

A. Upon final completion and acceptance of the Work in accordance with Paragraph 14.07 of the General Conditions, Owner shall pay the remainder of the Contract Price as recommended by Engineer as provided in said Paragraph 14.07.

ARTICLE 11 - INTEREST

11.01 All moneys not paid when due as provided in Article 14 of the General Conditions shall bear interest at the rate of _____ percent per annum.

ARTICLE 12 – CONTRACTOR'S REPRESENTATIONS

12.01 In order to induce Owner to enter into this Agreement Contractor makes the following representations:

A. Contractor has examined and carefully studied the Contract Documents and the other related data identified in the Bidding Documents.

B. Contractor has visited the Site and become familiar with and is satisfied as to the general, local, and Site conditions that may affect cost, progress, and performance of the Work.

C. Contractor is familiar with and is satisfied as to all federal, state, and local Laws and Regulations that may affect cost, progress, and performance of the Work.

D. Contractor has carefully studied all: (1) reports of explorations and tests of subsurface conditions at or contiguous to the Site and all drawings of physical conditions in or relating to existing surface or subsurface structures at or contiguous to the Site (except Underground Facilities) which have been identified in the Supplementary Conditions as provided in Paragraph 4.02 of the General Conditions and (2) reports and drawings of a Hazardous Environmental Condition, if any, at the Site which has been identified in the Supplementary Conditions as provided in Paragraph 4.06 of the General Conditions.

NOTE TO USER
1. *If the reports and/or drawings referred to in A-12.01.D do not exist, either modify A-12.01.D or delete A-12.01.D and renumber accordingly.*

E. Contractor has obtained and carefully studied (or assumes responsibility for having done so) all additional or supplementary examinations, investigations, explorations, tests, studies, and data concerning conditions (surface, subsurface, and Underground Facilities) at or contiguous to the Site which may affect cost, progress, or performance of the Work or which relate to any aspect of the means, methods, techniques, sequences, and procedures of construction to be employed by Contractor, including applying the specific means, methods, techniques, sequences, and procedures of construction, if any, expressly required by the Contract Documents to be employed by Contractor, and safety precautions and programs incident thereto.

SAMPLE

NOTES TO USER

1. If the reports and/or drawings referred to in A-12.01.D do not exist, delete the phrase "additional or supplementary" in the first sentence of Paragraph A-12.01.E.

F.   Contractor does not consider that any further examinations, investigations, explorations, tests, studies, or data are necessary for the performance of the Work at the Contract Price, within the Contract Times, and in accordance with the other terms and conditions of the Contract Documents.

G.   Contractor is aware of the general nature of work to be performed by Owner and others at the Site that relates to the Work as indicated in the Contract Documents.

H.   Contractor has correlated the information known to Contractor, information and observations obtained from visits to the Site, reports and drawings identified in the Contract Documents, and all additional examinations, investigations, explorations, tests, studies, and data with the Contract Documents.

I.   Contractor has given Engineer written notice of all conflicts, errors, ambiguities, or discrepancies that Contractor has discovered in the Contract Documents, and the written resolution thereof by Engineer is acceptable to Contractor.

J.   The Contract Documents are generally sufficient to indicate and convey understanding of all terms and conditions for performance and furnishing of the Work.

ARTICLE 13 - ACCOUNTING RECORDS

13.01 Contractor shall check all materials, equipment, and labor entering into the Work and shall keep such full and detailed accounts as may be necessary for proper financial management under this Agreement, and the accounting methods shall be satisfactory to Owner. Owner shall be afforded access to all Contractor's records, books, correspondence, instructions, drawings, receipts, vouchers, memoranda, and similar data relating to the Cost of the Work and Contractor's fee. Contractor shall preserve all such documents for a period of three years after the final payment by Owner.

ARTICLE 14 - CONTRACT DOCUMENTS

14.01 Contents

A.   The Contract Documents consist of the following:

1.   This Agreement (pages 1 to _____, inclusive).

2.   Performance bond (pages _____ to _____, inclusive).

3.   Payment bond (pages _____ to _____, inclusive).

4.   Other bonds (pages _____ to _____, inclusive).

a.   _____ (pages _____ to _____, inclusive).

b.   _____ (pages _____ to _____, inclusive).

c.   _____ (pages _____ to _____, inclusive).

5.   General Conditions (pages _____ to _____, inclusive).

6.   Supplementary Conditions (pages _____ to _____, inclusive).

7.   Specifications as listed in the table of contents of the Project Manual.

8.   Drawings consisting of a cover sheet and _____ sheets with each sheet bearing the following general title: _____ [or] the Drawings listed on attached sheet index.

9.   Addenda (numbers _____ to _____, inclusive).

10. Exhibits to this Agreement (enumerated as follows):

  a.   Contractor's Bid (pages _____ to _____, inclusive).

  b.   Documentation submitted by Contractor prior to Notice of Award (pages _____ to _____, inclusive).

  c.   _____

11. The following which may be delivered or issued on or after the Effective Date of the Agreement and are not attached hereto:

  a.   Notice to Proceed (pages _____ to _____, inclusive).

  b.   Work Change Directives.

  c.   Change Order(s).

B.   The documents listed in Paragraph 14.01.A are attached to this Agreement (except as expressly noted otherwise above).

C.   There are no Contract Documents other than those listed above in this Article 14.

D.   The Contract Documents may only be amended, modified, or supplemented as provided in Paragraph 3.04 of the General Conditions.

ARTICLE 15 - MISCELLANEOUS

15.01 Terms

A.   Terms used in this Agreement will have the meanings stated in the General Conditions and the Supplementary Conditions.

15.02 Assignment of Contract

A.   No assignment by a party hereto of any rights under or interests in the Contract will be binding on another party hereto without the written consent of the party sought to be bound; and, specifically but without limitation, moneys that may become due and moneys that are due may not be assigned without such consent (except to the extent that the effect of this restriction may be limited by law), and unless specifically stated to the contrary in any written consent to an assignment, no assignment will release or discharge the assignor from any duty or responsibility under the Contract Documents.

EJCDC C-525 Suggested Form of Agreement Between Owner and Contractor for Construction Contract (Cost-Plus)
Copyright © 2002 National Society of Professional Engineers for EJCDC. All rights reserved.
00525-9

15.03 Successors and Assigns

A.   Owner and Contractor each binds itself, its partners, successors, assigns, and legal representatives to the other party hereto, its partners, successors, assigns, and legal representatives in respect to all covenants, agreements, and obligations contained in the Contract Documents.

15.04 Severability

A.   Any provision or part of the Contract Documents held to be void or unenforceable under any Law or Regulation shall be deemed stricken, and all remaining provisions shall continue to be valid and binding upon Owner and Contractor, who agree that the Contract Documents shall be reformed to replace such stricken provision or part thereof with a valid and enforceable provision that comes as close as possible to expressing the intention of the stricken provision.

15.05 Other Provisions

NOTES TO USER
1. *If Owner intends to assign a procurement contract (for goods and services) to Contractor, see Note to User at Article 23 of Suggested Instructions to Bidders for Procurement Contracts (EJCDC No. P-200, 2000 Edition) for provisions to be inserted in this Article.*
2. *Insert other provisions here if applicable.*

EJCDC C-525 Suggested Form of Agreement Between Owner and Contractor for Construction Contract (Cost-Plus)
Copyright©2002 National Society of Professional Engineers for EJCDC. All rights reserved.
00525-10

IN WITNESS WHEREOF, Owner and Contractor have signed this Agreement in duplicate. One counterpart each has been delivered to Owner and Contractor. All portions of the Contract Documents have been signed or identified by Owner and Contractor or on their behalf.

NOTES TO USER
1. *See I-21 and correlate procedures for format and signing between the two documents.*

This Agreement will be effective on _____, _____ (which is the Effective Date of the Agreement).

OWNER: CONTRACTOR:

                                                _____

By:_____   By: _____

Title: _____   Title: _____

Address for giving notices: Address for giving notices:

_____   _____

_____   _____

_____   _____

(If Owner is a corporation, attach evidence of authority to sign. If Owner is a public body, attach evidence of authority to sign and resolution or other documents authorizing execution of Owner-Contractor Agreement.)

License No.: _____
              (Where applicable)

Agent for service of process: _____

(If Contractor is a corporation or a partnership, attach evidence of authority to sign)

EJCDC C-525 Suggested Form of Agreement Between Owner and Contractor for Construction Contract (Cost-Plus)
Copyright © 2002 National Society of Professional Engineers for EJCDC. All rights reserved.
00525-11

Appendix E

Arrow Notation Scheduling Calculations

E.1 CPM CALCULATIONS (ARROW NOTATION)

In making calculations with arrow notation, the arrow and its two associated nodes have attributes that are formally defined as symbols for mathematical purposes. This formal notation associated with the arrow is shown in Figure E.1.

The left-hand node on the arrow represents the event time at which the activity begins. It is referred to as the i node. The right-hand node represents the end time of the activity. It is referred to as the j node. Associated with each node is an earliest time, which is shown as T^E_i for the i node and T^E_j for the j node. Similarly, each node can have a latest event time, which is shown in the figure as T^L_i for the i node and T^L_j for the j node. This establishes four events, two associated with starting and two with ending nodes, which are of interest in calculating the critical path of the network. The duration of the activity (as shown in the figure) is given as t_{ij}. Because the starting and ending nodes in arrow notation are referred to as i and j, arrow notation is sometimes referred to as i-j notation.

A schematic diagram representing the application of the forward-pass algorithm using arrow notation is shown in Figure E.2. The objective of the forward-pass algorithm is to calculate the earliest point in time at which a given event can occur. That is, the algorithm calculates the earliest event time of a given node. The earliest event time for a given node is controlled by the earliest event times of each of the set of events that precede it. The algorithm is given as follows:

$$T^E_j = \max_{\substack{\text{All i} \\ i \in M}} [T^E_i + t_{ij}]$$

where M is the set of all i events that immediately precede j.

The earliest event time for a given node j is controlled by the earliest event times of each of the i nodes that precede it. Each i node plus the duration of the associated activity, t_{ij} which links it to the j node, must be investigated. The maximum of the preceding i node early event times plus the durations of the appropriate activity ij controls the earliest time at which a given event j, can occur.

To demonstrate this, consider Figure E.2. Node 30 is preceded by nodes 22, 25, and 26. The durations of the activities emanating from each of these nodes are as follows:

$$
\begin{array}{lll}
\text{Act } 22, 30 & t_{22,30} = & 7 \text{ days} \\
\text{Act } 25, 30 & t_{25,30} = & 2 \text{ days} \\
\text{Act } 26, 30 & t_{26,30} = & 6 \text{ days}
\end{array}
$$

Figure E.1 Arrow notation symbols.

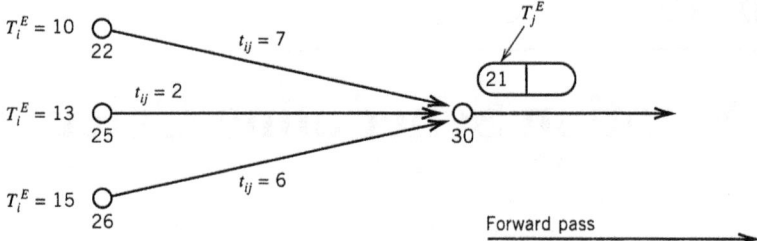

Figure E.2 Schematic of forward-pass calculation.

The earliest event times for each of the preceding nodes are as follows:

$$T^E_{22} = 10$$
$$T^E_{25} = 13$$
$$T^E_{26} = 15$$

The algorithm for the early event time for node 30 is then

$$T^E_{30} = \max(T^E_{22} + t_{22,30},\ T^E_{25} + t_{25,30},\ T^E_{26} + t_{26,30})$$

$$T^E_{30} = \max(10 + 7,\ 13 + 2,\ 15 + 6) = 21$$

E.2 CALCULATING THE EARLY EVENT TIMES (ARROW NOTATION)

In order to understand how the forward-pass algorithm is applied, consider the arrow notation network model of the small gas station in Figure E.3. In order to record the calculated values of the early event time (T^E_i), a partitioned oval is located above each node. The calculated early event time is recorded in the left side of the oval. During the backward pass, the late event times for each node will be recorded in the right side.

The forward-pass algorithm is applied repetitively starting with the source node (node A) and moving from left to right in a "bootstrapping" fashion. The starting node A is given an early event time of zero (0). Moving to node B, the set of preceding events consists of only one event. Therefore, T^E_B is max ($T^E_A + t_{AB}$) = max (0 + 10) = 10. Calculations for all of the nodes are shown in Table E.1. The values for each node are shown in Figure E.3.

The earliest time at which each activity can begin is given by the T^E_i value for the i node associated with the activity of interest. In addition to this information, it is now clear that the minimum duration of the project is 96 days since the earliest time at which node S can be realized has been calculated as 96 time units.

E.3 BACKWARD-PASS ALGORITHM (ARROW NOTATION)

A schematic diagram representing the application of the backward-pass algorithm is shown in Figure E.4. The backward-pass algorithm calculates the latest time at which each event can occur. The latest event time for a node i is controlled by the latest event times of the set of events that follow it. The late event time of each j node minus the duration of the associated activity, ij, must be investigated. The minimum of the following j node late event times minus the duration of activity ij controls the latest time at which the i event can occur. To demonstrate this, consider Figure E.4.

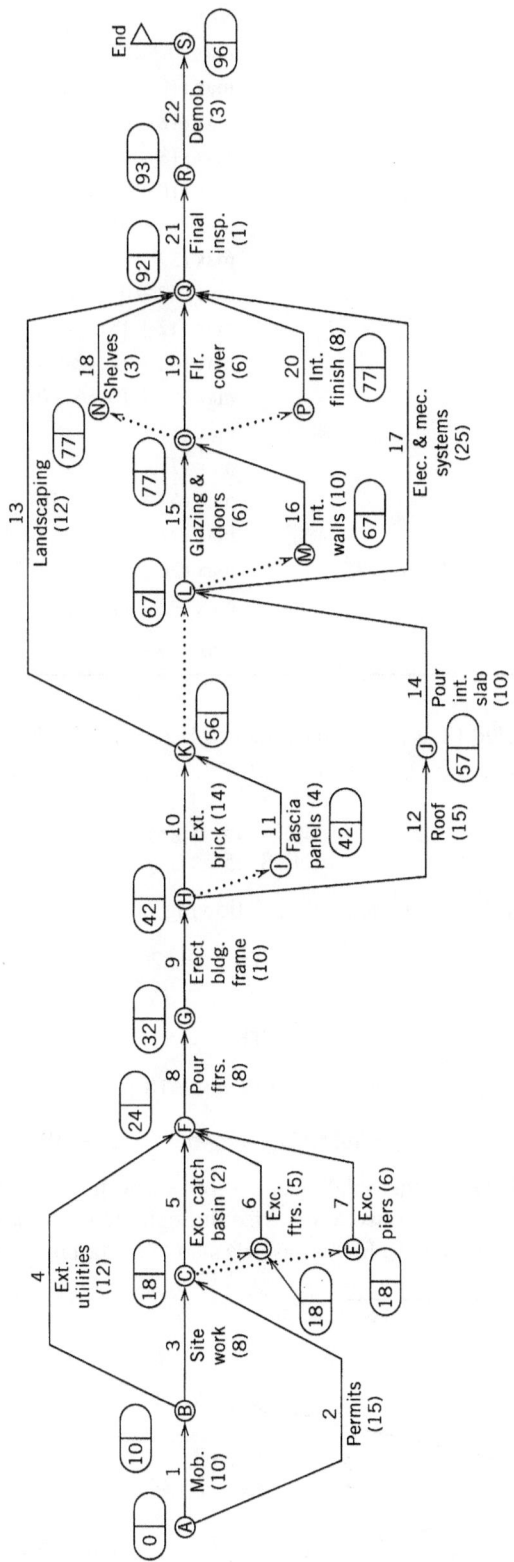

Note: Activity durations are shown in parentheses, e.g., (10).

Figure E.3 Expanded Project Model with Early Event Times. (The dotted lines represent so-called "dummy" logical connections and have a duration of zero).

Table E.1 Calculation of Early Event Times

Node	Formula	Numerical Value	T_i^E
A	N/A	N/A	0
B	$T_B^E = \max\left(T_A^E + t_{AB}\right)$	max (0 + 10)	10
C	$T_C^E = \max\left(T_B^E + t_{BC}, T_A^E + t_{AC}\right)$	max (10 + 8, 0 + 15)	18
D	$T_D^E = \max\left(T_C^E + t_{CD}\right)$	max (18 + 0)	18
E	$T_E^E = \max\left(T_C^E + t_{CE}\right)$	max (18 + 0)	18
F	$T_F^E = \max\left(T_B^E + t_{BF}, T_C^E + t_{CF}, T_D^E + t_{DF}, T_E^E + t_{EF}\right)$	max (10 + 12, 18 + 2, 18 + 5, 18 + 6)	24
G	$T_G^E = \max\left(T_F^E + t_{FG}\right)$	max (24 + 8)	32
H	$T_H^E = \max\left(T_G^E + t_{GH}\right)$	max (32 + 10)	42
I	$T_I^E = \max\left(T_H^E + t_{HI}\right)$	max (42 + 0)	42
J	$T_J^E = \max\left(T_H^E + t_{HJ}\right)$	max (42 + 15)	57
K	$T_K^E = \max\left(T_H^E + t_{HK}, T_I^E + t_{IK}\right)$	max (42 + 14, 42 + 4)	56
L	$T_L^E = \max\left(T_J^E + t_{JL}, T_K^E + t_{KL}\right)$	max (57 + 10, 56 + 0)	67
M	$T_M^E = \max\left(T_L^E + t_{LM}\right)$	max (67 + 0)	67
N	$T_N^E = \max\left(T_O^E + t_{ON}\right)$	max (77 + 0)	77
O	$T_O^E = \max\left(T_L^E + t_{LO}, T_M^E + t_{MO}\right)$	max (67 + 6, 67 + 10)	77
P	$T_P^E = \max\left(T_O^E + t_{OP}\right)$	max (77 + 0)	77
Q	$T_Q^E = \max\left(T_K^E + t_{KQ}, T_N^E + t_{NQ}, T_O^E + t_{OQ}, T_P^E + t_{PQ}, T_L^E + t_{LO}\right)$	max (56 + 12, 77 + 3, 77 + 6, 77 + 8, 67 + 25)	92
R	$T_R^E = \max\left(T_Q^E + t_{QR}\right)$	max (92 + 1)	93
S	$T_S^E = \max\left(T_R^E + t_S\right)$	max (93 + 3)	96

Node i, labeled 18, is followed by nodes 21, 23, and 25. The durations of the associated ij activities are:

$$\text{Act } 18, 21 \quad t_{18,21} = 12$$
$$\text{Act } 18, 23 \quad t_{18,23} = 3$$
$$\text{Act } 18, 25 \quad t_{18,25} = 10$$

The latest event times for each of the following nodes are as follows:

$$T_{21}^L = 52$$
$$T_{23}^L = 46$$
$$T_{25}^L = 53$$

The expression for the late event time of node 18 is:

$$T_{18}^L = \min(52 - 12, 46 - 3, 53 - 10) = 40$$

Again considering the small gas station arrow notation network, the same bootstrapping approach used to work stepwise through the network is used to determine late event times (see Figure E.5). In this case we start at the last node (right side) of the network

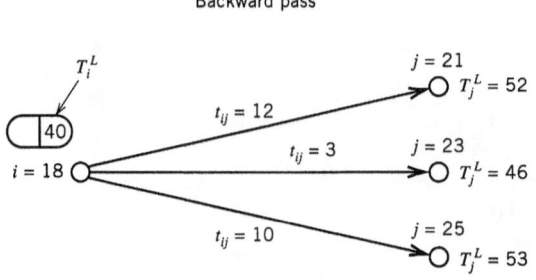

Figure E.4 Schematic of backward-pass algorithm.

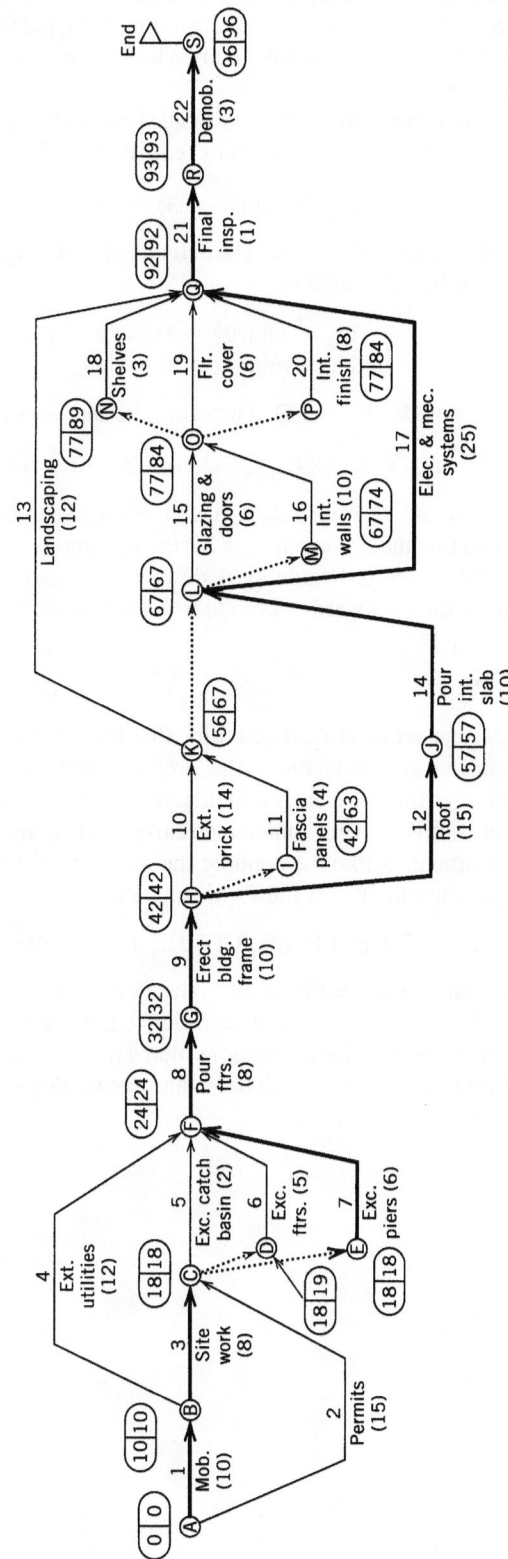

Note: Activity durations are shown in parentheses, e.g., (10).

Figure E.5 Expanded Project Arrow Notation Model with Early and Late Event Times.

and step from right to left. In order to commence calculations, a late event time is needed for node S. The late event time of S will be set to the early event time (i.e., 96). Since we wish to complete the project in the minimum period of time, it is logical to set the early and late event times of S to 96. If we were to use a greater number, the finish date for the project would be extended. Since 96 time units (e.g., days) is the minimum duration of the project, a lesser value is not feasible.

Walking the network from right to left, we would start with node R. Node R is followed only by node S. The backward-pass algorithm reduces to:

$$T^L_R = \min(96 - 3) = 93$$

Similarly, the late event time of Node Q can be calculated as 92. Again, the late event times for nodes N and P can be calculated as:

$$T^L_R = \min(92 - 3) = 89$$
$$T^L_P = \min(92 - 8) = 84$$

Node O is followed by nodes N, and Q. Therefore, its late event time is calculated as:

$$T^L_O = \min(89 - 0, 84 - 0, 92 - 6) = 84$$

All of the late event times are shown in the right side of the ovals above each node in Figure E.5. It should be noted that the late event time for the beginning node A is 0. The early and late event times of the source or beginning node must be equal (i.e., zero). Otherwise, a mistake in calculating the backward-pass values has occurred.

E.4   IDENTIFYING THE CRITICAL PATH

As previously stated, the set of critical activities that form the critical path(s) cannot be delayed without causing an extension of the project duration. Therefore, they can be identified as the activities that have early and late event times (associated with their i and j nodes) that are equal. Activities with i and j node early and late times that are not equal can be delayed a certain amount without extending the duration of the project. By looking at Figure E.5 it can be determined that the following activities are critical.

AB, BC, CE, EF, FG, GH, HJ, JL, LQ, QR, RS

This is both the critical and longest path through the network. The duration of this path must be equal to the minimum project duration calculated using the forward-pass algorithm. All other paths will have durations that are less than the minimum project duration. Check to see that the critical path has a duration of 96 and that all other paths have a duration less than 96 time units.

Appendix F

AGC Builders Association of Chicago: Typical Agent Job Descriptions

PROJECT MANAGER

A. General Functions

The project manager in the construction industry is usually an "inside" and "outside" man. The position may vary considerably from company to company. The project manager in some companies may be an estimator, and expeditor, and even handle some duties normally done by the job superintendent, while with other companies he may merely supervise superintendents.

B. Detailed Functions

1. May procure the invitation to bid on jobs.
2. May, when working as an estimator, prepare bids.
3. May handle the legal requirements for a contract.
4. May negotiate the specialty contractor's arrangements and agreements.
5. Set up completion schedules by bar graph or critical path method.
6. Supervise subcontractors and coordinate their material deliveries.
7. Arrange for sufficient manpower for the project.
8. Supervise superintendents on the job—"walk the job" each day to see progress being made and, during this time, review the work with a superintendent.
9. Control the movement of workers from one job to another.
10. Arrange for permits from the city, county, and so forth.
11. Hire and fire superintendents, foremen, engineers, and other personnel under his supervision.
12. Set up occupancy dates for buildings.
13. Act as public relations representative.
14. Coordinate with architect and owner requested revisions or errors found in drawings.

ESTIMATOR

A. General Functions

An estimator makes as close an estimate as possible of what the costs will be. In order to do so, he must itemize all of the building materials and calculate labor costs for the entire project—the cost estimate may also include a percentage for profit, though this may be done with or by top management.

B. Detailed Functions

1. Mail or telephone bid proposals to subcontractors.
2. Follow up with subcontractors on submission of their bids.
3. Review bid with subcontractors.
4. Prior to bid, inspect job site to determine access and that the land is the same as on the plans. Look for water conditions and other problems that might arise.
5. Analyze plans and specifications, that is, "learn the job."
6. Make a takeoff for each type of work to be done by general contractor forces.
7. Does takeoff for subcontractors when necessary.
8. May sit in on owner, architect, and contractor conferences.
9. May check on other estimator's work or have his work checked by another estimator.
10. Price the quantity takeoffs.
11. Read prints, noting discrepancies.
12. Make itemized lists of prices for materials.
13. Review and preview subcontractors' bids.
14. In some companies, purchase steel, lumber, and all other materials necessary for the job.
15. Compute a percentage for overhead and profit, which would be added to estimated cost.
16. Arrive at final bid price or cost price.
17. Prepare change order (estimates cost of changes) as needed or required—if major item and not handled by field personnel.
18. Expedite distribution of plans, including general and mechanical.
19. Serve in quality control capacity, because of position in purchasing, and plan review.
20. Make bar graphs, network, or CPM for scheduling.
21. Make cost breakdown of work performed by company forces for cost control purposes.

EXPEDITOR

A. General Functions

An expeditor may schedule or coordinate job material requirements. He serves as a troubleshooter when there is a breakdown in delivery schedule. He foresees problems by reviewing plans and specifications of the subcontractors and coordinating these with the plans and specifications of the architect.

B. Detailed Functions

1. In some companies receives the plans and specifications and breaks the specifications down by trade.

2. In some companies writes to all subcontractors advising them what is necessary to do on their plans.

3. Follows up on drawings, that is, shop drawings or the detailed drawings of project.

4. Submits drawings to the architect after having checked them to see if they match, that the job is correct, that the materials used are those specified, and analyzes the drawings.

5. Maintains constant follow-up on plans and drawings to ensure documents reach the proper place at the proper time.

6. Distributes approved plans to subs or to anyone else who should get them. Has to order enough plans from subcontractors so that entrusted parties will have sufficient documentation for project schedule.

7. In some companies establishes delivery time for materials, equipment, or labor, based on when they will be required and when they can be acquired, and determines the lead time required for acquisition.

8. May follow a CPM printout, make out delivery schedules, use a bar graph method or the critical path method. Makes sure the shop items are on the critical path method or the bar graph.

9. In some companies does small buying such as purchasing mailboxes, signs, and finish items.

10. Maintains constant follow-up to ensure that schedule is accomplished.

11. Checks all incoming tests to ensure they meet specifications.

12. In some companies checks the concrete design, that is, the mix or fixed formula of the concrete used.

13. Plans material delivery and schedules with job superintendents.

14. Keeps in contact each day with subcontractors.

15. Writes memos as needed to architects, superintendents, subcontractors, and so forth.

16. Follows up daily on trouble areas, that is, those places where delivery of materials may be lagging.

17. In some companies accumulates change order information.

18. Generally troubleshoots, especially for delivery problems.

EQUIPMENT SUPERINTENDENT

A.   General Functions

Maintains and repairs equipment owned by company. To do this, he supervises garage and yard personnel and coordinates delivery of equipment to the specific job sites and expedites repairs and deliveries.

B.   Detailed Functions

1. Supervises, maintains, and repairs.

2. Purchases parts for maintenance and repairs.

3. Expedites repairs of equipment that cannot be done in the company garage.

4. Keeps detailed records of equipment, including maintenance costs for each piece of equipment.

5. Keeps track of equipment, that is, which job is using it at any given time.

6. Makes recommendations concerning purchase of new equipment.

7. Provides delivery of equipment to job sites, helps to plan the time, provides means of delivery to the site, and provides equipment setup at site.

8. Keeps weekly repair costs on his crew.

9. Prepares an annual budget for operation.

FIELD SUPERINTENDENT

A.   General Functions

Builds the building. Manages men and materials on the job site so that the project is built for profit. Coordinates schedules so that men and materials are available to promote efficient erection of the building at a profit level.

B.   Detailed Functions

1. "Learns the building." Studies plans and specifications so that he can plan the work to be accomplished.

2. Tries to anticipate problems.

3. Studies the costs.

4. Arranges scheduling and manufacture of building parts or components.

5. Coordinates building when the manufactured items become available for the building.

6. Does survey and layout work or supervises technical or field engineer who does this.

7. Keeps constant check on all trades, overseeing workmanship and materials.

8. Hires and fires workmen.

9. Supplies information to accounting department so that records of costs can be maintained.

10. Supervises men directly or indirectly (i.e., through the foreman).

11. May be responsible for deliveries.

12. Is responsible for drawings and seeing that drawings are made of changes or incomplete items.

13. Arranges for plan changes as needed.

14. May be responsible for written schedules or physical schedules.

15. Does on-the-spot estimating (material or labor).

16. May price out extra items or charges.

17. Does limited buying (supplies and items missed by the purchasing department).

18. Makes daily safety inspections.

19. May record daily field activities in a log.

MECHANICAL SUPERINTENDENT

A.   General Functions

The mechanical superintendent coordinates subcontractor's work with that of the general contractor to ensure that project remains on schedule and quality is maintained.

B. Detailed Functions

1. Compiles listing of major mechanical electrical equipment required.
2. Expedites shop drawings and equipment deliveries.
3. Assists in preparation of project schedules.
4. Prepares weekly progress reports on electrical and mechanical work.
5. Coordinates subcontractors' work with general contractor.
6. Checks schedule to ensure project is on schedule.
7. Supervises general contractor's work done for subcontractors (equipment production, excavations, etc.).
8. Processes and distributes shop drawings.
9. Supervises, inspects, and evaluates work performed by subcontractors—ensures there is compliance with plans and specifications.
10. Supervises project closely to ensure that the owner is getting his money's worth on subcontractor work.

SCHEDULING ENGINEER (FIELD ENGINEER)

A. General Functions

Scheduling engineer schedules and coordinates. He serves as a troubleshooter when there is a breakdown in delivery schedule. He maintains a constant follow-up on the schedule to ensure progress as previously planned.

B. Detailed Functions

1. Receives plans and specifications and breaks them down by trade.
2. Writes to all subcontractors telling them when their work is necessary on the schedule.
3. Expedites follow-up for drawings, that is, shop drawings or detailed drawings. Checks with own staff for follow-up.
4. Keeps a close follow-up to ensure that plans and drawings reach the right people at the right time.
5. Establishes delivery times for materials, equipment, or labor, based on when they can be acquired, and determines the lead time required for acquisitions.
6. Makes out delivery schedules, using a bar graph method or critical path method. Makes sure that the shop items are on the bar graph or CPM.
7. Discusses material delivery and scheduling with job superintendent.
8. Keeps in touch with subcontractors as needed.
9. Writes memos as needed to superintendents, subcontractors, and so forth.
10. Follows up daily on trouble areas, where delivery of materials may be lagging.
11. Generally troubleshoots.

TIMEKEEPER

A. General Functions

A timekeeper is primarily concerned with maintaining cost control of labor force on a project. He maintains payroll records and may also maintain records on material deliveries.

B. Detailed Functions

1. Ensures that the men are on the job, checks what specific tasks they are performing, and checks this against job sheets given to him daily by the foreman.

2. Checks with the foreman to determine exact job and classification of work each man is doing so that the work can be coded and entered against the correct amount.

3. Walks the job a few times each day.

4. Computes previous day's work sheets to obtain costs.

5. Projects daily costs to determine if work was completed within the allocated budget.

6. Talks over costs with superintendent.

7. Posts workers' hours to the payroll on a daily basis.

8. Types a cost report each week. In some companies this may be done by central office staff.

9. Types payroll each week. In some companies this may be done by central office staff.

10. Types paychecks each week. In some companies this may be done by central office staff.

11. Types all back charges and time tickets.

12. Estimates costs of requests from subcontractors for sheds, shanties, carpenters, and concrete work performed for them.

13. May compile subcontractors' invoices for payment and discuss these with architect to determine accuracy.

14. Codes all delivery tickets to maintain costs on all building parts.

15. Keeps records of all reinforcing steel deliveries.

16. Records all concrete pours.

17. May assist superintendent by ordering labor, lumber, and other materials.

18. On certain big load days may call the union halls for extra men; will sign these men for the day and pay them by check at night.

19. Signs up all new workmen (W-4 forms, applications, etc.) and submits originals to central office.

20. Enters new employees' names and proper wage rate for the particular trade on payroll.

21. Types monthly report on welfare and pension. In some companies this may be done at the central office.

22. Balances the payroll and types it each Monday. Submits it to the main office so that checks can be made out and returned to the job site by Wednesday. In some companies this may be done at central office.

23. May travel to various job sites and perform same duties for each of the projects.

24. On projects involving federal funds, he collects payroll data from subcontractors for submission to the government in compliance with their regulations.

25. Maintains time record on company truck drivers when material deliveries are made.

26. May supervise "time checkers" on larger project.

Appendix G

Compound Interest Tables

TABLE G.1

0.25%	Compound Interest Factors						0.25%
	Single Payment		**Uniform Payment Series**				
n	Compound Amount Factor	Present Worth Factor	Sinking Fund Factor	Capital Recovery Factor	Compound Amount Factor	Present Worth Factor	n
	Find F given P F/P	Find P given F P/F	Find A given F A/F	Find A given P A/P	Find F given A F/A	Find P given A P/A	
1	1.00250	0.99751	1.00000	1.00250	1.00000	0.99751	1
2	1.00501	0.99502	0.49938	0.50188	2.00250	1.99252	2
3	1.00752	0.99254	0.33250	0.33500	3.00751	2.98506	3
4	1.01004	0.99006	0.24906	0.25156	4.01503	3.97512	4
5	1.01256	0.98759	0.19900	0.20150	5.02506	4.96272	5
6	1.01509	0.98513	0.16563	0.16813	6.03763	5.94785	6
7	1.01763	0.98267	0.14179	0.14429	7.05272	6.93052	7
8	1.02018	0.98022	0.12391	0.12641	8.07035	7.91074	8
9	1.02273	0.97778	0.11000	0.11250	9.09053	8.88852	9
10	1.02528	0.97534	0.09888	0.10138	10.11325	9.86386	10
11	1.02785	0.97291	0.08978	0.09228	11.13854	10.83677	11
12	1.03042	0.97048	0.08219	0.08469	12.16638	11.80725	12
13	1.03299	0.96806	0.07578	0.07828	13.19680	12.77532	13
14	1.03557	0.96565	0.07028	0.07278	14.22979	13.74096	14
15	1.03816	0.96324	0.06551	0.06801	15.26537	14.70420	15
16	1.04076	0.96084	0.06134	0.06384	16.30353	15.66504	16
17	1.04336	0.95844	0.05766	0.06016	17.34429	16.62348	17
18	1.04597	0.95605	0.05438	0.05688	18.38765	17.57953	18
19	1.04858	0.95367	0.05146	0.05396	19.43362	18.53320	19
20	1.05121	0.95129	0.04882	0.05132	20.48220	19.48449	20
21	1.05383	0.94892	0.04644	0.04894	21.53341	20.43340	21
22	1.05647	0.94655	0.04427	0.04677	22.58724	21.37995	22
23	1.05911	0.94419	0.04229	0.04479	23.64371	22.32414	23
24	1.06176	0.94184	0.04048	0.04298	24.70282	23.26598	24
25	1.06441	0.93949	0.03881	0.04131	25.76457	24.20547	25
30	1.07778	0.92783	0.03214	0.03464	31.11331	28.86787	30
31	1.08048	0.92552	0.03106	0.03356	32.19109	29.79339	31
32	1.08318	0.92321	0.03006	0.03256	33.27157	30.71660	32
33	1.08589	0.92091	0.02911	0.03161	34.35475	31.63750	33
34	1.08860	0.91861	0.02822	0.03072	35.44064	32.55611	34
35	1.09132	0.91632	0.02738	0.02988	36.52924	33.47243	35
36	1.09405	0.91403	0.02658	0.02908	37.62056	34.38647	36
42	1.11057	0.90044	0.02261	0.02511	44.22603	39.82300	42
48	1.12733	0.88705	0.01963	0.02213	50.93121	45.17869	48
60	1.16162	0.86087	0.01547	0.01797	64.64671	55.65236	60

TABLE G.2

0.50%	Compound Interest Factors						0.50%
	Single Payment		Uniform Payment Series				
n	Compound Amount Factor	Present Worth Factor	Sinking Fund Factor	Capital Recovery Factor	Compound Amount Factor	Present Worth Factor	n
	Find F given P F/P	Find P given F P/F	Find A given F A/F	Find A given P A/P	Find F given A F/A	Find P given A P/A	
1	1.00500	0.99502	1.00000	1.00500	1.00000	0.99502	1
2	1.01003	0.99007	0.49875	0.50375	2.00500	1.98510	2
3	1.01508	0.98515	0.33167	0.33667	3.01502	2.97025	3
4	1.02015	0.98025	0.24813	0.25313	4.03010	3.95050	4
5	1.02525	0.97537	0.19801	0.20301	5.05025	4.92587	5
6	1.03038	0.97052	0.16460	0.16960	6.07550	5.89638	6
7	1.03553	0.96569	0.14073	0.14573	7.10588	6.86207	7
8	1.04071	0.96089	0.12283	0.12783	8.14141	7.82296	8
9	1.04591	0.95610	0.10891	0.11391	9.18212	8.77906	9
10	1.05114	0.95135	0.09777	0.10277	10.22803	9.73041	10
11	1.05640	0.94661	0.08866	0.09366	11.27917	10.67703	11
12	1.06168	0.94191	0.08107	0.08607	12.33556	11.61893	12
13	1.06699	0.93722	0.07464	0.07964	13.39724	12.55615	13
14	1.07232	0.93256	0.06914	0.07414	14.46423	13.48871	14
15	1.07768	0.92792	0.06436	0.06936	15.53655	14.41662	15
16	1.08307	0.92330	0.06019	0.06519	16.61423	15.33993	16
17	1.08849	0.91871	0.05651	0.06151	17.69730	16.25863	17
18	1.09393	0.91414	0.05323	0.05823	18.78579	17.17277	18
19	1.09940	0.90959	0.05030	0.05530	19.87972	18.08236	19
20	1.10490	0.90506	0.04767	0.05267	20.97912	18.98742	20
21	1.11042	0.90056	0.04528	0.05028	22.08401	19.88798	21
22	1.11597	0.89608	0.04311	0.04811	23.19443	20.78406	22
23	1.12155	0.89162	0.04113	0.04613	24.31040	21.67568	23
24	1.12716	0.88719	0.03932	0.04432	25.43196	22.56287	24
25	1.13280	0.88277	0.03765	0.04265	26.55912	23.44564	25
30	1.16140	0.86103	0.03098	0.03598	32.28002	27.79405	30
31	1.16721	0.85675	0.02990	0.03490	33.44142	28.65080	31
32	1.17304	0.85248	0.02889	0.03389	34.60862	29.50328	32
33	1.17891	0.84824	0.02795	0.03295	35.78167	30.35153	33
34	1.18480	0.84402	0.02706	0.03206	36.96058	31.19555	34
35	1.19073	0.83982	0.02622	0.03122	38.14538	32.03537	35
36	1.19668	0.83564	0.02542	0.03042	39.33610	32.87102	36
42	1.23303	0.81101	0.02146	0.02646	46.60654	37.79830	42
48	1.27049	0.78710	0.01849	0.02349	54.09783	42.58032	48
60	1.34885	0.74137	0.01433	0.01933	69.77003	51.72556	60

TABLE G.3

0.75%	Compound Interest Factors						0.75%
	Single Payment		**Uniform Payment Series**				
	Compound Amount Factor	Present Worth Factor	Sinking Fund Factor	Capital Recovery Factor	Compound Amount Factor	Present Worth Factor	
n	Find F given P F/P	Find P given F P/F	Find A given F A/F	Find A given P A/P	Find F given A F/A	Find P given A P/A	n
1	1.00750	0.99256	1.00000	1.00750	1.00000	0.99256	1
2	1.01506	0.98517	0.49813	0.50563	2.00750	1.97772	2
3	1.02267	0.97783	0.33085	0.33835	3.02256	2.95556	3
4	1.03034	0.97055	0.24721	0.25471	4.04523	3.92611	4
5	1.03807	0.96333	0.19702	0.20452	5.07556	4.88944	5
6	1.04585	0.95616	0.16357	0.17107	6.11363	5.84560	6
7	1.05370	0.94904	0.13967	0.14717	7.15948	6.79464	7
8	1.06160	0.94198	0.12176	0.12926	8.21318	7.73661	8
9	1.06956	0.93496	0.10782	0.11532	9.27478	8.67158	9
10	1.07758	0.92800	0.09667	0.10417	10.34434	9.59958	10
11	1.08566	0.92109	0.08755	0.09505	11.42192	10.52067	11
12	1.09381	0.91424	0.07995	0.08745	12.50759	11.43491	12
13	1.10201	0.90743	0.07352	0.08102	13.60139	12.34235	13
14	1.11028	0.90068	0.06801	0.07551	14.70340	13.24302	14
15	1.11860	0.89397	0.06324	0.07074	15.81368	14.13699	15
16	1.12699	0.88732	0.05906	0.06656	16.93228	15.02431	16
17	1.13544	0.88071	0.05537	0.06287	18.05927	15.90502	17
18	1.14396	0.87416	0.05210	0.05960	19.19472	16.77918	18
19	1.15254	0.86765	0.04917	0.05667	20.33868	17.64683	19
20	1.16118	0.86119	0.04653	0.05403	21.49122	18.50802	20
21	1.16989	0.85478	0.04415	0.05165	22.65240	19.36280	21
22	1.17867	0.84842	0.04198	0.04948	23.82230	20.21121	22
23	1.18751	0.84210	0.04000	0.04750	25.00096	21.05331	23
24	1.19641	0.83583	0.03818	0.04568	26.18847	21.88915	24
25	1.20539	0.82961	0.03652	0.04402	27.38488	22.71876	25
30	1.25127	0.79919	0.02985	0.03735	33.50290	26.77508	30
31	1.26066	0.79324	0.02877	0.03627	34.75417	27.56832	31
32	1.27011	0.78733	0.02777	0.03527	36.01483	28.35565	32
33	1.27964	0.78147	0.02682	0.03432	37.28494	29.13712	33
34	1.28923	0.77565	0.02593	0.03343	38.56458	29.91278	34
35	1.29890	0.76988	0.02509	0.03259	39.85381	30.68266	35
36	1.30865	0.76415	0.02430	0.03180	41.15272	31.44681	36
42	1.36865	0.73065	0.02034	0.02784	49.15329	35.91371	42
48	1.43141	0.69861	0.01739	0.02489	57.52071	40.18478	48
60	1.56568	0.63870	0.01326	0.02076	75.42414	48.17337	60

TABLE G.4

1.00%	Compound Interest Factors						1.00%
	Single Payment		**Uniform Payment Series**				
	Compound Amount Factor	Present Worth Factor	Sinking Fund Factor	Capital Recovery Factor	Compound Amount Factor	Present Worth Factor	
n	Find F given P F/P	Find P given F P/F	Find A given F A/F	Find A given P A/P	Find F given A F/A	Find P given A P/A	n
1	1.01000	0.99010	1.00000	1.01000	1.00000	0.99010	1
2	1.02010	0.98030	0.49751	0.50751	2.01000	1.97040	2
3	1.03030	0.97059	0.33002	0.34002	3.03010	2.94099	3
4	1.04060	0.96098	0.24628	0.25628	4.06040	3.90197	4
5	1.05101	0.95147	0.19604	0.20604	5.10101	4.85343	5
6	1.06152	0.94205	0.16255	0.17255	6.15202	5.79548	6
7	1.07214	0.93272	0.13863	0.14863	7.21354	6.72819	7
8	1.08286	0.92348	0.12069	0.13069	8.28567	7.65168	8
9	1.09369	0.91434	0.10674	0.11674	9.36853	8.56602	9
10	1.10462	0.90529	0.09558	0.10558	10.46221	9.47130	10
11	1.11567	0.89632	0.08645	0.09645	11.56683	10.36763	11
12	1.12683	0.88745	0.07885	0.08885	12.68250	11.25508	12
13	1.13809	0.87866	0.07241	0.08241	13.80933	12.13374	13
14	1.14947	0.86996	0.06690	0.07690	14.94742	13.00370	14
15	1.16097	0.86135	0.06212	0.07212	16.09690	13.86505	15
16	1.17258	0.85282	0.05794	0.06794	17.25786	14.71787	16
17	1.18430	0.84438	0.05426	0.06426	18.43044	15.56225	17
18	1.19615	0.83602	0.05098	0.06098	19.61475	16.39827	18
19	1.20811	0.82774	0.04805	0.05805	20.81090	17.22601	19
20	1.22019	0.81954	0.04542	0.05542	22.01900	18.04555	20
21	1.23239	0.81143	0.04303	0.05303	23.23919	18.85698	21
22	1.24472	0.80340	0.04086	0.05086	24.47159	19.66038	22
23	1.25716	0.79544	0.03889	0.04889	25.71630	20.45582	23
24	1.26973	0.78757	0.03707	0.04707	26.97346	21.24339	24
25	1.28243	0.77977	0.03541	0.04541	28.24320	22.02316	25
30	1.34785	0.74192	0.02875	0.03875	34.78489	25.80771	30
31	1.36133	0.73458	0.02768	0.03768	36.13274	26.54229	31
32	1.37494	0.72730	0.02667	0.03667	37.49407	27.26959	32
33	1.38869	0.72010	0.02573	0.03573	38.86901	27.98969	33
34	1.40258	0.71297	0.02484	0.03484	40.25770	28.70267	34
35	1.41660	0.70591	0.02400	0.03400	41.66028	29.40858	35
36	1.43077	0.69892	0.02321	0.03321	43.07688	30.10751	36
42	1.51879	0.65842	0.01928	0.02928	51.87899	34.15811	42
48	1.61223	0.62026	0.01633	0.02633	61.22261	37.97396	48
60	1.81670	0.55045	0.01224	0.02224	81.66967	44.95504	60

TABLE G.5

1.25%	Compound Interest Factors						1.25%
	Single Payment		**Uniform Payment Series**				
n	Compound Amount Factor	Present Worth Factor	Sinking Fund Factor	Capital Recovery Factor	Compound Amount Factor	Present Worth Factor	n
	Find F given P F/P	Find P given F P/F	Find A given F A/F	Find A given P A/P	Find F given A F/A	Find P given A P/A	
1	1.01250	0.98765	1.00000	1.01250	1.00000	0.98765	1
2	1.02516	0.97546	0.49689	0.50939	2.01250	1.96312	2
3	1.03797	0.96342	0.32920	0.34170	3.03766	2.92653	3
4	1.05095	0.95152	0.24536	0.25786	4.07563	3.87806	4
5	1.06408	0.93978	0.19506	0.20756	5.12657	4.81784	5
6	1.07738	0.92817	0.16153	0.17403	6.19065	5.74601	6
7	1.09085	0.91672	0.13759	0.15009	7.26804	6.66273	7
8	1.10449	0.90540	0.11963	0.13213	8.35889	7.56812	8
9	1.11829	0.89422	0.10567	0.11817	9.46337	8.46234	9
10	1.13227	0.88318	0.09450	0.10700	10.58167	9.34553	10
11	1.14642	0.87228	0.08537	0.09787	11.71394	10.21780	11
12	1.16075	0.86151	0.07776	0.09026	12.86036	11.07931	12
13	1.17526	0.85087	0.07132	0.08382	14.02112	11.93018	13
14	1.18995	0.84037	0.06581	0.07831	15.19638	12.77055	14
15	1.20483	0.82999	0.06103	0.07353	16.38633	13.60055	15
16	1.21989	0.81975	0.05685	0.06935	17.59116	14.42029	16
17	1.23514	0.80963	0.05316	0.06566	18.81105	15.22992	17
18	1.25058	0.79963	0.04988	0.06238	20.04619	16.02955	18
19	1.26621	0.78976	0.04696	0.05946	21.29677	16.81931	19
20	1.28204	0.78001	0.04432	0.05682	22.56298	17.59932	20
21	1.29806	0.77038	0.04194	0.05444	23.84502	18.36969	21
22	1.31429	0.76087	0.03977	0.05227	25.14308	19.13056	22
23	1.33072	0.75147	0.03780	0.05030	26.45737	19.88204	23
24	1.34735	0.74220	0.03599	0.04849	27.78808	20.62423	24
25	1.36419	0.73303	0.03432	0.04682	29.13544	21.35727	25
30	1.45161	0.68889	0.02768	0.04018	36.12907	24.88891	30
31	1.46976	0.68038	0.02661	0.03911	37.58068	25.56929	31
32	1.48813	0.67198	0.02561	0.03811	39.05044	26.24127	32
33	1.50673	0.66369	0.02467	0.03717	40.53857	26.90496	33
34	1.52557	0.65549	0.02378	0.03628	42.04530	27.56046	34
35	1.54464	0.64740	0.02295	0.03545	43.57087	28.20786	35
36	1.56394	0.63941	0.02217	0.03467	45.11551	28.84727	36
42	1.68497	0.59348	0.01825	0.03075	54.79734	32.52132	42
48	1.81535	0.55086	0.01533	0.02783	65.22839	35.93148	48
60	2.10718	0.47457	0.01129	0.02379	88.57451	42.03459	60

TABLE G.6

1.50%	Compound Interest Factors						1.50%
	Single Payment		**Uniform Payment Series**				
	Compound Amount Factor	Present Worth Factor	Sinking Fund Factor	Capital Recovery Factor	Compound Amount Factor	Present Worth Factor	
n	Find F given P F/P	Find P given F P/F	Find A given F A/F	Find A given P A/P	Find F given A F/A	Find P given A P/A	n
1	1.01500	0.98522	1.00000	1.01500	1.00000	0.98522	1
2	1.03023	0.97066	0.49628	0.51128	2.01500	1.95588	2
3	1.04568	0.95632	0.32838	0.34338	3.04522	2.91220	3
4	1.06136	0.94218	0.24444	0.25944	4.09090	3.85438	4
5	1.07728	0.92826	0.19409	0.20909	5.15227	4.78264	5
6	1.09344	0.91454	0.16053	0.17553	6.22955	5.69719	6
7	1.10984	0.90103	0.13656	0.15156	7.32299	6.59821	7
8	1.12649	0.88771	0.11858	0.13358	8.43284	7.48593	8
9	1.14339	0.87459	0.10461	0.11961	9.55933	8.36052	9
10	1.16054	0.86167	0.09343	0.10843	10.70272	9.22218	10
11	1.17795	0.84893	0.08429	0.09929	11.86326	10.07112	11
12	1.19562	0.83639	0.07668	0.09168	13.04121	10.90751	12
13	1.21355	0.82403	0.07024	0.08524	14.23683	11.73153	13
14	1.23176	0.81185	0.06472	0.07972	15.45038	12.54338	14
15	1.25023	0.79985	0.05994	0.07494	16.68214	13.34323	15
16	1.26899	0.78803	0.05577	0.07077	17.93237	14.13126	16
17	1.28802	0.77639	0.05208	0.06708	19.20136	14.90765	17
18	1.30734	0.76491	0.04881	0.06381	20.48938	15.67256	18
19	1.32695	0.75361	0.04588	0.06088	21.79672	16.42617	19
20	1.34686	0.74247	0.04325	0.05825	23.12367	17.16864	20
21	1.36706	0.73150	0.04087	0.05587	24.47052	17.90014	21
22	1.38756	0.72069	0.03870	0.05370	25.83758	18.62082	22
23	1.40838	0.71004	0.03673	0.05173	27.22514	19.33086	23
24	1.42950	0.69954	0.03492	0.04992	28.63352	20.03041	24
25	1.45095	0.68921	0.03326	0.04826	30.06302	20.71961	25
30	1.56308	0.63976	0.02664	0.04164	37.53868	24.01584	30
31	1.58653	0.63031	0.02557	0.04057	39.10176	24.64615	31
32	1.61032	0.62099	0.02458	0.03958	40.68829	25.26714	32
33	1.63448	0.61182	0.02364	0.03864	42.29861	25.87895	33
34	1.65900	0.60277	0.02276	0.03776	43.93309	26.48173	34
35	1.68388	0.59387	0.02193	0.03693	45.59209	27.07559	35
36	1.70914	0.58509	0.02115	0.03615	47.27597	27.66068	36
42	1.86885	0.53509	0.01726	0.03226	57.92314	30.99405	42
48	2.04348	0.48936	0.01437	0.02937	69.56522	34.04255	48
60	2.44322	0.40930	0.01039	0.02539	96.21465	39.38027	60

TABLE G.7

1.75%	Compound Interest Factors						1.75%
	Single Payment		**Uniform Payment Series**				
n	Compound Amount Factor	Present Worth Factor	Sinking Fund Factor	Capital Recovery Factor	Compound Amount Factor	Present Worth Factor	n
	Find F given P F/P	Find P given F P/F	Find A given F A/F	Find A given P A/P	Find F given A F/A	Find P given A P/A	
1	1.01750	0.98280	1.00000	1.01750	1.00000	0.98280	1
2	1.03531	0.96590	0.49566	0.51316	2.01750	1.94870	2
3	1.05342	0.94929	0.32757	0.34507	3.05281	2.89798	3
4	1.07186	0.93296	0.24353	0.26103	4.10623	3.83094	4
5	1.09062	0.91691	0.19312	0.21062	5.17809	4.74786	5
6	1.10970	0.90114	0.15952	0.17702	6.26871	5.64900	6
7	1.12912	0.88564	0.13553	0.15303	7.37841	6.53464	7
8	1.14888	0.87041	0.11754	0.13504	8.50753	7.40505	8
9	1.16899	0.85544	0.10356	0.12106	9.65641	8.26049	9
10	1.18944	0.84073	0.09238	0.10988	10.82540	9.10122	10
11	1.21026	0.82627	0.08323	0.10073	12.01484	9.92749	11
12	1.23144	0.81206	0.07561	0.09311	13.22510	10.73955	12
13	1.25299	0.79809	0.06917	0.08667	14.45654	11.53764	13
14	1.27492	0.78436	0.06366	0.08116	15.70953	12.32201	14
15	1.29723	0.77087	0.05888	0.07638	16.98445	13.09288	15
16	1.31993	0.75762	0.05470	0.07220	18.28168	13.85050	16
17	1.34303	0.74459	0.05102	0.06852	19.60161	14.59508	17
18	1.36653	0.73178	0.04774	0.06524	20.94463	15.32686	18
19	1.39045	0.71919	0.04482	0.06232	22.31117	16.04606	19
20	1.41478	0.70682	0.04219	0.05969	23.70161	16.75288	20
21	1.43954	0.69467	0.03981	0.05731	25.11639	17.44755	21
22	1.46473	0.68272	0.03766	0.05516	26.55593	18.13027	22
23	1.49036	0.67098	0.03569	0.05319	28.02065	18.80125	23
24	1.51644	0.65944	0.03389	0.05139	29.51102	19.46069	24
25	1.54298	0.64810	0.03223	0.04973	31.02746	20.10878	25
30	1.68280	0.59425	0.02563	0.04313	39.01715	23.18585	30
31	1.71225	0.58403	0.02457	0.04207	40.69995	23.76988	31
32	1.74221	0.57398	0.02358	0.04108	42.41220	24.34386	32
33	1.77270	0.56411	0.02265	0.04015	44.15441	24.90797	33
34	1.80372	0.55441	0.02177	0.03927	45.92712	25.46238	34
35	1.83529	0.54487	0.02095	0.03845	47.73084	26.00725	35
36	1.86741	0.53550	0.02018	0.03768	49.56613	26.54275	36
42	2.07227	0.48256	0.01632	0.03382	61.27236	29.56780	42
48	2.29960	0.43486	0.01347	0.03097	74.26278	32.29380	48
60	2.83182	0.35313	0.00955	0.02705	104.67522	36.96399	60

TABLE G.8

2.00%	Compound Interest Factors						2.00%
	Single Payment		Uniform Payment Series				
n	Compound Amount Factor	Present Worth Factor	Sinking Fund Factor	Capital Recovery Factor	Compound Amount Factor	Present Worth Factor	n
	Find F given P F/P	Find P given F P/F	Find A given F A/F	Find A given P A/P	Find F given A F/A	Find P given A P/A	
1	1.02000	0.98039	1.00000	1.02000	1.00000	0.98039	1
2	1.04040	0.96117	0.49505	0.51505	2.02000	1.94156	2
3	1.06121	0.94232	0.32675	0.34675	3.06040	2.88388	3
4	1.08243	0.92385	0.24262	0.26262	4.12161	3.80773	4
5	1.10408	0.90573	0.19216	0.21216	5.20404	4.71346	5
6	1.12616	0.88797	0.15853	0.17853	6.30812	5.60143	6
7	1.14869	0.87056	0.13451	0.15451	7.43428	6.47199	7
8	1.17166	0.85349	0.11651	0.13651	8.58297	7.32548	8
9	1.19509	0.83676	0.10252	0.12252	9.75463	8.16224	9
10	1.21899	0.82035	0.09133	0.11133	10.94972	8.98259	10
;11	1.24337	0.80426	0.08218	0.10218	12.16872	9.78685	11
12	1.26824	0.78849	0.07456	0.09456	13.41209	10.57534	12
13	1.29361	0.77303	0.06812	0.08812	14.68033	11.34837	13
14	1.31948	0.75788	0.06260	0.08260	15.97394	12.10625	14
15	1.34587	0.74301	0.05783	0.07783	17.29342	12.84926	15
16	1.37279	0.72845	0.05365	0.07365	18.63929	13.57771	16
17	1.40024	0.71416	0.04997	0.06997	20.01207	14.29187	17
18	1.42825	0.70016	0.04670	0.06670	21.41231	14.99203	18
19	1.45681	0.68643	0.04378	0.06378	22.84056	15.67846	19
20	1.48595	0.67297	0.04116	0.06116	24.29737	16.35143	20
21	1.51567	0.65978	0.03878	0.05878	25.78332	17.01121	21
22	1.54598	0.64684	0.03663	0.05663	27.29898	17.65805	22
23	1.57690	0.63416	0.03467	0.05467	28.84496	18.29220	23
24	1.60844	0.62172	0.03287	0.05287	30.42186	18.91393	24
25	1.64061	0.60953	0.03122	0.05122	32.03030	19.52346	25
30	1.81136	0.55207	0.02465	0.04465	40.56808	22.39646	30
31	1.84759	0.54125	0.02360	0.04360	42.37944	22.93770	31
32	1.88454	0.53063	0.02261	0.04261	44.22703	23.46833	32
33	1.92223	0.52023	0.02169	0.04169	46.11157	23.98856	33
34	1.96068	0.51003	0.02082	0.04082	48.03380	24.49859	34
35	1.99989	0.50003	0.02000	0.04000	49.99448	24.99862	35
36	2.03989	0.49022	0.01923	0.03923	51.99437	25.48884	36
42	2.29724	0.43530	0.01542	0.03542	64.86222	28.23479	42
48	2.58707	0.38654	0.01260	0.03260	79.35352	30.67312	48
60	3.28103	0.30478	0.00877	0.02877	114.05154	34.76089	60

TABLE G.9

2.50%	Compound Interest Factors						2.50%
	Single Payment		**Uniform Payment Series**				
	Compound Amount Factor	Present Worth Factor	Sinking Fund Factor	Capital Recovery Factor	Compound Amount Factor	Present Worth Factor	
n	Find F given P F/P	Find P given F P/F	Find A given F A/F	Find A given P A/P	Find F given A F/A	Find P given A P/A	n
1	1.02500	0.97561	1.00000	1.02500	1.00000	0.97561	1
2	1.05063	0.95181	0.49383	0.51883	2.02500	1.92742	2
3	1.07689	0.92860	0.32514	0.35014	3.07563	2.85602	3
4	1.10381	0.90595	0.24082	0.26582	4.15252	3.76197	4
5	1.13141	0.88385	0.19025	0.21525	5.25633	4.64583	5
6	1.15969	0.86230	0.15655	0.18155	6.38774	5.50813	6
7	1.18869	0.84127	0.13250	0.15750	7.54743	6.34939	7
8	1.21840	0.82075	0.11447	0.13947	8.73612	7.17014	8
9	1.24886	0.80073	0.10046	0.12546	9.95452	7.97087	9
10	1.28008	0.78120	0.08926	0.11426	11.20338	8.75206	10
11	1.31209	0.76214	0.08011	0.10511	12.48347	9.51421	11
12	1.34489	0.74356	0.07249	0.09749	13.79555	10.25776	12
13	1.37851	0.72542	0.06605	0.09105	15.14044	10.98318	13
14	1.41297	0.70773	0.06054	0.08554	16.51895	11.69091	14
15	1.44830	0.69047	0.05577	0.08077	17.93193	12.38138	15
16	1.48451	0.67362	0.05160	0.07660	19.38022	13.05500	16
17	1.52162	0.65720	0.04793	0.07293	20.86473	13.71220	17
18	1.55966	0.64117	0.04467	0.06967	22.38635	14.35336	18
19	1.59865	0.62553	0.04176	0.06676	23.94601	14.97889	19
20	1.63862	0.61027	0.03915	0.06415	25.54466	15.58916	20
21	1.67958	0.59539	0.03679	0.06179	27.18327	16.18455	21
22	1.72157	0.58086	0.03465	0.05965	28.86286	16.76541	22
23	1.76461	0.56670	0.03270	0.05770	30.58443	17.33211	23
24	1.80873	0.55288	0.03091	0.05591	32.34904	17.88499	24
25	1.85394	0.53939	0.02928	0.05428	34.15776	18.42438	25
30	2.09757	0.47674	0.02278	0.04778	43.90270	20.93029	30
31	2.15001	0.46511	0.02174	0.04674	46.00027	21.39541	31
32	2.20376	0.45377	0.02077	0.04577	48.15028	21.84918	32
33	2.25885	0.44270	0.01986	0.04486	50.35403	22.29188	33
34	2.31532	0.43191	0.01901	0.04401	52.61289	22.72379	34
35	2.37321	0.42137	0.01821	0.04321	54.92821	23.14516	35
36	2.43254	0.41109	0.01745	0.04245	57.30141	23.55625	36
42	2.82100	0.35448	0.01373	0.03873	72.83981	25.82061	42
48	3.27149	0.30567	0.01101	0.03601	90.85958	27.77315	48
60	4.39979	0.22728	0.00735	0.03235	135.99159	30.90866	60

TABLE G.10

3.00%	Compound Interest Factors						3.00%
	Single Payment		Uniform Payment Series				
n	Compound Amount Factor	Present Worth Factor	Sinking Fund Factor	Capital Recovery Factor	Compound Amount Factor	Present Worth Factor	n
	Find F given P F/P	Find P given F P/F	Find A given F A/F	Find A given P A/P	Find F given A F/A	Find P given A P/A	
1	1.03000	0.97087	1.00000	1.03000	1.00000	0.97087	1
2	1.06090	0.94260	0.49261	0.52261	2.03000	1.91347	2
3	1.09273	0.91514	0.32353	0.35353	3.09090	2.82861	3
4	1.12551	0.88849	0.23903	0.26903	4.18363	3.71710	4
5	1.15927	0.86261	0.18835	0.21835	5.30914	4.57971	5
6	1.19405	0.83748	0.15460	0.18460	6.46841	5.41719	6
7	1.22987	0.81309	0.13051	0.16051	7.66246	6.23028	7
8	1.26677	0.78941	0.11246	0.14246	8.89234	7.01969	8
9	1.30477	0.76642	0.09843	0.12843	10.15911	7.78611	9
10	1.34392	0.74409	0.08723	0.11723	11.46388	8.53020	10
11	1.38423	0.72242	0.07808	0.10808	12.80780	9.25262	11
12	1.42576	0.70138	0.07046	0.10046	14.19203	9.95400	12
13	1.46853	0.68095	0.06403	0.09403	15.61779	10.63496	13
14	1.51259	0.66112	0.05853	0.08853	17.08632	11.29607	14
15	1.55797	0.64186	0.05377	0.08377	18.59891	11.93794	15
16	1.60471	0.62317	0.04961	0.07961	20.15688	12.56110	16
17	1.65285	0.60502	0.04595	0.07595	21.76159	13.16612	17
18	1.70243	0.58739	0.04271	0.07271	23.41444	13.75351	18
19	1.75351	0.57029	0.03981	0.06981	25.11687	14.32380	19
20	1.80611	0.55368	0.03722	0.06722	26.87037	14.87747	20
21	1.86029	0.53755	0.03487	0.06487	28.67649	15.41502	21
22	1.91610	0.52189	0.03275	0.06275	30.53678	15.93692	22
23	1.97359	0.50669	0.03081	0.06081	32.45288	16.44361	23
24	2.03279	0.49193	0.02905	0.05905	34.42647	16.93554	24
25	2.09378	0.47761	0.02743	0.05743	36.45926	17.41315	25
30	2.42726	0.41199	0.02102	0.05102	47.57542	19.60044	30
31	2.50008	0.39999	0.02000	0.05000	50.00268	20.00043	31
32	2.57508	0.38834	0.01905	0.04905	52.50276	20.38877	32
33	2.65234	0.37703	0.01816	0.04816	55.07784	20.76579	33
34	2.73191	0.36604	0.01732	0.04732	57.73018	21.13184	34
35	2.81386	0.35538	0.01654	0.04654	60.46208	21.48722	35
36	2.89828	0.34503	0.01580	0.04580	63.27594	21.83225	36
42	3.46070	0.28896	0.01219	0.04219	82.02320	23.70136	42
48	4.13225	0.24200	0.00958	0.03958	104.40840	25.26671	48
60	5.89160	0.16973	0.00613	0.03613	163.05344	27.67556	60

TABLE G.11

4.00%	Compound Interest Factors						4.00%
	Single Payment		**Uniform Payment Series**				
n	Compound Amount Factor	Present Worth Factor	Sinking Fund Factor	Capital Recovery Factor	Compound Amount Factor	Present Worth Factor	n
	Find F given P F/P	Find P given F P/F	Find A given F A/F	Find A given P A/P	Find F given A F/A	Find P given A P/A	
1	1.04000	0.96154	1.00000	1.04000	1.00000	0.96154	1
2	1.08160	0.92456	0.49020	0.53020	2.04000	1.88609	2
3	1.12486	0.88900	0.32035	0.36035	3.12160	2.77509	3
4	1.16986	0.85480	0.23549	0.27549	4.24646	3.62990	4
5	1.21665	0.82193	0.18463	0.22463	5.41632	4.45182	5
6	1.26532	0.79031	0.15076	0.19076	6.63298	5.24214	6
7	1.31593	0.75992	0.12661	0.16661	7.89829	6.00205	7
8	1.36857	0.73069	0.10853	0.14853	9.21423	6.73274	8
9	1.42331	0.70259	0.09449	0.13449	10.58280	7.43533	9
10	1.48024	0.67556	0.08329	0.12329	12.00611	8.11090	10
11	1.53945	0.64958	0.07415	0.11415	13.48635	8.76048	11
12	1.60103	0.62460	0.06655	0.10655	15.02581	9.38507	12
13	1.66507	0.60057	0.06014	0.10014	16.62684	9.98565	13
14	1.73168	0.57748	0.05467	0.09467	18.29191	10.56312	14
15	1.80094	0.55526	0.04994	0.08994	20.02359	11.11839	15
16	1.87298	0.53391	0.04582	0.08582	21.82453	11.65230	16
17	1.94790	0.51337	0.04220	0.08220	23.69751	12.16567	17
18	2.02582	0.49363	0.03899	0.07899	25.64541	12.65930	18
19	2.10685	0.47464	0.03614	0.07614	27.67123	13.13394	19
20	2.19112	0.45639	0.03358	0.07358	29.77808	13.59033	20
21	2.27877	0.43883	0.03128	0.07128	31.96920	14.02916	21
22	2.36992	0.42196	0.02920	0.06920	34.24797	14.45112	22
23	2.46472	0.40573	0.02731	0.06731	36.61789	14.85684	23
24	2.56330	0.39012	0.02559	0.06559	39.08260	15.24696	24
25	2.66584	0.37512	0.02401	0.06401	41.64591	15.62208	25
30	3.24340	0.30832	0.01783	0.05783	56.08494	17.29203	30
31	3.37313	0.29646	0.01686	0.05686	59.32834	17.58849	31
32	3.50806	0.28506	0.01595	0.05595	62.70147	17.87355	32
33	3.64838	0.27409	0.01510	0.05510	66.20953	18.14765	33
34	3.79432	0.26355	0.01431	0.05431	69.85791	18.41120	34
35	3.94609	0.25342	0.01358	0.05358	73.65222	18.66461	35
36	4.10393	0.24367	0.01289	0.05289	77.59831	18.90828	36
42	5.19278	0.19257	0.00954	0.04954	104.81960	20.18563	42
48	6.57053	0.15219	0.00718	0.04718	139.26321	21.19513	48
60	10.51963	0.09506	0.00420	0.04420	237.99069	22.62349	60

TABLE G.12

5.00%	Compound Interest Factors						5.00%
	Single Payment		Uniform Payment Series				
n	Compound Amount Factor	Present Worth Factor	Sinking Fund Factor	Capital Recovery Factor	Compound Amount Factor	Present Worth Factor	n
	Find F given P F/P	Find P given F P/F	Find A given F A/F	Find A given P A/P	Find F given A F/A	Find P given A P/A	
1	1.05000	0.95238	1.00000	1.05000	1.00000	0.95238	1
2	1.10250	0.90703	0.48780	0.53780	2.05000	1.85941	2
3	1.15763	0.86384	0.31721	0.36721	3.15250	2.72325	3
4	1.21551	0.82270	0.23201	0.28201	4.31013	3.54595	4
5	1.27628	0.78353	0.18097	0.23097	5.52563	4.32948	5
6	1.34010	0.74622	0.14702	0.19702	6.80191	5.07569	6
7	1.40710	0.71068	0.12282	0.17282	8.14201	5.78637	7
8	1.47746	0.67684	0.10472	0.15472	9.54911	6.46321	8
9	1.55133	0.64461	0.09069	0.14069	11.02656	7.10782	9
10	1.62889	0.61391	0.07950	0.12950	12.57789	7.72173	10
11	1.71034	0.58468	0.07039	0.12039	14.20679	8.30641	11
12	1.79586	0.55684	0.06283	0.11283	15.91713	8.86325	12
13	1.88565	0.53032	0.05646	0.10646	17.71298	9.39357	13
14	1.97993	0.50507	0.05102	0.10102	19.59863	9.89864	14
15	2.07893	0.48102	0.04634	0.09634	21.57856	10.37966	15
16	2.18287	0.45811	0.04227	0.09227	23.65749	10.83777	16
17	2.29202	0.43630	0.03870	0.08870	25.84037	11.27407	17
18	2.40662	0.41552	0.03555	0.08555	28.13238	11.68959	18
19	2.52695	0.39573	0.03275	0.08275	30.53900	12.08532	19
20	2.65330	0.37689	0.03024	0.08024	33.06595	12.46221	20
21	2.78596	0.35894	0.02800	0.07800	35.71925	12.82115	21
22	2.92526	0.34185	0.02597	0.07597	38.50521	13.16300	22
23	3.07152	0.32557	0.02414	0.07414	41.43048	13.48857	23
24	3.22510	0.31007	0.02247	0.07247	44.50200	13.79864	24
25	3.38635	0.29530	0.02095	0.07095	47.72710	14.09394	25
30	4.32194	0.23138	0.01505	0.06505	66.43885	15.37245	30
31	4.53804	0.22036	0.01413	0.06413	70.76079	15.59281	31
32	4.76494	0.20987	0.01328	0.06328	75.29883	15.80268	32
33	5.00319	0.19987	0.01249	0.06249	80.06377	16.00255	33
34	5.25335	0.19035	0.01176	0.06176	85.06696	16.19290	34
35	5.51602	0.18129	0.01107	0.06107	90.32031	16.37419	35
36	5.79182	0.17266	0.01043	0.06043	95.83632	16.54685	36
42	7.76159	0.12884	0.00739	0.05739	135.23175	17.42321	42
48	10.40127	0.09614	0.00532	0.05532	188.02539	18.07716	48
60	18.67919	0.05354	0.00283	0.05283	353.58372	18.92929	60

TABLE G.13

6.00%	Compound Interest Factors						6.00%
	Single Payment		Uniform Payment Series				
n	Compound Amount Factor	Present Worth Factor	Sinking Fund Factor	Capital Recovery Factor	Compound Amount Factor	Present Worth Factor	n
	Find F given P F/P	Find P given F P/F	Find A given F A/F	Find A given P A/P	Find F given A F/A	Find P given A P/A	
1	1.06000	0.94340	1.00000	1.06000	1.00000	0.94340	1
2	1.12360	0.89000	0.48544	0.54544	2.06000	1.83339	2
3	1.19102	0.83962	0.31411	0.37411	3.18360	2.67301	3
4	1.26248	0.79209	0.22859	0.28859	4.37462	3.46511	4
5	1.33823	0.74726	0.17740	0.23740	5.63709	4.21236	5
6	1.41852	0.70496	0.14336	0.20336	6.97532	4.91732	6
7	1.50363	0.66506	0.11914	0.17914	8.39384	5.58238	7
8	1.59385	0.62741	0.10104	0.16104	9.89747	6.20979	8
9	1.68948	0.59190	0.08702	0.14702	11.49132	6.80169	9
10	1.79085	0.55839	0.07587	0.13587	13.18079	7.36009	10
11	1.89830	0.52679	0.06679	0.12679	14.97164	7.88687	11
12	2.01220	0.49697	0.05928	0.11928	16.86994	8.38384	12
13	2.13293	0.46884	0.05296	0.11296	18.88214	8.85268	13
14	2.26090	0.44230	0.04758	0.10758	21.01507	9.29498	14
15	2.39656	0.41727	0.04296	0.10296	23.27597	9.71225	15
16	2.54035	0.39365	0.03895	0.09895	25.67253	10.10590	16
17	2.69277	0.37136	0.03544	0.09544	28.21288	10.47726	17
18	2.85434	0.35034	0.03236	0.09236	30.90565	10.82760	18
19	3.02560	0.33051	0.02962	0.08962	33.75999	11.15812	19
20	3.20714	0.31180	0.02718	0.08718	36.78559	11.46992	20
21	3.39956	0.29416	0.02500	0.08500	39.99273	11.76408	21
22	3.60354	0.27751	0.02305	0.08305	43.39229	12.04158	22
23	3.81975	0.26180	0.02128	0.08128	46.99583	12.30338	23
24	4.04893	0.24698	0.01968	0.07968	50.81558	12.55036	24
25	4.29187	0.23300	0.01823	0.07823	54.86451	12.78336	25
30	5.74349	0.17411	0.01265	0.07265	79.05819	13.76483	30
31	6.08810	0.16425	0.01179	0.07179	84.80168	13.92909	31
32	6.45339	0.15496	0.01100	0.07100	90.88978	14.08404	32
33	6.84059	0.14619	0.01027	0.07027	97.34316	14.23023	33
34	7.25103	0.13791	0.00960	0.06960	104.18375	14.36814	34
35	7.68609	0.13011	0.00897	0.06897	111.43478	14.49825	35
36	8.14725	0.12274	0.00839	0.06839	119.12087	14.62099	36
42	11.55703	0.08653	0.00568	0.06568	175.95054	15.22454	42
48	16.39387	0.06100	0.00390	0.06390	256.56453	15.65003	48
60	32.98769	0.03031	0.00188	0.06188	533.12818	16.16143	60

TABLE G.14

7.00%	Compound Interest Factors						7.00%
	Single Payment		Uniform Payment Series				
n	Compound Amount Factor	Present Worth Factor	Sinking Fund Factor	Capital Recovery Factor	Compound Amount Factor	Present Worth Factor	n
	Find F given P F/P	Find P given F P/F	Find A given F A/F	Find A given P A/P	Find F given A F/A	Find P given A P/A	
1	1.07000	0.93458	1.00000	1.07000	1.00000	0.93458	1
2	1.14490	0.87344	0.48309	0.55309	2.07000	1.80802	2
3	1.22504	0.81630	0.31105	0.38105	3.21490	2.62432	3
4	1.31080	0.76290	0.22523	0.29523	4.43994	3.38721	4
5	1.40255	0.71299	0.17389	0.24389	5.75074	4.10020	5
6	1.50073	0.66634	0.13980	0.20980	7.15329	4.76654	6
7	1.60578	0.62275	0.11555	0.18555	8.65402	5.38929	7
8	1.71819	0.58201	0.09747	0.16747	10.25980	5.97130	8
9	1.83846	0.54393	0.08349	0.15349	11.97799	6.51523	9
10	1.96715	0.50835	0.07238	0.14238	13.81645	7.02358	10
11	2.10485	0.47509	0.06336	0.13336	15.78360	7.49867	11
12	2.25219	0.44401	0.05590	0.12590	17.88845	7.94269	12
13	2.40985	0.41496	0.04965	0.11965	20.14064	8.35765	13
14	2.57853	0.38782	0.04434	0.11434	22.55049	8.74547	14
15	2.75903	0.36245	0.03979	0.10979	25.12902	9.10791	15
16	2.95216	0.33873	0.03586	0.10586	27.88805	9.44665	16
17	3.15882	0.31657	0.03243	0.10243	30.84022	9.76322	17
18	3.37993	0.29586	0.02941	0.09941	33.99903	10.05909	18
19	3.61653	0.27651	0.02675	0.09675	37.37896	10.33560	19
20	3.86968	0.25842	0.02439	0.09439	40.99549	10.59401	20
21	4.14056	0.24151	0.02229	0.09229	44.86518	10.83553	21
22	4.43040	0.22571	0.02041	0.09041	49.00574	11.06124	22
23	4.74053	0.21095	0.01871	0.08871	53.43614	11.27219	23
24	5.07237	0.19715	0.01719	0.08719	58.17667	11.46933	24
25	5.42743	0.18425	0.01581	0.08581	63.24904	11.65358	25
30	7.61226	0.13137	0.01059	0.08059	94.46079	12.40904	30
31	8.14511	0.12277	0.00980	0.07980	102.07304	12.53181	31
32	8.71527	0.11474	0.00907	0.07907	110.21815	12.64656	32
33	9.32534	0.10723	0.00841	0.07841	118.93343	12.75379	33
34	9.97811	0.10022	0.00780	0.07780	128.25876	12.85401	34
35	10.67658	0.09366	0.00723	0.07723	138.23688	12.94767	35
36	11.42394	0.08754	0.00672	0.07672	148.91346	13.03521	36
42	17.14426	0.05833	0.00434	0.07434	230.63224	13.45245	42
48	25.72891	0.03887	0.00283	0.07283	353.27009	13.73047	48
60	57.94643	0.01726	0.00123	0.07123	813.52038	14.03918	60

TABLE G.15

8.00%	Compound Interest Factors						8.00%
	Single Payment		**Uniform Payment Series**				
n	Compound Amount Factor	Present Worth Factor	Sinking Fund Factor	Capital Recovery Factor	Compound Amount Factor	Present Worth Factor	n
	Find F given P F/P	Find P given F P/F	Find A given F A/F	Find A given P A/P	Find F given A F/A	Find P given A P/A	
1	1.08000	0.92593	1.00000	1.08000	1.00000	0.92593	1
2	1.16640	0.85734	0.48077	0.56077	2.08000	1.78326	2
3	1.25971	0.79383	0.30803	0.38803	3.24640	2.57710	3
4	1.36049	0.73503	0.22192	0.30192	4.50611	3.31213	4
5	1.46933	0.68058	0.17046	0.25046	5.86660	3.99271	5
6	1.58687	0.63017	0.13632	0.21632	7.33593	4.62288	6
7	1.71382	0.58349	0.11207	0.19207	8.92280	5.20637	7
8	1.85093	0.54027	0.09401	0.17401	10.63663	5.74664	8
9	1.99900	0.50025	0.08008	0.16008	12.48756	6.24689	9
10	2.15892	0.46319	0.06903	0.14903	14.48656	6.71008	10
11	2.33164	0.42888	0.06008	0.14008	16.64549	7.13896	11
12	2.51817	0.39711	0.05270	0.13270	18.97713	7.53608	12
13	2.71962	0.36770	0.04652	0.12652	21.49530	7.90378	13
14	2.93719	0.34046	0.04130	0.12130	24.21492	8.24424	14
15	3.17217	0.31524	0.03683	0.11683	27.15211	8.55948	15
16	3.42594	0.29189	0.03298	0.11298	30.32428	8.85137	16
17	3.70002	0.27027	0.02963	0.10963	33.75023	9.12164	17
18	3.99602	0.25025	0.02670	0.10670	37.45024	9.37189	18
19	4.31570	0.23171	0.02413	0.10413	41.44626	9.60360	19
20	4.66096	0.21455	0.02185	0.10185	45.76196	9.81815	20
21	5.03383	0.19866	0.01983	0.09983	50.42292	10.01680	21
22	5.43654	0.18394	0.01803	0.09803	55.45676	10.20074	22
23	5.87146	0.17032	0.01642	0.09642	60.89330	10.37106	23
24	6.34118	0.15770	0.01498	0.09498	66.76476	10.52876	24
25	6.84848	0.14602	0.01368	0.09368	73.10594	10.67478	25
30	10.06266	0.09938	0.00883	0.08883	113.28321	11.25778	30
31	10.86767	0.09202	0.00811	0.08811	123.34587	11.34980	31
32	11.73708	0.08520	0.00745	0.08745	134.21354	11.43500	32
33	12.67605	0.07889	0.00685	0.08685	145.95062	11.51389	33
34	13.69013	0.07305	0.00630	0.08630	158.62667	11.58693	34
35	14.78534	0.06763	0.00580	0.08580	172.31680	11.65457	35
36	15.96817	0.06262	0.00534	0.08534	187.10215	11.71719	36
42	25.33948	0.03946	0.00329	0.08329	304.24352	12.00670	42
48	40.21057	0.02487	0.00204	0.08204	490.13216	12.18914	48
60	101.25706	0.00988	0.00080	0.08080	1253.21330	12.37655	60

TABLE G.16

9.00%	Compound Interest Factors						9.00%
	Single Payment		Uniform Payment Series				
n	Compound Amount Factor	Present Worth Factor	Sinking Fund Factor	Capital Recovery Factor	Compound Amount Factor	Present Worth Factor	n
	Find F given P F/P	Find P given F P/F	Find A given F A/F	Find A given P A/P	Find F given A F/A	Find P given A P/A	
1	1.09000	0.91743	1.00000	1.09000	1.00000	0.91743	1
2	1.18810	0.84168	0.47847	0.56847	2.09000	1.75911	2
3	1.29503	0.77218	0.30505	0.39505	3.27810	2.53129	3
4	1.41158	0.70843	0.21867	0.30867	4.57313	3.23972	4
5	1.53862	0.64993	0.16709	0.25709	5.98471	3.88965	5
6	1.67710	0.59627	0.13292	0.22292	7.52333	4.48592	6
7	1.82804	0.54703	0.10869	0.19869	9.20043	5.03295	7
8	1.99256	0.50187	0.09067	0.18067	11.02847	5.53482	8
9	2.17189	0.46043	0.07680	0.16680	13.02104	5.99525	9
10	2.36736	0.42241	0.06582	0.15582	15.19293	6.41766	10
11	2.58043	0.38753	0.05695	0.14695	17.56029	6.80519	11
12	2.81266	0.35553	0.04965	0.13965	20.14072	7.16073	12
13	3.06580	0.32618	0.04357	0.13357	22.95338	7.48690	13
14	3.34173	0.29925	0.03843	0.12843	26.01919	7.78615	14
15	3.64248	0.27454	0.03406	0.12406	29.36092	8.06069	15
16	3.97031	0.25187	0.03030	0.12030	33.00340	8.31256	16
17	4.32763	0.23107	0.02705	0.11705	36.97370	8.54363	17
18	4.71712	0.21199	0.02421	0.11421	41.30134	8.75563	18
19	5.14166	0.19449	0.02173	0.11173	46.01846	8.95011	19
20	5.60441	0.17843	0.01955	0.10955	51.16012	9.12855	20
21	6.10881	0.16370	0.01762	0.10762	56.76453	9.29224	21
22	6.65860	0.15018	0.01590	0.10590	62.87334	9.44243	22
23	7.25787	0.13778	0.01438	0.10438	69.53194	9.58021	23
24	7.91108	0.12640	0.01302	0.10302	76.78981	9.70661	24
25	8.62308	0.11597	0.01181	0.10181	84.70090	9.82258	25
30	13.26768	0.07537	0.00734	0.09734	136.30754	10.27365	30
31	14.46177	0.06915	0.00669	0.09669	149.57522	10.34280	31
32	15.76333	0.06344	0.00610	0.09610	164.03699	10.40624	32
33	17.18203	0.05820	0.00556	0.09556	179.80032	10.46444	33
34	18.72841	0.05339	0.00508	0.09508	196.98234	10.51784	34
35	20.41397	0.04899	0.00464	0.09464	215.71075	10.56682	35
36	22.25123	0.04494	0.00424	0.09424	236.12472	10.61176	36
42	37.31753	0.02680	0.00248	0.09248	403.52813	10.81337	42
48	62.58524	0.01598	0.00146	0.09146	684.28041	10.93358	48
60	176.03129	0.00568	0.00051	0.09051	1944.79213	11.04799	60

TABLE G.17

10.00%	Compound Interest Factors						10.00%
	Single Payment		**Uniform Payment Series**				
	Compound Amount Factor	Present Worth Factor	Sinking Fund Factor	Capital Recovery Factor	Compound Amount Factor	Present Worth Factor	
n	Find F given P F/P	Find P given F P/F	Find A given F A/F	Find A given P A/P	Find F given A F/A	Find P given A P/A	n
1	1.10000	0.90909	1.00000	1.10000	1.00000	0.90909	1
2	1.21000	0.82645	0.47619	0.57619	2.10000	1.73554	2
3	1.33100	0.75131	0.30211	0.40211	3.31000	2.48685	3
4	1.46410	0.68301	0.21547	0.31547	4.64100	3.16987	4
5	1.61051	0.62092	0.16380	0.26380	6.10510	3.79079	5
6	1.77156	0.56447	0.12961	0.22961	7.71561	4.35526	6
7	1.94872	0.51316	0.10541	0.20541	9.48717	4.86842	7
8	2.14359	0.46651	0.08744	0.18744	11.43589	5.33493	8
9	2.35795	0.42410	0.07364	0.17364	13.57948	5.75902	9
10	2.59374	0.38554	0.06275	0.16275	15.93742	6.14457	10
11	2.85312	0.35049	0.05396	0.15396	18.53117	6.49506	11
12	3.13843	0.31863	0.04676	0.14676	21.38428	6.81369	12
13	3.45227	0.28966	0.04078	0.14078	24.52271	7.10336	13
14	3.79750	0.26333	0.03575	0.13575	27.97498	7.36669	14
15	4.17725	0.23939	0.03147	0.13147	31.77248	7.60608	15
16	4.59497	0.21763	0.02782	0.12782	35.94973	7.82371	16
17	5.05447	0.19784	0.02466	0.12466	40.54470	8.02155	17
18	5.55992	0.17986	0.02193	0.12193	45.59917	8.20141	18
19	6.11591	0.16351	0.01955	0.11955	51.15909	8.36492	19
20	6.72750	0.14864	0.01746	0.11746	57.27500	8.51356	20
21	7.40025	0.13513	0.01562	0.11562	64.00250	8.64869	21
22	8.14027	0.12285	0.01401	.0.11401	71.40275	8.77154	22
23	8.95430	0.11168	0.01257	0.11257	79.54302	8.88322	23
24	9.84973	0.10153	0.01130	0.11130	88.49733	8.98474	24
25	10.83471	0.09230	0.01017	0.11017	98.34706	9.07704	25
30	17.44940	0.05731	0.00608	0.10608	164.49402	9.42691	30
31	19.19434	0.05210	0.00550	0.10550	181.94342	9.47901	31
32	21.11378	0.04736	0.00497	0.10497	201.13777	9.52638	32
33	23.22515	0.04306	0.00450	0.10450	222.25154	9.56943	33
34	25.54767	0.03914	0.00407	0.10407	245.47670	9.60857	34
35	28.10244	0.03558	0.00369	0.10369	271.02437	9.64416	35
36	30.91268	0.03235	0.00334	0.10334	299.12681	9.67651	36
42	54.76370	0.01826	0.00186	0.10186	537.63699	9.81740	42
48	97.01723	0.01031	0.00104	0.10104	960.17234	9.89693	48
60	304.48164	0.00328	0.00033	0.10033	3034.81640	9.96716	60

TABLE G.18

11.00%	Compound Interest Factors						11.00%
	Single Payment		Uniform Payment Series				
n	Compound Amount Factor	Present Worth Factor	Sinking Fund Factor	Capital Recovery Factor	Compound Amount Factor	Present Worth Factor	n
	Find F given P F/P	Find P given F P/F	Find A given F A/F	Find A given P A/P	Find F given A F/A	Find P given A P/A	
1	1.11000	0.90090	1.00000	1.11000	1.00000	0.90090	1
2	1.23210	0.81162	0.47393	0.58393	2.11000	1.71252	2
3	1.36763	0.73119	0.29921	0.40921	3.34210	2.44371	3
4	1.51807	0.65873	0.21233	0.32233	4.70973	3.10245	4
5	1.68506	0.59345	0.16057	0.27057	6.22780	3.69590	5
6	1.87041	0.53464	0.12638	0.23638	7.91286	4.23054	6
7	2.07616	0.48166	0.10222	0.21222	9.78327	4.71220	7
8	2.30454	0.43393	0.08432	0.19432	11.85943	5.14612	8
9	2.55804	0.39092	0.07060	0.18060	14.16397	5.53705	9
10	2.83942	0.35218	0.05980	0.16980	16.72201	5.88923	10
11	3.15176	0.31728	0.05112	0.16112	19.56143	6.20652	11
12	3.49845	0.28584	0.04403	0.15403	22.71319	6.49236	12
13	3.88328	0.25751	0.03815	0.14815	26.21164	6.74987	13
14	4.31044	0.23199	0.03323	0.14323	30.09492	6.98187	14
15	4.78459	0.20900	0.02907	0.13907	34.40536	7.19087	15
16	5.31089	0.18829	0.02552	0.13552	39.18995	7.37916	16
17	5.89509	0.16963	0.02247	0.13247	44.50084	7.54879	17
18	6.54355	0.15282	0.01984	0.12984	50.39594	7.70162	18
19	7.26334	0.13768	0.01756	0.12756	56.93949	7.83929	19
20	8.06231	0.12403	0.01558	0.12558	64.20283	7.96333	20
21	8.94917	0.11174	0.01384	0.12384	72.26514	8.07507	21
22	9.93357	0.10067	0.01231	0.12231	81.21431	8.17574	22
23	11.02627	0.09069	0.01097	0.12097	91.14788	8.26643	23
24	12.23916	0.08170	0.00979	0.11979	102.17415	8.34814	24
25	13.58546	0.07361	0.00874	0.11874	114.41331	8.42174	25
30	22.89230	0.04368	0.00502	0.11502	199.02088	8.69379	30
31	25.41045	0.03935	0.00451	0.11451	221.91317	8.73315	31
32	28.20560	0.03545	0.00404	0.11404	247.32362	8.76860	32
33	31.30821	0.03194	0.00363	0.11363	275.52922	8.80054	33
34	34.75212	0.02878	0.00326	0.11326	306.83744	8.82932	34
35	38.57485	0.02592	0.00293	0.11293	341.58955	8.85524	35
36	42.81808	0.02335	0.00263	0.11263	380.16441	8.87859	36
42	80.08757	0.01249	0.00139	0.11139	718.97790	8.97740	42
48	149.79695	0.00668	0.00074	0.11074	1352.69958	9.03022	48
60	524.05724	0.00191	0.00021	0.11021	4755.06584	9.07356	60

TABLE G.19

12.00%	Compound Interest Factors						12.00%
	Single Payment		**Uniform Payment Series**				
n	Compound Amount Factor	Present Worth Factor	Sinking Fund Factor	Capital Recovery Factor	Compound Amount Factor	Present Worth Factor	n
	Find F given P F/P	Find P given F P/F	Find A given F A/F	Find A given P A/P	Find F given A F/A	Find P given A P/A	
1	1.12000	0.89286	1.00000	1.12000	1.00000	0.89286	1
2	1.25440	0.79719	0.47170	0.59170	2.12000	1.69005	2
3	1.40493	0.71178	0.29635	0.41635	3.37440	2.40183	3
4	1.57352	0.63552	0.20923	0.32923	4.77933	3.03735	4
5	1.76234	0.56743	0:15741	0.27741	6.35285	3.60478	5
6	1.97382	0.50663	0.12323	0.24323	8.11519	4.11141	6
7	2.21068	0.45235	0.09912	0.21912	10.08901	4.56376	7
8	2.47596	0.40388	0.08130	0.20130	12.29969	4.96764	8
9	2.77308	0.36061	0.06768	0.18768	14.77566	5.32825	9
10	3.10585	0.32197	0.05698	0.17698	17.54874	5.65022	10
11	3.47855	0.28748	0.04842	0.16842	20.65458	5.93770	11
12	3.89598	0.25668	0.04144	0.16144	24.13313	6.19437	12
13	4.36349	0.22917	0.03568	0.15568	28.02911	6.42355	13
14	4.88711	0.20462	0.03087	0.15087	32.39260	6.62817	14
15	5.47357	0.18270	0.02682	0.14682	37.27971	6.81086	15
16	6.13039	0.16312	0.02339	0.14339	42.75328	6.97399	16
17	6.86604	0.14564	0.02046	0.14046	48.88367	7.11963	17
18	7.68997	0.13004	0.01794	0.13794	55.74971	7.24967	18
19	8.61276	0.11611	0.01576	0.13576	63.43968	7.36578	19
20	9.64629	0.10367	0.01388	0.13388	72.05244	7.46944	20
21	10.80385	0.09256	0.01224	0.13224	81.69874	7.56200	21
22	12.10031	0.08264	0.01081	0.13081	92.50258	7.64465	22
23	13.55235	0.07379	0.00956	0.12956	104.60289	7.71843	23
24	15.17863	0.06588	0.00846	0.12846	118.15524	7.78432	24
25	17.00006	0.05882	0.00750	0.12750	133.33387	7.84314	25
30	29.95992	0.03338	0.00414	0.12414	241.33268	8.05518	30
31	33.55511	0.02980	0.00369	0.12369	271.29261	8.08499	31
32	37.58173	0.02661	0.00328	0.12328	304.84772	8.11159	32
33	42.09153	0.02376	0.00292	0.12292	342.42945	8.13535	33
34	47.14252	0.02121	0.00260	0.12260	384.52098	8.15656	34
35	52.79962	0.01894	0.00232	0.12232	431.66350	8.17550	35
36	59.13557	0.01691	0.00206	0.12206	484.46312	8.19241	36
42	116.72314	0.00857	0.00104	0.12104	964.35948	8.26194	42
48	230.39078	0.00434	0.00052	0.12052	1911.58980	8.29716	48
60	897.59693	0.00111	0.00013	0.12013	7471.64111	8.32405	60

TABLE G.20

13.00%	Compound Interest Factors						13.00%
	Single Payment		Uniform Payment Series				
	Compound Amount Factor	Present Worth Factor	Sinking Fund Factor	Capital Recovery Factor	Compound Amount Factor	Present Worth Factor	
n	Find F given P F/P	Find P given F P/F	Find A given F A/F	Find A given P A/P	Find F given A F/A	Find P given A P/A	n
1	1.13000	0.88496	1.00000	1.13000	1.00000	0.88496	1
2	1.27690	0.78315	0.46948	0.59948	2.13000	1.66810	2
3	1.44290	0.69305	0.29352	0.42352	3.40690	2.36115	3
4	1.63047	0.61332	0.20619	0.33619	4.84980	2.97447	4
5	1.84244	0.54276	0.15431	0.28431	6.48027	3.51723	5
6	2.08195	0.48032	0.12015	0.25015	8.32271	3.99755	6
7	2.35261	0.42506	0.09611	0.22611	10.40466	4.42261	7
8	2.65844	0.37616	0.07839	0.20839	12.75726	4.79877	8
9	3.00404	0.33288	0.06487	0.19487	15.41571	5.13166	9
10	3.39457	0.29459	0.05429	0.18429	18.41975	5.42624	10
11	3.83586	0.26070	0.04584	0.17584	21.81432	5.68694	11
12	4.33452	0.23071	0.03899	0.16899	25.65018	5.91765	12
13	4.89801	0.20416	0.03335	0.16335	29.98470	6.12181	13
14	5.53475	0.18068	0.02867	0.15867	34.88271	6.30249	14
15	6.25427	0.15989	0.02474	0.15474	40.41746	6.46238	15
16	7.06733	0.14150	0.02143	0.15143	46.67173	6.60388	16
17	7.98608	0.12522	0.01861	0.14861	53.73906	6.72909	17
18	9.02427	0.11081	0.01620	0.14620	61.72514	6.83991	18
19	10.19742	0.09806	0.01413	0.14413	70.74941	6.93797	19
20	11.52309	0.08678	0.01235	0.14235	80.94683	7.02475	20
21	13.02109	0.07680	0.01081	0.14081	92.46992	7.10155	21
22	14.71383	0.06796	0.00948	0.13948	105.49101	7.16951	22
23	16.62663	0.06014	0.00832	0.13832	120.20484	7.22966	23
24	18.78809	0.05323	0.00731	0.13731	136.83147	7.28288	24
25	21.23054	0.04710	0.00643	0.13643	155.61956	7.32998	25
30	39.11590	0.02557	0.00341	0.13341	293.19922	7.49565	30
31	44.20096	0.02262	0.00301	0.13301	332.31511	7.51828	31
32	49.94709	0.02002	0.00266	0.13266	376.51608	7.53830	32
33	56.44021	0.01772	0.00234	0.13234	426.46317	7.55602	33
34	63.77744	0.01568	0.00207	0.13207	482.90338	7.57170	34
35	72.06851	0.01388	0.00183	0.13183	546.68082	7.58557	35
36	81.43741	0.01228	0.00162	0.13162	618.74933	7.59785	36
42	169.54876	0.00590	0.00077	0.13077	1296.52895	7.64694	42
48	352.99234	0.00283	0.00037	0.13037	2707.63342	7.67052	48
60	1530.05347	0.00065	0.00009	0.13009	11761.94979	7.68728	60

TABLE G.21

14.00%	Compound Interest Factors						14.00%
	Single Payment		**Uniform Payment Series**				
n	Compound Amount Factor	Present Worth Factor	Sinking Fund Factor	Capital Recovery Factor	Compound Amount Factor	Present Worth Factor	n
	Find F given P F/P	Find P given F P/F	Find A given F A/F	Find A given P A/P	Find F given A F/A	Find P given A P/A	
1	1.14000	0.87719	1.00000	1.14000	1.00000	0.87719	1
2	1.29960	0.76947	0.46729	0.60729	2.14000	1.64666	2
3	1.48154	0.67497	0.29073	0.43073	3.43960	2.32163	3
4	1.68896	0.59208	0.20320	0.34320	4.92114	2.91371	4
5	1.92541	0.51937	0.15128	0.29128	6.61010	3.43308	5
6	2.19497	0.45559	0.11716	0.25716	8.53552	3.88867	6
7	2.50227	0.39964	0.09319	0.23319	10.73049	4.28830	7
8	2.85259	0.35056	0.07557	0.21557	13.23276	4.63886	8
9	3.25195	0.30751	0.06217	0.20217	16.08535	4.94637	9
10	3.70722	0.26974	0.05171	0.19171	19.33730	5.21612	10
11	4.22623	0.23662	0.04339	0.18339	23.04452	5.45273	11
12	4.81790	0.20756	0.03667	0.17667	27.27075	5.66029	12
13	5.49241	0.18207	0.03116	0.17116	32.08865	5.84236	13
14	6.26135	0.15971	0.02661	0.16661	37.58107	6.00207	14
15	7.13794	0.14010	0.02281	0.16281	43.84241	6.14217	15
16	8.13725	0.12289	0.01962	0.15962	50.98035	6.26506	16
17	9.27646	0.10780	0.01692	0.15692	59.11760	6.37286	17
18	10.57517	0.09456	0.01462	0.15462	68.39407	6.46742	18
19	12.05569	0.08295	0.01266	0.15266	78.96923	6.55037	19
20	13.74349	0.07276	0.01099	0.15099	91.02493	6.62313	20
21	15.66758	0.06383	0.00954	0.14954	104.76842	6.68696	21
22	17.86104	0.05599	0.00830	0.14830	120.43600	6.74294	22
23	20.36158	0.04911	0.00723	0.14723	138.29704	6.79206	23
24	23.21221	0.04308	0.00630	0.14630	158.65862	6.83514	24
25	26.46192	0.03779	0.00550	0.14550	181.87083	6.87293	25
30	50.95016	0.01963	0.00280	0.14280	356.78685	7.00266	30
31	58.08318	0.01722	0.00245	0.14245	407.73701	7.01988	31
32	66.21483	0.01510	0.00215	0.14215	465.82019	7.03498	32
33	75.48490	0.01325	0.00188	0.14188	532.03501	7.04823	33
34	86.05279	0.01162	0.00165	0.14165	607.51991	7.05985	34
35	98.10018	0.01019	0.00144	0.14144	693.57270	7.07005	35
36	111.83420	0.00894	0.00126	0.14126	791.67288	7.07899	36
42	245.47301	0.00407	0.00057	0.14057	1746.23582	7.11376	42
48	538.80655	0.00186	0.00026	0.14026	3841.47534	7.12960	48
60	2595.91866	0.00039	0.00005	0.14005	18535.13328	7.14011	60

TABLE G.22

15.00%	Compound Interest Factors						15.00%
	Single Payment		Uniform Payment Series				
n	Compound Amount Factor	Present Worth Factor	Sinking Fund Factor	Capital Recovery Factor	Compound Amount Factor	Present Worth Factor	n
	Find F given P F/P	Find P given F P/F	Find A given F A/F	Find A given P A/P	Find F given A F/A	Find P given A P/A	
1	1.15000	0.86957	1.00000	1.15000	1.00000	0.86957	1
2	1.32250	0.75614	0.46512	0.61512	2.15000	1.62571	2
3	1.52088	0.65752	0.28798	0.43798	3.47250	2.28323	3
4	1.74901	0.57175	0.20027	0.35027	4.99338	2.85498	4
5	2.01136	0.49718	0.14832	0.29832	6.74238	3.35216	5
6	2.31306	0.43233	0.11424	0.26424	8.75374	3.78448	6
7	2.66002	0.37594	0.09036	0.24036	11.06680	4.16042	7
8	3.05902	0.32690	0.07285	0.22285	13.72682	4.48732	8
9	3.51788	0.28426	0.05957	0.20957	16.78584	4.77158	9
10	4.04556	0.24718	0.04925	0.19925	20.30372	5.01877	10
11	4.65239	0.21494	0.04107	0.19107	24.34928	5.23371	11
12	5.35025	0.18691	0.03448	0.18448	29.00167	5.42062	12
13	6.15279	0.16253	0.02911	0.17911	34.35192	5.58315	13
14	7.07571	0.14133	0.02469	0.17469	40.50471	5.72448	14
15	8.13706	0.12289	0.02102	0.17102	47.58041	5.84737	15
16	9.35762	0.10686	0.01795	0.16795	55.71747	5.95423	16
17	10.76126	0.09293	0.01537	0.16537	65.07509	6.04716	17
18	12.37545	0.08081	0.01319	0.16319	75.83636	6.12797	18
19	14.23177	0.07027	0.01134	0.16134	88.21181	6.19823	19
20	16.36654	0.06110	0.00976	0.15976	102.44358	6.25933	20
21	18.82152	0.05313	0.00842	0.15842	118.81012	6.31246	21
22	21.64475	0.04620	0.00727	0.15727	137.63164	6.35866	22
23	24.89146	0.04017	0.00628	0.15628	159.27638	6.39884	23
24	28.62518	0.03493	0.00543	0.15543	184.16784	6.43377	24
25	32.91895	0.03038	0.00470	0.15470	212.79302	6.46415	25
30	66.21177	0.01510	0.00230	0.15230	434.74515	6.56598	30
31	76.14354	0.01313	0.00200	0.15200	500.95692	6.57911	31
32	87.56507	0.01142	0.00173	0.15173	577.10046	6.59053	32
33	100.69983	0.00993	0.00150	0.15150	664.66552	6.60046	33
34	115.80480	0.00864	0.00131	0.15131	765.36535	6.60910	34
35	133.17552	0.00751	0.00113	0.15113	881.17016	6.61661	35
36	153.15185	0.00653	0.00099	0.15099	1014.34568	6.62314	36
42	354.24954	0.00282	0.00042	0.15042	2354.99693	6.64785	42
48	819.40071	0.00122	0.00018	0.15018	5456.00475	6.65853	48
60	4383.99875	0.00023	0.00003	0.15003	29219.99164	6.66515	60

TABLE G.23

20.00%	Compound Interest Factors						20.00%
	Single Payment		Uniform Payment Series				
n	Compound Amount Factor	Present Worth Factor	Sinking Fund Factor	Capital Recovery Factor	Compound Amount Factor	Present Worth Factor	n
	Find F given P F/P	Find P given F P/F	Find A given F A/F	Find A given P A/P	Find F given A F/A	Find P given A P/A	
1	1.20000	0.83333	1.00000	1.20000	1.00000	0.83333	1
2	1.44000	0.69444	0.45455	0.65455	2.20000	1.52778	2
3	1.72800	0.57870	0.27473	0.47473	3.64000	2.10648	3
4	2.07360	0.48225	0.18629	0.38629	5.36800	2.58873	4
5	2.48832	0.40188	0.13438	0.33438	7.44160	2.99061	5
6	2.98598	0.33490	0.10071	0.30071	9.92992	3.32551	6
7	3.58318	0.27908	0.07742	0.27742	12.91590	3.60459	7
8	4.29982	0.23257	0.06061	0.26061	16.49908	3.83716	8
9	5.15978	0.19381	0.04808	0.24808	20.79890	4.03097	9
10	6.19174	0.16151	0.03852	0.23852	25.95868	4.19247	10
11	7.43008	0.13459	0.03110	0.23110	32.15042	4.32706	11
12	8.91610	0.11216	0.02526	0.22526	39.58050	4.43922	12
13	10.69932	0.09346	0.02062	0.22062	48.49660	4.53268	13
14	12.83918	0.07789	0.01689	0.21689	59.19592	4.61057	14
15	15.40702	0.06491	0.01388	0.21388	72.03511	4.67547	15
16	18.48843	0.05409	0.01144	0.21144	87.44213	4.72956	16
17	22.18611	0.04507	0.00944	0.20944	105.93056	4.77463	17
18	26.62333	0.03756	0.00781	0.20781	128.11667	4.81219	18
19	31.94800	0.03130	0.00646	0.20646	154.74000	4.84350	19
20	38.33760	0.02608	0.00536	0.20536	186.68800	4.86958	20
21	46.00512	0.02174	0.00444	0.20444	225.02560	4.89132	21
22	55.20614	0.01811	0.00369	0.20369	271.03072	4.90943	22
23	66.24737	0.01509	0.00307	0.20307	326.23686	4.92453	23
24	79.49685	0.01258	0.00255	0.20255	392.48424	4.93710	24
25	95.39622	0.01048	0.00212	0.20212	471.98108	4.94759	25
30	237.37631	0.00421	0.00085	0.20085	1181.88157	4.97894	30
31	284.85158	0.00351	0.00070	0.20070	1419.25788	4.98245	31
32	341.82189	0.00293	0.00059	0.20059	1704.10946	4.98537	32
33	410.18627	0.00244	0.00049	0.20049	2045.93135	4.98781	33
34	492.22352	0.00203	0.00041	0.20041	2456.11762	4.98984	34
35	590.66823	0.00169	0.00034	0.20034	2948.34115	4.99154	35
36	708.80187	0.00141	0.00028	0.20028	3539.00937	4.99295	36
42	2116.47106	0.00047	0.00009	0.20009	10577.35529	4.99764	42
48	6319.74872	0.00016	0.00003	0.20003	31593.74358	4.99921	48
60	56347.51435	0.00002	0.00000	0.20000	281732.57177	4.99991	60

TABLE G.24

25.00%	Compound Interest Factors						25.00%
	Single Payment		**Uniform Payment Series**				
n	Compound Amount Factor	Present Worth Factor	Sinking Fund Factor	Capital Recovery Factor	Compound Amount Factor	Present Worth Factor	n
	Find F given P F/P	Find P given F P/F	Find A given F A/F	Find A given P A/P	Find F given A F/A	Find P given A P/A	
1	1.25000	0.80000	1.00000	1.25000	1.00000	0.80000	1
2	1.56250	0.64000	0.44444	0.69444	2.25000	1.44000	2
3	1.95313	0.51200	0.26230	0.51230	3.81250	1.95200	3
4	2.44141	0.40960	0.17344	0.42344	5.76563	2.36160	4
5	3.05176	0.32768	0.12185	0.37185	8.20703	2.68928	5
6	3.81470	0.26214	0.08882	0.33882	11.25879	2.95142	6
7	4.76837	0.20972	0.06634	0.31634	15.07349	3.16114	7
8	5.96046	0.16777	0.05040	0.30040	19.84186	3.32891	8
9	7.45058	0.13422	0.03876	0.28876	25.80232	3.46313	9
10	9.31323	0.10737	0.03007	0.28007	33.25290	3.57050	10
11	11.64153	0.08590	0.02349	0.27349	42.56613	3.65640	11
12	14.55192	0.06872	0.01845	0.26845	54.20766	3.72512	12
13	18.18989	0.05498	0.01454	0.26454	68.75958	3.78010	13
14	22.73737	0.04398	0.01150	0.26150	86.94947	3.82408	14
15	28.42171	0.03518	0.00912	0.25912	109.68684	3.85926	15
16	35.52714	0.02815	0.00724	0.25724	138.10855	3.88741	16
17	44.40892	0.02252	0.00576	0.25576	173.63568	3.90993	17
18	55.51115	0.01801	0.00459	0.25459	218.04460	3.92794	18
19	69.38894	0.01441	0.00366	0.25366	273.55576	3.94235	19
20	86.73617	0.01153	0.00292	0.25292	342.94470	3.95388	20
21	108.42022	0.00922	0.00233	0.25233	429.68087	3.96311	21
22	135.52527	0.00738	0.00186	0.25186	538.10109	3.97049	22
23	169.40659	0.00590	0.00148	0.25148	673.62636	3.97639	23
24	211.75824	0.00472	0.00119	0.25119	843.03295	3.98111	24
25	264.69780	0.00378	0.00095	0.25095	1054.79118	3.98489	25
30	807.79357	0.00124	0.00031	0.25031	3227.17427	3.99505	30
31	1009.74196	0.00099	0.00025	0.25025	4034.96783	3.99604	31
32	1262.17745	0.00079	0.00020	0.25020	5044.70979	3.99683	32
33	1577.72181	0.00063	0.00016	0.25016	6306.88724	3.99746	33
34	1972.15226	0.00051	0.00013	0.25013	7884.60905	3.99797	34
35	2465.19033	0.00041	0.00010	0.25010	9856.76132	3.99838	35
36	3081.48791	0.00032	0.00008	0.25008	12321.95164	3.99870	36
42	11754.94351	0.00009	0.00002	0.25002	47015.77403	3.99966	42
48	44841.55086	0.00002	0.00001	0.25001	179362.20343	3.99991	48
60	652530.44680	0.00000	0.00000	0.25000	2610117.78720	3.99999	60

TABLE G.25

30.00%	Compound Interest Factors						30.00%
	Single Payment		**Uniform Payment Series**				
n	Compound Amount Factor	Present Worth Factor	Sinking Fund Factor	Capital Recovery Factor	Compound Amount Factor	Present Worth Factor	n
	Find F given P F/P	Find P given F P/F	Find A given F A/F	Find A given P A/P	Find F given A F/A	Find P given A P/A	
1	1.30000	0.76923	1.00000	1.30000	1.00000	0.76923	1
2	1.69000	0.59172	0.43478	0.73478	2.30000	1.36095	2
3	2.19700	0.45517	0.25063	0.55063	3.99000	1.81611	3
4	2.85610	0.35013	0.16163	0.46163	6.18700	2.16624	4
5	3.71293	0.26933	0.11058	0.41058	9.04310	2.43557	5
6	4.82681	0.20718	0.07839	0.37839	12.75603	2.64275	6
7	6.27485	0.15937	0.05687	0.35687	17.58284	2.80211	7
8	8.15731	0.12259	0.04192	0.34192	23.85769	2.92470	8
9	10.60450	0.09430	0.03124	0.33124	32.01500	3.01900	9
10	13.78585	0.07254	0.02346	0.32346	42.61950	3.09154	10
11	17.92160	0.05580	0.01773	0.31773	56.40535	3.14734	11
12	23.29809	0.04292	0.01345	0.31345	74.32695	3.19026	12
13	30.28751	0.03302	0.01024	0.31024	97.62504	3.22328	13
14	39.37376	0.02540	0.00782	0.30782	127.91255	3.24867	14
15	51.18589	0.01954	0.00598	0.30598	167.28631	3.26821	15
16	66.54166	0.01503	0.00458	0.30458	218.47220	3.28324	16
17	86.50416	0.01156	0.00351	0.30351	285.01386	3.29480	17
18	112.45541	0.00889	0.00269	0.30269	371.51802	3.30369	18
19	146.19203	0.00684	0.00207	0.30207	483.97343	3.31053	19
20	190.04964	0.00526	0.00159	0.30159	630.16546	3.31579	20
21	247.06453	0.00405	0.00122	0.30122	820.21510	3.31984	21
22	321.18389	0.00311	0.00094	0.30094	1067.27963	3.32296	22
23	417.53905	0.00239	0.00072	0.30072	1388.46351	3.32535	23
24	542.80077	0.00184	0.00055	0.30055	1806.00257	3.32719	24
25	705.64100	0.00142	0.00043	0.30043	2348.80334	3.32861	25
30	2619.99564	0.00038	0.00011	0.30011	8729.98548	3.33206	30
31	3405.99434	0.00029	0.00009	0.30009	11349.98112	3.33235	31
32	4427.79264	0.00023	0.00007	0.30007	14755.97546	3.33258	32
33	5756.13043	0.00017	0.00005	0.30005	19183.76810	3.33275	33
34	7482.96956	0.00013	0.00004	0.30004	24939.89853	3.33289	34
35	9727.86043	0.00010	0.00003	0.30003	32422.86808	3.33299	35
36	12646.21855	0.00008	0.00002	0.30002	42150.72851	3.33307	36
42	61040.88153	0.00002	0.00000	0.30000	203466.27175	3.33328	42
48	294632.67632	0.00000	0.00000	0.30000	982105.58773	3.33332	48
60	6864377.17274	0.00000	0.00000	0.30000	22881253.90915	3.33333	60

Appendix H

Plans for Small Gas Station

SERVICE BAY SIDE ELEVATION

SHOW ROOM SIDE ELEVATION

REAR ELEVATION

FRONT ELEVATION

DRAWING SCHEDULE:

CONSTRUCTIONAL DRAWINGS

DWG. NO.	TITLE
SHT. 1	ELEVATIONS
SHT. 2	FOUNDATION & FLOOR PLAN & ANCHOR BOLTS
SHT. 3	FDN. & WALL SECTIONS & DETAILS
SHT. 4	INTERIOR DETAILS & HEATING
SHT. 5	AIR & WATER PIPING
SHT. 6	ELECTRICAL
SHT. 7	SHELVING

METAL BLDG.

SPECIFICATION 200 A-160

GENERAL NOTES:

1. THE CONTRACTOR SHALL REFER TO THE "PLOT PLAN" INSTRUCTIONS TO THE BIDDERS AND/OR ADDENDUM TO THE SPECIFICATIONS FOR SUPPLEMENTARY INFORMATION IN CONNECTION WITH THE WORK SHOWN ON THESE DRAWINGS.

2. DESIGN LOADS & GENERAL REQUIREMENTS:

	LIVE LOAD	WIND PRESSURE	SOIL BEARING PRESSURE
ROOF	40 PSF	EXT. WALLS 15 PSF	ASSUMED TO BE 3000 PSF
FLOOR	100 PSF	ROOF 20 PSF	

ADDITIONAL REQUIREMENTS ARE LISTED UNDER THE APPLICABLE SECTIONS OF THE CONSTRUCTION SHEETS.

3. SEE PLOT PLAN FOR SIZE, TYPE, LOCATION & NUMBER OF PUMP ISLANDS, SIGNS, BASES & ALL OTHER WORK TO BE CONSTRUCTED TO WITHIN PLUG OR MINUS 1/4".

4. ALL SERVICE BAY FLOOR LOADS SHALL BE
5. SEE PLOT PLAN FOR NUMBER OF LIFTS TO BE INSTALLED.

SERVICE STATION

ELEVATIONS

DRAWING NO.	20-200
SCALE 1/8" = 1'-0"	SHEET NO. 1 of 7

1 of 7

REV. 3

REVISIONS

NO.	DATE BY
1	GENERAL FL. TO PANELS/BLK.
2	ALTERED FL. TO PANELS/BLK.
3	RELATED BRICK TEXT. (TOP.)

437

Appendix I

Site Reconnaissance Checklist

GENERAL CONSIDERATIONS

    A.   What features are native to topography and climate?

    B.   What is required for construction method selected?

    C.   What features are needed to support construction force?

    D.   What features might encroach on local society or environment?

    A.   Features native to topography and climate

        1. Actual topography (excessive grades, etc.)

        2. Elevation

        3. Geology (soil characteristics, rock, etc.)

        4. Ground cover

        5. Excessive seasonal effects

        6. Wind direction

        7. Natural defenses

        8. Drainage

        9. Subsurface water conditions

        10. Seismic zones

    B.   Features required that contribute to construction method

        1. Accessibility to site (rail, road, water)

        2. Labor availability (skill, cost, attitude)

        3. Material availability (salvage, cost, attitude)

        4. Locate borrow pits (gravel, sand, base, fill)

        5. Locate storage areas, plant sites

        6. Alternate building, campsites

        7. General working room about site

        8. Location of existing structures and utilities

        9. Conflicts with existing structures and utilities

        10. Overhead

        11. Disposal areas

        12. Land usage

        13. Local building practices

C.   Features to support construction force

1. Billeting/shelter
2. Food (also on-job meals)
3. Special equipment
4. Clothing
5. Communications
6. Local hazards
7. Fire/security protection available
8. Local customs/culture
9. Potable H_2O
10. Sanitary facilities (also for job)
11. Entertainment
12. Small stores
13. Medical
14. Banking, currency
15. Transportation
16. Local maintenance available

D.   Features that might encroach on local society or environment

1. Noise
2. Dust
3. Blasting
4. Hauling over roads
5. Use of water
6. Burning (smoke)
7. Drainage (create problems)
8. Flight operations
9. Disposal areas
10. Utility disruption
11. Relocation problems
12. Work hours
13. Economy impact
14. Community attitude
15. Security
16. Political

Appendix $J(1)$

The Cumulative Normal Distribution Function[†]

$$\Phi(z)=\int_{-\infty}^{z}\phi(t)\,dt \qquad \text{for } (-\infty < z \leqq 0)$$

z	·00	·01	·02	·03	·04	·05	·06	·07	·08	·09
− ·0	·5000	·4960	·4920	·4880	·4840	·4801	·4761	·4721	·4681	·4641
− ·1	·4602	·4562	·4522	·4483	·4443	·4404	·4364	·4325	·4286	·4247
− ·2	·4207	·4168	·4129	·4090	·4052	·4013	·3974	·3936	·3897	·3859
− ·3	·3821	·3783	·3745	·3707	·3669	·3632	·3594	·3557	·3520	·3483
− ·4	·3446	·3409	·3372	·3336	·3300	·3264	·3228	·3192	·3156	·3121
− ·5	·3085	·3050	·3015	·2981	·2946	·2912	·2877	·2843	·2810	·2776
− ·6	·2743	·2709	·2676	·2643	·2611	·2578	·2546	·2514	·2483	·2451
− ·7	·2420	·2389	·2358	·2327	·2297	·2266	·2236	·2206	·2177	·2148
− ·8	·2119	·2090	·2061	·2033	·2005	·1977	·1949	·1922	·1894	·1867
− ·9	·1841	·1814	·1788	·1762	·1736	·1711	·1685	·1660	·1635	·1611
−1·0	·1587	·1562	·1539	·1515	·1492	·1469	·1446	·1423	·1401	·1379
−1·1	·1357	·1335	·1314	·1292	·1271	·1251	·1230	·1210	·1190	·1170
−1·2	·1151	·1131	·1112	·1093	·1075	·1056	·1038	·1020	·1003	·09853
−1·3	·09680	·09510	·09342	·09176	·09012	·08851	·08691	·08534	·08379	·08226
−1·4	·08076	·07927	·07780	·07636	·07493	·07353	·07215	·07078	·06944	·06811
−1·5	·06681	·06552	·06426	·06301	·06178	·06057	·05938	·05821	·05705	·05592
−1·6	·05480	·05370	·05262	·05155	·05050	·04947	·04846	·04746	·04648	·04551
−1·7	·04457	·04363	·04272	·04182	·04093	·04006	·03920	·03836	·03754	·03673
−1·8	·03593	·03515	·03438	·03362	·03288	·03216	·03144	·03074	·03005	·02938
−1·9	·02872	·02807	·02743	·02680	·02619	·02559	·02500	·02442	·02385	·02330
−2·0	·02275	·02222	·02169	·02118	·02068	·02018	·01970	·01923	·01876	·01831
−2·1	·01786	·01743	·01700	·01659	·01618	·01578	·01539	·01500	·01463	·01426
−2·2	·01390	·01355	·01321	·01287	·01255	·01222	·01191	·01160	·01130	·01101
−2·3	·01072	·01044	·01017	$\cdot0^2$9903	$\cdot0^2$9642	$\cdot0^2$9387	$\cdot0^2$9137	$\cdot0^2$8894	$\cdot0^2$8656	$\cdot0^2$8424
−2·4	$\cdot0^2$8198	$\cdot0^2$7976	$\cdot0^2$7760	$\cdot0^2$7549	$\cdot0^2$7344	$\cdot0^2$7143	$\cdot0^2$6947	$\cdot0^2$6756	$\cdot0^2$6569	$\cdot0^2$6387
−2·5	$\cdot0^2$6210	$\cdot0^2$6037	$\cdot0^2$5868	$\cdot0^2$5703	$\cdot0^2$5543	$\cdot0^2$5386	$\cdot0^2$5234	$\cdot0^2$5085	$\cdot0^2$4940	$\cdot0^2$4799
−2·6	$\cdot0^2$4661	$\cdot0^2$4527	$\cdot0^2$4396	$\cdot0^2$4269	$\cdot0^2$4145	$\cdot0^2$4025	$\cdot0^2$3907	$\cdot0^2$3793	$\cdot0^2$3681	$\cdot0^2$3573
−2·7	$\cdot0^2$3467	$\cdot0^2$3364	$\cdot0^2$3264	$\cdot0^2$3167	$\cdot0^2$3072	$\cdot0^2$2980	$\cdot0^2$2890	$\cdot0^2$2803	$\cdot0^2$2718	$\cdot0^2$2635
−2·8	$\cdot0^2$2555	$\cdot0^2$2477	$\cdot0^2$2401	$\cdot0^2$2327	$\cdot0^2$2256	$\cdot0^2$2186	$\cdot0^2$2118	$\cdot0^2$2052	$\cdot0^2$1988	$\cdot0^2$1926
−2·9	$\cdot0^2$1866	$\cdot0^2$1807	$\cdot0^2$1750	$\cdot0^2$1695	$\cdot0^2$1641	$\cdot0^2$1589	$\cdot0^2$1538	$\cdot0^2$1489	$\cdot0^2$1441	$\cdot0^2$1395
−3·0	$\cdot0^2$1350	$\cdot0^2$1306	$\cdot0^2$1264	$\cdot0^2$1223	$\cdot0^2$1183	$\cdot0^2$1144	$\cdot0^2$1107	$\cdot0^2$1070	$\cdot0^2$1035	$\cdot0^2$1001
−3·1	$\cdot0^3$9676	$\cdot0^3$9354	$\cdot0^3$9043	$\cdot0^3$8740	$\cdot0^3$8447	$\cdot0^3$8164	$\cdot0^3$7888	$\cdot0^3$7622	$\cdot0^3$7364	$\cdot0^3$7114
−3·2	$\cdot0^3$6871	$\cdot0^3$6637	$\cdot0^3$6410	$\cdot0^3$6190	$\cdot0^3$5976	$\cdot0^3$5770	$\cdot0^3$5571	$\cdot0^3$5377	$\cdot0^3$5190	$\cdot0^3$5009
−3·3	$\cdot0^3$4834	$\cdot0^3$4665	$\cdot0^3$4501	$\cdot0^3$4342	$\cdot0^3$4189	$\cdot0^3$4041	$\cdot0^3$3897	$\cdot0^3$3758	$\cdot0^3$3624	$\cdot0^3$3495
−3·4	$\cdot0^3$3369	$\cdot0^3$3248	$\cdot0^3$3131	$\cdot0^3$3018	$\cdot0^3$2909	$\cdot0^3$2803	$\cdot0^3$2701	$\cdot0^3$2602	$\cdot0^3$2507	$\cdot0^3$2415
−3·5	$\cdot0^3$2326	$\cdot0^3$2241	$\cdot0^3$2158	$\cdot0^3$2078	$\cdot0^3$2001	$\cdot0^3$1926	$\cdot0^3$1854	$\cdot0^3$1785	$\cdot0^3$1718	$\cdot0^3$1653
−3·6	$\cdot0^3$1591	$\cdot0^3$1531	$\cdot0^3$1473	$\cdot0^3$1417	$\cdot0^3$1363	$\cdot0^3$1311	$\cdot0^3$1261	$\cdot0^3$1213	$\cdot0^3$1166	$\cdot0^3$1121
−3·7	$\cdot0^3$1078	$\cdot0^3$1036	$\cdot0^4$9961	$\cdot0^4$9574	$\cdot0^4$9201	$\cdot0^4$8842	$\cdot0^4$8496	$\cdot0^4$8162	$\cdot0^4$7841	$\cdot0^4$7532
−3·8	$\cdot0^4$7235	$\cdot0^4$6948	$\cdot0^4$6673	$\cdot0^4$6407	$\cdot0^4$6152	$\cdot0^4$5906	$\cdot0^4$5669	$\cdot0^4$5442	$\cdot0^4$5223	$\cdot0^4$5012
−3·9	$\cdot0^4$4810	$\cdot0^4$4615	$\cdot0^4$4427	$\cdot0^4$4247	$\cdot0^4$4074	$\cdot0^4$3908	$\cdot0^4$3747	$\cdot0^4$3594	$\cdot0^4$3446	$\cdot0^4$3304
−4·0	$\cdot0^4$3167	$\cdot0^4$3036	$\cdot0^4$2910	$\cdot0^4$2789	$\cdot0^4$2673	$\cdot0^4$2561	$\cdot0^4$2454	$\cdot0^4$2351	$\cdot0^4$2252	$\cdot0^4$2157
−4·1	$\cdot0^4$2066	$\cdot0^4$1978	$\cdot0^4$1894	$\cdot0^4$1814	$\cdot0^4$1737	$\cdot0^4$1662	$\cdot0^4$1591	$\cdot0^4$1523	$\cdot0^4$1458	$\cdot0^4$1395
−4·2	$\cdot0^4$1335	$\cdot0^4$1277	$\cdot0^4$1222	$\cdot0^4$1168	$\cdot0^4$1118	$\cdot0^4$1069	$\cdot0^4$1022	$\cdot0^5$9774	$\cdot0^5$9345	$\cdot0^5$8934
−4·3	$\cdot0^5$8540	$\cdot0^5$8163	$\cdot0^5$7801	$\cdot0^5$7455	$\cdot0^5$7124	$\cdot0^5$6807	$\cdot0^5$6503	$\cdot0^5$6212	$\cdot0^5$5934	$\cdot0^5$5668
−4·4	$\cdot0^5$5413	$\cdot0^5$5169	$\cdot0^5$4935	$\cdot0^5$4712	$\cdot0^5$4498	$\cdot0^5$4294	$\cdot0^5$4098	$\cdot0^5$3911	$\cdot0^5$3732	$\cdot0^5$3561
−4·5	$\cdot0^5$3398	$\cdot0^5$3241	$\cdot0^5$3092	$\cdot0^5$2949	$\cdot0^5$2813	$\cdot0^5$2682	$\cdot0^5$2558	$\cdot0^5$2439	$\cdot0^5$2325	$\cdot0^5$2216
−4·6	$\cdot0^5$2112	$\cdot0^5$2013	$\cdot0^5$1919	$\cdot0^5$1828	$\cdot0^5$1742	$\cdot0^5$1660	$\cdot0^5$1581	$\cdot0^5$1506	$\cdot0^5$1434	$\cdot0^5$1366
−4·7	$\cdot0^5$1301	$\cdot0^5$1239	$\cdot0^5$1179	$\cdot0^5$1123	$\cdot0^5$1069	$\cdot0^5$1017	$\cdot0^6$9680	$\cdot0^6$9211	$\cdot0^6$8765	$\cdot0^6$8339
−4·8	$\cdot0^6$7933	$\cdot0^6$7547	$\cdot0^6$7178	$\cdot0^6$6827	$\cdot0^6$6492	$\cdot0^6$6173	$\cdot0^6$5869	$\cdot0^6$5580	$\cdot0^6$5304	$\cdot0^6$5042
−4·9	$\cdot0^6$4792	$\cdot0^6$4554	$\cdot0^6$4327	$\cdot0^6$4111	$\cdot0^6$3906	$\cdot0^6$3711	$\cdot0^6$3525	$\cdot0^6$3348	$\cdot0^6$3179	$\cdot0^6$3019

Example: $\Phi(-3\cdot57) = \cdot0^3 1785 = 0\cdot0001785.$

† By permission from A. Hald, *Statistical Tables, and Formulas*, John Wiley & Sons, Inc., New York, 1952.

Appendix J(2)

The Cumulative Normal Distribution Function[†]

$$\Phi(z) = \int_{-\infty}^{z} \phi(t)\,dt \qquad \text{for } (0 \leqq z \lneq \infty)$$

z	·00	·01	·02	·03	·04	·05	·06	·07	·08	·09
·0	·5000	·5040	·5080	·5120	·5160	·5199	·5239	·5279	·5319	·5359
·1	·5398	·5438	·5478	·5517	·5557	·5596	·5636	·5675	·5714	·5753
·2	·5793	·5832	·5871	·5910	·5948	·5987	·6026	·6064	·6103	·6141
·3	·6179	·6217	·6255	·6293	·6331	·6368	·6406	·6443	·6480	·6517
·4	·6554	·6591	·6628	·6664	·6700	·6736	·6772	·6808	·6844	·6879
·5	·6915	·6950	·6985	·7019	·7054	·7088	·7123	·7157	·7190	·7224
·6	·7257	·7291	·7324	·7357	·7389	·7422	·7454	·7486	·7517	·7549
·7	·7580	·7611	·7642	·7673	·7703	·7734	·7764	·7794	·7823	·7852
·8	·7881	·7910	·7939	·7967	·7995	·8023	·8051	·8078	·8106	·8133
·9	·8159	·8186	·8212	·8238	·8264	·8289	·8315	·8340	·8365	·8389
1·0	·8413	·8438	·8461	·8485	·8508	·8531	·8554	·8577	·8599	·8621
1·1	·8643	·8665	·8686	·8708	·8729	·8749	·8770	·8790	·8810	·8830
1·2	·8849	·8869	·8888	·8907	·8925	·8944	·8962	·8980	·8997	·90147
1·3	·90320	·90490	·90658	·90824	·90988	·91149	·91309	·91466	·91621	·91774
1·4	·91924	·92073	·92220	·92364	·92507	·92647	·92785	·92922	·93056	·93189
1·5	·93319	·93448	·93574	·93699	·93822	·93943	·94062	·94179	·94295	·94408
1·6	·94520	·94630	·94738	·94845	·94950	·95053	·95154	·95254	·95352	·95449
1·7	·95543	·95637	·95728	·95818	·95907	·95994	·96080	·96164	·96246	·96327
1·8	·96407	·96485	·96562	·96638	·96712	·96784	·96856	·96926	·96995	·97062
1·9	·97128	·97193	·97257	·97320	·97381	·97441	·97500	·97558	·97615	·97670
2·0	·97725	·97778	·97831	·97882	·97932	·97982	·98030	·98077	·98124	·98169
2·1	·98214	·98257	·98300	·98341	·98382	·98422	·98461	·98500	·98537	·98574
2·2	·98610	·98645	·98679	·98713	·98745	·98778	·98809	·98840	·98870	·98899
2·3	·98928	·98956	·98983	·$9^{2}0097$	·$9^{2}0358$	·$9^{2}0613$	·$9^{2}0863$	·$9^{2}1106$	·$9^{2}1344$	·$9^{2}1576$
2·4	·$9^{2}1802$	·$9^{2}2024$	·$9^{2}2240$	·$9^{2}2451$	·$9^{2}2656$	·$9^{2}2857$	·$9^{2}3053$	·$9^{2}3244$	·$9^{2}3431$	·$9^{2}3613$
2·5	·$9^{2}3790$	·$9^{2}3963$	·$9^{2}4132$	·$9^{2}4297$	·$9^{2}4457$	·$9^{2}4614$	·$9^{2}4766$	·$9^{2}4915$	·$9^{2}5060$	·$9^{2}5201$
2·6	·$9^{2}5339$	·$9^{2}5473$	·$9^{2}5604$	·$9^{2}5731$	·$9^{2}5855$	·$9^{2}5975$	·$9^{2}6093$	·$9^{2}6207$	·$9^{2}6319$	·$9^{2}6427$
2·7	·$9^{2}6533$	·$9^{2}6636$	·$9^{2}6736$	·$9^{2}6833$	·$9^{2}6928$	·$9^{2}7020$	·$9^{2}7110$	·$9^{2}7197$	·$9^{2}7282$	·$9^{2}7365$
2·8	·$9^{2}7445$	·$9^{2}7523$	·$9^{2}7599$	·$9^{2}7673$	·$9^{2}7744$	·$9^{2}7814$	·$9^{2}7882$	·$9^{2}7948$	·$9^{2}8012$	·$9^{2}8074$
2·9	·$9^{2}8134$	·$9^{2}8193$	·$9^{2}8250$	·$9^{2}8305$	·$9^{2}8359$	·$9^{2}8411$	·$9^{2}8462$	·$9^{2}8511$	·$9^{2}8559$	·$9^{2}8605$
3·0	·$9^{2}8650$	·$9^{2}8694$	·$9^{2}8736$	·$9^{2}8777$	·$9^{2}8817$	·$9^{2}8856$	·$9^{2}8893$	·$9^{2}8930$	·$9^{2}8965$	·$9^{2}8999$
3·1	·$9^{3}0324$	·$9^{3}0646$	·$9^{3}0957$	·$9^{3}1260$	·$9^{3}1553$	·$9^{3}1836$	·$9^{3}2112$	·$9^{3}2378$	·$9^{3}2636$	·$9^{3}2886$
3·2	·$9^{3}3129$	·$9^{3}3363$	·$9^{3}3590$	·$9^{3}3810$	·$9^{3}4024$	·$9^{3}4230$	·$9^{3}4429$	·$9^{3}4623$	·$9^{3}4810$	·$9^{3}4991$
3·3	·$9^{3}5166$	·$9^{3}5335$	·$9^{3}5499$	·$9^{3}5658$	·$9^{3}5811$	·$9^{3}5959$	·$9^{3}6103$	·$9^{3}6242$	·$9^{3}6376$	·$9^{3}6505$
3·4	·$9^{3}6631$	·$9^{3}6752$	·$9^{3}6869$	·$9^{3}6982$	·$9^{3}7091$	·$9^{3}7197$	·$9^{3}7299$	·$9^{3}7398$	·$9^{3}7493$	·$9^{3}7585$
3·5	·$9^{3}7674$	·$9^{3}7759$	·$9^{3}7842$	·$9^{3}7922$	·$9^{3}7999$	·$9^{3}8074$	·$9^{3}8146$	·$9^{3}8215$	·$9^{3}8282$	·$9^{3}8347$
3·6	·$9^{3}8409$	·$9^{3}8469$	·$9^{3}8527$	·$9^{3}8583$	·$9^{3}8637$	·$9^{3}8689$	·$9^{3}8739$	·$9^{3}8787$	·$9^{3}8834$	·$9^{3}8879$
3·7	·$9^{3}8922$	·$9^{3}8964$	·$9^{4}0039$	·$9^{4}0426$	·$9^{4}0799$	·$9^{4}1158$	·$9^{4}1504$	·$9^{4}1838$	·$9^{4}2159$	·$9^{4}2468$
3·8	·$9^{4}2765$	·$9^{4}3052$	·$9^{4}3327$	·$9^{4}3593$	·$9^{4}3848$	·$9^{4}4094$	·$9^{4}4331$	·$9^{4}4558$	·$9^{4}4777$	·$9^{4}4988$
3·9	·$9^{4}5190$	·$9^{4}5385$	·$9^{4}5573$	·$9^{4}5753$	·$9^{4}5926$	·$9^{4}6092$	·$9^{4}6253$	·$9^{4}6406$	·$9^{4}6554$	·$9^{4}6696$
4·0	·$9^{4}6833$	·$9^{4}6964$	·$9^{4}7090$	·$9^{4}7211$	·$9^{4}7327$	·$9^{4}7439$	·$9^{4}7546$	·$9^{4}7649$	·$9^{4}7748$	·$9^{4}7843$
4·1	·$9^{4}7934$	·$9^{4}8022$	·$9^{4}8106$	·$9^{4}8186$	·$9^{4}8263$	·$9^{4}8338$	·$9^{4}8409$	·$9^{4}8477$	·$9^{4}8542$	·$9^{4}8605$
4·2	·$9^{4}8665$	·$9^{4}8723$	·$9^{4}8778$	·$9^{4}8832$	·$9^{4}8882$	·$9^{4}8931$	·$9^{4}8978$	·$9^{5}0226$	·$9^{5}0655$	·$9^{5}1066$
4·3	·$9^{5}1460$	·$9^{5}1837$	·$9^{5}2199$	·$9^{5}2545$	·$9^{5}2876$	·$9^{5}3193$	·$9^{5}3497$	·$9^{5}3788$	·$9^{5}4066$	·$9^{5}4332$
4·4	·$9^{5}4587$	·$9^{5}4831$	·$9^{5}5065$	·$9^{5}5288$	·$9^{5}5502$	·$9^{5}5706$	·$9^{5}5902$	·$9^{5}6089$	·$9^{5}6268$	·$9^{5}6439$
4·5	·$9^{5}6602$	·$9^{5}6759$	·$9^{5}6908$	·$9^{5}7051$	·$9^{5}7187$	·$9^{5}7318$	·$9^{5}7442$	·$9^{5}7561$	·$9^{5}7675$	·$9^{5}7784$
4·6	·$9^{5}7888$	·$9^{5}7987$	·$9^{5}8081$	·$9^{5}8172$	·$9^{5}8258$	·$9^{5}8340$	·$9^{5}8419$	·$9^{5}8494$	·$9^{5}8566$	·$9^{5}8634$
4·7	·$9^{5}8699$	·$9^{5}8761$	·$9^{5}8821$	·$9^{5}8877$	·$9^{5}8931$	·$9^{5}8983$	·$9^{6}0320$	·$9^{6}0789$	·$9^{6}1235$	·$9^{6}1661$
4·8	·$9^{6}2067$	·$9^{6}2453$	·$9^{6}2822$	·$9^{6}3173$	·$9^{6}3508$	·$9^{6}3827$	·$9^{6}4131$	·$9^{6}4420$	·$9^{6}4696$	·$9^{6}4958$
4·9	·$9^{6}5208$	·$9^{6}5446$	·$9^{6}5673$	·$9^{6}5889$	·$9^{6}6094$	·$9^{6}6289$	·$9^{6}6475$	·$9^{6}6652$	·$9^{6}6821$	·$9^{6}6981$

Example: $\Phi(3 \cdot 57) = \cdot 9^{3}8215 = 0 \cdot 9998215$.

[†] By permission from A. Hald, *Statistical Tables, and Formulas*, John Wiley & Sons, Inc., New York, 1952.

Bibliography

1. Adrian, J. and D. J. Adrian. (2006). *Construction Accounting: Financial, Managerial, Auditing, and Tax.* Champaign IL: Stipes Publishing LLC.
2. Ahuja, H. N., S. P. Dozzi, and S. M. Abourizk (1994). *Project Management-Techniques in Planning and Controlling Construction Projects,* 2nd edition, Hoboken, NJ: John Wiley & Sons.
3. Allensworth, W. and American Bar Association. Forum on the Construction Industry. (2009). *Construction law.* Chicago, Forum on the Construction Industry, American Bar Association.
4. Angerame, M. (2002). *Engineering & Construction Project Management.* Denver, CO: Hampton Group.
5. Antill, J. M. and R. W. Woodhead (1982). *Critical Path Methods in Construction Practice,* 3rd edition, Hoboken, NJ: John Wiley & Sons.
6. Au, T. and T. P. Au (1992). *Engineering Economics for Capital Investment Analysis,* 2nd edition, Upper Saddle River, NJ: Prentice Hall.
7. Barrie, D. S., & Paulson, B. C. (1992). *Professional construction management: including CM, design-construct, and general contracting* (3rd ed). New York, NY: McGraw-Hill.
8. Bonny, John B., and Joseph P. Frein (1973). *Handbook of Construction Management and Organization,* New York, NY: Van Nostrand Reinhold Co.
9. *Building (The) Estimator's Reference Book,* 28th edition (2009). Chicago, IL: The Frank R. Walker Co.
10. Callahan, M. T. (2005). *Procurement of construction and design contracts.* New York, NY: Aspen Publishers.
11. Caterpillar Performance Handbook, 39th edition (2009). Peoria, IL: Caterpillar Tractor Co.
12. Collier, C. A. and W. B. Ledbetter (1988). *Engineering economic and cost analysis.* New York, Harper & Row.
13. Collier, K. (2001). *Construction Contracts,* 3rd. edition, Upper Saddle River, NJ: Prentice Hall.
14. Collier, N. S., Collier, C. A. and Halperin, D. A. (2002). *Construction Funding,* The Process of Real Estate Development, 3rd Edition, Hoboken, NJ: John Wiley & Sons.
15. Clough, R.H. and G.A. Sears (1994). *Construction Contracting,* 6th edition. New York: John Wiley & Sons.
16. *Credit Reports* (individual subscription), Building Construction Division, New York, NY: Dunn and Bradstreet, Inc.
17. Culp, G. and R. Anne Smith (1992). *Managing People (Including Yourself) for Project Success,* New York, NY: Van Nostrand Reinhold Co.
18. Dallavia, L. (1957). *Estimating General Construction Costs,* 2nd edition, New York, NY: F. W. Dodge Co.
19. De Kruif, P. (1932). *Microbe Hunters,* New York, NY: Harcourt, Brace and Co.
20. *Dodge Reports* (daily publication), New York, NY: F. W. Dodge Corp. (Division of McGraw-Hill, Inc.).
21. Dorsey, Robert W. (1997). *Project Delivery Systems for Building Construction,* Arlington, VA: Associated General Contractors of America.
22. Engineering News Record (weekly publication), New York, NY: McGraw-Hill, Inc.
23. Ernstrom, M., Hanson, D. et al. (2006). *The Contractors' Guide to BIM.* Arlington, VA: The Associated General Contractors of America.
24. Eschenbach, T. (2003). *Engineering economy: applying theory to practice.* New York, Oxford University Press.
25. Estey, M. (1967). *The Unions,* New York, NY: Harcourt, Brace Jovanovich.
26. Fellow, R., Langford D., et al. (2002) *Construction management in practice,* 2nd edition. Oxford, UK: Blackwell Science.
27. Fisk, E. R. and W. D. Reynolds (2006). *Construction project administration.* Upper Saddle River, N.J., Pearson Prentice Hall.
28. Friedrich, A. J., editor (1989). *Sons of Martha,* New York, NY: American Society of Civil Engineers.
29. Fryer, B. G., Fryer, M., Egbu, C. O., Ellis, R., & Gorse, C. A. (2004). *The practice of construction management: people and business performance* (4th ed.). Oxford, U.K.; Malden, Mass.: Blackwell Pub.
30. Gibb, T. W., Jr. (1975). *Building Construction in the Southeastern United States,* Report presented to the School of Civil Engineering, Georgia Institute of Technology, Atlanta GA.
31. Gould, F. E. and N. Joyce (2009). *Construction project management.* Upper Saddle River, N.J., Pearson Prentice Hall.
32. Halpin, D. and Senior, B. (2009). *Financial Management and Accounting Fundamentals for Construction.* Hoboken, N.J.: John Wiley & Sons.
33. Halpin, D. W. (1977). *"CYCLONE -Method for Modeling Job Site Processes,"* Journal of the Construction Division ASCE, 103 (CO3), 489–499.
34. Halpin, D. W. (1992). *MicroCYCLONE User's Manual,* West Lafayette, IN: Learning Systems, Inc.
35. Halpin, D. W. and L. S. Riggs (1992). *Planning and Analysis of Construction Operations,* Hoboken, NJ: John Wiley & Sons.
36. Halpin, D. W. and R. D. Neathammer (1973). *"Construction Time Overruns,"* Technical Report P-16, Champaign, IL: Construction Engineering Research Laboratory.
37. Haltenhoff, C. E. (1999). *The CM contracting system: fundamentals and practices.* Upper Saddle River, N.J.: Prentice Hall.
38. Hardin, B. (2009). *BIM and construction management: proven tools, methods, and workflows* (1st ed.) Indianapolis, Ind.: Wiley Technology Pub.

39. Harris, F., McCaffer, R., & Edum-Fotwe, F. (2006). *Modern construction management* (6th ed.). Oxford; Malden, MA: Blackwell.

40. Harris, R. B. (1978). Precedence and Arrow Networking Techniques for Construction, Hoboken, NJ: John Wiley & Sons.

41. Hinze, J. (2008). *Construction planning and scheduling.* Upper Saddle River, N.J., Pearson Prentice Hall.

42. Hinze, Jimmie (2001). *Construction Contracts*, 2nd Edition, New York, NY: McGraw-Hill.

43. Hoover, Herbert (1951). *The Memoirs of Herbert Hoover - Years of Adventure 1874-1920*, New York, NY: The MacMillan Company.

44. Jackson, B. J. (2004). *Construction management jumpstart.* San Francisco, Calif.; London: SYBEX.

45. Jackson, John Howard (1973). *Contract Law in Modern Society*, St. Paul, MN: West Publishing Co.

46. Kelly, J. E. and Walker, M. R. (1959). *Critical Path Planning and Scheduling*, Proceedings of the Eastern Joint Computer Conference, pp. 160–173.

47. Kenig, M., Coordinator (2004). *Project Delivery Systems for Construction*, Arlington, VA: Associated General Contractors of America.

48. Kennedy Venoit, W. (2009). *International construction law: a guide for cross-border transactions and legal disputes.* Chicago, IL: American Bar Association.

49. Kerzner, H. (1989). *Project Management—A System's Approach to Planning, Scheduling and Controlling*, 3rd edition, New York, NY: Van Nostrand Reinhold Co.

50. Knutson, K., & Schexnayder, C. J. (2009). *Construction management fundamentals* (2nd ed.). Boston: McGraw-Hill Higher Education.

51. Kubba, S. *Green construction project management and cost oversight.* Burlington, MA, Architectural Press.

52. Kymmell, W. (2008). *Building information modeling: planning and managing construction projects with 4D CAD and simulations.* New York, McGraw-Hill.

53. Lang, H. J. and M. De Coursey (1983). *Profitability Accounting and Bidding Strategy for Engineering and Construction Management*, Van Nostrand Reinhold Co. New York.

54. Levitt, R. E. and N. M. Samelson (1993). *Construction Safety Management*, 2nd edition, Hoboken, NJ: John Wiley & Sons.

55. Levy, S. M. (1994). *Project Management in Construction*, 2nd edition, New York, NY: McGraw-Hill, Inc.

56. Levy, S. M. (1996). *Build Operate Transfer*, Hoboken, NJ: John Wiley & Sons.

57. Loulakis, M. C. (1995). *Construction management: law and practice.* New York: Wiley Law Publications.

58. Maor, E., *The Story of a Number*, Princeton, N.J.: Princeton University Press.

59. McCullough, D. (1972). *The Great Bridge*, New York, NY: Touchstone Books, Simon and Schuster.

60. McCullough, D. (1977). *The Path Between the Seas*, Touchstone Books, New York, NY: Simon and Schuster.

61. McGeorge, W. D., Palmer, A., & London, K. (2002). *Construction management: new directions* (2nd ed.). Osney Mead, Oxford UK; Malden, Mass: Blackwell Science.

62. Menheere, S. C. M. and Pollais, S. N. (1996). *Case Studies on Build Operate Transfer*, Delft, The Netherlands: Delft University of Technology.

63. Mills, D. Q. (1994). *Labor Management Relations*, 5th edition, New York, NY: McGraw-Hill.

64. Moder, J. J., C. R. Phillips, and E. W. Davis (1983). *Project Management with CPM, PERT and Precedence Diagramming*, 3rd edition, New York, NY: Van Nostrand Reinhold Co.

65. National Research Council (U.S.). Committee on Advancing the Productivity and Competitiveness of the U.S. Construction Industry (2009). *Advancing the competitiveness and efficiency of the U.S. construction industry.* Washington, D.C., National Academies Press.

66. National Research Council (U.S.). Transportation Research Board. (1992). *Construction quality and construction management.* Washington, D.C.: Transportation Research Board National Academy Press.

67. Navarette, Pablo F. (1995). *Planning, Estimating, and Control of Chemical Construction Projects*, New York, Basel, Hong Kong: Marcel Dekker, Inc.

68. Neale, R. H. (1995). *Managing international construction projects: an overview.* Geneva: International Labor Office.

69. Neil, J. M. (1983). *Construction Cost Estimating for Project Control*, Upper Saddle River, NJ: Prentice Hall.

70. *New Manual of Accident Prevention in Construction* (2009). Publication #0101, Arlington, VA: The Associated General Contractors of America.

71. Newhouse, Elizabeth, L., editor (1992). *The Builders—Marvels of Engineering*, Washington, DC: National Geographic Society.

72. Nunnally, S. W. (2001). *Construction Methods and Management*, 5th edition, Upper Saddle River, NJ: Prentice Hall.

73. O'Brien, W.J. (2000). Implementing issues in project web sites: a practitioner's viewpoint, Journal of Management in Engineering ASCE, 16(3), 34–39.

74. Oberlender, G. D. (1993). *Project Management for Engineering and Construction*, New York, NY: McGraw-Hill.

75. O'Brien, J. J. (1994). *Preconstruction Estimating*, New York, NY: McGraw-Hill.

76. O'Brien, J. J., & Plotnick, F. L. (2005). *CPM in construction management* (6th ed.). New York: McGraw-Hill.

77. Oglesby, C. H., H. W. Parker, et al. (1989). *Productivity improvement in construction.* New York, McGraw-Hill.

78. Peurifoy, R. L. and G. D. Oberlender (2002). *Estimating Construction Costs*, 5th Edition, New York, NY: McGraw-Hill.

79. Peurifoy, R. L., C. J. Schexnayder, et al. (2006). *Construction planning, equipment, and methods.* Boston, McGraw-Hill Higher Education.

80. Pilcher, R. (1992). *Principles of construction management* (3rd ed.). London; New York: McGraw-Hill Book Co.

81. Poage, W. S. (1990). *The Building Professional's Guide to Contract Documents*, Kingston, MA: R. S. Means Company, Inc.

82. Price, A. D. F. (1995). *Financing international projects.* Geneva: International Labour Office.

83. R. S. Means Company, *Building Construction Cost Data*, 67th annual edition (2009), Kingston, MA: R. S. Means Company, Inc. (published annually).

84. Ranns, R. H. B., & Ranns, E. J. M. (2005). *Practical construction management.* London; New York: Taylor & Francis.

85. *Richardson General Construction Estimating Standards*, Mesa, AZ: Richardson Engineering Service, Inc., published annually.

86. Ritz, G. J. (1994). *Total Construction Project Management*, New York, NY: McGraw-Hill.

87. Sears, S. K., Clough, R. H., & Sears, G. A. (2008). *Construction project management: a practical guide to field construction management* (5th ed.). Hoboken, N.J.: John Wiley & Sons.

88. Senthilkumar, V., Koshy, V. and Chandran, A. (2010). Web-based system for design interface management of construction projects. Automation in Construction, 19(2), 197–212.

89. Shuette, S. D. and R. W. Liska (1994). *Building Construction Estimating*, New York, NY: McGraw-Hill.

90. Spinner, M. P. (1997). *Project Management Principles and Practices*, Upper Saddle River, NJ: Prentice Hall.

91. Sullivan, W. G., E. M. Wicks, et al. (2009). *Engineering economy*. Upper Saddle River, N.J., Pearson Prentice Hall.

92. Twomey, T. R. (1989). *Understanding the Legal Aspects of Design/Build*, Kingston, MA: R. S. Means Company, Inc.

93. University of Texas at Austin. Construction Industry Institute. Design-Build Research Team. (1997). *Project delivery systems: CM at risk, design-build, design-bid-build*. Austin, Tex., The Institute, University of Texas at Austin.

94. Vaughn, R. C. and S. R. Borgmann (1993). *Legal Aspects of Engineering*, 5th edition, Dubuque, IA: Kendall/Hunt Publishing Co.

95. Williams, T. (2009). *Construction management: emerging trends and technologies*. New York, NY: Delmar.

Index